Santa Fe Community
College Library
6401 Richards Ave.
Santa Fe, NM 87508

Taphonomy
A Process Approach

Taphonomy: A Process Approach is the first book to review the entire field of taphonomy, the science of fossil preservation. It describes the formation of plant and animal fossils in oceanic, terrestrial and river settings and how this effects deciphering the ecology and extinction of past lifeforms and the environments in which they lived.

This volume emphasizes a process approach to taphonomy and reviews the taphonomic behaviour of all important taxa, plant and animal. Taphonomic behaviour is described at a range of scales, from the formation of fossil Lagerstätten to cyclic and secular trends in preservation over hundreds of millions of years. The author discusses applications of taphonomy to the solution of both academic and practical problems, and describes mathematical models of bioturbation, fossil assemblage formation, and stratigraphic completeness.

This book will be useful to anyone interested in the preservation of fossils and the formation of fossil assemblages; it is aimed primarily at advanced students and professionals working in paleontology, stratigraphy, sedimentology, climate modeling and biogeochemistry.

RON MARTIN is Associate Professor of Geology at the University of Delaware. He holds a doctorate in protozoology from the University of California at Berkeley (1981). Prior to going to Delaware he worked as an operations micropaleontologist for UNOCAL (Houston). He has published numerous articles on foraminiferal biostratigraphy and taphonomy, and received the 1996 best paper award in the journal *Palaios*. He is author of *One Long Experiment: Scale and Process in Earth History* (Columbia University Press), editor of the forthcoming book *Environmental Micropaleontology* (Plenum Press), and is an associate editor for *Palaios* and the *Journal of Foraminiferal Research*.

Cambridge Paleobiology Series

Series editors
D. E. G. Briggs *University of Bristol*
P. Dodson *University of Pennsylvania*
B. J. Macfadden *University of Florida*
J. J. Sepkoski *University of Chicago* (deceased)
R. A. Spicer *Open University*

Cambridge Paleobiology Series is a new collection of books in the multidisciplinary area of modern paleobiology. The series will provide accessible and readable reviews of the exciting and topical aspects of paleobiology. The books will be written to appeal to advanced students and to professional earth scientists, paleontologists and biologists who wish to learn more about the developments in the subject.

Books in the Series:
1. *The Enigma of Angiosperm Origins* Norman F. Hughes
2. *Patterns and Processes of Vertebrate Evolution* Robert L. Carroll
3. *The Fossils of the Hunsrück Slate: Marine Life in the Devonian* Christoph Bartels, Derek E. G. Briggs and Günther Brassel
4. *Taphonomy: A Process Approach* Ronald E. Martin

Taphonomy
A Process Approach

Ronald E. Martin

PUBLISHED BY THE PRESS SYNDICATE OF THE UNIVERSITY OF CAMBRIDGE
The Pitt Building, Trumpington Street, Cambridge CB2 1RP, United Kingdom

CAMBRIDGE UNIVERSITY PRESS
The Edinburgh Building, Cambridge CB2 2RU, UK http://www.cup.cam.ac.uk
40 West 20th Street, New York, NY 10011-4211, USA http://www.cup.org
10 Stamford Road, Oakleigh, Melbourne 3166, Australia

© Cambridge University Press 1999

This book is in copyright. Subject to statutory exception
and to the provisions of relevant collective licensing agreements,
no reproduction of any part may take place without
the written permission of Cambridge University Press.

First published 1999

Printed in the United Kingdom at the University Press, Cambridge

Typeface Adobe Garamond 12/15pt   *System* 3B2 [wv]

A catalogue record for this book is available from the British Library

Library of Congress Cataloging in Publication data
Martin, Ronald E.
    Taphonomy: a process approach / Ronald E. Martin.
        p.   cm. – (Cambridge paleobiology series; 4)
    Includes bibliographical references.
    ISBN 0 521 59171 6. – ISBN 0 521 59833 8 (pbk.)
    1. Taphonomy.   I. Title.   II. Series.
    QE721.2.F6M37   1999
    560–dc21   98-32341 CIP

ISBN 0 521 59171 6 hardback
ISBN 0 521 59833 8 paperback

For Merle Selander, W.D.L., and D.C.S.

"... I have had arguments at Oxford with my friend William of Occam ... He has sown doubts in my mind. Because if only the sense of the individual is just, the proposition that identical causes have identical effects is difficult to prove. A single body can be hot or cold, sweet or bitter, wet or dry, in one place — and not in another place. How can I discover the universal bond that orders all things if I cannot lift a finger without creating an infinity of new entities? For with such a movement all the relations of position between my finger and all other objects change. The relations are the ways in which my mind perceives the connections between single entities, but what is the guarantee that this is universal and stable?"

"But you know that a certain thickness of glass corresponds to a certain power of vision, and it is because you know this that now you can make lenses like the ones you have lost: otherwise, how could you?"

"An acute reply, Adso. In fact, I have worked out this proposition: equal thickness corresponds necessarily to equal power of vision. I have posited it because on other occasions I have had individual insights of the same type. To be sure, anyone who tests the curative property of herbs knows that individual herbs of the same species have equal effects of the same nature on the patient, and therefore the investigator formulates the proposition that every herb of a given type helps the feverish, or that every lens of such a type magnifies the eye's vision to the same degree. The science Bacon spoke of rests unquestionably on these propositions. You understand, Adso, I must believe that my proposition works, because I learned it by experience: but to believe it I must assume there are universal laws. ..."

"And so, if I understand you correctly, you act, and you know why you act, but you don't know why you know that you know what you do?"

"... Perhaps that's it. In any case, this tells you why I feel so uncertain of my truth, even if I believe in it. ..."

Umberto Eco, *The Name of the Rose*

Contents

Preface xiii

1 Introduction: the science of taphonomy 1
1.1 The foundations of taphonomy 1
1.2 Methodology in historical sciences 5
1.3 Laws, rules, and hierarchy 11
1.4 Rules of taphonomy 12
1.5 Models and classifications of fossil assemblages 14
    1.5.1 *Johnson's models of assemblage formation* 15
    1.5.2 *Biostratinomic classification* 16
    1.5.3 *The R-sediment model* 22
    1.5.4 *Taphofacies* 24
1.6 Facts or artifacts? 25

2 Biostratinomy I: necrolysis, transport, and abrasion 27
2.1 Introduction 27
2.2 Fundamentals of fluid and sediment movement 28
2.3 Microfossils 35
    2.3.1 *Mainly foraminifera and other calcareous microfossils* 35
    2.3.2 *Mainly non-calcareous microfossils* 43
2.4 Cnidaria and associated biota 45
2.5 Mainly bivalved shells: brachiopods and pelecypods 51
2.6 Arthropods and annelids 55
2.7 Echinoderms 65
2.8 Vertebrates: bones as stones 72
    2.8.1 *Mammals* 72
    2.8.2 *Reptiles* 85
    2.8.3 *Amphibians, fish, and birds* 92

	2.9	Macroflora and pollen 93
		2.9.1 *Plant macrofossils* 93
		2.9.2 *Pollen* 102
	2.10	Techniques of enumeration 104

3 Biostratinomy II: dissolution and early diagenesis 110

3.1 Introduction 110
3.2 CaCO$_3$ dissolution and precipitation 110
3.3 Shell mineralogy, architecture, microstructure, and size 116
    3.3.1 *Mineralogy* 116
    3.3.2 *Architecture, microstructure, and size* 118
3.4 Pyritization 133
3.5 Silicification 139
3.6 Phosphatization 145
3.7 Concretions 147
3.8 Soils 152

4 Bioturbation 161

4.1 Introduction 161
4.2 Bioturbation in terrestrial environments 162
4.3 Diffusion models of bioturbation 162
    4.3.1 *The random walk* 162
    4.3.2 *Derivation of Fick's equations* 169
    4.3.3 *Solution of the Guinasso–Schink equation* 172
    4.3.4 *Estimation of sedimentary parameters* 175
4.4 Caveats of diffusion-based bioturbation models 179
4.5 Box models 182
4.6 Deconvolution 183

5 Time-averaging of fossil assemblages: taphonomy and temporal resolution 186

5.1 Introduction 186
5.2 Consequences of time-averaging 191
    5.2.1 *Abundance* 191
    5.2.2 *Taxonomic composition and diversity* 192
    5.2.3 *Trophic and life habits* 194
    5.2.4 *Stratigraphic disorder* 194
5.3 Types of time-averaged assemblages 199

Contents

- 5.4 Recognition of time-averaging 202
  - 5.4.1 *Actualistic criteria* 202
  - 5.4.2 *Criteria for ancient settings* 203
- 5.5 Durations of time-averaging 206
  - 5.5.1 *Macroinvertebrates* 206
  - 5.5.2 *Microfossils* 212
  - 5.5.3 *Vertebrates* 220
  - 5.5.4 *Plant macrofossils and pollen* 227

6 Exceptional preservation 235

- 6.1 Introduction 235
- 6.2 Genesis of Konservat-Lagerstätten 237
- 6.3 Recurrent associations of Konservat-Lagerstätten 244
  - 6.3.1 *Ediacaran association* 244
  - 6.3.2 *Burgess Shale association* 247
  - 6.3.3 *Beecher's Trilobite Bed* 251
  - 6.3.4 *Hunsrückschiefer (Hunsrück Slate)* 251
  - 6.3.5 *Orsten association* 252
  - 6.3.6 *Posidonienschiefer (Holzmaden)* 253
  - 6.3.7 *Solnhofen Limestone* 253
  - 6.3.8 *Mazon Creek association* 256
  - 6.3.9 *Coal balls* 258
  - 6.3.10 *Lacustrine association* 259
- 6.4 Hot springs 263
- 6.5 Traps 264
  - 6.5.1 *Amber* 265
  - 6.5.2 *Peat bogs* 266

7 Sedimentation and stratigraphy 268

- 7.1 Introduction 268
- 7.2 Stratigraphic maturity 269
- 7.3 Stratigraphic completeness 271
  - 7.3.1 *Fractals* 271
  - 7.3.2 *The marine record* 275
  - 7.3.3 *The terrestrial record* 279
- 7.4 Sequence stratigraphy 284
- 7.5 The stratigraphy of shell concentrations 289
- 7.6 The Signor–Lipps effect 296

7.7	Graphic correlation 301	
7.8	The hierarchy of taphonomic processes 306	

8 Megabiases I: cycles of preservation and biomineralization 309
- 8.1 Introduction 309
- 8.2 Cycles of sedimentation and climate 310
  - 8.2.1 *Sea-level, CO_2, lithology, and biotic response* 310
  - 8.2.2 *Cementation and diagenesis* 317
  - 8.2.3 *Storms* 319
  - 8.2.4 *Fluctuations in the CCD* 320
  - 8.2.5 *The latitudinal lysocline* 322
- 8.3 The origin and biomineralization of skeletons 322

9 Megabiases II: secular trends in preservation 330
- 9.1 Introduction 330
- 9.2 Types of secular megabiases 330
- 9.3 Stratigraphic completeness 334
  - 9.3.1 *The pull of the Recent* 334
  - 9.3.2 *Geochemical uniformitarianism?* 337
- 9.4 Pyritization 338
  - 9.4.1 *Modern analogs and ancient settings* 338
  - 9.4.2 *Secular trends in pyritization* 341
- 9.5 Seafood through time: energy and evolution 348
  - 9.5.1 *The Cambro-Devonian* 348
  - 9.5.2 *The Permo-Carboniferous* 353
  - 9.5.3 *The Meso-Cenozoic* 359
  - 9.5.4 *Secular increase in biomass and diversity* 362
  - 9.5.5 *Alternative interpretations* 365
  - 9.5.6 *Evolution of the biogeochemical cycles of carbon and silica* 366

10 Applied taphonomy 369
- 10.1 Introduction 369
- 10.2 Stratigraphic completeness: rates and patterns of evolution 371
- 10.3 Extinction 373
  - 10.3.1 *The terrestrial record* 373
  - 10.3.2 *The marine record* 375
- 10.4 Stasis and community unity? 378
- 10.5 Disturbance and alternative community states 380

Contents

10.6 Population dynamics and extinction in the fossil record 382
10.7 Holocene sea-level change 383
10.8 Paleophysiology 385

11 Taphonomy as a historical science 387
11.1 Major themes 387
11.2 Some more rules 389
11.3 Final thoughts 392

References 396

Index 479

Preface

Taphonomy is frequently defined as the science of fossil preservation. To those only vaguely acquainted with it, it deals with death, decay, and disintegration, and is the science of dead, rotting things accompanied by a terrific stench. It is what one is often glibly treated to in the first laboratory of a historical geology or paleontology course about molds, casts, and carbon films, before the course moves on, presumably to more important things.

But taphonomy is much more than that. Taphonomy is concerned with the information content of the fossil record and the processes by which fossils are incorporated into the fossil record. Traditionally, taphonomists and non-taphonomists alike have emphasized information loss, but with the publication of Behrensmeyer and Kidwell's (1985) seminal paper, there has been a groundswell of research about information *gain*. The fossil record is a rich source of information about phenomena that occur over temporal scales that far exceed those of a human generation, and that often occur so slowly that they appear constant to us, if we are cognizant of them at all; moreover, the stratigraphic record suggests that we cannot simply scale upward from ecological to geological scales. The fossil record is, then, a rich source of *environmental* information, including that on scales of decades to centuries – scales that encompass multiple biological surveys from which the "noise" of short-term fluctuations has been filtered by taphonomic processes (Martin, 1991, 1995; Kidwell and Flessa, 1995). But if this information is to be understood meaningfully in a (paleo)environmental context (whatever the spatiotemporal scale), we must be cognizant of the filters involved in recording it.

Since publication of Behrensmeyer and Kidwell's (1985) paper, there have been a number of symposia and associated publications on taphonomy (e.g., Behrensmeyer and Kidwell, 1988; Allison and Briggs, 1991; Donovan, 1991; Kidwell and Behrensmeyer, 1993). Despite the increased prominence of the science, however, I am not quite unconvinced that it has yet come of age (Allison, 1991): the consequences of preservation and the quality of the stratigraphic record are still too often ignored before sampling occurs (although ecologists and paleontologists began a constructive dialog at the 1998 Geological Society of America Penrose conference on linking spatial and temporal scales). I think the reason for this state of affairs is that taphonomy is really a young (in terms of the intensity of the research in the past decade or so) and highly interdisciplinary science, taxonomically, sedimentologically, and geochemically. Moreover, given the burgeoning literature, it is impossible for taphonomists to *really* pay attention to what each one of us is doing. If that is the case, how can we expect non-specialists to pay any attention? Lastly, taphonomy has – until only fairly recently – emphasized information *loss*.

Perhaps it is time, then, to assess the accomplishments and future directions of taphonomy and the pertinence of the field to the earth sciences in general. In writing this synthesis, I maintained a process approach so that general themes emerged, and highlighted the positive aspects of taphonomy, not just loss. I have tried to present a balanced treatment that gives due consideration to all disciplines and that attempts to present important generalizations supported by relevant examples. I have also concentrated on more recent developments in the field because they would seem to be most useful in assessing future research directions; my fear was that if I had attempted to present an exhaustive treatment, the text would have become so large that readers would have been deterred by its length. Chapter 1 is probably what most readers would expect: a brief review of the history of the field and the scientific method; the discussion of the scientific method is somewhat unconventional, however, and serves as a central theme of the book, so it should not be skipped. Chapter 2 follows with basic results of recent biostratinomic studies, while Chapter 3 follows with a synopsis of information about early diagenesis. Chapter 4 derives mathematical models for bioturbation, which I feel is one of the foremost problems hindering interpretation of the fossil record and its application to the environmental arena; I have typically derived equations so that they are not enshrined, and enshrouded, in gobbledegook. Chapter 5 is concerned with time-averaging and is followed by the exceptions: Lagerstätten. Although perhaps not falling under the purview of taphonomy *per se* but because of its tremendous importance, stratigraphic completeness is discussed in Chapter 7. Chapters 8 and 9 are devoted to cyclic and secular megabiases, which may limit the application of strict actualistic and steady-state approaches to the fossil record over geological scales of time; portions of these chapters are speculative. In a number of cases, I have presented contradictory data or interpretations and have let readers decide for themselves.

Throughout the text, I have emphasized the application of taphonomy to interpretation of the fossil record, rather than isolating taphonomy as a separate discipline. Given the present economic situation, if paleontology and taphonomy are to persist as viable disciplines with realistic employment opportunities, I feel we must do a better job of communicating the importance of taphonomy to the environmental sciences, and so Chapter 10 is devoted to the application of taphonomy both in and out of the ivory tower. Chapter 11 ends where the book begins: taphonomy as a historical science and the importance of historical sciences to the current environmental predicament. The text is extensively referenced throughout in the hope that the book will serve as both synthesis and source.

I wish to express my sincere gratitude to numerous colleagues for their support over the years and to Catherine Flack for her invitation to write the book. My thanks also to those who were kind enough to send reprints and preprints, and to the National Science Foundation and Exxon U.S.A. for their support of my research. The bulk of the manuscript was written while I was on a half-year sabbatical from the University of Delaware. Molly Chappel and Barbara Broge, of University of Delaware Media Design, drafted figures; my thanks to Molly, especially, for her work ethic, skill, and patience. My thanks also to various scientific organizations and publishers for their kind permission to use their drawings. Cheryl Doherty (Geology Department, University of Delaware) patiently retyped a large number of the tables and Howard Farrell carefully copy-edited the manuscript. Last, but not least, my wife, Carol, provided financial assistance for figure preparation ("How many copies of this book did you say you're going to sell?"). After all, one should be content to do things like this for the glory...

Ron Martin, Department of Geology, University of Delaware, Newark, DE 19716, U.S.A.; e-mail:Daddy@Strauss.Udel.Edu

References

Allison, P. A. 1991. Taphonomy has come of age! *Palaios* 6: 345–346.

Allison, P. A. and Briggs, D. E. G. 1991. *Taphonomy: Releasing the Data Locked in the Fossil Record.* New York: Plenum Press.

Behrensmeyer, A. K. and Kidwell, S. M. 1985. Taphonomy's contributions to paleobiology. *Paleobiology* 11: 105–119.

Behrensmeyer, A. K. and Kidwell, S. M. (eds.) 1988. Ecological and evolutionary implications of taphonomic processes. *Palaeogeography, Palaeoclimatology, Palaeoecology* 63: 1–291.

Donovan, S. K. (ed.) 1991. *The Processes of Fossilization*. London: Belhaven Press.

Kidwell, S. M. and Behrensmeyer, A. K. (eds.) 1993. *Taphonomic Approaches to Time Resolution in Fossil Assemblages*. Pittsburgh: Paleontological Society Short Courses in Paleontology No. 6.

Kidwell, S. M. and Flessa, K. W. 1995. The quality of the fossil record: populations, species, and communities. *Annual Review of Ecology and Systematics* 26: 269–299.

Martin, R. E. 1991. Beyond biostratigraphy: micropaleontology in transition? *Palaios* 6: 437–438.

Martin, 1995. The once and future profession of micropaleontology. *Journal of Foraminiferal Research* 25: 372–373.

1 Introduction: the science of taphonomy

Nature is full of infinite causes that have never occurred in experience.
Leonardo da Vinci

1.1 The foundations of taphonomy[1]

Taphonomy is the science of the "laws of burial" (from the Greek *taphos* + *nomos*). It is the study of the transition of organic remains from the biosphere into the lithosphere or the processes of "fossilization" from death to diagenesis. Although the term "taphonomy" was first coined by Efremov (1940), the science of taphonomy has been practiced for centuries (Cadée, 1991). Taphonomic investigations were first conducted by Leonardo da Vinci (1452–1519), who used observations on living and dead bivalves to infer that fossils found in nearby mountains had not been transported there by the Biblical Deluge, but had actually lived and died *in situ* (see excerpts from da Vinci's notebooks in Bolles, 1997; see also Chapters 2, 3). Subsequent taphonomic inferences were made by none other than Steno, who concluded that so-called tonguestones or glossopetrae were actually shark's teeth (Albritton, 1986); Robert Hooke, who compared the cellular structure of cork to that of petrified wood, thereby supporting Steno's assertion that fossils were of organic origin and not the result of the "plastic virtue" of the surrounding rocks (Albritton, 1986); the vertebrate paleontologist and anatomist, Cuvier; Alcide d'Orbigny, who erected the first detailed biostratigraphic zonations; and Armand Gressly, who formulated the concept of "facies."

Near the end of the nineteenth century and continuing into the twentieth, German paleontologists came to dominate the science (Cadée, 1991). Johannes Walther (1904, 1910) studied marine environments in the vicinity of Naples and the Jurassic Solnhofen, and Abel (1912, 1927, 1935), who published several books on taphonomy, carried out initial studies of *"fossil-Lagerstätten"* or "fossil

---

[1] This section is based on the excellent summaries of Behrensmeyer and Kidwell (1985) and Cadée (1991).

2 Introduction: the science of taphonomy

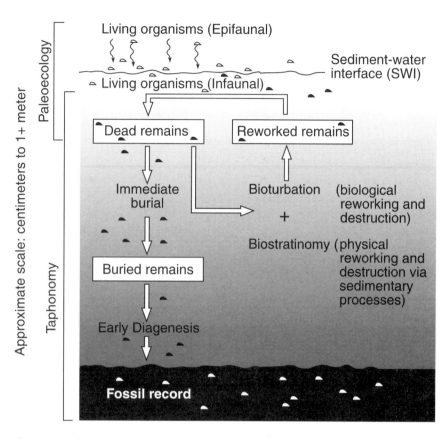

Figure 1.1. The processes of fossilization. Note the *dynamic* aspects of taphonomy, especially the recycling of fossils before final destruction or burial. Efremov (1940) included "fossil diagenesis," or the chemical and mechanical alterations of fossils within sediment, as the final stage of taphonomy (Cadée, 1991), but diagenesis may begin as soon as hardparts enter the surface layer of sediment. (Based on Lawrence, 1968; Behrensmeyer and Kidwell, 1985; and Newton and Laporte 1989.)

mother lodes" of spectacularly preserved fossil biotas, which have received increasing attention (e.g., Seilacher, 1970, 1976; Whittington and Conway Morris, 1985; Allison and Briggs, 1991a,b; Chapter 6). Weigelt (1927; translated in 1989) is best known for his careful description of the decomposition, transport and burial of carcasses at the edge of a lake in the U.S. Gulf Coast after a severe "norther" in December, 1924, caused massive mortality of cattle and other vertebrates (he also described other vertebrates from modern and ancient deposits). Weigelt's study was the first well-documented investigation of the *biostratonomy* (*biostratinomy*) or sedimentary history of fossils from necrolysis (death and decomposition; Figure 1.1) through final burial (most workers prefer the term *taphonomy*, perhaps because it is more inclusive and euphonious; cf. Figure 1.1).

1.1 The foundations of taphonomy

Weigelt (1928) also studied the biostratinomy of plants preserved in the Kupferschiefer (Upper Permian, Zechstein, Germany). Somewhat earlier, Chaney (1924) studied the correspondence between plant fossil assemblages and original vegetation, and Potonié (cited in Cadée, 1991) reviewed the formation of peats and coals. Taphonomic studies of fossil pollen were also beginning about this time (Cadée, 1991). Unfortunately, German investigations were largely ignored in other countries because of language difficulties and anti-German sentiment associated with the rise of the National Socialist party (Cadée, 1991). Consequently, taphonomy was not recognized as a distinct discipline outside of Europe until after World War II by such workers as Olson (1952), working on Permian vertebrates, and Johnson (1957, 1960, 1962), who studied modern and Pleistocene shallow marine invertebrates of the northern California coast (Behrensmeyer and Kidwell, 1985).

Unlike the German taphonomists, who were concerned primarily with paleoenvironmental interpretation, Efremov, who was a vertebrate paleontologist, emphasized the incompleteness of the fossil record. As a result, to this day, taphonomy has come to be associated – even by many taphonomists – with the documentation of "information loss" and "bias" in the fossil record. Lawrence (1968), for example, compared modern and Oligocene oyster communities and concluded that 75% of the macroinvertebrates were not preserved. Numerous studies followed among a diversity of specialists, who largely worked independently of one another and who did not recognize taphonomy as a distinct discipline of the Earth sciences.

In an attempt to unify the discipline, Behrensmeyer and Kidwell (1985, p. 105) defined taphonomy as "the study of processes of preservation and how they affect information in the fossil record." As the science of taphonomy has emerged as a separate – albeit highly interdisciplinary – entity, especially in the past decade, its body of theory has become sufficient to begin predicting the utility of the fossil record in ecological and evolutionary studies (Chapter 7). For example, ancient river channel accumulations of bone and plant fossils are likely to represent regional, rather than local, samples of the biota, given the distances that the remains have been transported (Behrensmeyer, 1982; Behrensmeyer and Hook, 1992). In contrast, oxbows and other abandoned channels will likely fill with biota from the immediate vicinity and may record ecological succession from aquatic to terrestrial habitats (Behrensmeyer, 1982; Behrensmeyer and Hook, 1992). Although deltas may offer well-preserved fossil assemblages (because of rapid burial), their record will likely be highly discontinuous because of lobe-switching (Schindel, 1980, 1982); such a record may be adequate for studying short-term ecological phenomena, but the record may prove too discontinuous to assess evolutionary lineages. Because of rapid sedimentation and subsidence (burial), active continental margins (e.g., California) may be characterized by taphonomically less complex shell beds than those of passive margins (Maryland coastal plain; Kidwell, 1988). Even along active margins, however, the degree of complexity and *fidelity* of shell beds to living communities will vary according to rates of uplift, erosion, and sedimentation (Meldahl and Cutler, 1992; Meldahl, 1993).

Behrensmeyer and Kidwell (1985) also emphasized the positive contributions of taphonomy to our understanding of the fossil record. For example, etching or breakage of shells contribute not only to information loss, but also to recognition of the biostratinomic agents involved in the formation of fossil assemblages (such as waves, currents, predators and scavengers, hermit crabs, and birds), length of exposure of hardparts in the surface mixed layer (where early diagenetic phenomena are concentrated; Chapters 3, 4), rates of sedimentation, and pore water chemistry.

Typically, "time-averaging" of fossil assemblages results because rates of sedimentation are too slow to prevent mixing of "ecological" signals into accumulations of longer duration and lower temporal resolution (Chapters 4, 5). Although viewed negatively by most workers, time-averaging is actually an advantage, since short-term "noise" is damped and longer-term signals from a biological community are preserved (Behrensmeyer and Kidwell, 1985; Wilson, 1988*b*; Kidwell and Flessa, 1995); in fact, modern death assemblages from soft-bottom habitats are perhaps comparable to repeated (and expensive!) biological surveys in assessing the long-term dynamics of biological communities (Peterson, 1977; Kidwell and Bosence, 1991; Kidwell and Flessa, 1995).

Taphonomy has, however, only recently begun to assess "megabiases" in the fossil record (Behrensmeyer and Kidwell, 1985; Behrensmeyer and Hook, 1992), although Efremov (1940) was certainly aware of them (Chapters 8, 9). These include, but are not restricted to (1) the outcrop (sampling) area of particular environments, which reflects the influence of sea-level, continental configuration, and other climatic influences on patterns of sedimentation (Efremov, 1940; Signor, 1985); (2) cycles of preservation, which again appear to reflect the influence of plate tectonics, sea-level, and atmospheric CO_2 (e.g., Walker and Diehl, 1985; Martin, 1995*a*, 1996*a,b*); and (3) secular changes in the Earth's biota, which have affected the cycling (through predation, scavenging, bioturbation) of biogeochemically important elements between – and their storage in – various reservoirs, as foreshadowed by Efremov's (1940) definition of taphonomy (e.g.,

Vermeij, 1987; Boss and Wilkinson, 1991; Kidwell, 1991; Bambach, 1993; Martin, 1995a, 1996a,b; cf. Figure 1.1).

1.2 Methodology in historical sciences

"In a profession more observational and comparative than experimental, the ordering of diverse objects into sensible categories becomes a *sine qua non* of causal interpretation" because it represents a *causal ordering* (Gould, 1986). Taphonomy, and its sister disciplines of paleoecology, sedimentology, and stratigraphy, and geology itself for that matter, are historical sciences because, fundamentally, they are concerned with the history of Earth and its Life as they are recorded in the rocks. Other scientists regularly construct historical hypotheses based on observations arranged in stages, from the life histories of species and stars, to the development of atolls: similarly, taphonomists often refer to "taphonomic histories" and "pathways" of formation of fossil assemblages.

One may also infer history from single, *unique objects* by looking at anomalous features or imperfections (e.g., Lagerstätten; Chapter 6). Although each "singularity" is unique because of historical constraints, collections of singularities (fossil assemblages, especially Lagerstätten) may exhibit certain "nomothetic" (general or universal) relationships that can be predicted and tested (e.g., durations of time-averaging based on radiocarbon dates of hardparts from different depositional settings; Chapter 5).

By far and away the most common approach to interpreting the fossil record involves *upward scaling* from short-term observations to geological phenomena. The foundation of the Earth sciences, *and of all other sciences*, is the Principle of Uniformitarianism: "the present is the key to the past." The origins of uniformitarianism are found in James Hutton's *Theory of the Earth*, but uniformitarianism probably received its greatest impetus from Sir Charles Lyell's incredibly influential *Principles of Geology, Being An Attempt to Explain the Former Changes of the Earth's Surface, by Reference to Causes Now in Operation*, the first volume of which was published in 1830 and which went through 11 editions in ~50 years. Because of his religious outlook, Lyell was strongly committed to a steady-state view of the Earth (a view easily rationalized with a divine presence), and uniformitarianism was an attempt by him to deny any form of catastrophism or directionalism (progressionism) in the inorganic and organic worlds, such as that recognized by Cuvier and certain other early nineteenth century scientists (Bowler, 1976; Ruse, 1979); indeed, "jettison steady-statism, and you jeopardize ... uniformitarianism" (Ruse, 1979, p. 79). The influence of Lyell –

and later Darwin (who read the *Principles* and subscribed to its tenets) – has been such that even to the present day, progressionism in the fossil record is viewed suspiciously as teleology: after all, "progress is not inevitable by the canons of natural selection" (Desmond, 1982, p. 101; see also Gould, 1996). Geochemical models are typically conceived, for example, under steady-state conditions primarily for the sake of simplicity (the shorter the interval considered the less likely conditions will change; cf. Chapters 8, 9; see also Bowler, 1976).

Lyell's philosophical view consisted of two parts: the explanation of past geological phenomena by processes that (1) are *observable* today ("actualism") and (2) are of the same *rate* as those observed today (uniformitarianism *sensu stricto*; Ruse, 1979).[2] Like many of his contemporaries (including Darwin), Lyell was probably influenced by the then (and still!) prevalent view of the astronomer Sir John Herschel's *A Preliminary Discourse on the Study of Natural Philosophy* (1831), which espoused physics as the paradigm of a mature, quantitative science (Ruse, 1979). According to Herschel (1831), there are two kinds of laws: empirical (those that state relationships without stating their causes; e.g., Kepler's laws) and *verae causae* or true causes (e.g., Newton's laws of motion and gravitation, from which Kepler's laws may be deduced; Ruse, 1979). According to Herschel (1831), *verae causae can only be determined by analogy (comparison) with our own experience*.

Lyell's views were not without controversy, however (especially uniformitarianism *sensu stricto*), and were attacked by none other than Adam Sedgwick (who taught Darwin field geology in Wales shortly before his departure on H.M.S. *Beagle*) and the Reverend William Whewell (who also had extensive contact with Darwin) because phenomena that could not be observed (such as the intervention of a divine Creator) were automatically ruled out by Lyell (see Ruse, 1979, for further discussion).

Gould (1965; see also Lyman, 1994a[3]) also distinguished two types of uniformitarianism: methodological and substantive. According to methodological uniformitarianism, no unknown *processes* need be invoked if historical records can be explained by processes observed in the present (a form of Occam's

---

[2]Both are referred to as uniformitarianism in England and the U.S.A., whereas in Russia uniformitarianism has meant "the specific Lyellian hypothesis ... while actualism is a method" (Hooykaas, 1963, p. v).

[3]Lyman (1994a) gives a detailed – and often personal – view of the history and methodology of taphonomy, especially with respect to zooarcheology. The following discussion of methodology is based on his work, and also that of Salmon (1967), Gould (1986), Frodeman (1995), and Martin (1998a).

1.2 Methodology in historical sciences

Razor); whereas, according to the substantive doctrine, *rates* of change have always been uniform and gradual and catastrophic changes are not caused by sudden changes in rates (cf. uniformitarianism *sensu stricto*). Methodological uniformitarianism is essentially identical to actualism and has been a mainstay of taphonomic research. Based on this tenet, modern fossil assemblages and their taphonomic settings, as well as field and laboratory experiments, can be used to make inferences about the processes — and their rates — that formed ancient assemblages; the actualistic method, then, involves argument by analogy (Lyman, 1994a). Although this approach began as early as da Vinci, it received tremendous support from German workers, among them Richter (1928), who founded the institute at Senckenberg am Meer in Wilhelmshaven (Germany) along the Wadden Sea, to study "aktuopaläontologie" (Cadée, 1991). But the actualistic approach, at least in marine environments, probably received its greatest impetus from the translation (1972) of Schäfer's (1962) actuopaleontological studies in the North Sea (Behrensmeyer and Kidwell, 1985; Cadée, 1991). Schäfer documented the death, decay, and disintegration of modern vertebrate and invertebrate remains; the traces of these and other animals ("Lebensspuren" or ichnofossils); and the transition from *biocoenosis* (living community) to *thanatocoenosis* or fossil (death) assemblage ("any group of fossils from a suitably restricted stratigraphic interval and geographic locality"; Fagerstrom, 1964). These terms had been used earlier by Wasmund (1926) in his study of lakes (see Kidwell and Bosence, 1991, their table 1, for review of the usage of these terms by different investigators; some workers have adopted the term *taphocoenosis* for a taphonomically modified thanatocoenosis).

It is in uniformity, however, where the rub lies. First, the *assumption* of uniformity cannot be tested because we cannot actually observe the past (Hubbert, 1967; Kitts, 1977); therefore, we are forced to assume an actualistic stance. And second, appeal to processes not observable in the present, especially unusual ones such as the classic view of "catastrophism," is, strictly speaking, precluded (Kitts, 1977; Lyman, 1994a; see also Ruse, 1979). With regard to the first criticism, since we cannot demonstrate natural laws to be invariant in the past (even in "hard" — and largely ahistorical — sciences like chemistry and physics), conclusions are arrived at through the process of induction (Salmon, 1967; Lyman, 1994a). Unlike deduction, inductively derived conclusions contain inferences not present in the premises and can never be shown to be absolutely true (inductive arguments are therefore said to be ampliative); nevertheless, inductively derived generalizations are extremely useful as premises of deductive arguments (i.e., prediction) and are the basis of the hypothetico-deductive method of scientific inquiry

(Salmon, 1967). If inductively derived generalizations – or deductions based upon them – turn out to be incorrect, we seek new ones.

But logically, we cannot conclude that historical phenomena are explained by our actualistic generalizations if those generalizations were used to infer the historical phenomena in the first place. This represents *the* fundamental criticism of the process of scientific induction by the Scottish philospher David Hume ("Hume's paradox"): namely, how *does* one acquire knowledge of the unobserved (Salmon, 1967)? "If we attempt to rationally justify scientific induction by use of an inductively strong argument, we ... [must] *assume* that scientific induction is reliable in order to prove that scientific induction is reliable; we are reduced to begging the question. Thus, we cannot use an inductively strong argument to rationally justify scientific induction" (Skyrms, 1966, p. 25; see also Salmon, 1967). Nevertheless, we use induction because the approach works and there is no alternative (Bridgman, 1959; Lyman, 1994a). We are, in effect, engaged in a pragmatic (some might say circular) form of reasoning that seems to work most of the time.

Unlike induction, deduction is non-ampliative: it "purchases ... truth preservation by sacrificing any extension of content" (Salmon, 1967, p. 8). It is partly for this reason that scientists (unwittingly) subscribe to the principle of the "uniformity of nature": such a principle is a "synthetic *a priori*" statement that, when part of the premises of an inductive argument – consciously or otherwise – attempts to make an inductive argument deductive in nature; synthetic statements are arrived at inductively, however, and their accuracy is indeterminate (Salmon, 1967). Moreover, Hume asked, how can we know *a priori* that Nature is uniform?

For this and a number of other reasons, Karl Popper – who has probably influenced scientific methodology more than any other philosopher this century – rejected induction, and proposed instead a *hypothetico*-deductive method that he claimed avoided the problems of induction by making statements (hypotheses) that could be falsified (Popper, 1959, and later works; see also Woodward and Goodstein, 1996). According to Popper, the more likely it is that a hypothesis can be falsified, the better it is. Hypotheses must run as great a risk as possible of being overturned, because the more falsifiable a hypothesis is, the more it tells us; therefore, the more falsifiable a hypothesis is, the more it excludes extraneous possibilities and the greater the risk it runs of being false (Salmon, 1967). This approach differs from that assumed by most scientists (who attribute it to Popper): that the more a hypothesis is corroborated by *positive* support, the more likely it is to be *confirmed*. Unfortunately, this often leaves too many hypotheses to explain the same phenomenon, and the "hypothetico-deductive theorist"

1.2 Methodology in historical sciences

will likely choose the most probable one, whereas Popper would pick the least likely one because of the inverse relation between falsifiability and probability. According to Popper, a highly falsifiable hypothesis which is stringently and repeatedly tested and left unfalsified becomes "highly corroborated," which is not the same thing as being confirmed. Popper considered his method to be strictly deductive; nevertheless, inductive argument still creeps in because without it conclusions (hypotheses) would only confirm the premises (observations), and science would "amount to [no] more than a mere collection of ... observations and various reformulations thereof" (Salmon, 1967, p. 24).

Perhaps we can take some comfort in the fact that deduction may arrive at absurd conclusions if the premises are false. Understanding Nature by deduction from "indubitable" first principles grounded in *pure reason* was championed by rationalists such as Descartes and Leibniz, who wanted to reason *to* nature not *from* it (Ruse, 1979), and "who were impressed by the power of the mathematics they had helped to create" but "which failed to account for ... observational and experimental aspect[s]" (Salmon, 1967, pp. 1–2). An early typical example of deductive reasoning was to accept the Judeo-Christian god as the Creator of the universe and its occupants, and to deduce what were thought to be the necessary consequences, such as that the creation occurred only a few thousand years ago, all species are immutable, and so on (Moore, 1993). Empiricist philosophers such as Sir Francis Bacon (1561–1626) found this approach repugnant, and emphasized that one should begin with data based on observation and experiment, not faith (Moore, 1993; see also Martin, 1998*a*). Not surprisingly, given the times, James Hutton – like Isaac Newton before him – believed that natural phenomena demonstrated the existence of a divine plan; according to Greene (1982), Hutton's approach was more deductive than inductive. In taphonomy, one example of such a deduction, which would seem to be obvious based on reason alone, is that small bones should vastly outnumber larger ones in a vertebrate fossil assemblage because small animals vastly outnumber larger ones in living populations ("*Law of Numbers*"; see also Kidwell and Flessa, 1995; Chapter 2). Although *intuitively* this prediction makes perfect sense, it is exactly the opposite of observations made on mammalian remains of Amboseli Basin (East Africa) by Behrensmeyer and Boaz (1980).

With regard to the second criticism of the assumption of uniformity, modern usage of the principle does accept that the rates and intensities of processes have varied during the Earth's history (even Hutton accepted this; Albritton, 1986). When historical phenomena cannot be explained using the actualistic approach we must concede that either our knowledge of modern processes is incomplete or that there are processes that we have not yet observed or that are no longer operative in the modern realm (e.g., "megabiases"; cf. substantive uniformitarianism). Ironically, it was rationalists such as Descartes who argued that since our senses (or in modern terms, our scale of observation) often deceive us, only deductive arguments are valid.

The problem of scale strikes at the heart of actualism and is why historical sciences are so important (Martin, 1998a). If we do not observe a process over the typical span of a grant proposal of 3–5 years or a scientific career of several decades, much less a human life span of say 70 years, does that mean that the process does not occur? How many times has human civilization, which has spanned thousands of years, recorded the collision of an extraterrestrial body with the Earth (the closest in recent memory was the Tunguska event in Siberia in 1908), much less the impact of a comet with Jupiter's atmosphere? Because we observe streams to erode gradually downward, are we always justified in extrapolating these rates to all river valleys? The Lake Missoula floods argue otherwise (Parfit, 1995). Mass extinctions may have occurred over considerable spans of geological time and may be preceded by gradual climate change that would be undetectable over many human generations but which nevertheless culminates in biological catastrophe in the fossil record (e.g., Martin, 1998a).

In the case of fossilization, even if we observe, for example, certain features on shells or bones to be produced at certain rates in laboratory or field experiments, we cannot blindly extrapolate those rates to the past. Similar features of fossils may have been produced at rates different from those observed in modern analogs and experiments (Behrensmeyer, 1982; Kotler *et al.*, 1992). Moreover, taphonomic features of a fossil or an assemblage that may appear to be diagnostic of a particular taphonomic agent may arise for different reasons (*equifinality* of Lyman, 1994a); i.e., the same features may result from different taphonomic pathways or histories.

Nevertheless, "data derived from actualistic research are ... commonly used as a source of empirical generalizations or *formal analogies* rather than to build *relational analogies* and postulate diagnostic criteria" (Lyman, 1994a, p. 69). In formal analogies, two or more objects are said to be similar because they share certain attributes; such analogies are weak because the properties may have arisen by chance (Hodder, 1982). For example, just because two shells possess a similar *taphonomic grade* (surface appearance) does not mean that they have identical taphonomic histories: young shells, for example, may *appear* to be quite old and old shells may appear to be quite young (Flessa *et al.*, 1993; Kidwell, 1993a; Martin *et al.*, 1996; Chapter 5), so that shells of similar appearance may actually have quite different taphonomic histories (contrary to intuition; cf. Brandt, 1989).

In relational analogies, attributes are interdependent and causally related (Hodder, 1982). In the case of shell grade and age, shell grade is not a function of shell age itself, but of the shell's residence time near the sediment–water interface (SWI; cf. Figure 1.1) and accompanying exposure to bioeroders and dissolution before final burial (Flessa *et al.*, 1993; Cutler, 1995; Martin *et al.*, 1996). Exposure at the SWI is in turn a function of factors such as rates of sedimentation and bioturbation and reworking by storms (e.g., Meldahl, 1987; Flessa *et al.*, 1993). Thus, relational analogies can either weaken or strengthen formal analogies and result from *context*, which in this case is the taphonomic (depositional) setting.

Thus, taphonomic histories may be quite complex, and even deceptive, and the term "taphonomy" is itself a misnomer, as Efremov (1940) was no doubt aware (Cadée, 1991). In historical sciences we do not necessarily seek laws that "apply to all parts of space and time without restriction" (Salmon, 1967, p. 5) so much as principles or "rules of thumb" that can guide us, although not necessarily unerringly, in interpreting the history of fossil assemblages (Weigelt, 1989, also used the term "law" but in a much more restrictive sense, and it is clear from his discussion that his "laws" are really principles).

1.3 Laws, rules, and hierarchy

So-called laws, and even principles, are constrained by context or history (Olson, 1980; Allen and Starr, 1982; Martin, 1998*a*). If we interpret the past strictly in terms of anthropocentric laws, we will never truly understand what the fossil record has to tell us (Martin, 1998*a*; cf. Chapter 11). The Principle of Superposition, for example, states that younger sedimentary rocks lie on top of older ones, but not if they have been overturned by folding. Our inferences about ancient sediment and soils are based on the laws of physics and chemistry, but the exact chemical conditions that pertained to the formation of a particular fossil assemblage depend upon the contextual relations of bedrock, climate (arid, wet), type and amount of vegetation (especially in the case of soil), rates of weathering, intensity of bioturbation (including trampling), and so on.

Much of nature consists of hierarchies, which consist of discrete levels called "holons." Each holon has three aspects: (1) its interior, which consists of (2) its parts (which may in turn be separate holons with their own parts), and (3) its surrounding environment (which may be another holon surrounded by its environment). Thus, holons are both parts and wholes simultaneously (Salthe, 1985). A taxonomic holon, for example, is a level that contains other objects (e.g., the species of a genus) and is in turn subsumed by a higher taxon (a family of genera in the taxonomic hierarchy). If the holon is repeatedly recognized using different techniques, then it is robust (Salthe, 1985, delves into these topics from a philosophical approach; Allen and Starr, 1982, suggest multivariate statistical methods that can be used to detect holons; see also Ahl and Allen, 1996). A holon also exhibits spatiotemporal continuity: whatever it is that we recognize, it is sufficiently stable to persist over some area and last for some recognizable interval of time.

Holons have a history, and they have unique properties that have been determined in part by unique configurations of historical contingency. Hierarchy works by recognizing differences (history); it is what Salthe (1985) calls an *idiographic* approach, which emphasizes particularities. On the other hand, reductionist science works by using observational regularities or similarities (laws) discovered by comparing measurements (Salthe, 1985); reductionism is a nomothetic approach because it seeks general or universal laws. The processes (and their rates) studied using idiographic and nomothetic approaches may not interact directly (Salthe, 1985), thereby isolating the disciplines from one another.

Hierarchies may also be viewed as systems of constraint. When examining a holon we must consider the holon immediately above and that immediately below (which forms the constituent parts of the holon in question; i.e., integration). Higher (larger) holons tend to constrain the behavior of their constituent lower holons because the higher holons provide the environment (boundary conditions) within which lower holons must operate; conversely, lower holons provide the "initiating conditions" or "possibilities," which, depending upon the boundary conditions, may or may not be realized. The greater the number of boundary conditions, the fewer the possibilities that are realized. Higher-level constraints produce boundaries that are historical in nature, whereas lower-level processes act in a more "lawful" manner (Salthe, 1985). The dynamics of lower levels are rate-dependent because they are dependent upon laws, which are "inexorable … incorporeal … and universal," but they are constrained by rules at higher levels, which are independent of lower level rates because they are "arbitary, … structure-dependent, … [and] … local. In other words, we can never alter or evade laws of nature; we can always evade and change rules" (Pattee, 1978, in Allen and Starr, 1982, p. 42). History certainly has.

1.4 Rules of taphonomy

Considering that the taphonomy of fossil assemblages has been investigated for at least 500 years, taphonomists ought to have developed some empirical

1.4 Rules of taphonomy

generalizations – principles or rules – by now. Wilson (1988*b*) lists a number of them, which I have modified or supplemented

(1) Organisms are more likely to be preserved if they have hardparts.
(2) Preservation is greatly enhanced by rapid burial, especially in fine-grained sediment (low turbulence) or in the absence of decay and scavenging.
(3) During the transition from biocoenosis to thanatocoenosis, disarticulation and chemical alteration resulting from decay, abrasion, transportation, predation, scavenging, or dissolution cause loss of information about species abundances and community diversity and structure.
(4) Fossil assemblages consist of (a) *autochthonous* remains, which represent organisms that lived in the community and may have been preserved in life positions; (b) *parautochthonous* remains, which are autochthonous components that have been moved (disarticulated, reoriented, concentrated) from their original position by bioturbators, predators, or scavengers, but not transported from another community; and (c) *allochthonous* or foreign remains that have been derived from other communities (Kidwell *et al.*, 1986).
(5) Taphonomic loss, especially through dissolution and bioerosion, is typically most severe in shallow-water marine environments. Perhaps this "rule" also results from the attention these environments have received from "actuopaleontologists" because of their greater accessibility: salt-marshes, for example, are largely characterized by autochthonous remains (Scott and Medioli, 1980*b*; Behrensmeyer and Hook, 1992; Chapter 10), whereas complete unmixed deep-sea marine records are by no means the norm, despite the "optimistic assessments" of many biostratigraphers and paleoceanographers (Schiffelbein, 1984).
(6) Information loss in terrestrial and fluvial biotas results largely from transport, disarticulation, sorting, and breakage by water, predators, scavengers, and trampling.
(7) Bioturbation and physical reworking also cause time-averaging (temporal mixing) of different communities and may lead to *increased* diversity and variation in morphological features of fossil lineages. Temporal mixing often goes unrecognized in fossil assemblages.
(8) Thus, false First and Last Appearance Datums (FADs and LADs) may result from bioturbation and physical reworking. False LADs are most serious because bioturbation and reworking preferentially mix sediment upward.

(9) Nevertheless, information *gain* about taphonomic settings and long-term community dynamics may result from the actions of taphonomic agents (this point remains largely unappreciated by those outside the field).

(10) Furthermore, catastrophic burial or smothering (*obrution*) may result in Lagerstätten that serve as "snapshots" of population dynamics. These "fossil censuses" may not, however, be truly representative of the long-term dynamics of the population, and so multiple snapshots of a fossil biota probably better represent the temporal variation in populations.

As Cadée (1991, p. 16) notes, after half a century of intensive investigation, much of it concentrated in the past decade or two, the number of taphonomic rules "seems rather meager" (see also Olson, 1980). But given the nature of laws and the role of history, this is to be expected (section 1.3; see also Chapter 11). Although we must begin with the assumption of the uniformity of Nature, any application of principles or rules must be done in a *comparative* (case-by-case) manner because each historical entity bears the imprint of the unique (or nearly so) circumstances that led up to it (Olson, 1980; Martin, 1998a).

Not surprisingly, most of these generalizations deal with information loss, lack any true predictive ability, and in hindsight, seem like so much common sense or intuition. But these rules, along with numerous corroborative studies, have served as a foundation for inductive *models* ("an intellectual construct for organizing experience"; Allen and Starr, 1982) and classifications, both of which may be considered "working hypotheses" and which hold the greatest promise for deductive or predictive approaches.

1.5 Models and classifications of fossil assemblages

In this section, I discuss several classifications and models of formation of marine fossil assemblages that emphasize *environmental gradients* of taphonomic processes. The review of these models serves as a foundation for much of the rest of this book: similar classifications and models of fossil vertebrate and plant assemblage formation are explored in succeeding chapters. These sorts of models are of heuristic value because they demonstrate that, although fossil assemblage formation is complex, the character and utility of fossil assemblages can be predicted. Although at first glance the models emphasize information loss, they demonstrate how much paleoenvironmental information can be *gained* through careful analysis of taphonomic pathways and agents (Behrensmeyer and Kidwell, 1985; Wilson, 1988b), and how any and all criteria – paleontological, sedimentological, and stratigraphic

1.5 Models and classifications of fossil assemblages

– should be brought to bear in paleoenvironmental interpretation. Some of the models also demonstrate the value of *comparing* fossil assemblages of greatly different ages but of similar preservational histories.

1.5.1 Johnson's models of assemblage formation

Johnson's (1960) models of assemblage formation are among the first – if not the very first – to emphasize taphonomic gradients, and other models of marine assemblage formation can be viewed as outgrowths of his work. Johnson (1960) plotted three theoretical *taphonomic modes* of formation of fossil concentrations (especially for bivalves) according to "exposure effects" (a function of residence time at the SWI) versus "transportation effects" (energy + shell import; Figure 1.2). A taphonomic mode is a "recurring pattern of preservation of organic remains in a particular sedimentary context, accompanied by characteristic taphonomic features" (Behrensmeyer, 1988, p. 183). Each assemblage mode is characterized by certain taphonomic criteria (Figure 1.2). Model I represents a *census assemblage* which is rapidly buried so that there is little or no chance of transportation: remains are largely autochthonous and some may still be in life position. Model II represents a *low-energy assemblage* (*within-habitat time-averaged*) dominated by parautochthonous hardparts that mostly exhibit some degree of wear and movement through such agents as waves, currents, and bioturbation. Model III also consists largely of parautochthonous remains, but includes allochthonous hardparts as well, and represents a *high-energy* version of Model II. Although not

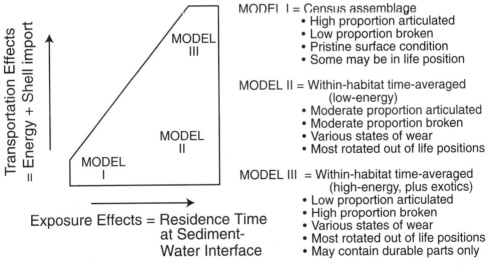

Figure 1.2. Plot of Johnson's (1960) models of fossil assemblage formation according to transportation versus exposure effects. (Redrawn from Kidwell, 1993a, after Johnson, 1960.)

stated explicitly, these models are in effect "end-members" and all gradations of assemblages conceivably occur between these extremes, as indicated by Johnson's (1960) initial evaluation of the Millerton Formation (Pleistocene, Tomales Bay, California). Johnson's approach to analyzing fossil assemblages has exerted a tremendous influence on subsequent taphonomic models, such as those that follow.

1.5.2 Biostratinomic classification

Kidwell *et al.* (1986) developed a *descriptive* nomenclature and a *genetic* classification for level (soft)-bottom fossil concentrations along modern and ancient onshore–offshore bathymetric transects. The descriptive procedure uses four features – taxonomic composition, bioclastic packing (biofabric), geometry, and internal structure – that can be used in the field to assess the genetic significance of biostratinomic factors (Figure 1.3), and is intended to "facilitate systematic characterization of local sections in terms of their skeletal concentrations, which are at present underexploited in the differentiation and mapping of sedimentary facies" (p. 236). According to *taxonomic composition*, concentrations may be monotypic or polytypic according to whether they consist of one or more types of skeleton; these terms apply to any taxonomic category appropriate to a study (monotypic accumulations may, for example, be said to consist of bivalves, oysters,

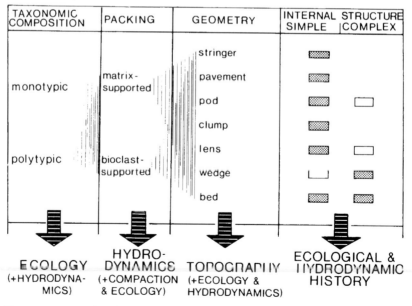

Figure 1.3. Procedure for describing skeletal concentrations proposed by Kidwell *et al.* (1986). (Reprinted with permission of SEPM [Society for Sedimentary Geology].)

1.5 Models and classifications of fossil assemblages

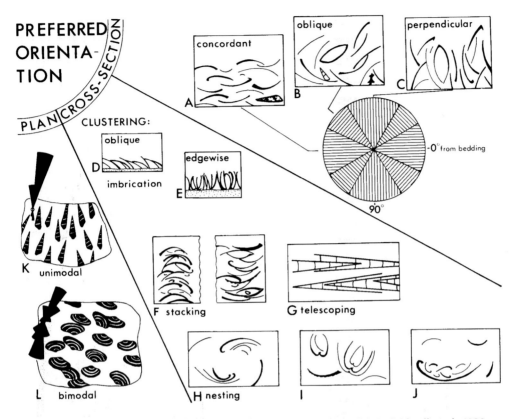

Figure 1.4. Terminology for hardpart orientation and biofabric. (Kidwell *et al.*, 1986; reprinted with permission of SEPM.)

or *Crassostrea virginica*), but the lower the taxonomic level, the greater the ecological or hydrodynamic significance of monotypy.

Biofabric refers to the three-dimensional arrangement of skeletal remains, including orientation, sorting by size and shape, and close-packing, which may range anywhere between matrix and bioclast-supported (Figure 1.3). Biofabric depends mainly on hydrodynamics but may also reflect ecology (life position), necrology (decay), predation, scavenging, bioturbation, and rotation and disarticulation during compaction. Kidwell *et al.* (1986) proposed descriptive terms for hardpart orientation and biofabric (Figure 1.4).

The *geometry* of a fossil deposit depends on a number of factors (Figures 1.3, 1.5), among them antecedent topography (including burrows and crevices); mode of life of the hardpart producers (e.g., whether they lived in clumps, such as oysters or archaeocyathids), biological activity (e.g., bioturbation, selective deposit feeding); and physical processes that produce syngenetic topography (shell lags, channels, etc.).

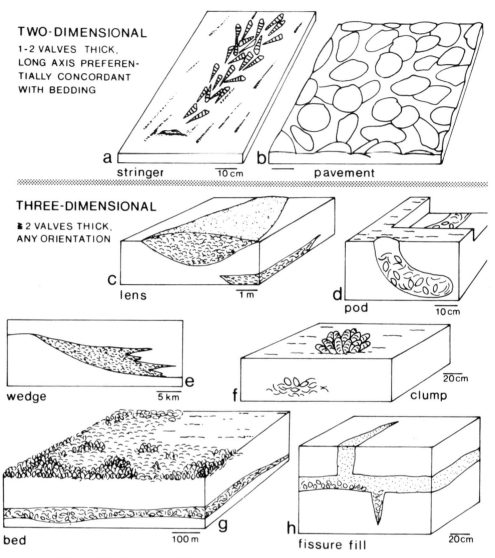

Figure 1.5. Geometry of skeletal accumulations. (Kidwell *et al.*, 1986; reprinted with permission of SEPM.)

The fourth criterion is that of *internal structure* of assemblages. *Simple* concentrations are internally homogeneous or exhibit some monotonic trend, such as upward fining of matrix or bioclasts (e.g., shelly turbidites, "tempestites" or storm deposits). *Complex* concentrations, on the other hand, include assemblages that consist of alternating horizons of articulated and disarticulated hardparts or concentrations that consist of lateral or vertical amalgamations of smaller-scale concentrations. Stringers and pavements are almost always simple, whereas both simple and complex internal structures occur in thicker beds (Figure 1.3).

1.5 Models and classifications of fossil assemblages

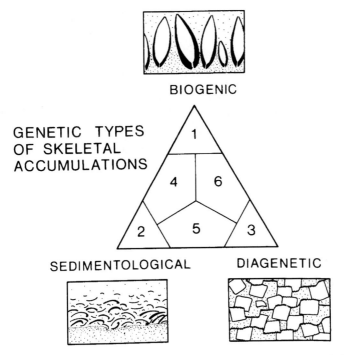

Figure 1.6. Genetic classification of hardpart concentrations consisting of endmember (1–3) and mixed assemblages (4–6). (Kidwell *et al.*, 1986; reprinted with permission of SEPM.)

The genetic classification is represented by a ternary diagram of lithological, biological, and diagenetic end-members (1–3; Figure 1.6) that includes three mixed concentration types (4–6). Any of the six assemblages may be autochthonous, parautochthonous, or allochthonous. *Intrinsic biogenic* concentrations are produced by the gregarious behavior of organisms in life through, for example, preferential colonization of sites already occupied by adults (e.g., brachiopods, vermetid gastropods, oysters) or their remains in death, and are usually autochthonous or parautochthonous. *Extrinsic biogenic* concentrations result from the interactions of organisms with other skeletonized organisms or their hardparts, and are typically parautochthonous or allochthonous; such assemblages include subsurface layers produced by Conveyor Belt ("head-down") Deposit Feeders (CDFs; van Straaten, 1952; Rhoads and Stanley, 1965; Cadée, 1976; Meldahl, 1987; Boudreau, 1997), shell-filled shallow excavations produced by skates and rays (Gregory *et al.*, 1979; Fürsich and Flessa, 1987), accumulations produced by birds (Teichert and Serventy, 1947; Lindberg and Kellogg, 1982; Meldahl and Flessa, 1990), and *Diopatra* burrows lined by shells (Schäfer, 1972). In either case, live–dead interactions can change the physical nature of the substratum and influence the structure of benthic communities via *taphonomic feedback* (Kidwell and Jablonski, 1983; Kidwell, 1986*b*).

Sedimentological concentrations result primarily from hydraulic processes of hardpart concentration. Such accumulations include (1) winnowed, parautochthonous fair-weather or storm lags (cf. model II of Johnson, 1960); (2) gradual accumulations of autochthonous–parautochthonous hardparts during intervals of low net sedimentation (cf. model II of Johnson, 1960); and (3) transport of allochthonous hardparts into otherwise autochthonous–parautochthonous assemblages (model III of Johnson, 1960).

Diagenetic concentrations result from physical and chemical processes that significantly concentrate shells after burial, including compaction (Fürsich and Kauffman, 1984), selective pressure solution, which concentrates fossils along stylolites in limestones, or the destruction of hardparts in adjacent beds (Fürsich, 1982; Haszeldine, 1984).

Mixed concentrations result from the interaction of two or more end-members, one of which may strongly overprint the other. Oyster biostromes formed by gregarious settlement or pavements of wave or tidal current-oriented shells of the high-spired gastropod *Turritella* (unimodal or telescoped orientations of Figure 1.4) may be further concentrated by hydraulic sorting (Figure 1.6, area 4). In some cases, hydraulic reworking may be sufficient to obliterate any evidence of biogenic accumulation, but if the reworked shells are judged to be strictly allochthonous, the accumulation is classified as sedimentological rather than overprinted biogenic; such overprinting may be indicated by lenses of hydraulically oriented specimens of species that are also found in surrounding or adjacent biogenic accumulations. Early cementation of hydraulically sorted shell pavements or of concretions following mass mortality (Brett and Baird, 1986) results in mixed assemblages of types 5 and 6, respectively.

Kidwell *et al.* (1986) suggested that the ternary genetic classification was applicable to environmental gradients across many soft-bottom environments (Figure 1.7), such as those preserved in the Miocene Calvert, Choptank, and St Mary's Formations of Maryland; Triassic Muschelkalk (Aigner, 1982*a*, 1985); and Pliocene Purisima Formation of California (Norris, 1986). Intertidal and supratidal flats are characterized by both biogenic accumulations (e.g., oyster bars, subsurface CDF-generated shell beds, ray pits, bird nests, hermit crab-generated concentrations) and winnowed lags. Biogenic accumulations also occur in lagoons, but sedimentological concentrations are likely to be represented by storm washovers and flood deposits. Similarly, beaches and shallow subtidal shoals consist almost entirely of sedimentological concentrations formed through

1.5 Models and classifications of fossil assemblages

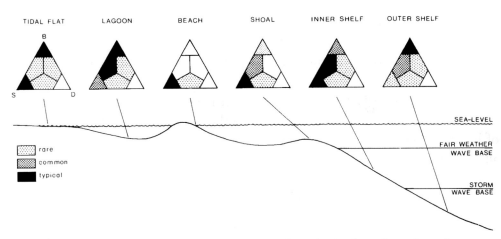

Figure 1.7. Distribution of shell bed types along the depth gradient of a terrigenous marine shelf. Sedimentation rate is assumed to be constant along the transect. Frequency of sedimentological and biogenic concentrations decrease and increase, respectively, across the gradient. (cf. Figure 1.6; Kidwell *et al.*, 1986; reprinted with permission of SEPM.)

storm and fair-weather wave activity. Nevertheless, shell hashes in shallow water are not necessarily produced by abrasion-related processes; for example, the occurrence of fine sediment matrix that is not hydraulically equivalent to shell remains is a contraindication of high-energy conditions and suggests that fragmentation probably resulted primarily from dissolution (Kidwell, 1989; Martin, personal observations at Calvert Cliffs, Maryland; see also Chapter 3). On open soft-bottom shelves above storm wave base, individual and amalgamated storm lags and reworked and recolonized biogenic concentrations predominate. Further offshore, biogenic beds dominate, although rare storm beds may occur, and where sediment bypassing or starvation may produce phosphate-rich "bone beds" (Kidwell, 1989).

In the case of the Maryland Miocene, durations of time-averaging of major shell beds are of the order of a few thousands to a few tens of thousands of years and 10^5 years at most (Kidwell, 1989). Major shell beds can consist of minor shell layers condensed over ecological time scales (10^0–10^2 years). At the other extreme, major shell beds may form over longer intervals that approximate biostratigraphic zones based on molluscs, diatoms, or planktonic foraminifera or transgressive–regressive depositional sequences of approximately 1 million years duration (Kidwell, 1989). The longer the duration of formation, however, the greater the likelihood of shell concentrations of mixed origins (Kidwell *et al.*, 1986).

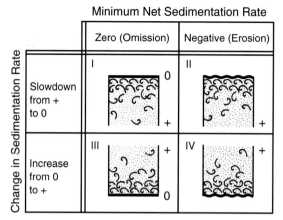

Figure 1.8. Shell bed types based on the R-sediment model. (Redrawn from Kidwell, 1986a.)

1.5.3 The R-sediment model

Based on her work in the Miocene of Maryland, Kidwell (1986a) developed an "R (rate)-sediment" model for the production of marine hardpart concentrations. In this model, she held hardpart (mainly mollusc) input constant ([shell production *in situ* + transport from other environments] − destruction) and allowed sediment accumulation rates to vary from positive (net accumulation) to negative (net erosion). *Type I shell beds* grade upward from positive accumulation to zero and exhibit an upward increase in the density of shells so that the top of the bed represents a shell pavement (Figure 1.8). The upward decrease in accumulation can result either from (1) reduction in sediment supply (sediment starvation) or (2) attainment of baselevel equilibrium and sedimentary bypassing either through aggradation or a downward shift in baselevel. *Type II shell beds* are similar to type I, but are capped by an erosional surface (negative sediment accumulation) consisting of reworked shells. *Types III and IV* are the inverse of types I and II (Figure 1.8). Kidwell (1986a) successfully tested the R-sediment model by varying both sediment and hardpart input and found that it was remarkably robust to changes in both variables (Figure 1.9).

The shell bed types are predicted to differ with respect to post-mortem bias (Figure 1.10). Because of slow sediment accumulation rates, assemblages from the tops of type I and II beds should exhibit greater degrees of time-averaging (enhanced mixing of successive living populations), greater abrasion and fragmentation, and greater degrees of encrustation, bioerosion, and taxa that prefer shell gravels (taphonomic feedback). The concentration of hardparts may also buffer against dissolution so that hardparts are better preserved at the tops of beds than during normal burial, when hardparts are dispersed in the sediment

1.5 Models and classifications of fossil assemblages

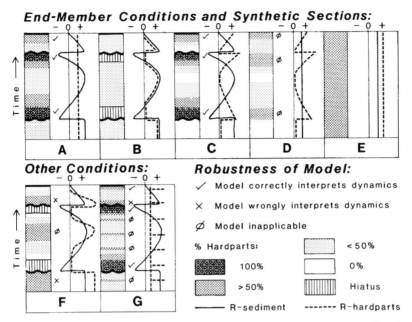

Figure 1.9. Synthetic stratigraphic sections generated using the R-sediment model. (Reprinted with permission of the editors of *Paleobiology*.)

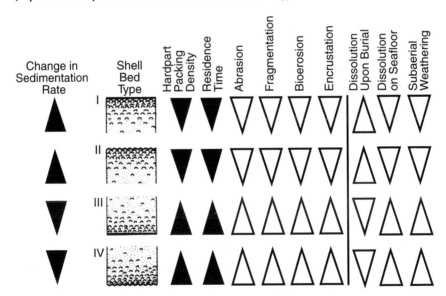

Figure 1.10. Post-mortem bias predicted for each of the shell bed types generated with the R-sediment model. (Redrawn from Kidwell, 1986a.)

matrix, although this effect may be counteracted by enhanced dissolution by undersaturated waters as the shells lie exposed to the SWI or subaerially. Because type II and IV beds are characterized by net erosion at the top and base, respectively, they should exhibit somewhat greater post-mortem bias within the shell

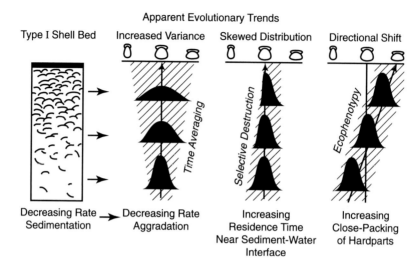

Figure 1.11. *Apparent* intraspecific (evolutionary) morphometric trends generated by the R-sediment model. (Redrawn from Kidwell, 1986a.)

concentrations than types I and III. Bioerosion should, however, *decrease* in types II and IV because they are associated with increased turbidity and physical disturbance. These temporal changes in taphonomic conditions can have important effects on measured morphometric trends within fossil lineages (Figure 1.11).

1.5.4 Taphofacies

The differential preservation of fossils among facies reflects the differential activity of taphonomic processes in different depositional settings. *Taphofacies* ("taphonomic facies") consist of "suites of sedimentary rock characterized by particular combinations of preservational features of the contained fossils" (Brett and Speyer, 1990). They are similar to, but distinct from, *biofacies*, which are distinguished by recurring groups of organisms that presumably lived together (biocoenosis). Taphofacies are really the end products of many of the factors portrayed in the models and classifications discussed above.

I wish only to briefly review the relationship of taphofacies to environmental gradients, the main point being that taphofacies represent both information loss *and gain* along paleoenvironmental gradients, and are of tremendous predictive value in paleoenvironmental intepretation (e.g., Speyer and Brett, 1988). In the case of marine environments, taphofacies reflect such factors as background and episodic (storm) sedimentation rates, current intensity and direction, and levels of oxygenation in water above the SWI and in sediment, as well as geochemical features such as degree of pyritization of fossils. Based on such features in the Middle Devonian Hamilton Group of New York State, for example, Brett and

Baird (1986) and Speyer and Brett (1986) distinguished nine taphofacies along a paleobathymetric (energy) gradient; based on this *inductive* model, Speyer and Brett (1988) developed a *deductive* model of taphofacies for Paleozoic epeiric seas (Chapters 2, 3).

Taphonomic modes and taphofacies were early recognized by vertebrate taphonomists. In one of the more recent studies, Behrensmeyer (1988) recognized two basic taphonomic modes – akin to taphofacies – in her study of attritional vertebrate assemblages of the Siwalik deposits (Miocene) of northern Pakistan and the Lower Permian Belle Plains and Arroyo Formations (Wichita Group) of Texas. She emphasized that the two modes represent end-members of a *spectrum* of assemblages (cf. classifications above), that the modes are defined primarily on overall sedimentary context and secondarily on taphonomic features of the bone assemblages, and that rate and mode of sedimentation is critical to preservation (see also Chapters 2, 3, 7).

1.6 Facts or artifacts?

All earth scientists have been indoctrinated into the dogma of reductionist science: the simplest system is the most easily understood and the simplest explanation is the most likely (Occam's Razor or the Principle of Parsimony). Besides, the geological record is incomplete: Earth history and paleontology can't *prove* anything.

Nevertheless, the fossil record not only allows us to assess the impact of ecological processes over periods of time much longer than those normally considered by an ecologist, but also reveals that there are likely to be processes (as indicated by patterns in the fossil record) that occur only on long time scales. To corrupt Marshall McLuhan's famous phrase, the pattern may be the process. Moreover, time-averaging of fossil assemblages actually enhances expression of ecological signals because it damps (filters) out short-term noise. That is not to say that understanding how ecological signals are "filtered" or "smoothed" is not important; they are if we are to understand what gets incorporated into the fossil record and *how*.

These patterns, and presumably the processes that produced them, are by no means strictly academic (Chapter 10). They bear strongly upon man's impact on the health and biotic diversity of the planet, not to mention the future of the profession of paleontology. Paleontology and stratigraphy bear upon such environmental phenomena as population dynamics; the dynamics of speciation and extinction; biotic recovery from extinction; the organization and resilience of biological communities to disturbance; and the occurrence of alternative community states in response to environmental disturbance, both natural and anthropogenic.

Besides attempting to answer these broad questions and posing new questions in their stead, paleontology and stratigraphy continue to affect us directly on the short time scales of our everyday lives. Ironically, while humankind has the means to answer questions about the state of the planet's health, it has historically concentrated on finding resources that will alter it (oil and gas), and will continue to do so as long as civilization remains dependent on fossil fuels. On the other hand, we can also answer such questions as those about the state of the environment prior to the spread of humans to an area. The best way to assess anthropogenic effects on ecosystems – and to develop truly effective management strategies for them – is to study those ecosystems before human settlement and industrialization (e.g., Greenstein *et al.*, 1995; see also Russell, 1997). What constitutes a pristine environment and what doesn't? What have been the effects of land use and runoff (including fertilizers, sewage, and nutrients released from deforested soils) on coastal and estuarine ecosystems over the past few decades and the past few centuries (e.g., Anderson, 1994)? How might such studies affect legislation regarding coastal ecosystems? How fast is sea-level rising? Such studies might also contribute to ecological and evolutionary theory regarding, say, disturbance. Thus, paleontology and taphonomy have tremendous potential as *environmental sciences*. But to realize this potential, we *really must know how fossil and subfossil assemblages are generated*: no matter how good the stratigraphic record *looks*, it cannot necessarily be taken at face value (Figures 1.10, 1.11).

2 Biostratinomy I: necrolysis, transport, and abrasion

Alas, poor Yorick ... William Shakespeare, *Hamlet*

2.1 Introduction

Because of a large body of theory regarding sediment erosion, transport, and deposition, and the relative ease of flume and settling experiments, the movement of individual shells and bones as sediment particles has probably received more attention than any other aspect of taphonomy. Nevertheless, the extent of transport in natural assemblages may still be difficult to discern. Ideally, fossil assemblages consist of skeletal remains that are *autochthonous*. Normally, however, shells and bones are moved to a minor extent from their source (*parautochthonous*) or, in some cases, hardparts are transported considerable distances and mixed with other assemblages and comprise *allochthonous* components of an assemblage.

The enclosing sediment (*matrix*) is no less important than fossils in discerning operative taphonomic processes. The sediment may be of two basic types: (1) *terrigenous*, which is derived from land (often from outside the basin), is almost invariably siliciclastic, and is transported to some degree before deposition; and (2) *carbonate*, which normally consists of the autochthonous or parautochthonous remains of organisms living within the depositional basin (Maiklem, 1968). Thus, there is typically greater opportunity for hydraulic factors to modify the size distributions and surface features of biogenic and non-biogenic particles of terrigenous deposits than in pure carbonate sediments. Smith (1977) published a dimensionless parameter (H) for *physical* mixing that represents the ratio of the depth of physical erosion to the frequency of erosion relative to sediment accumulation rate (see also Chapter 4). Because fine sediment is last deposited and first eroded, large values of H (intense physical reworking) indicate that only coarse sediment and homogenous (or massive) sands will accumulate, whereas finer – and more heterogenous – sediments tend to accumulate as H decreases (see also Nittrouer and Sternberg, 1981). On the Washington shelf (c. 75 m water depth), for example, Nittrouer and Sternberg (1981) estimated that the seabed is eroded for approximately 75 days each year.

Besides the duration and distance of transport, other factors in terrigenous regimens include *sediment supply* and the *nature of the bioclasts* themselves; the *mode of transportation* (e.g., *bottom traction, saltation, suspension, or colloids*); the *shape, angularity, size, and density of the sediment grains* (including shell and bone); and the *competency* and *capacity* of the transporting agent. By contrast, the size distributions of bioclasts in carbonate environments are largely inherited from the producers and mainly reflect the types, abundance, and size distribution of source organisms and their shell architecture and microstructure.

2.2 Fundamentals of fluid and sediment movement[1]

Shells and bones are transported by water, whereas leaves and pollen are typically transported by air. The basic physical properties of fluids are determined by their *density* and *viscosity* (Boggs, 1987). Density (ρ or mass per unit fluid volume) affects the magnitude of the forces that act within a fluid, on a substrate, and on the settling velocity of particles. The density of water is more than 700 times that of air and increases with decreasing temperature; thus, the *competence* and *capacity* of water is much greater than that of wind.

Fluid viscosity is a measure of the ability of fluids to flow (Boggs, 1987). Fluids of low viscosity (such as air) flow readily, whereas fluids of high viscosity (e.g., ice) flow slowly. *Dynamic viscosity* (μ) is a measure of the resistance of a fluid to change in shape caused by a force (*shear stress* or force per unit area, τ) acting parallel to the surface of the fluid:

$$\mu = \frac{\tau}{du/dy} \tag{2.1}$$

where du/dy is the rate of deformation across the fluid (i.e., du/dy is a velocity gradient perpendicular to the direction of shear; Figure 2.1; cf. Figure 2.2). Shear stress can be generated at the boundaries between moving fluids and is a function of how much a slower fluid retards the velocity of a faster moving fluid; the greater the viscosity, the greater the shear stress must be to produce the same rate of deformation (du/dy). Shear stress at the sediment–water interface (SWI) is called *shear velocity* (in cm s^{-1}) and is critical to the erosion, entrainment, and transport of sediment.

---

[1] This section is based on the excellent summary in Boggs (1987).

2.2 Fundamentals of fluid and sediment movement

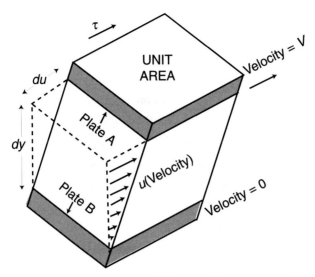

Figure 2.1. Movement of a fluid encased between two rigid plates (A and B). See text for further discussion. (Redrawn from Boggs, 1987.)

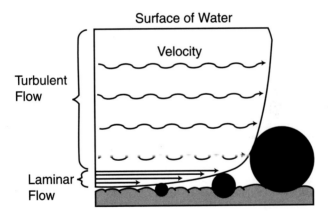

Figure 2.2. Schematic cross-section of a current, showing distribution of velocity and turbulence. Note projection of larger sediment particles above laminar layer into faster turbulent layers above. (Redrawn from Kuenen, 1950.)

Fluid movement is of two basic types (Boggs, 1987). *Laminar* flow is flow in slowly moving fluids (e.g., ice or water just above a stream bed) and consists of parallel sheets of fluid moving past one another (Figure 2.2). If fluid velocity is increased or viscosity decreased, fluid movement becomes *turbulent*, in which *eddies* move normal to the mean direction of flow. Laminar flow near a stream bed tends to be destroyed by sediment grains that project above bottom (Figure 2.2); thus, turbulent flow is also partly a function of bed roughness. Most water and air flow is turbulent, although water has a much greater ability to move sediment than air.

Because fluid viscosity and density both decrease with increasing temperature, they are often combined into a single parameter called *kinematic viscosity* (ν), which affects the production of eddies:

$$\nu = \frac{\mu}{\rho} \qquad (2.2)$$

The upward motion of eddies slows the settling of sediment particles and allows them potentially to be transported farther than they would be otherwise. Laminar flow near a stream bed tends to remain more or less constant and turbulent flow varies instantaneously about an average value, but mean fluid velocity tends to increase with the logarithm of distance above bottom (Figure 2.2). Moreover, a fluid undergoing turbulent flow resists deformation more than fluid undergoing laminar flow, so that turbulent fluids have a higher apparent viscosity (*eddy viscosity*), which is normally several orders of magnitude higher than dynamic viscosity. Thus, turbulent flow erodes and transports larger and denser particles than does laminar flow.

The main difference between laminar and turbulent flow therefore arises from the ratio of forces that cause turbulence (which tends to move particles) and viscosity (which tends to dampen fluid deformation), respectively. This ratio is expressed as a dimensionless *Reynolds number* (*Re*):

$$Re = \frac{UL}{\nu} \qquad (2.3)$$

where U is mean velocity of fluid flow, L is length (typically water depth), and ν is kinematic viscosity. When Reynolds numbers are small (low velocity, shallow water depth, or high viscosity such as in mudflows), fluid flow past a particle is laminar and the particle descends in a stable orientation. As flow velocity increases, Reynolds numbers also increase and turbulence dominates (as with wind and most rivers) and a wake is generated that causes the particle to assume unstable orientations (e.g., lateral oscillations). The transition from laminar to turbulent flow takes place for $Re \simeq 500\text{--}2000$, and depends on channel depth and geometry (Futterer, 1978c). Because *Re* is dimensionless it can be used to predict the extent of turbulence in both natural and scaled-down (experimental) settings, although to my knowledge this has not been done in taphonomic studies.

In order for sediment grains to be picked up and moved by fluids (*entrainment*), a *critical threshold or traction velocity* must be reached by the transporting medium. The threshold velocity is a function of fluid velocity (which increases above bottom), *boundary (bed) shear stress* (τ_0), fluid viscosity, and particle size,

2.2 Fundamentals of fluid and sediment movement

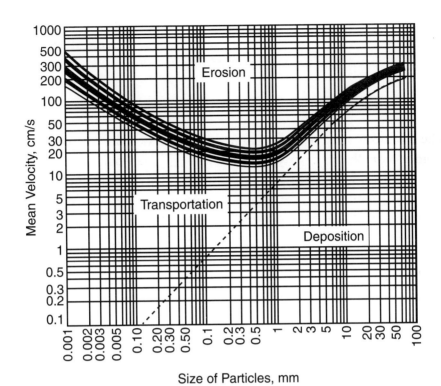

Figure 2.3. Hjulström diagram showing relation between sediment grain size and current velocities required for erosion, transport, and deposition. See text for further discussion. (Redrawn from Kuenen, 1950.)

shape, and density. Sediment particles resist movement because of the downward force of gravity (weight) and friction between sediment particles. In order for entrainment to occur, fluid velocity must generally increase with increasing grain size above approximately 0.5 mm (medium-grained sand). Clays have greater resistance than expected based on grain size alone because of electrochemical bonds between clay particles, as indicated by the *Hjulström diagram* (Figure 2.3). To initiate transport, a fluid must exert a *drag force* (F_D) parallel to the bottom (related to boundary shear stress) and a lift force due to the *Bernouilli effect* (Figure 2.4), where

$$F_D = \tau_0/N \qquad (2.4)$$

and N is number of exposed grains per unit area.

A plot similar to the Hjulström diagram (*Shields diagram*) may be generated using two dimensionless parameters, Re_g (*grain Reynolds number*), which is analogous to the Reynolds number but is for sediment grains and is a measure of

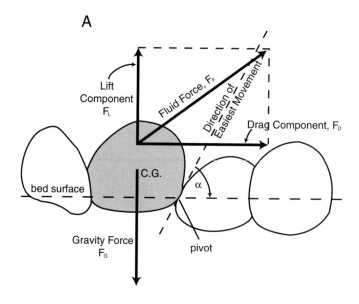

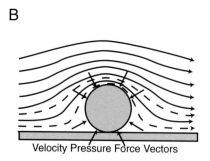

Figure 2.4. (A) Forces acting during fluid flow on a sediment grain. (B) Fluid flow over individual grain showing Bernouilli effect. C.G., center of gravity. (Redrawn from Boggs, 1987.)

turbulence at the grain–fluid boundary, and *dimensionless shear stress* (τ^*):

$$Re_g = \frac{U^* D}{\nu} \qquad (2.5)$$

where U^* is "*friction velocity*" (a measure of turbulence), D is sediment grain size, and ν is kinematic viscosity, and

$$\tau^* = \frac{\tau_0}{(\gamma_s - \gamma_f)D} \qquad (2.6)$$

2.2 Fundamentals of fluid and sediment movement

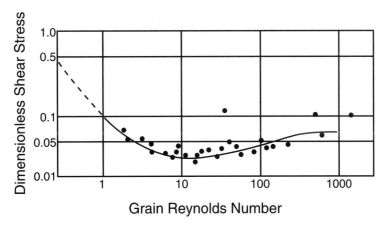

Figure 2.5. Shields diagram showing relation between grain Reynolds number and dimensionless shear stress. (Redrawn from Boggs, 1987.)

in which γ_s is specific gravity of sediment particles and γ_f is specific gravity of fluid, respectively (Figure 2.5). An increase in Re_g indicates either an increase in grain size, increased friction velocity and turbulence, or decreased kinematic viscosity, whereas an increase in τ^* indicates an increase in shear stress flow velocity or a decrease in grain size or density. Points above the curve in the Shields diagram represent grains on the bed that are in motion, whereas points below the curve indicate grains at rest. τ_0 increases slightly as Re_g increases above ~5–10; at lower Re_g, however, τ_0 increases steadily because when sediment grains are up to fine sand in size (~0.125–0.25 mm), sediment particles do not project up into the turbulent layers (cf. Figure 2.2). Although it is more difficult to interpret, the Shields diagram is preferred over the Hjulström diagram because it is more general and is applicable to transport by wind. Sediment movement in both diagrams can, however, be affected by cohesiveness of clays and instantaneous fluctuations in eddy velocity.

If fluid velocity decreases sufficiently, sediment grains settle out of suspension. Initially, the particle velocity is quite rapid as the particle accelerates downward, but eventually the particle reaches a constant terminal *settling velocity*. The smaller the particle, the faster it reaches terminal velocity, which is a function of fluid viscosity, and the size, shape, and density of the particle. As particles settle, both upwardly directed forces of fluid buoyancy and viscous drag act to retard downward movement caused by gravity. The upwardly directed drag force is proportional – for spherical grains – to fluid density (ρ_f), grain diameter (d), and fall velocity (V), where

$$C_d \pi \frac{d^2}{4} \frac{\rho_f V^2}{2} \tag{2.7}$$

and C_d is the *drag coefficient* that is related to Re_g and particle shape (curves of C_d versus Re_g are given in most fluid mechanics texts). The upward buoyancy force is given by:

$$\frac{4}{3}\pi\left(\frac{d}{2}\right)^3 \rho_f g \tag{2.8}$$

where g is gravitational acceleration, and the downward gravitational force is given by:

$$\frac{4}{3}\pi\left(\frac{d}{2}\right)^3 \rho_s g \tag{2.9}$$

where ρ_s is particle density. After the grain has achieved terminal velocity, the downward drag force equals the downward gravitational force minus the upward force of buoyancy:

$$C_d \pi \frac{d^2}{4} \frac{\rho_f V^2}{2} = \frac{4}{3}\pi\left(\frac{d}{2}\right)^3 \rho_s g - \frac{4}{3}\pi\left(\frac{d}{2}\right)^3 \rho_f g \tag{2.10}$$

Equation 2.10 may be rearranged so that

$$V^2 = \frac{4gd}{3C_d}\frac{(\rho_s - \rho_f)}{\rho_f} \tag{2.11}$$

For slow laminar flow with low sediment concentrations and low Re_g, $C_d = 24/Re_g$ or $24/U^*D/\nu$, which equals $24/U^*D/(\mu/\rho_f)$; substituting into equation 2.11:

$$V = \frac{1}{18}\frac{(\rho_s - \rho_f)gd^2}{\mu} \tag{2.12}$$

which is known as *Stokes Law* and is usually simplified to

$$V = CD^2 \tag{2.13}$$

where C is a constant $[= (\rho_s - \rho_f)g/18\mu]$ and D is particle diameter of a sphere (cm). Stokes Law accurately predicts settling behavior only for particles less than about 0.2 mm in size; larger grains have slower settling velocities than those predicted by Stokes Law owing to turbulence produced by the faster rates of fall of larger particles. V also decreases with decreased fluid temperature and increased suspended load (increased fluid viscosity), decreased particle density, and departure from spherical shape. For spherical particles outside the size range of Stokes Law, the following empirical relationship is used (Gibbs *et al.*, 1971, as modified

2.3 Microfossils

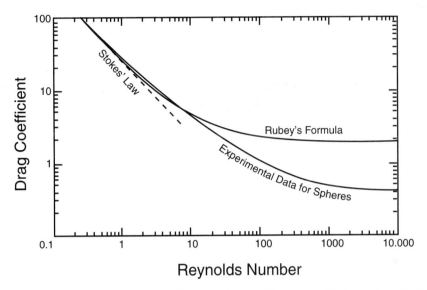

Figure 2.6. Variation of drag coefficient with Reynolds number. (Redrawn from Blatt et al., 1972.)

from Rubey, 1933):

$$V = \frac{-3\mu + [9\mu^2 + gr^2\rho(\rho_s - \rho)(0.015476 + 0.19841r)]^{1/2}}{\rho(0.011607 + 0.14881r)} \quad (2.14)$$

where V is settling velocity (cm sec^{-1}); μ is dynamic viscosity of water (equation 2.1); ρ_s and ρ are sediment grain and water densities, respectively; g is gravitational acceleration; and r is sphere radius (in cm; Figure 2.6).

2.3 Microfossils

2.3.1 Mainly foraminifera and other calcareous microfossils

Because of their relatively small size, foraminiferal tests and other microfossils are potentially very susceptible to post-mortem transport. Consequently, post-mortem transport of microfossils has received a great deal of attention from micropaleontologists. Mackenzie *et al.* (1965), Martin (1988), and Brunner and Culver (1992), for example, have used foraminifera as sediment tracers in turbidites and other types of debris flows, in which foraminifera may be hydraulically sorted and reflect different depositional processes on the slope (see also Brunner and Normark, 1985; Brunner and Ledbetter, 1987; Lundquist *et al.*, 1997).

Foraminifera may also be put into suspension in water, especially in shallow high-energy regimes of estuaries and tidal flats. Murray (1965) detected empty tests of modern shallow-water benthic foraminifera in plankton samples, while Grabert (1971) used the occurrence of shallow-water trochammines (relatively simple agglutinated tests) to infer the transport of sand from near the shoreline into a lagoon (see also Culver, 1980; Murray et al., 1982; Gao and Collins, 1995). Grabert determined through settling velocity experiments that trochammine tests and quartz grains with the same hydraulic-equivalent diameter have a size ratio of about 3:1. Field studies indicated, however, that finer sediment grains and tests were deposited together, so that hydraulically equivalent grains were not equivalent in terms of transport. She concluded that size, shape, and density were important in determining transport and settling velocity.

Kontrovitz et al. (1978) conducted flume studies of the suspension and transport of benthic foraminifera. Twenty-five specimens of each of 12 shallow shelf-to-slope species were measured for volume, weight, *operational sphericity* ($\sqrt[3]{V_1/V_2}$, where V_1 is volume of the test and V_2 is volume of a circumscribing sphere), *maximum projection sphericity* ($MPS = \sqrt[3]{S^2/L \cdot I}$, where S is the shortest axis of the particle and L and I are the longest and intermediate axes, respectively), *effective density* (ratio of test weight in water to test volume), and *nominal diameter* or D_n (the diameter of a sphere having the same volume as the test), in which

$$D_n = (D_s D_i D_l)^{1/3} \qquad (2.15)$$

and where D_s, D_i, and D_l are the short, intermediate, and long axes of the grain, respectively (when $D_s = D_i = D_l$, the ellipsoid becomes a sphere and D_n becomes the sphere's diameter; Fok-Pun and Komar, 1983). Mean traction (threshold transport) velocities ranged from 5.1 to 18.7 cm s^{-1} on a bed of subangular to rounded fine sand (cf. equations 2.5, 2.6 and Figures 2.4, 2.5). Although maximum projection sphericity is, according to Kontrovitz et al. (1978), a better indicator of hydraulic behavior than operational sphericity, only operational sphericity and weight exhibited significant correlations with traction velocity. After removing two species as outliers (most notably *Archaias angulatus*, which is a large reef-dwelling foraminifer in West Indian shallow-water carbonate sediments), they determined through multiple regression that mean traction velocity (V_t in cm s^{-1}) is related to maximum projection sphericity (MPS) and weight (W in mg) by the equation

$$V_t = 18.4 - 11.4 MPS - 38.9 W \qquad (2.16)$$

Similarly, Zhang et al. (1993) examined the hydraulic behavior of five of the most common (~0.25 mm in size) foraminifera from tidal flats of Bahia la Choya,

Sonora, Mexico (northern Gulf of California). They found that, in general, test shape, initial orientation (a matter of investigator control), and nature of the substrate all influenced traction velocities (range 1–22 cm s^{-1}), and that the type of movement varied from rolling or sliding to saltation and suspension. Once in suspension, tests were subject to lateral transport.

Despite reports of test transport, however, assemblages would appear to remain largely parautochthonous in most shelf environments below wave base. Snyder et al. (1990a,b) conducted extensive field and experimental studies of transport of benthic foraminifera on the Washington continental shelf. They determined the traction velocities for 31 benthic foraminiferal species that were dominant components (90–100% of natural assemblages). They found that traction velocities ranged from approximately 4.1 to 13.3 cm s^{-1} and that

$$V_t = 22.3 - 19.8 MPS \qquad (2.17)$$

Snyder et al. (1990a,b) recognized three traction velocity groups: (1) four species with roughly equidimensional tests and low traction velocities ($\sim$4.0–6.1 cm s^{-1}); (2) 19 species with elongate, inflated or coiled, moderately inflated tests exhibiting intermediate traction velocities ($\sim$6.4–9.4 cm s^{-1}); and (3) eight species with elongate, highly compressed or coiled, discoidal tests and high traction velocities ($\sim$10.5–13.3 cm s^{-1}). Kontrovitz et al. (1978) and Snyder et al. (1990b) predicted that, depending upon bottom current velocities, surface sediment assemblages ought to be dominated by species belonging to one of the three traction velocity groups; instead, they found bottom assemblages to consist of mixtures of traction velocity groups and concluded that bottom current velocities were too low at the depths sampled (31–185 m) and too erratic to sort foraminiferal assemblages effectively. Possibly this is also because the tests do not project above the SWI (cf. Figures 2.2, 2.4); foraminifera tend to occur and reproduce in patches of various sizes; and tests are mixed via bioturbation (Chapter 4).

Current velocities in marshes also appear to be too low to significantly affect foraminiferal assemblages. In marshes, normal flood tides transport only mud (particle size less than 62 μm; Redfield, 1972) and are therefore typically too weak to transport significant numbers of tests (normally larger than 62 μm; Scott and Medioli, 1980b). Stumpf (1983), for example, found that after a severe northeaster, layers of aluminum glitter were buried in stream-side and high Delaware marsh sites by several millimeters of mud and not eroded. Hippensteel and Martin (unpublished observations) concluded that, initially, significant amounts of surficial layers of sand-sized glass beads are probably suspended through surface tension and lost to tidal currents, especially in the low marsh;

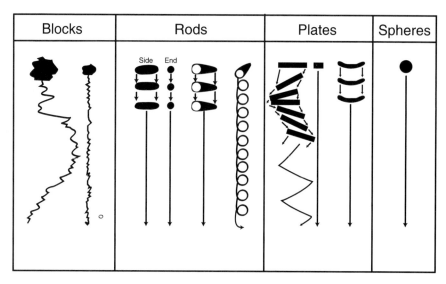

Figure 2.7. Effect of biogenic grain shape on settling path. (Redrawn from Maiklem, 1968.)

however, most foraminifera no doubt remain buried in surficial sediment and attached to surrounding sediment by their pseudopodia.

Tests also vary in their settling behavior. Based on smaller tidal flat foraminifera, Zhang *et al.* (1993) found that both test weight and size influenced settling velocities of foraminifera (0.8–4.5 cm s^{-1}). They determined that settling motion was mainly straight and not substantially influenced by shape. However, Maiklem (1968) determined that the shape of *larger* foraminifera (≥1–2 mm) and other biogenic carbonate clasts (classified as blocks, rods, spheres, and plates) can significantly influence settling velocities (Figure 2.7). For the same Re_g, disks, for example, have the highest C_ds, whereas spheres have the lowest (see also Stringham *et al.*, 1969). "Spheres" (spinose tests of the common Pacific reef-dwelling foraminifera *Baculogypsina* and *Calcarina*, sometimes with the spines worn off) settled smoothly, whereas rough, approximately equidimensional blocks of coral and coralline algae approximately 3–4 mm in size had an unpredictable settling path. The settling path of equidimensional grains became more regular as grain size decreased. Plates (tests of the large, disk-shaped foraminifer *Marginopora*, platelets of the calcareous green alga *Halimeda*, disarticulated valves of unidentified pelecypods) tended to settle with their short axis approximately vertical. Large flat plates zig-zagged back and forth while settling (cf. equations 2.3, 2.5), whereas small flat plates settled smoothly. Large curved plates (pelecypods) settled smoothly with their convex side down (i.e., concave-up). Rods (gastropods, the foraminifer *Alveolinella*, coral sticks, alyconarian and echinoid spines) of all sizes

2.3 Microfossils

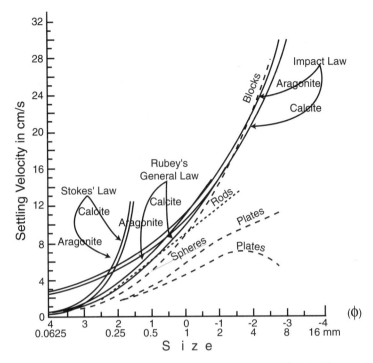

Figure 2.8. Settling curves for biogenic carbonate particles of different shapes versus Stokes, Rubey, and Impact curves for aragonite and calcite grains (cf. Figure 2.9). See text for explanation of Stokes' and Rubey's Laws. Size is intermediate grain diameter. (Redrawn from Maiklem, 1968.)

typically settled smoothly with their long axis horizontal, or if one end of the rod was heavier than the other (e.g., gastropods with air bubbles trapped at the apex), the shell spiraled downward with the heavy end leading (see also sections 2.4, 2.7). In general, calculated settling velocities for grains of calcite and aragonite were up to four times greater than velocities for shallow-water calcareous biogenic grains (Maiklem, 1968; Figure 2.8). These differences disappeared for larger blocks, but became more pronounced for other grain shapes with increasing grain size.

Among all grain shapes, plates (especially *Marginopora*, tests of which consist of highly subdivided chambers that result in low bulk density and specific gravity) exhibited the slowest settling velocities (Figure 2.9; Maiklem, 1968; foraminifera may also be transported on floating vegetation; see, for example, Anderson *et al.*, 1997; Chapter 5). The curves for settling velocities of plates have two parts: one for grain sizes smaller than 2 mm, which appears to correspond to Rubey's (1933) General Settling Curve (cf. equation 2.14), and a flatter portion for particles larger than 2 mm. The divergence in settling velocities of *Marginopora* versus

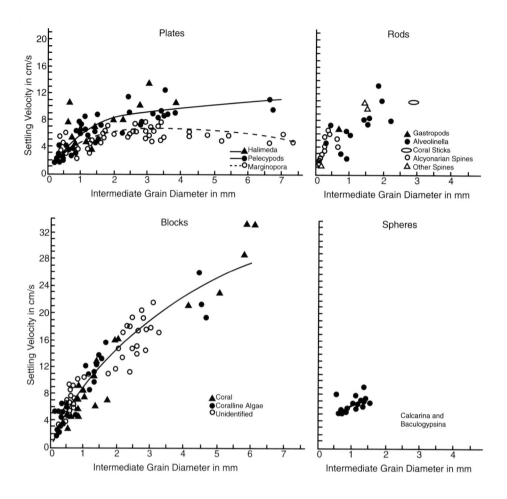

Figure 2.9. Settling velocities of biogenic carbonate grains according to shape. Curves fitted visually. (Redrawn from Maiklem, 1968.)

Halimeda and pelecypods suggests that specific gravity is of greater importance to settling velocity than grain shape for larger particles. The decrease in settling velocity for *Marginopora* larger than about 3 mm in size reflects the greater amplitude of the zig-zag motion of its tests during settling. Not surprisingly, spheres of *Calcarina* and *Baculogypsina*, which have approximately the same specific gravity as *Marginopora* (~2.0–2.4), settled approximately 1.3–1.5 times faster than *Marginopora* of the same size. Block-shaped particles (e.g., corals, coralline algae) less than about 2 mm in size settled relatively slowly, but blocks larger than this settled at velocities close to calculated values. Apparently, organic matter content, roughness, and specific gravity have a greater effect on settling velocities of blocks less than 2 mm in size.

2.3 Microfossils

Besides the potential for post-mortem transport, foraminiferal tests deposited in shallow water are also potentially subject to abrasion and fragmentation. Martin (1986) and Martin and Wright (1988) found that large living populations of the fragile, scab-like foraminifer *Planorbulina* spp., which encrusts plants and other surfaces, were greatly reduced in sediment thanatocoenoses of wave-swept, outer back-reef environments off Key Largo, Florida, and that sediment assemblages were dominated by the robust *Archaias angulatus*, the tests of which were typically highly degraded. Only small living populations of *Archaias* were found on vegetation, however, so that sediment assemblages must represent time-averaged authochthonous-to-parautochthonous assemblages (Chapter 5).

Later, Martin and Liddell (1988, 1991) and Liddell and Martin (1989) adopted a taphofacies approach (based primarily on water energy, abrasion, and dissolution; see Chapter 3) for fringing reefs of Discovery Bay, Jamaica. Based on these field studies, Kotler *et al.* (1992) conducted laboratory experiments on abrasion resistance (using shaker tables) of the 12 most common reef-associated foraminifera. They found that after abrasion in carbonate sand and buffered artificial seawater for up to 1000 hours (which corresponds to a calculated distance of approximately 35 km), most tests exhibited relatively few surface pits, scratches, etc., and only the most fragile species (e.g., *Planorbulina*) were completely destroyed.

Peebles and Lewis (1991) also conducted abrasion experiments on foraminifera, but used tumblers in which the foraminiferal tests themselves served as the abrasive. Although the calculated transport distances in both sets of experiments were similar, their tests exhibited much greater rates of weight loss, which was undoubtedly maximized by their experimental protocol (cf. protocol of Kotler *et al.*, 1992). Shroba (1993) produced significant surficial test degradation via abrasion with quartz sand, but after 10 days surface degradation of tests from natural thanatocoenoses was still much more pronounced than on laboratory specimens. Since these experiments resulted in *maximum* rates of abrasive reduction (Kidwell and Bosence, 1991), they cannot be directly extrapolated to natural settings, but they indicate that once produced, foraminiferal tests may persist for long periods of time in carbonate settings (see also Chapters 3, 5).

Foraminifera are also subject to bioerosion in shallow-water settings. Freiwald (1995) documented the breakdown of the calcareous foraminifer, *Cibicides lobatulus*, via bacteria; he concluded that test destruction was dependent upon the microstructure of the test and the distribution of organic matter in the shell. Martin and Liddell (1991) reported substantial damage to reef-dwelling foraminifera exposed on stakes above the SWI, but the degradation did not resemble that typically seen on tests from natural assemblages; possibly it was the result of predators or grazing fish (e.g., Lipps, 1988). Foraminifera have also been reported as bioeroders themselves, forming small pits in shells, but the quantitative significance appears to be substantially less than that of other microbes (Vénec-Peyré, 1996).

In deeper shelf settings, foraminiferal assemblages are probably largely autochthonous because they form below the storm wave base (cf. Figure 1.7). Further offshore, predominantly planktonic foraminiferal assemblages generally reflect the composition of the plankton of the water column above bottom (Bé, 1977), which indicates that significant test transport can occur only in the upper few hundred meters of the water column (Berger and Piper, 1972). Even smaller microfossils such as coccolithophorids and diatoms, which are smaller than about 50 μm, can potentially be transported large distances (e.g., Sherrod, 1995), but normally settle very rapidly to the bottom (a few days) after their ingestion and incorporation into fecal pellets (Schrader, 1971; Honjo, 1977; see also Chapters 3, 9). Diatoms are also sufficiently small that they can be transported long distances by wind (Stroeven *et al.*, 1996), which is, for example, of no small consequence in determining the stability of the Antarctic ice sheet in the late Neogene (cf. Stroeven *et al.*, 1998 and Harwood and Webb, 1998).

Bottom scour, erosion, and potentially microfossil transport can also occur in the deep sea (e.g., Keller and Barron, 1983; see also Chapter 7). Antarctic diatoms transported by Antarctic Bottom Water (AABW) have been reported as far as 30° N latitude in the Atlantic (Kennett, 1982). Kontrovitz *et al.* (1979) determined the traction velocities of 15 species of planktonic foraminifera in a flume with a substrate of fine sand. They found that threshold velocities ranged from 12.9 to 22.6 cm s^{-1} (within the range reported for currents with 0.5 km of the seafloor; Kennett, 1982), and that intercept sphericity and *MPS* were the only variables signficantly correlated with velocity:

$$V_t = 29.8 - 15.1 MPS \tag{2.18}$$

They could not distinguish between velocities of species of the same genera, and concluded that threshold velocity is a feature of the genus level. The different equations for benthic and planktonic foraminifera reflect basic differences in size and shape of benthic versus planktonic tests. Moreover, the relationship between settling velocity and threshold velocity for foraminifera remains unclear.

Settling behavior of planktonic tests is critical in calculating rates of planktonic foraminiferal production and sedimentation from sediment trap data. Based on Stokes Law (cf. equations 2.12, 2.13), Berger and Piper (1972) determined that for 19 species of planktonic foraminifera (62–250 μm in size), the modal settling velocity of empty tests was 0.3 cm s^{-1} for tests of 62–125 μm size to 2.3 cm s^{-1} for tests bigger than 250 μm, which was about 2.4 times less than that of a quartz sphere. For an ocean basin of 5000 m depth, this translates into a descent time of about 19 days for tests of 62–125 μm size to 2.5 days for tests larger than 250 μm, depending upon test weight and spinosity (Bé, 1977). Non-spinose thick-walled tests may settle twice as fast as spinose thin-walled tests of another species (Bé, 1977). Berger and Piper (1972) also determined that the sequence of test settling approximated that found in foraminiferal assemblages of graded beds (see also Brunner and Normark, 1985; Brunner and Ledbetter, 1987).

Nevertheless, Berger and Piper (1972) found that large variations in settling velocity occurred as a result of test shape and wall thickness (effective density). The cone-shaped species, *Globorotalia truncatulinoides*, for example, settles preferentially with its spiral (flattened portion of cone) downward (Bé et al., 1977). Fok-Pun and Komar (1983) examined the effects of shape and density in greater detail, and found that tests (all larger than 210 μm) did not settle in the Stokes region because the grain Reynolds numbers were always greater than 1.0, where

$$Re_g = V_s D / \nu \qquad (2.19)$$

and V_s is test settling velocity, D is nominal test diameter, and ν is kinematic viscosity (cf. equations 2.2, 2.3, 2.5). In the case of *Orbulina universa*, for example, which has a spherical terminal chamber, the Re_g ranged from 7.3 to 40, whereas the Stokes settling relationship is valid for $Re_g < 0.5$ (Fok-Pun and Komar, 1983). In order for the Stokes relation to apply, the diameter of *O. universa* would have to be larger than about 100–200 μm (Fok-Pun and Komar, 1983; Boggs, 1987).

Fok-Pun and Komar (1983) calculated test shape according to the *Corey Shape Factor* (*CSF*), where

$$CSF = D_s / (D_i / D_l)^{1/2} \qquad (2.20)$$

For a spherical shape, $CSF = 1.0$, and the lower the *CSF*, the less spherical the particle. As test shape departed from the spherical (e.g., *Globigerinoides ruber*), C_d increased and settling velocity decreased (cf. equation 2.11), and measured settling velocities ranged from 0.736 to 0.915 times predicted velocities.

2.3.2 Mainly non-calcareous microfossils

The transport and abrasion of ostracodes (which are sometimes impregnated with $CaCO_3$ or other salts and are very sensitive to bottom conditions), conodonts, radiolarians, and dinoflagellate cysts have also been investigated. Kontrovitz

(1975) studied the behavior of whole carapaces and separate valves of different ostracodes in a flume; he found that whole carapaces were entrained at current velocities of 6.3–12.7 cm s^{-1}, whereas separate valves were entrained at velocities of 17.5–26.3 cm s^{-1}. Although separate valves tended to settle concave-up, in the flume valves tended to flip over (convex-up), whereupon current velocities required to move them approximately doubled. Kontrovitz (1975) concluded that shape was most important in ostracode transport: whole carapaces moved at lower velocities than did separate valves because whole carapaces (essentially biconvex in shape) typically exhibited less contact with the substrate. This may explain the observations by other workers that larger ostracodes are sometimes moved before smaller ones, and that size is more important than shape in determining ostracode transport (Kontrovitz, 1987). Overrepresentation of juvenile molts versus adults in transported assemblages has also been attributed to size-based sorting, whereas in certain asymmetric genera, selective sorting of left/right valves on the basis of shape may occur (van Harten, 1986, 1987). The same factors also affect carapace preservation during sediment compaction (Kontrovitz et al., 1998; see Wetmore, 1987, for similar studies of foraminifera).

In asymmetric species, left and right valves may also exhibit differential preservation according to strength, thereby mimicking differential sorting (Whatley et al., 1982; van Harten, 1986, 1987), so that fragile species may be lost during burial and compaction (Izuka and Kaesler, 1986). Shell size, shape, ornamentation, and thickness alone are relatively unimportant in determining ostracode shell strength; shell strength is instead the result of a combination of these factors along with shell chemistry, structure, and ultrastructure (Whatley et al., 1982).

Ornamentation may markedly affect settling velocities of dinoflagellates, however. Modern dinoflagellate cysts (*Gyrodinium uncatenum*, *Gonyaulax tamarensis*, and *Scrippsiella trochoidea*) settle faster than vegetative cells (6–11 m d^{-1}). Unlike spinose foraminifera, though, short spines on the cysts of *S. trochoidea* *increased* sinking rates by approximately one-third. Thus, dinoflagellate assemblages may be affected by density differences and current regimes of water masses (Wall et al., 1977; Anderson et al., 1985). Eddies associated with upwelling have also been shown to affect the orientation of radiolarians (Dozen and Ishiga, 1997); based on flume experiments, current velocities were higher and test orientations more variable in Jurassic red-bedded cherts than in black (stagnant) ones.

Post-mortem transport of conodonts has also been studied in laboratory settings. Unlike most other microfossil taxa, conodonts are members of multi-element skeletons in life, and are studied from recurrent groups of dissociated elements. Broadhead et al. (1990) and McGoff (1991) found that in settling experiments, conodonts are hydraulically equivalent to phosphate spheres of coarse silt to sand size; they settle in the Stokesian range with a wide range of velocities ($\sim 1-10$ cm s^{-1}) depending on shape, and spin about a central axis as they settle (cf. Figure 2.7). McGoff (1991) found that the settling velocity of a particular conodont element is of about the same magnitude as the current velocity necessary to entrain the particle. She also found that elements with equal Re_gs may have different C_ds, implying that shape is a primary control on conodont hydrodynamic behavior. Length was found to be a satisfactory proxy for overall element shape (McGoff, 1991). Individual conodont elements have a high specific gravity and are largely resistant to abrasion and bioerosion (cf. Argast *et al.*, 1987). Thus, conodonts may form lags (Ellison, 1987), even under low-energy conditions (given suitable shapes) or be reworked (Metzger, 1989) with little or no surface alteration (McGoff, 1991; Broadhead and Driese, 1994). Ramiform conodonts are thought, however, to be more prone to fracturing and destruction (McGoff, 1991). Like foraminifera, conodonts and ostracodes may also be transported by debris flows, which may be recognized by mixing of faunas of different depths and concentrations of hydrodynamically similar elements (van Harten, 1986; McGoff, 1991).

Conodonts have been used extensively in zonation and correlation across Paleozoic mass extinction intervals (Frasnian–Famennian, Late Permian). Despite their potential for reworking, however, Sweet (1992) dismissed reworking of conodonts as a complicating factor in placement of the Permo-Triassic boundary (see also Chapter 10). He found that conodont taxa that extended into the Triassic were probably not reworked because the species in question all disappeared in approximately the same order in different boundary sections (cf. Tozer, 1988).

2.4 Cnidaria and associated biota

The sediment grains of carbonate environments are mostly unlike those of terrigenous settings because the grains are almost purely biogenic calcareous skeletons, the durability of which differs markedly from inorganic sediment grains and from each other (Table 2.1). Based on tumbling experiments using chert pebbles and silica sand (cf. section 2.3.1), Chave (1964) found that the least durable forms were filamentous calcareous algae (*Corallina*) and bryozoans, which were reduced to unrecognizable particles in less than an hour. Destruction rates were slower in quartz sand than in chert pebbles, but the relative rates remained about the same. Gastropods, pelecypods, echinoids, and scleractinian corals (*Acropora cervicornis*, which is a prominent member of West Indian shallow-water tropical carbonate ecosystems) were of intermediate durability, although at the end of 40 hours,

Table 2.1. *Types of carbonate skeletons*

Type	Main Phyletic groups	Examples
Spicules	Sponges	*Leucetta floridana* (calcisponge)
	Alcyonarians	*Gorgonia flabellum* (sea fan), *Briareum asbestinum* (sea whip)
	Tunicates	*Didemnum candidum*
	Holothurians	*Holothuria floridana*
Sheath	Codiacean green algae	*Penicillus, Udotea, Rhipocephalus*
Segments	Codiacean green algae	*Halimeda*
	Dasycladacean green algae	*Chalmasia, Acetabularia*
	Red algae	*Jania, Amphiroa*
	Echinoid spines	*Diadema antillarum, Lytechinus variegatus*
	Asteroids, Ophiuroids	*Linckia guildingi*
Branches	Red algae	*Goniolithon*, some *Lithothamnium*
	Corals	*Acropora* spp., *Porites divaricata*
	Bryozoans	*Schizoporella floridana*
	Hydrozoans	*Millepora complanata*
Chambers	Gastropods	Cerithids
	Foraminifera	Miliolids, Peneroplids
	Pelecypod valves	*Chione cancellata*
	Echinoderm tests	*Lytechinus variegatus, Clypeaster rosaceus*
	Worms	Spirorbids
Crusts	Red algae	*Lithothamnium, Melobesia*
	Corals	*Diploria clivosa*
	Hydrozoans	*Millepora alcicornis*
	Bryozoans	*Schizoporella floridana*
	Annelid worms	Spirorbids, serpulids
	Foraminifera	*Homotrema rubra, Planorbulina acervalis*
Massive	Corals	*Montastrea* spp., *Diploria* spp., *Siderastrea sidera*

Grouping is in order of increasing resistance to mechanical breakdown.
Enos and Perkins (1977) after Ginsburg *et al.* (1963).

echinoids (*Strongylocentrotus*), starfish (*Pisaster*), and *Corallina* were present only as fragments less than 2 mm in size, whereas peleycpods were only slightly worn. Skeletal mineralogy (aragonite versus calcite or mixed mineralogies) did not appear to be related to durability, but size did, even within the same genus.

Chave (1964) concluded that despite the multiple sources and traits of carbonate skeletons, assemblages formed in high-energy regimes would all be

2.4 Cnidaria and associated biota

simplified to thanatocoenoses of more durable skeletons. He concluded that the primary factors in controlling the durability of carbonate skeletons were microarchitecture of the shell and the dispersion of organic matrix among the $CaCO_3$ crystallites, as did Sorby (1879) much earlier (*Sorby Principle* of Folk and Robles, 1964; see also Chapter 3 on dissolution). Dense, fine-grained skeletons of pelecypods and gastropods were the most durable; oysters and corals with more porous skeletons were intermediate; and echinoderms, bryozoans, and algae – all with very open, relatively lightly calcificied skeletons – least durable.

More recent studies have demonstrated considerable variation in the durability of bryozoans, which may be significant contributors of hardparts in temperate-water settings (see Smith and Nelson, 1994, and Smith, 1995, for reviews). Smith and Nelson (1996; see also Cheetham and Thomsen, 1981) found that free-living and erect-rigid fenestrate forms (which are strongly calcified and reinforced by cross-bars) are most resistant to abrasion in tumbling experiments, whereas multilaminar encrusting forms are much weaker. High-energy faunas are largely destroyed, leaving only unidentifiable fragments of erect-flexible articulated and encrusting unilaminar growth forms; elongate robust genera (e.g., *Archimedes*) may be oriented biomodally in ancient deposits (Wulff, 1990). By contrast, moderate and low-energy faunas are characterized by more diverse growth forms and are better preserved; breakdown of skeletons may also produce significant quantities of carbonate mud in low-energy environments. Smith and Nelson (1996) also suggested that degree of surface wear may be useful in assessing abrasion and hydraulic regime (but see Chapters 3 and 5). Like Chave (1964), Smith *et al.* (1992) found that mineralogy was a poor predictor of relative rates of dissolution of bryozoan skeletons: small delicate skeletons with non-compact morphologies dissolved most rapidly in laboratory acid baths (see also Chapter 3). The paucity of bryozoans in modern assemblages and the general increase in bryozoan abundance through the Cenozoic (Horowitz and Pachut, 1994) may therefore be partially a function of the general fragility of the skeleton (see also Chapter 9).

Based on field studies of tropical carbonate sediments off Alacran Reef (Yucatan, Mexico), Folk and Robles (1964) similarly found that the mechanical breakdown and grain size distribution of *Acropora cervicornis* and *Halimeda* was controlled by the internal geometry of the crystallites. In the case of *Halimeda*, complete breakdown resulted in the production of a bimodal distribution of grains of relatively coarse (250–400 µm) flakes and microscopic (~1 µm) aragonite needles, which is typical of carbonate sediments (Force, 1969; Scoffin, 1992). Thus, although platelets of *Halimeda* and other calcareous algae may be preserved in ancient carbonate sediments (e.g., Boss and Liddell, 1987), in other cases they may be completely reduced to carbonate sand and mud, leaving little or no hint as to their presence (Moberly, 1968; Neumann and Land, 1975). Well-sorted carbonate sands may result under high-energy regimes if resistant grains are originally sand-size (e.g., Martin, 1986; Martin and Wright, 1988) or if the contributing taxa produce sand-size particles upon breakdown (Scoffin, 1992). On the other hand, poorly sorted carbonate sediments may result from the trapping action of marine angiosperms (e.g., *Thalassia* or "turtle grass") and cyanobacteria (Neumann *et al.*, 1970; Scoffin, 1970), neither of which may leave any record.

Carbonate grains are also subject to extensive encrustation and bioerosion in shallow-water settings. Clionid sponges remove up to approximately $7\,\mathrm{kg\,m^{-2}}$ a year from experimental substrates (Neumann, 1966) and produce from about 20 to 80 μm size scalloped chips that may make up a substantial portion of silt-size carbonate sediment (Futterer, 1974; Rützler, 1975; Scoffin, 1992). Microboring fungi and filamentous green algae and cyanobacteria, which are most prevalent in quiet-water sediments of lagoons (Swinchatt, 1965; Alexandersson, 1972; May *et al.*, 1982; Peebles and Lewis, 1988) and often infest preferred substrates (Perkins and Halsey, 1971), also cause considerable damage to carbonate grains from the SWI up to 1.6 m below the surface (Perkins and Halsey, 1971; Golubic *et al.*, 1975; Kobluk and Kahle, 1977; May and Perkins, 1979). These microboring taxa have been reported to remove up to about $0.4\,\mathrm{kg\,m^{-2}}$ a year in lagoonal sediments of Davies Reef, Australia (Tudhope and Risk, 1985). Microbes also outranked chitons and echinoids, and then sponges, on loose carbonate blocks placed in the shallow subtidal zone (−3 to −17 m) off Rhodes, Greece; conversely, grazing by chitons and block disturbance (tumbling) slowed bioerosion, although availability of larvae of bioeroders may have also played a role (Bromley *et al.*, 1990).

Chamberlain (1978) found that only minor amounts of boring can reduce the strength of coral skeleton by up to 50%. The activities of boring, microboring and encrusting taxa were further investigated by Greenstein and Moffat (1996) to estimate rates of burial of corals. They found that *Acropora cervicornis* and *A. palmata* from sediments off San Salvador (Bahamas) were more heavily abraded, encrusted (coralline algae, worm tubes, bryozoans, corals), and bored (clionid sponges, lithophagid peleycpods) than their Pleistocene counterparts. They concluded that the Pleistocene corals had been rapidly buried (<10 yr), whereas the modern corals lay exposed at the SWI for much longer intervals of time. Ketcher and Allmon (1993) described the formation of a Late Pliocene coral bed in Florida that consisted of broken, "sand-blasted" colonies of *Septastrea crassa* formed by

2.4 Cnidaria and associated biota

storms. As the dead corals lay at the SWI, they were attacked by bioeroders and encrusting organisms for an estimated 25–30 years before being buried relatively rapidly, possibly by a migrating sand wave. Like Greenstein and Moffat (1996; see also Greenstein *et al.*, 1997), Ketcher and Allmon (1993) cautioned that coral beds may have different taphonomic histories. Lirman and Fong (1997), for example, found that standing colonies protect regenerating fragments of *Acropora palmata* from removal and that mortality of transported fragments is affected by the nature of the post-transport substrate.

Nevertheless, there may be some depth-related taphonomic gradients that may be useful in interpreting ancient reef assemblages. Pandolfi and Minchin (1995) found that the taxonomic composition of coral death assemblages from low-energy sites in Madang Lagoon (Papua New Guinea) more closely resembled their corresponding living communities than did death assemblages from high-energy reef crest sites. Pandolfi and Greenstein (1997*a*) found, as expected, that greater biological degradation occurred in low energy (leeward) and deeper (6–7 m) sites off Orpheus Island (Great Barrier Reef, Australia) than at high-energy (windward) sites, but that a gradient in taphonomic alteration was greatest at windward sites. Massive corals suffered greater alteration than free-living corals (*Fungia*), which suffered greater damage than branching corals such as *Acropora*. They suggested that greater bioerosion of massive corals was related to greater skeletal surface area and density and the lower amount of skeleton not covered by living coral tissue. Also, fragile corals such as *Acropora* grow faster than massive species and may contribute more hardparts to sediment. Pandolfi and Greenstein (1997*b*) and Greenstein and Pandolfi (1997) came to similar results in the Florida Keys, although they found that the Keys death assemblage was more diverse than that of Indo-Pacific assemblages, perhaps because Indo-Pacific life assemblages are represented by more fragile species.

Pandolfi and Greenstein (1997*a*) hypothesized that because massive corals may suffer more alteration before final disintegration than fragile ones, massive coral hardparts may reside longer in the surface mixed layer ("Taphonomically Active Zone" or TAZ; Davies *et al.*, 1989) than fragile hardparts (see also discussion of *taphonomic grades* and *taphonomic clocks* in Chapters 3, 5). Similarly, Perry (1996) found that in Late Miocene reefs of Mallorca (Spain), massive corals were always more extensively bored and contained a higher diversity of borers than platy or branched corals, but that massive corals were more likely to survive boring; boring was most extensive in corals from the lagoonal facies. Reefs, or portions thereof, constructed of more fragile species were found to aggrade more slowly because structurally weakened corals were more likely to be destroyed by storms (Perry, 1996). If further confirmed, this hypothesis might explain the "all-or-none" preservation of fragile reef corals observed in the fossil record and would have important implications for coral reef management and conservation efforts (e.g., Jackson, 1992; Brown, 1995; Greenstein and Curran, in press; see also Chapter 10).

The extent of bioerosion and encrustation in reef settings appears to be influenced by nutrient levels. Although extremely productive, scleractinian reefs prefer oligotrophic (nutrient-poor) waters. On fringing reefs located on the west side of Barbados (West Indies), Tomascik and Sander (1985; 1987a,b) found that growth rates of the common frame-building coral *Montastrea annularis* increased with rising concentrations of nutrients up to a maximum, and then decreased in response to reduced levels of light and photosymbiosis. They (1987a) concluded that anthropogenic eutrophication (sewage) stressed Barbados reefs via its effect on feeding strategies, and that *Porites astreoides*, *P. porites*, *Siderastrea radians*, and *Agaricia agaricites* were the most abundant species on polluted reefs because they are all less susceptible to fouling. Elevated nutrient levels also caused increased densities of plankton populations and turbidity and allowed the invasion of bioeroders and opportunistic species that overwhelm reefs (Highsmith, 1980; Smith *et al.*, 1981; Birkeland, 1987; Hallock 1987, 1988; Martin, 1998b; see also Chapters 9, 10). Hermatypic reef corals adapted to low light levels (e.g., *Agaricia*) may move upward into shallower waters under these conditions (Acevedo and Morelock, 1988).

Similar processes appear to occur naturally on regional scales. Margalef (1968, 1971) noted the segregation of well-developed hermatypic coral reefs and highly productive pelagic ecosystems in the Caribbean and Pacific. Wilkinson (1987) suggested that nutrient levels in the Caribbean are higher than in the Indo-West Pacific based on the much greater sponge (suspension feeder) biomass of West Indian reefs. Higher nutrient levels in the Caribbean may result from localized upwelling (Margalef, 1971) and river input to this basin, which is much more landlocked than the Indo-West Pacific (Highsmith, 1980; Birkeland, 1982; maps in Berger, 1989). Higher nutrient levels as a result of upwelling have also been cited to partially explain the poorly developed reefs of the eastern equatorial Pacific (Glynn, 1996). Differences in regional nutrient levels may also affect life history strategies, rates of speciation, and grazing by invertebrates, all of which affect spat survival, species diversity and preservation potential of reefs (Martin, 1998b). For example, Risk *et al.* (1995) recently reported a decline in rates of bioerosion of *Acropora* across the continental shelf of the Great Barrier Reef.

2.5 Mainly bivalved shells: brachiopods and pelecypods

Post-mortem transport and destruction of macrobiota has received considerable attention. Gastropods, for example, may be reoriented by currents or concentrated by other means (cf. Figure 1.4) or transported because of air trapped inside the shell, whereas cephalopod conchs typically settle to the bottom and undergo minimal transport out of the habitat of living populations (Reyment, 1958, 1980; Chamberlain et al., 1981; Chamberlain, 1987; Boston and Mapes, 1991). Based on comparison of modern *Nautilus* and fossil assemblages (despite difficulties of direct application; cf. Saunders and Ward, 1994, and Jacobs and Landman, 1994; see also Jacobs, 1996), negative buoyancy of necroplanktonic conchs occurs rapidly via water infiltration into empty buoyant shells (depending on morphology and strength of internal shell structures), through punctures from (usually rapid) predation, and possibly epizoan encrustation. If gas or softparts remain in some portion of negatively buoyant conchs, implosion and fragmentation can also occur (Boston and Mapes, 1991) and may make dissolution more likely, leaving behind more resistant aptychi (Chapter 3).

Among macroinvertebrates, however, post-mortem disarticulation and movement of brachiopods and pelecypods has received far and away the most attention. One of the earliest actualistic studies of bivalved shell transport was the experimental work of Menard and Boucot (1951), who studied the behavior of terebratuloid brachiopods in a flume. They found that brachiopod transport is primarily dependent on effective density and secondarily on shape and size. Although the shells weighed as much as about 0.6 g, they were much more easily moved than sand grains, each of which weighed only a few thousandths of a gram. For example, a terebratuloid shell filled with water had a density in air of $1.28\,\text{cm}^{-3}$, but when submersed, the buoyancy of water reduced the density to $0.28\,\text{g}\,\text{cm}^{-3}$ (effective density). Thus, a "water-filled shell moved by a slow current is analogous to an inflated balloon moved by the faintest breeze" (Menard and Boucot, 1951, p. 134). Single valves moved by both sliding and rolling, so that valves occurred both convex upward and downward. Competent velocities ($\sim$5–15 cm s^{-1}) were not strictly a function of size, but also of shape (sphericity), size, and weight, and the smaller the effective density, the more important the shape. Also, shells were more likely to be moved on gravel than on sand substrata. Unlike foraminifera (section 2.3), the shells were of sufficient size (nominal diameter 0.41–5.5 mm) to project into the turbulent layers above bottom, so that the turbulent wake downstream from the shells exerted a drag on the shells; in some cases, shells were buried in the resulting scour mark (Figure 2.10; see also Johnson, 1957; Futterer, 1978a,b).

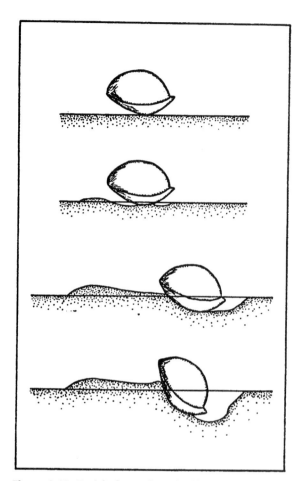

Figure 2.10. Burial of a terebratuloid brachiopod via current scour. Current direction is from right to left and is of constant velocity. (From Menard and Boucot, 1951; reprinted by permission of the *American Journal of Science*.)

Single valves that were buried in the flume were always convex up (Menard and Boucot, 1951). Concave-up orientations require higher flow velocities than convex-up shells in laboratory flumes (Savarese, 1994); moreover, convex-up orientations are common under traction flow, but concave-up orientations are found in wave-swept, shallow marine settings, and turbidites (Clifton and Boggs, 1970); convex-up orientations have also been produced in settling studies (e.g., Middleton, 1967; Maiklem, 1968; Futterer, 1978c; Allen, 1984).

However, decreases in flume current velocities can result in mixtures of concave up and concave down shells (Menard and Boucot, 1951). Thus, concave-up or concave-down orientations of valves are not infallible indicators of environmental (water) energy or the tops and bottoms of strata (e.g., Menard and Boucot, 1951; Kidwell and Bosence, 1991). Moreover, shell orientations of pelecypods and brachiopods, as well as the orientations of other shells, may be modified by birds (Cadée, 1989, 1994; Meldahl and Flessa, 1990), scavengers and predators (e.g., skates and rays; Fürsich and Flessa, 1987), hermit crabs (Walker, 1988, 1989, 1992; Walker and Carlton, 1995), and bioturbators (e.g., Salazar-Jiminéz et al., 1982; Chapter 4). Much fragmentation of shells may also result from these sorts of biological agents and may strongly resemble that produced by physicochemical processes (Cate and Evans, 1994).

Other workers have used disarticulated valves as current direction indicators (e.g., Nagle, 1967; Alexander, 1986). Based on the work of Futterer (1978a,b,c) and Allen (1984), Kidwell and Bosence (1991) emphasized that in order for such paleocurrent direction indicators to be reliable, the seafloor must be sufficiently firm for the shells to move freely (threshold velocity of sediment is greater than the threshold velocity of shells), such as on dewatered surfaces and where shells or other obstacles do not interfere with movement. Also, Brenchley and Newall (1970) found that the number of shells transported increased with increasing current velocity up to approximately $25-30 \, \text{cm s}^{-1}$ on both sand and mud substrates, but that above this velocity so many objects were being transported that they began to interfere with one another and were progressively redeposited as current velocity increased up to a maximum of about $40-45 \, \text{cm s}^{-1}$; if shells interfere with one another during transport, they may also assume unusual equilibrium orientations with respect to the current. Shells began transport at lower velocities and were transported farther on sand than on mud, but were more easily inverted on mud surfaces than on sand, as sand is more easily eroded (cf. Figure 2.3). Nevertheless, current scour around objects and movement of sand into concave-up shells often resulted in burial of shells at high angles to the SWI.

Despite such variation, Brenchley and Newall (1970) found that under low-flow regimes in flume studies of real shells and models of pelecypods (and gastropods), shells and models assumed similar final positions. Cross-current orientations prevailed except for the gastropod *Turritella* (models were made of modelling clay embedded with homogenized lead dust to yield a specific gravity of approximately 2.6, which is near that of shell calcite). Since low-flow regimes are more likely than high-flow regimes (e.g., rivers, beaches), Brenchley and Newall (1970) concluded that cross-current patterns would dominate in the fossil record. This is in contrast to the studies of Nagle (1967), among others, who often found current-parallel orientations. Nevertheless, Brenchley and Newall (1970, p. 219) concluded that "under different experimental conditions and with differently shaped objects, a variety of orientation patterns can result," depending upon life orientation (e.g., Alexander, 1984, 1986), hydrodynamic properties, and orientation of shells by investigators in experiments (e.g., concave-up and current-parallel orientations were most stable in their experiments).

The behavior of whole shells versus separate valves also raises the question of the role of disarticulation in shell transport and destruction. According to some of the basic rules of taphonomy (Chapter 1), shells are more likely to remain articulated under low-energy conditions and rapid burial. Such conditions are more likely to prevail on open shelves, especially those below fair-weather wave base, than in nearshore environments (e.g., Henderson and Frey, 1986; Kidwell and Bosence, 1991; Figure 1.7). Boucot (1953) and Boucot et al. (1958) suggested using the ratio of pedicle to brachial valves as an index of disarticulation to discriminate between life and death assemblages. Relatively high percentages of articulated shells may, however, result from storms or turbidity currents (*obrution* or "smothering"; Fürsich and Flessa, 1987; Seilacher et al., 1985; Brett and Seilacher, 1991). Low temperature and anoxia also prevent disarticulation; anoxia does not prevent organic decay *per se*, but minimizes bioturbation and promotes early mineralization of soft tissues (Chapters 3, 6). Disarticulation may also result from the activities of the same agents that reorient shells (see above). Other factors affecting skeletal articulation in bivalved shells include: (1) epifaunal versus infaunal habit (infaunal taxa are more likely to remain articulated than epifaunal ones because they are already "buried"); (2) amount of connective tissue and extent of organic decay prior to reworking (e.g., Schäfer, 1972); and (3) the fit of skeletal elements (size of ligaments, hinge teeth, etc.; Sheehan, 1977; Alexander, 1990).

Whether articulated or not, however, shells may still decrease in mechanical strength after death, thereby accelerating their disintegration. Modern inarticulate brachiopods (*Glottidia palmeri*) appear to undergo rapid disintegration and require high sedimentation rates to be preserved (Kowalweski, 1996b; see Glover and Kidwell, 1993, for bivalves). Daley (1993) found rapid loss of strength of the hinge and shell walls of the modern articulate *Terebratalia transversa*. In other cases, articulated valves appear to persist for considerable periods of time: Holland (1988) found that valves of the Ordovician articulate brachiopod *Platystrophia ponderosa* remain articulated long after death, as evidenced by repeated sediment infillings and unidentified cryptobiota that could only have formed if the valves had remained articulated but gaping (other workers have concluded that the degree of encrustation is not a reliable indicator of residence time at the SWI; e.g., Bordeaux and Brett, 1990). Holland (1988) also found that the pedicle:brachial valve ratio increased with increasing abrasion and breakage in disarticulated assemblages; the pedicle valve is substantially thicker than the brachial, which suggests preferential destruction of the brachial valve during burial (Velbel and Brandt, 1989). Thus, alteration of pedicle:brachial ratios need not be purely a function of post-mortem transport and sorting (cf. Boucot, 1953; Boucot et al., 1958).

Driscoll (1967, 1970) and Driscoll and Weltin (1973) conducted field and laboratory studies of abrasion of selected pelecypods (and gastropods). They found that rates of abrasion were greatest in the surf zone and least in the shallow sublittoral, and that rates of abrasion were greater using larger-grained (gravel versus sand) and more poorly sorted abrasives. Shells with greater surface area per unit weight were more rapidly destroyed on a percentage basis, whereas more robust shells (less surface area per unit weight) lost shell material at a slower rate in absolute terms (i.e., although robust shells lost more weight, they also had more shell to lose, so they were relatively unaffected). Driscoll (1967) also found that abrasion in natural environments is much slower than laboratory (tumbling) experiments (cf. section 2.3.1).

Moreover, heavier shells were buried more rapidly, whereas lighter shells remained close to the SWI, where they were extensively bored (Driscoll, 1970). Smith (1992), Aller (1995), and Cutler (1995), also concluded that bioerosion of molluscs is their principal means of destruction, which may be completed in in hundreds to thousands of years. Shell destruction by bioerosion may be most pronounced in higher latitudes, such as temperate-water settings (Smith, 1992; Cutler and Flessa, 1995; but see Walter and Burton, 1990). Nevertheless, such destructive mechanisms may lead to "case-hardening" of shells and deceleration of their destruction (Cutler and Flessa, 1995).

One caveat to the study of taphonomy of bivalved shells (and other taxa) is the assumption that shells of the same species come from similar populations (a form of "taxonomic uniformitarianism"). Peterson and Black (1988) demonstrated that mortality (and therefore shell input) of populations of suspension-feeding sand-flat venerid bivalves (*Katelysia scalarina* and *K. rhytiphora*) in response to sedimentation was increased by crowding (i.e., a physical agent acts in a density-*dependent* manner). Thus, historical conditioning to one factor (in this case, crowding) can alter a population's susceptibility to other stresses (e.g., mortality in response to sedimentation) and perhaps rates of shell input to assemblages (cf. Kidwell, 1986a; Chapter 1).

2.6 Arthropods and annelids

The fossil record of arthropods and annelids is largely the result of exceptional preservation. Unlike most other taxa, arthropods and annelids are mainly

Table 2.2. *Fossilization potential in a modern oyster community*

Taxa	Total species	Soft-bodied	With preservable hardparts				Possible redundancy
			Ca	Ch	Si	Ph	
Porifera	5	—	—	—	5	—	3
Coelenterata	6	5	1	—	—	—	—
Platyhelminthes	1	1	—	—	—	—	—
Nemertea	2	2	—	—	—	—	—
Bryozoa							
Ectoprocta	7	4	3	—	—	—	—
Annelida							
Polychaeta	13	13	—	—	—	—	4
Mollusca							
Gastropoda	9	—	9	—	—	—	1
Pelecypoda	13	—	13	—	—	—	2
Arthropoda							
Crustacea	19	10	4	5	—	—	5
Arachnida (?)	1	1	—	—	—	—	—
Insecta	1	1	—	—	—	—	—
Chordata							
Tunicata	2	2	—	—	—	—	—
Vertebrata	1	—	—	—	—	1	1
Totals	80	39	30	5	5	1	15
Percentages of total community	100	49	38	6	6	1	19

Among the arthropods, only decapod crabs with relatively well-calcified and/or well-tanned exoskeletons have been included with the organisms with hardparts. Ca, calcareous; Ch, chitinous; Si, siliceous; Ph, phosphatic.

Lawrence, 1968.

represented by soft-bodied species that vary considerably in the degree of mineralization of an outer cuticle composed of chitin (a mucopolysaccharide) and protein. Lawrence (1968) documented substantial loss of arthropods and annelids during the preservation of oyster communities (Table 2.2). Staff *et al.* (1986) observed similar changes during the formation of shallow-water molluscan thanatocoenoses of shallow Texas bays.

    Other workers have documented the decay of individual soft-bodied taxa. Schäfer (1972) described the disintegration of crabs, shrimp, ostracodes, barnacles,

2.6 Arthropods and annelids

insects, and polychaetes. Much of the early disarticulation and disintegration of benthic arthropods is the result of predation and scavenging by various invertebrates and vertebrates, both at the SWI and within sediment (Schäfer, 1972; see Plotnick, 1986a, for numerous references). Chitinoclastic bacteria are extremely common in the guts of organisms, in seawater and sediment, and on zooplankton (see Plotnick, 1986a, for references). Soft tissues are quickly attacked by bacteria, which may completely obliterate the carcass in a matter of days depending on the size of the carcass and water temperature (Chapters 3, 6). Harding (1973), for example, documented the complete destruction of marine copepods by chitinoclastic bacteria within 11 days at 4 °C and within 3 days at 22 °C, and concluded that collection of surface corpses by net was highly improbable after the first day of death in subtropical waters and beyond the sixth day in temperate coastal waters. Rates of bacterial degradation of carcasses are accelerated by decreased substrate particle size (increased surface area) that results from predation and scavenging.

As Plotnick (1986a) pointed out, however, most studies of arthropod preservation are *a posteriori* interpretations of fossil occurrences that explain only fossilization events and are therefore of somewhat limited value. More recently, a number of workers have used an experimental approach to determine why cases of exceptional preservation of these soft-bodied taxa have occurred in an attempt to change arthropod taphonomy from an anecdotal to a predictive science. In laboratory (jar) experiments maintained at 12 °C in darkness, Plotnick (1986a) documented different stages of decay of the freshly killed (frozen) shrimp, *Pandalus danae*, which were reduced to tiny fragments after about 4 weeks (see also Briggs and Kear, 1994b). Decomposition was accompanied by overall softening of the cuticle, breakdown of membranes, loss of soft tissues, and decreased physical strength of the carcass. All specimens were easily disarticulated by slight physical disturbance after the first week, although individual pieces of cuticle remained intact; the relative size (weight), substrate (mud or sand), and oxygen content of water (oxic/anoxic) made little difference in the sequence and timing of shrimp decay (Plotnick, 1986a, table 3).

Field experiments involving shrimp buried both in sandy sediment oxygenated (bioturbated) by *Callianassa* and *Upogebia* and in muddy anoxic sediment (Puget Sound), indicated that bioturbation quickly disarticulated experimental carcasses (Plotnick, 1986a). Short-term preservation was best in muddy anoxic sediments, and survival time increased with depth of burial, especially below approximately 10 cm, which is probably beneath the zone of the most intense mixing (TAZ). Shell (mollusc) layers at anoxic sites also inhibited bioturbation (taphonomic feedback of Kidwell and Jablonski, 1983; Kidwell,

1986b; Chapter 1). Enclosure of experimental shrimp in wire mesh cages greatly increased survival time of carcasses. Plotnick (1986a) concluded that arthropod preservation is best in sediments in which bioturbation is inhibited (e.g., anoxia or movement of carcasses well below the TAZ via rapid burial).

Plotnick *et al.* (1988) conducted further experiments using the crab *Panopeus* buried in carbonate sediments of Bermuda. In contrast to the shrimp studies, *Panopeus* remained relatively well preserved. They attributed this to low organic matter content of the sediment, low density of infauna, and to the carbonate sediment, which probably buffered against acids produced by soft tissue decay. Unfortunately, the relative importance of these factors was not apparent from the field experiments.

Rapid burial and anoxia have traditionally been viewed as the main prerequisites for preservation of soft-bodied taxa (*Konservat-Lagerstätten*; e.g., Seilacher *et al.*, 1985) because rapid burial inhibits scavenging and bioturbation and promotes anoxia by restricting oxygen supply (Allison, 1988a; see also Chapter 6). But, in laboratory (airtight jar) experiments, Allison (1988a) found that freshly killed (via asphyxiation) carcasses of the polychaete *Nereis* sp. and the crustaceans *Nephrops norvegicus* (heavily calcified) and *Palaemon adspersus* all decayed despite anoxia. Like Plotnick (1986a), Allison (1988a) described a sequence of decay over about 25 weeks at 10 °C. Most importantly, *Nephrops* decayed more rapidly than did *Palaemon* despite heavier calcification; thin sections of *Nephrops* demonstrated the removal of chitin that bound calcite layers together, thereby accelerating decay of the exoskeleton. Rates of decay were substantially faster in marine waters than fresh because of oxidation of organic matter by sulfate (SO_4^{2-})-reducing bacteria (Chapter 3; see Hof and Briggs, 1997, for discussion of necrolysis of mantis shrimps).

Allison (1988a) did not recover any remains of *Nereis* in his jar experiments. Briggs and Kear (1993a), however, described a sequence of decay of *Nereis virens* at 20 °C over about 30 days: whole/shriveled → flaccid → unsupported gut → cuticle sac → jaws and setae. Thus, only sclerotized tissues (e.g., jaws) are likely to survive beyond 30 days in the absence of early diagenetic mineralization (Chapter 3). All stages of decay are represented in the fossil record, however, and this suggested to Briggs and Kear (1993a) that the stages could be used to estimate taphonomic thresholds of how far decay had proceeded before it was halted.

Because of the strong possibility of rapid decay induced disarticulation of soft-bodied taxa, the preservation of complete soft-bodied fossils is viewed as reflecting little or no transport. Nevertheless, based on tumbling experiments, Allison (1986) demonstrated that soft-bodied carcasses of *Nephrops* and *Palaemon*

2.6 Arthropods and annelids

can potentially be transported significant distances either by currents or via flotation caused by decay gases. The implications of Allison's (1986, 1988a) results for the formation of Lagerstätten are assessed in Chapter 6.

Among arthropods, only some trilobites, crustaceans, and millipedes have heavily mineralized skeletons (Briggs and Kear, 1993a, 1994b; Hof and Briggs, 1997). Although soft tissues decay rapidly, the exoskeleton of millipedes and crickets may persist for years (Seastedt and Tate, 1981; Seastedt and Crossley, 1984). The extent of preservation of molted elements varies with the degree of calcification of juvenile skeletons (e.g., chelae of crabs, lobsters, and shrimps are most likely to be preserved). Trilobites are preserved in almost all marine facies of the Paleozoic by virtue of a cuticle in which are embedded primarily calcitic, and perhaps phosphatic, minerals (Speyer, 1991). Intact outstretched specimens most likely formed under conditions of rapid burial without disturbance, whereas preburial disturbance more likely produced enrolled forms (Table 2.3).

Besides disarticulation, individual trilobites repeatedly contributed multiple elements to sediment through the process of molting (ecdysis). Individual elements of disarticulated or molted trilobites were, in turn, subject to size and shape sorting along energy gradients (Figure 2.11). Some studies have indicated differences in hydrodynamic behavior of trilobite parts. In flume studies of *Flexicalymene meeki*, individual artificial (epoxy) cephala and pygidia exhibited concave-down behavior (no matter what the initial orientation), but settled concave-up (Lask, 1993). Like bivalves, individual sclerites were buried by scour pits at current velocities up to $50\,\text{cm}\,\text{s}^{-1}$. Surprisingly, enrolled specimens were transported at current speeds much lower (approximately $13\text{–}18\,\text{cm}\,\text{s}^{-1}$) than those needed to move individual sclerites from stable positions (Lask, 1993; cf. Kontrovitz, 1975). Lask (1993) concluded that in unidirectional currents, concentrations of pygidia may indicate somewhat higher current velocities than mixed assemblages of sclerites, but in wave-swept environments both cephala and pygidia occur concave up. Similarly, Westrop (1986) concluded that unusual concentrations of disarticulated trilobite hardparts caused by tempestites must be ruled out before ecological causes are considered as concentration agents.

The settling and flume results of Lask (1993) differed from those reported for *Dikelocephalus* based on models made from dental acrylic (Hesslebo, 1987; e.g., concave-down settling of pygidia), and were thought to have resulted from different morphologies or densities of the models (Lask, 1993). Both concave-up and concave-down behavior are found in Devonian rocks, however (Table 2.3): predominantly concave-down orientations reflect reorientation due to surface currents; equal proportions of concave-up and concave-down orientations

Table 2.3. *Modes of occurrence and taphonomic conditions among trilobite remains and assemblages*

Disarticulated remains
General. Disarticulated remains indicate periods of exposure associated with surface scavenging and/or current agitation. Post-burial disarticulation may result from intrastratal bioturbation.

Manner of occurrence	*Taphonomic conditions*
1. Disarticulated skeletons with preferred convex-up orientation among concavo-convex elements	1. Persistent surface currents: not usually associated with burial
2. Disarticulated skeletons with preferred concave-up orientations among concavo-convex elements	2. Settling postures after rapid, turbid deposition: scavenger reorientation
3. Disarticulated skeletons show no preferred orientation	3. Rapid, mass deposition: long-term accumulation in low-energy environments
4. Fragmentation of disarticulated sclerites	4. Episodic or persistent agitation in high-energy environments

Articulated remains
General. Articulated trilobite remains indicate rapid burial and conservation from destructive agent (currents, bioturbation). Important index for recognizing event-deposited strata.

Manner of occurrence	*Taphonomic conditions*
1. Outstretched carcass remains	1. Rapid burial without pre-burial disturbance: overturned bodies indicate settling and/or behavior
2. Enrolled carcass remains	2. Pre-burial disturbance (toxicity, turbidity) followed by deep burial
3. Intact molt remains: molt ensembles and configurations	3. Burial without current agitation: predominance indicates shallow burial

Other considerations
General. Various other conditions of preservation reflect specific taphonomic circumstances. These are also useful in evaluating the history of preservation of trilobite assemblages.

Manner of occurrence	*Taphonomic conditions*
1. Diagenetic mineralization: silica, pyrite, carbonate, phosphate	1. Decay initiates chemical pathway according to background circumstances: indicates carcass remains
2. Cuticle color and thickness	2. Degree to which cuticle carbonate matrix has been dissolved and/or modified indicates ambient pH and level of carbonate saturation

Speyer, 1991.

2.6 Arthropods and annelids

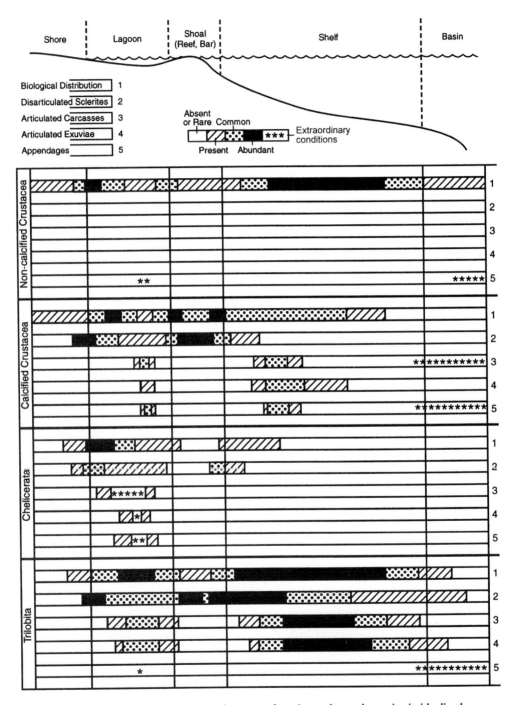

Figure 2.11. Comparative taphonomy of marine arthropod remains in idealized environments. (Redrawn from Speyer, 1991.)

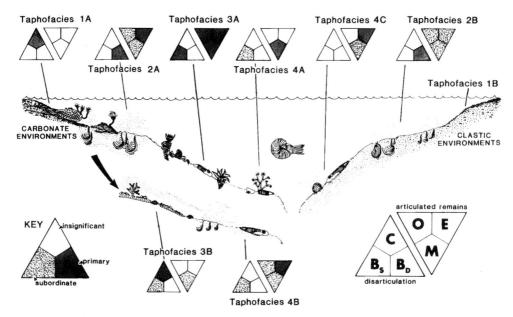

Figure 2.12. Trilobite taphofacies for the Middle Devonian Hamilton Group (New York State). C, current-related processes; B_s, surficial bioturbation; B_D, deep bioturbation; O, outstretched specimens; E, enrolled specimens; M, molts. (Speyer and Brett, 1986; reprinted with permission of SEPM.)

suggest deep bioturbation; and predominantly concave-up orientations indicate limited bioturbation (Speyer and Brett, 1986). The differing experimental results again indicate the pitfalls of laboratory protocols when coupled to strict analogical reasoning (Chapters 1, 11).

Based on phacopid trilobites, Brett and Baird (1986) and Speyer and Brett (1986) distinguished nine taphofacies in the Middle Devonian Hamilton Group of New York State along a paleobathymetric (energy) gradient (Figure 2.12; Table 2.4; cf. Figure 1.7), irrespective of clastic or carbonate regimes. Taphofacies 1 is characterized by disarticulated, fragmented, and sorted trilobite remains, and is interpreted as consisting of reworked and lag deposits formed under a high-energy regime near shore. Taphofacies 2 formed in slightly deeper water with moderately high sedimentation rates, as indicated by relatively deep (*Zoophycos*) burrows and mainly disarticulated or enrolled specimens. Taphofacies 3 and 4 have high proportions of complete outstretched specimens and molt assemblages. Because of low sediment influx, taphofacies 3A also has numerous, but unfragmented, remains that accumulated as thin, laterally persistent layers, whereas taphofacies 3B is characterized by prominent trilobite layers that accumulated as a result of sediment bypass. In taphofacies 4A, trilobite debris accumulated under conditions of low rates of sedimentation in deep, quiet water, and shell

Table 2.4. *Relationships of biostratinomic features to sedimentation rate and environmental energy*

		Sedimentation rates		
Environmental energy	Skeletal type	Episodic, very rapid (1–50 cm/10^2 yr)	Intermediate to rapid (10–100 m/10^3 yr)	Low to intermediate (1–100 cm/10^3 yr)
High	Fragile Ramose Bivalved Shells	Minor fragmentation Mostly articulated; rarely *in situ*	Strong fragmentation Partially articulated; some fragmented	Absent Disarticulated; fragmented; abraded
	Multi-element Skeletons	Mostly articulated; rarely *in situ*	Partially articulated; pieces sorted	Disarticulated; pieces sorted
Low	Fragile Ramose	Intact; not fragmented	Some fragmentation	Strong fragmentation; corrosion
	Bivalved Shells	Articulated; some *in situ*	Mostly disarticulated; complete valves	Disarticulated; minor fragmentation; corrosion
	Multi-element Skeletons	Completely articulated; some *in situ*; intact molts	Partially articulated; non-sorted	Disarticulated; non-sorted

Brett and Baird, 1986.

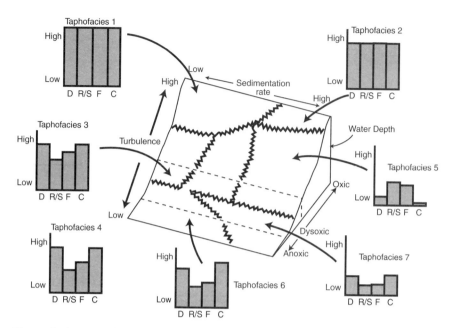

Figure 2.13. General taphofacies model for epeiric seas. Seven taphofacies are recognized on the basis of the degree of disarticulation (D), reorientation and sorting (R/S), fragmentation (F), and "corrasion" (C; corrosion + abrasion). Sedimentation rate decreases from right to left, whereas turbulence and seafloor oxygen (oxic, dysoxic, anoxic) decrease downslope from top to bottom. (Redrawn from Brett and Speyer, 1990.)

layers may have served as substrates for other benthos (taphonomic feedback), whereas in taphofacies 4B sediment accumulated that had bypassed shallower sites. Early pyrite (FeS) formation in taphofacies 4B suggests that dysoxic–anoxic conditions also prevailed; pyrite precipitation also occurred in localized microenvironments around decaying carcasses (see Chapter 3 for caveats). Oxygen levels were presumably lower in taphofacies 4C, in which low diversity faunas and bypassed sediment rapidly accumulated. Based on this inductive model, Speyer and Brett (1988) developed a deductive model of taphofacies for Paleozoic epeiric seas (Figure 2.13).

Besides scavenging and predation, certain arthropods are important taphonomic agents themselves. In addition to the shrimp, *Callianassa* and *Upogebia*, fiddler and other types of crabs produce extensive burrows in shallow marine settings, especially marshes, that affect porewater chemistry (Chapter 3) and destroy temporal resolution (Chapter 5). Walker (1988, 1989, 1992; Walker and Carlton, 1995) described how hermit crabs produce various anomalies in the gastropod fossil record: (1) between-habitat anomalies (transport between bathymetric zones); (2) within-habitat anomalies (infaunal–epifaunal displacement); (3) shell selection and resulting abundance and size–frequency anomalies; (4) wear and destruction anomalies; and (5) temporal anomalies (reworking between strata).

2.7 Echinoderms

Unlike arthropods, the echinoderm skeleton consists of highly calcified plates or ossicles (except for the spicules of holothurians) that are each individual calcite crystals (Donovan, 1991). The crystals consist, however, of a meshwork (*sterom*) of calcite that is highly porous and formed of calcite rods (*trabeculae*). Like arthropods, though, the echinoderm skeleton typically disarticulates almost completely after death, and taxonomic and taphonomic studies are therefore largely dependent on "preservational rarities."

Two main features of the echinoderm test determine its preservability: the articulation of the plates and the durability of the tissues that hold them together (Donovan, 1991; see also Kidwell and Baumiller, 1989; Allison, 1990; LeClair, 1993). In a flexible echinoid corona, the ossicles overlap and slide past one another; echinoderms with flexible joints between ossicles are especially dependent on the durability of ligaments (*mutable collagenous tissues* or MCTs), which are found in certain pore spaces of the sterom and which vary in rate of decay (Donovan, 1991). Most echinoid coronas are of the inflexible type, however (e.g., cidaroids), in which adjacent ossicles join at inflexible sutures that are perpendicular to the test surface. In other echinoids, trabeculae of adjacent plates penetrate each other's sterom so that the test remains rigid even after ligaments have decayed, and in still others, interlocking socket and peg structures occur. The extreme in test strength is found in the clypeastroids (sand dollars), in which internal pillars reinforce the test between oral and aboral surfaces; the genus *Clypeaster* also secretes a nearly continuous layer of calcite on the inner surface of the corona.

Rapid disarticulation of arms, pinnules, and cirri has been reported in modern comatulid crinoids (even under anaerobic conditions; Blyth Cain, 1968), although calyces may remain intact in the absence of scavengers or under low-energy conditions (Meyer, 1971; Liddell, 1975). On the other hand, Meyer *et al.* (1989) found that the rigidly constructed calyces of most monobathrid camerates and disparid inadunates are relatively insensitive to long-term exposure, and some fossil crinoids persisted long enough to be encrusted or dragged along the bottom (Donovan, 1991). Meyer and Meyer (1986) concluded that the burial of crinoid calyces with intact arms is *normally* rapid, but that the preservation of calyces alone does not necessarily require rapid burial.

Individual echinoderms may contribute multiple generations of ossicles to sediment through autotomy, which is a response to adverse external stimuli such as disease, old age, desiccation (reef flat crinoids) and predation (fish; seabirds in the case of reef flat crinoids) via irreversible degeneration of MCTs (Donovan, 1991). In the case of modern isocrinid crinoids, autotomy occurs preferentially between cirri-bearing columnals (nodals) and columnals lacking cirri (internodals; Baumiller et al., 1995).

Thus, predation does not have to be lethal for echinoderms, which may live up to about 20 years depending upon the taxon (Donovan, 1991), to contribute ossicles to sediment. Meyer and Meyer (1986), for example, noted large numbers of crinoid loose arms, damaged individuals, and ossicles in sediments at the base of the escarpment of Lizard Island (Great Barrier Reef, Australia) where attacks on crinoids by triggerfish were observed. Moreover, scavengers may preferentially attack the calyces because they contain the gonads, a highly nutritious food source (Maples and Archer, 1989). Both fresh and worn ossicles were found by Meyer and Meyer (1986) in the same sediment samples, suggesting that ossicle accumulation is a time-averaged phenomenon (cf. Seilacher, 1973).

A number of authors have described the disarticulation and reorientation of disarticulated ossicles in various echinoderms. Schäfer (1972) described the disarticulation of regular echinoid and starfish (Table 2.5); he also noted current-orientation and disintegration of ophiuroid arms after only 15 hours. Lewis (1986, 1987) used modern ophiuroids as analogs for the progressive disarticulation of Paleozoic crinoids (Table 2.6); unlike Schäfer (1972), however, he found that ophiuroids disarticulated over a period of weeks. Based on field and flume studies, Schwarzacher (1963) demonstrated that elongate crinoid remains are oriented cross-current (cf. Nagle, 1967, and Brenchley and Newall, 1970).

Donovan (1991) suggested that the different results of Lewis (1986, 1987) and Schäfer (1972) may be related to the level of wave energy and ambient water temperature. Schäfer's (1972) observations were primarily from the cool-water North Sea region, whereas Lewis's (1986) observations were conducted in the laboratory. Kidwell and Baumiller (1989) also suggested that temperature-related latitudinal (and bathymetric) gradients in echinoid preservation may exist. They found that in the laboratory, low temperature retards MTC decomposition of the regular urchins *Strongylocentrotus purpuratus* and *S. droebachiensis* more effectively than anoxia, and that the primary role of anoxia in echinoderm preservation may be to exclude scavengers and slow disarticulation. Based on tumbling experiments (cf. section 2.3.1), Kidwell and Baumiller (1989) also found that specimens that had decayed for 2 days at 30 °C disintegrated about six

2.7 Echinoderms

Table 2.5. *Patterns of disarticulation in certain Recent echinoderms*

Asteroid *Asterias rubens* (after Schäfer, 1972)	Time (days)	Regular echinoid (after Smith, 1984; Schäfer, 1972)
Arms flexible, but stiffen soon after death		Radioles and pedicellariae droop; moribund
DEATH	0	DEATH
Red coloration soon fades	1	Radioles drooping
	2	Pedicellariae then radioles detach
Decomposition gases arch body and inflate carcass	3	
	4	
	5	
Dorsal integument starts to lift	6	
	7	
	8	Peristomial and periproctal membranes disintegrate
Dorsal skin loosens and drifts away as pieces	9	
	10	
	11	
Decomposition has removed water vascular system, digestive system and musculature from ventral skeleton	12	Apical system largely detached
	13	
	14	
	15	Aristotle's Lantern completely disarticulated
	16	
	17	
Complete disarticulation		Corona fragments

Donovan, 1991.

Table 2.6 *Post-mortem disarticulation of ophiuroids*

Stage	Effect
1	Gradual loss of color and body fluids
	Release of decomposition gases
	Some stiffening of proximal arms
2	Arms flexible, with little or no disarticulation
	Dorsal integument of disk may become detached
3	Partial decomposition, continuing to completeness

Donovan (1991), after Lewis (1987).

Table 2.7. *Preservation scale of crinoids in Fort Payne Formation*

Lithofacies	Lithology	Bedding	Geometry
Crinoidal packstone buildups	Crinoidal packstone, green shale	Massive, thickens toward flanks	Broad dome, flat top
Wackestone buildups	Wackestone, packstone	Massive, thickens toward flanks	Drape over shale core; domal
Sheetlike packstones	Packstone siltstone	Graded, lenticular to tabular	Sheetlike
Channelform packstones	Packstone	Graded, thin, tabular	Channel-filling
Green shales	Green shale packstone	None	Moundlike, flanks of carbonates

Meyer *et al.*, 1989.

times faster than specimens that decayed in cool water (11 °C), and that the longer specimens were allowed to decay before tumbling, the faster they disintegrated, up to a threshold time that appears to represent the decay of MTCs. The threshold time was a few days at 30 °C, about 2 weeks at 23 °C, and more than 4 weeks at 11 °C. In tumbling experiments, specimens that decayed beyond the threshold time decayed almost instantaneously along sutures, whereas prior to the threshold time, tests disintegrated by both sutural separation and cross-plate fractures. Thus, a whole suite of preservational styles may obtain in a seemingly uniform depositional environment (Allison, 1990). The implications of Kidwell and Baumiller's (1989) results for echinoderm Lagerstätten are considered in Chapter 6 (see also Allison, 1988*a*).

Obviously, the interplay between echinoderm skeletal morphology, environment, and preservation can be quite complex. Meyer *et al.* (1989) studied the preservation of pelmatozoans (crinoids and blastoids) in the Fort Payne Formation (Early Mississippian of Kentucky and Tennessee). They concluded that complete pelmatozoans with arms and stalks can endure some transport close to the living site, so that completely articulated pelmatozoans are not necessarily indicative of rapid burial. Meyer *et al.* (1989) used sedimentological evidence, degree of articulation (Table 2.7), and taphonomic effects of morphology (Table 2.8) to distinguish several taphofacies (Table 2.9) that were indicative of degree of transport and burial (Figure 2.14). Chiefly autochthonous facies included (1) crinoidal packstone buildups showing a full range of preservation including articulated

2.7 Echinoderms

Table 2.8. *Morphological effects on pelmatozoan taphonomy in the Fort Payne Formation*

Constructional morphology	Preservation with gradual burial	Examples from Ft Payne
Multiplated calyxes with fixed postradials, rigid tegmen forming unitized calyx, arms with little flexibility; rigid blastoid thecae with delicate, inflexible brachioles	Intact calyxes lacking arms, stem	Monobathrid camerates: *Alloprosallocrinus, Agaricocrinus. Uperocrinus*; some blastoids
Multiplated calyxes as above but with larger plates less rigidly united	Disarticulated or partial calyxes	Monobathrid camerates: *Actinocrinites*; some blastoids
Compact calyx with reduced number of plates, plates often fused, anal sac not rigidly plated, arms moderately flexible	Intact aboral cups with no anal sac, arms, or stem, many disarticulated	Disparid inadunates: *Synbathocrinus, Halysiocrinus, Catillocrinus*; Flexibles: *Gaulocrinus*
As above but with larger plates, less well sutured	Disarticulated or partial calyxes	Cyathocrine inadunates, Poteriocrine inadunates, Monobathrid camerates: *Platycrinites, Paradichocrinus*
Calyx and arms with flexible articulations	Disarticulated or partial calyxes	Flexibles: *Metichthyocrinus*

Meyer *et al.*, 1989.

specimens; (2) wackestone buildups dominated by complete calyces lacking arms and stalks; and (3) green shales dominated by disarticulated specimens but including complete calyces and arm segments, suggesting slow background deposition (relatively quiet water) punctuated by rapid influxes. Allochthonous assemblages included sheetlike packstones dominated by complete and partial calyces, and channelform packstones dominated by disarticulated debris with occasional fully articulated crinoids (probably a crinoidal turbidite or tempestite). These taphofacies were found to be in general agreement with depositional gradients, and it appeared that more complete fossils were found upsection because of the progradation of taphofacies with more autochthonous assemblages over those composed of more allochthonous ones.

Table 2.9. *Crinoidal taphofacies of the Fort Payne Formation*

	No transport		Transport	
	Gradual accumulation	Rapid burial	Gradual accumulation	Rapid burial
Articulation	Disarticulated; calyx can be intact	Articulated; long stems	Disarticulated; calyx can be intact; short stems	Articulated; calyx can be separated from stem
Posture	None	"Trauma"	None or "splay"	"Trauma"
Orientation	None	Horizontal; can be parallel	Little or none	Horizontal or chaotic; can be parallel
Abrasion	Possible	None	Possible	Possible
Holdfasts	*In situ*	*In situ*; stem attached	Not *in situ*; lack cirri, roots	Not *in situ*

Meyer *et al.* 1989.

2.7 Echinoderms

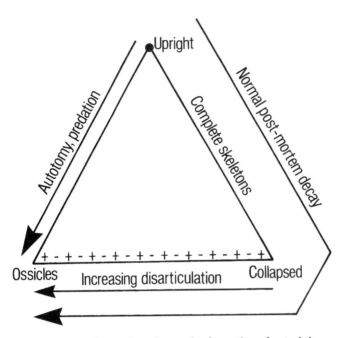

Figure 2.14. Taphonomic pathways for formation of autochthonous and parautochthonous crinoid assemblages. (Donovan and Pickerell, 1995; reprinted with permission of SEPM.)

Just because numerous disarticulated echinoderm elements are contributed to the sediment does not mean, however, that they will be detectable. For example, populations of the Caribbean echinoid *Diadema antillarum* were decimated by an unidentified pathogen during 1983–1984; this mass mortality event produced a geologically instantaneous input of disarticulated elements to the sediment. Nevertheless, Greenstein (1989) was unable to locate a corresponding spike of *Diadema* elements in the sediment.

These sorts of studies are not just academic, as they hold important implications for ecosystem conservation and management. Bell (1992), for example, suggested that rising nutrient levels may have contributed to outbreaks of the "Crown-of-Thorns" starfish, *Acanthaster planci*, on the Great Barrier Reef. The starfish then attack the corals and decimate the reefs. Like *Diadema*, when the starfish die, they contribute significant numbers of skeletal elements to the sediment. One way to test for the occurrence of similar outbreaks in the past (*prior* to anthropogenic nutrient input) is to examine the sedimentary record, the interpretation of which has been disputed (see Greenstein *et al.*, 1995, for review). In order to clarify the situation, Greenstein *et al.* (1995) conducted field experiments that simulated outbreaks and mortality of the starfish. They found that after 4 years, the signature of the simulated outbreaks was best recorded by starfish ossicles of 1–4 mm in size.

Ossicles in the 0.5–1 mm and greater than 4 mm size ranges increased and decreased, respectively, because of "taphonomic bias" (based on tumbling experiments) that influences any estimates of past population sizes. Clearly, taphonomy must play a significant role in environmental management decisions based on the sedimentary record (see also Chapter 10).

2.8 Vertebrates: bones as stones

2.8.1 Mammals

The biostratinomy of vertebrate remains has received substantial attention both in the field and laboratory. Weigelt (1929; translated 1989) documented the disintegration of modern fish, reptile, and mammalian carcasses, including those resulting from mass death. He described (1989, p. 17) a sequence of disarticulation for cow carcasses rotting on the warm coastal plain of the Gulf of Mexico:

> *The slowly dying animals almost always fall to one side with legs and neck outstretched. In a short time the carcass swells enormously due to intestinal gas..., causing the hind leg lying uppermost during rigor mortis to rise to the horizontal or even higher. In extreme cases, gas pressure can force the carcass to roll onto its back or even completely over. The foreleg is only slightly raised because the area around the lungs and pectoral girdle deflates more quickly than the abdominal area. The anus protrudes considerably, facilitating the efforts of vultures, which always attack there first... The vultures' feeding quickly releases the gas pressure and the whole thing collapses, the jutting hind leg falling usually a little in front of the one lying beneath it...*
>
> *It is astonishing how the carcasses, so enormous and conspicuous when bloated, seem almost to melt into the ground... Under the burning sun, part of the flesh dries onto the bones and does not decompose until much later. How much is eaten by scavengers [including beetles and maggots; Weiglet, 1989] depends almost entirely on how much food is available to them... The scapula and forelegs are especially easy to remove, the hind legs much more difficult... The bones are still held together rather well by the tendons.*

Weigelt (1989) also noted that although elephant herds were quite large at the time of his writing (c. 1925), skeletons were only rarely encountered (cf. Figure 2.15), and Hill (1980) noted that despite their size, large animals such as elephants are essentially no more resistant to carnivore damage than small ones.

Weigelt (1989) suggested that there were definite sequences of mammalian disarticulation. His *Law of the Lower Jaw*, for example, states that because of the ease of detachment of the lower jaw from the rest of the skeleton, the ease of detachment of the symphysis, and preferential attack of the mouth by scavengers

2.8 Vertebrates: bones as stones

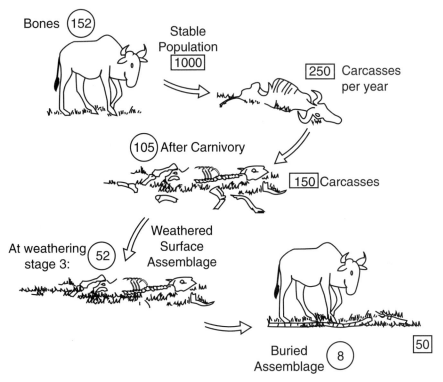

Figure 2.15. Progressive loss of wildebeest carcasses and bones via taphonomic processes. Circled numbers refer to bones per carcass, whereas boxed numbers to carcasses per year from a stable population of 1000 animals. Numbers are based on field studies in Amboseli Basin. See text for weathering stages. (Redrawn from Behrensmeyer, 1982.)

(second only to the anus), the lower jaw is often splayed or missing altogether. The *Law of the Ribs*, on the other hand, refers to the splaying or convergence of ribs in carcasses lying on their sides.

Toots (1965) later studied the sequence of disarticulation in antelope, domestic sheep, and coyotes. Unlike Weigelt (1989), he concluded that the sequence of disarticulation varies from one taxon to the next (see also Hill, 1980), and depends on the type of joint, the degree of interlocking at articulating surfaces, and the relative amounts of resistant versus decomposable tissue holding the skeleton together. In larger herbivores, for example, the lower jaw has a high ascending ramus that restricts the mobility of the jaw and slows its disarticulation. Toots (1965) also found that the relative mobility of the skull on the atlas promoted its separation from the rest of the skeleton. On the other hand, vertebrae interlock strongly, so the vertebral column is among the last portions of the skeleton to disarticulate. Hide and ligaments, both of which are relatively resistant, also resist disarticulation.

Toots (1965) concluded that most scattering of disconnected bones occurs during the late stages of disarticulation after most digestible tissues have been lost, but Hill (1980) found that medium-sized bovid skeletons in East Africa were largely disarticulated after only a few weeks. Carnivores and scavengers, including birds and ants, may, however, concentrate certain bones and fragments (depending upon the desired portion of the carcass) in their caves, dens, and lairs (e.g., Brain, 1980; Shipman and Walker, 1980; Hoffman, 1988). Consequently the bones often exhibit tooth (gnaw) marks and scratches produced not only by carnivores but also by such seemingly unlikely agents as porcupines (Brain, 1980). Pellets (regurgitated or defecated) are important sources of bones (usually disarticulated and broken) for determining the diets of predators and scavengers (Dodson and Wexlar, 1979; Wilson, 1987). In the case of porcupines, the bones were probably collected after they had dried in the sun, as the bones did not exude fats (Brain, 1980).

In arid habitats, individual bones may weather while lying exposed on the ground or partially buried at the soil–air interface. Behrensmeyer (1978) recognized six weathering stages for modern mammalian bones in Amboseli Basin (southern Kenya), which is a relatively arid region with alkaline soils (see also Hill, 1980; and Chapters 3, 10). These stages are:

Stage 0: no obvious weathering, bone may still be greasy (cf. above), with marrow, skin, and ligaments attached

Stage 1: cracking normally parallel to fiber structure (i.e., longitudinal in long bones), soft tissues may or may not be present

Stage 2: outer concentric layers of bone flaking off until most outer layers lost; remnants of ligaments, etc., may remain

Stage 3: bone surface covered by rough, homogeneously weathered patches of compact bone, tissue rarely present

Stage 4: bone surface coarse and rough, splinters, cracks and deep weathering into bone interior

Stage 5: bone falling apart *in situ*, large splinters.

In the case of whole carcasses, she examined a number of bones before estimating weathering stage. Based on field studies, she estimated that most bones decompose beyond recognition in 10–15 years, and that bones of small mammals (<100 kg) and juveniles weather more rapidly than large bones. Small bones are also more

2.8 Vertebrates: bones as stones

likely to be destroyed by carnivores and scavengers and trampling ("bioturbation"), especially in swamps (Behrensmeyer and Boaz, 1980).

Behrensmeyer (1978) found no consistent relation, however, between habitat (woodland, bush, plain, lakebed, swamp) and weathering stage, and that there is substantial variation in weathering stages within the same habitat. Also, weathering stages varied on the same bone: Amboseli bones were weathered more on exposed than buried surfaces. But portions of bones extending more than about 10 cm above ground were often less weathered than portions close to the ground, suggesting that diurnal fluctuations in temperature and moisture at the soil–air interface are of utmost importance in accelerating breakdown of the organic matrix of bone (Hare, 1974, 1980). Although the highly alkaline Amboseli soils are conducive to preservation (Behrensmeyer, 1978; Behrensmeyer and Boaz, 1980), they may also promote cracking and splintering on buried bone surfaces through crystallization of salts (cf. Retallack, 1984). This suggests that localized conditions of temperature and moisture (shade versus sun) are more important to bone weathering than overall characteristics of the habitat. Nevertheless, more equable habitats are more likely to be represented by bones of earlier weathering stages (Behrensmeyer, 1978; Hill, 1980).

Shotwell (1955) attempted to determine the proximity of bone source area to the site of bone deposition for mammals excavated from quarries in late Tertiary sediments, and in so doing, to distinguish between autochthonous and allochthonous remains of assemblages. He assumed that the minimum number of individuals of each genus in an assemblage can be calculated from the element most often duplicated, and from these data he calculated the number of specimens per individual. Then, using the corrected number of specimens per individual, he determined the "completeness" of the representation of each genus. Genera with higher than average completeness were considered "proximal" and to have lived near the site of deposition, whereas "distant" genera were transported from elsewhere.

Voorhies (1969) seriously criticized Shotwell's (1955) methodology. First, the number of small mammal bones (e.g., rodents, insectivores) is much smaller than expected based on the availability of bones in modern populations (cf. Figure 2.15 and *Law of Numbers*, Chapter 1). Second, small bones are more likely to be winnowed or destroyed, which would affect the apparent distance of transport and Shotwell's (1955) "completeness" (Figure 2.16; see also Dodson, 1973; Behrensmeyer, 1978). Although broad similarities in depositional environments, and therefore taphonomic processes, may exist, even within the same (tapho)facies, microvertebrate remains may be differentially concentrated because

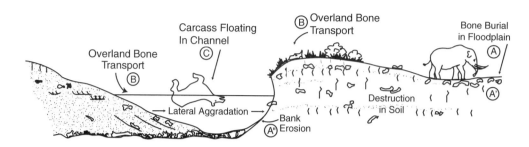

Figure 2.16. Input of bones to fluvial systems with channels that contemporaneously erode floodplain sediments. Letters denote various taphonomic pathways: (A) autochthonous burial on floodplain, followed in some cases by A′ (alteration by soil processes) or A″ (erosion from channel banks into active channel); (B) overland transport into channel; (C) origination in channel as part of a carcass. (Redrawn from Behrensmeyer, 1982.)

of size and shape differences or entrapment by larger bones (Behrensmeyer, 1990; Eberth, 1990; Blob and Fiorillo, 1996). Martill (1991) pointed out that bones vary in their strength because they are used for different purposes (e.g., large bones and the inner ear bones of cetaceans are relatively dense for bearing loads and sound transmission, respectively); even the same bone may vary in its resistance to stress (human bones will break if a stress is applied in the wrong place). Third, most of the sediments examined by Shotwell (1955) were fluvial in origin, but Shotwell inferred the proximity of grasslands. As Voorhies (1969) points out, however, at the time of deposition (approximately Late Miocene–Early Pliocene), the rivers in the study area probably had relatively wide floodplains, as evidenced by peculiar widespread lithologies such as green clay (soil?) horizons (Chapter 3). Voorhies (1969) explained the abundant remains of grazers as being the result of the mobility of grazers that probably lived in grassy to wooded interstream areas and often visited the rivers to drink. Moreover, floodplain deposits are conducive to preservation, but are also often reworked by meandering rivers and streams (see also Chapters 5, 7).

Voorhies (1969) conducted a series of flume experiments on the transport of bones using velocities comparable to those in natural streams ($\sim$20–150 cm s^{-1}). He found that the orientation of the same bones from the same taxon varied somewhat with water depth: in "shallow" water of about 4 cm depth (i.e., partially submerged), bones were more likely to be oriented cross-current (although most were oriented current-parallel), whereas in "deep" ($\sim$7–10 cm) water bones were largely oriented current-parallel. He recognized three separate transport groups for coyote and sheep bones (Table 2.10) and came to several generalizations: (1) long bones and complete pelves were the most reliable mammalian skeletal

2.8 Vertebrates: bones as stones

Table 2.10. *Susceptibility of mammalian bones to transport in a flume. Based on 15 runs each with disarticulated coyote and sheep skeletons*

Group I Immediately removed, transported by saltation or flotation	Group II Removed gradually, transported by traction	Group III Lag deposit
Ribs	Femur	Skull
Vertebrae	Tibia	Mandible
Sacrum	Humerus	
Sternum	Metapodia	*Ramus*
	Pelvis	
Scapula	Radius	
Phalanges		
Ulna	*Scapula*	
	Ramus	
	Phalanges	
	Ulna	

Elements in italic type are intermediate between the two groups in which they appear. Voorhies, 1969.

remains for inferring current direction; (2) large ends of bones are normally oriented downstream; (3) a majority of cross-current-oriented long bones indicates deposition in very shallow water; and (4) strongly convex lower jaws should usually lie convex up (most hydraulically stable orientation) provided the current is not strong enough to orient them randomly (cf. discussion of bivalved shells).

Behrensmeyer (1975) also conducted flume studies and settling experiments that compared the density of bone soaked in water ($<1.0–2.0$ g cm^{-1}; for teeth $\sim 1.7–2.3$ g cm^{-1}) with quartz spheres. She found that bones are generally hydraulically equivalent to quartz particles of smaller nominal diameter (Figure 2.17), and that in fluvial environments associations of bones with sediments of much smaller hydraulic equivalence or vice versa (e.g., hippopotamus skull in silt) may indicate other modes of origin for the association (e.g., flotation of carcasses or death *in situ*; cf. Figure 2.17). Hydraulic equivalence between skeletal remains and sediment matrix does not, however, prove that the bone assemblage was transported with the sediment, but a lack of equivalence between skeletal remains and matrix identifies remains that could not have been transported with the sediment (Badgley, 1986*b*). Behrensmeyer (1975) concluded that Voorhies Group I bones would move at normal flow velocities, but that Groups II and III required floods

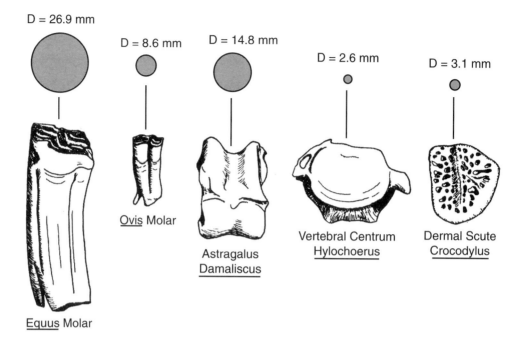

Figure 2.17. Hydraulic equivalents of modern bones determined by settling velocity experiments in terms of diameter (D) of quartz spheres. Variation in hydraulic equivalence of bones is primarily the result of density variation in bone. (Redrawn from Behrensmeyer, 1975.)

for significant transport. Thus, if bones with a wide range of transport potentials are found together, they indicate that *selective* sorting did *not* occur. She cautioned, however, that such relationships should not be assumed for bones with high surface/volume ratios (e.g., ribs, scapulae, etc.) because of differential transport (Figure 2.18).

Behrensmeyer (1975) then applied her experimental results to interpretation of vertebrate assemblages of the Kooba Fora Formation (Plio-Pleistocene) of Kenya. She found that teeth were more abundant in channel and floodplain samples, whereas vertebrae and phalanges (Voorhies Group I) were more abundant in delta margin deposits farther downstream (see also Hanson, 1980, pp. 178–179). Based on the relative abundance of teeth, Behrensmeyer (1975) concluded that lighter remains may be removed during reworking, and she suggested that the tooth/vertebra ratio be used as an indicator of winnowing.

Hanson (1980) conducted the first study of bone transport using a mathematical model. Based on the concepts developed in section 2.2, he found that the predicted *transportability of bones* (R') was

$$R' \approx 75 C_d A c^{1/3} / W_s \mu \tag{2.21}$$

2.8 Vertebrates: bones as stones

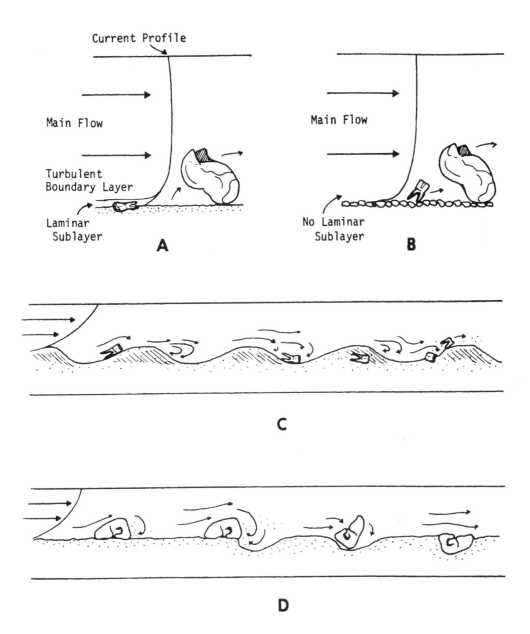

Figure 2.18. Effects of bone size and shape and bottom morphology on bone transport. (A) Small tooth remains at rest in low-velocity laminar layer, while an astragalus is moved by turbulent layer above. (B) On a coarser bottom with no laminar sublayer, tooth and astragalus are moved. (C) Tooth is dropped on downstream side of migrating ripples, buried, and re-excavated, losing its roots via erosion. (D) A metapodial (shown in end view) creates turbulence and scour around itself, rolls into the resulting pit and is buried (cf. Figures 2.2, 2.10). (Reprinted from Behrensmeyer, 1975, with permission of the Museum of Comparative Zoology, Harvard University.)

where C_d is drag coefficient, A is cross-sectional area of the bone (minimum when bone is parallel to current direction), c is bone height, W_s is submerged weight of bone ($=V[\rho_b - \rho]g$, where V is volume, ρ_b is bulk density of bone, ρ is bulk density of water, g is gravitational acceleration), and μ is sliding or rolling coefficient). When the fluid force (F_f or shear stress; section 2.2) just exceeds the maximum static force of resistance (F_S) due to friction between bone and substrate, the bone begins to move:

$$F_S = \mu F_N \tag{2.22}$$

where F_N is normal or gravitational force. If the bed is horizontal, then $F_N = W_s$, and

$$F_S = \mu W_s \tag{2.23}$$

At the point of incipient movement, $F_f = F_S$ and their ratio (R) is therefore

$$R = F_f/F_S = F_f/W_s\mu = 1 \tag{2.24}$$

Equations 2.21–2.23 are for rolling bones, but Hanson (1980) developed similar equations for sliding bones.

Hanson (1980) then plotted his empirical flume data against R and c to define several areas of bone transport (movement is predicted for $R > 1$; Figure 2.19). He attributed the widening of the zone of discontinuous movement (category C) with decreasing bone height (c) to obstruction by other bones, minor bed irregularities, and to the sensitivity of small bones to small-scale turbulence. Sliding was more common for larger bones. Hanson (1980) pointed out that instantaneous velocities that exceed the long-term mean velocity will still transport bones with $R < 1$ for short distances (cf. above).

Hanson (1980), along with A.K. Behrensmeyer, also conducted field experiments of bone transport in the East Fork River of western Wyoming in which measured and numbered bones were placed and subsequently relocated following snowmelts (floods). Hanson (1980) plotted the resulting data versus *estimated transportability* (R') and distance transported (Figure 2.20). He found that most bone transport occurred during floods (as did Behrensmeyer, 1975), which are characterized by greater traction and suspended loads that increase the density and force of fluids (cf. section 2.2). He (1980, pp. 175–176) attributed the scatter in the data (Figure 2.20) to

> *randomly distributed obstructions and unpredictable temporal and spatial variations in flow and bed conditions along individual transport paths. These factors would*

2.8 Vertebrates: bones as stones

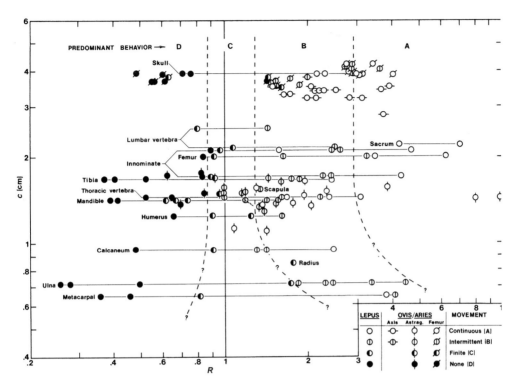

Figure 2.19. Flume observations of bone transport versus transportability (*R*) and element height (*c*; see text for explanation). Movement is predicted for elements with *R* > 1. *Lepus*, jackrabbit; *Ovis/Capra*, sheep/goat. Different values of *R* and *c* for a given kind of element reflect different orientations and flow conditions as well as individual variation. "Finite" movement (category C), element stopped after short initial movement. See text for further discussion. (Reprinted from Hanson, 1980, in *Fossils in the Making*, edited by A.K. Behrensmeyer and A.P. Hill, with permission of University of Chicago Press.)

cause an original association, even of identical elements, to become spread over progressively longer stream segments as their mean transport distance increases (this dispersion is masked by the logarithmic distance scale.) . . . Conversely, if an association of disarticulated elements does occur in a transported assemblage, the elements probably originated at different times and distances upstream, and hence represent different individuals.

This last point is essentially the same conclusion reached by Aslan and Behrensmeyer (1996) in subsequent field experiments in the East Fork River. Aslan and Behrensmeyer (1996) also found that (1) assemblages of unsorted bones develop from combinations of individual point sources *in less than 10 years*; (2) long transport distances did not necessarily result in significantly greater abrasion than short distances; (3) that light and porous bones such as vertebrae, patellae, and phalanges were transported farther than heavy and dense bones (limb bones

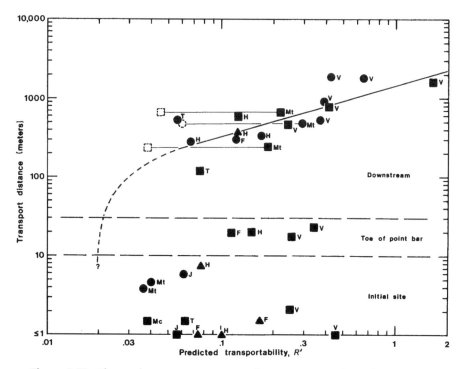

Figure 2.20. Observed one-season transport distance versus estimated transportability (R'; see text for explanation) for domestic cattle bones visually relocated after exposure to 1975 flood on East Fork River, Wyoming. Horizontal dashed lines separate natural assemblages of elements: those which remained on the bar where initially placed or within 10 m along the channel; those trapped at the toe of the point bar; and those which moved beyond the site of initial placement. Solid line is regression curve fitted to downstream data points. Little or no transport is expected for elements with $R' < \sim 0.02$ (dashed curve). Open, dashed symbols are plotted for R' values corresponding to parallel-to-current orientation for downstream metatarsals; solid symbols indicate more probable perpendicular orientation during transport. Triangular symbols denote elements initially placed at site A; square symbols, site B; circles, site C. Note increase in proportion of transported elements (downstream) versus untransported elements (initial site) with increasing R'. F, femur; H, humerus; J, jaw; Mc, Metacarpal; Mt, Metatarsal; T, tibia; V, Thoracic vertebra. (Reprinted from Hanson, 1980, in *Fossils in the Making*, edited by A.K. Behrensmeyer and A.P. Hill, with permission of University of Chicago Press.)

and mandibles; cf. Table 2.10), but that (4) bone size and density do not solely control bone transport; and that (5) low flow velocities and low rates of sediment transport through wide, shallow channels favored sorting of bones by size and shape within 1–2 years. Interestingly, Hanson (1980) found virtually no abrasion or other transport-related damage on bones carried more than 2 km, but minor damage did occur during bank storage, weathering, and trampling. He also concluded from an exponential mathematical model that bones transported from upstream sources may dominate an assemblage over those from communities

2.8 Vertebrates: bones as stones

near a sample site (cf. Shotwell, 1955). Other workers have documented abrasion of bones in the field (e.g., Boaz in Lyman, 1994a) and in tumblers (e.g., Korth, 1979), but Shipman and Rose (1983, 1988) expressed caution about interpreting both tumbler-induced features (scratches and grooves were rarely produced no matter what the conditions) and rates of bone abrasion, which appear to be accelerated in tumblers (cf. section 2.3.1).

Based on these sorts of studies, Behrensmeyer (1988) recognized two basic taphonomic modes – akin to taphofacies – in her study of attritional vertebrate assemblages of the Siwalik deposits (Miocene) of northern Pakistan and the Lower Permian Belle Plains and Arroyo Formations (Wichita Group) of Texas (Table 2.11). She emphasized that the two modes represent end-members of a spectrum of assemblages; that the modes are defined primarily on overall sedimentary context and secondarily on taphonomic features of the bone assemblages; and that rate and mode of sedimentation is critical to preservation. Given a constant input of bones (cf. Chapter 1), if sedimentation is too slow, bones will be destroyed faster than they can be buried, whereas if sedimentation is too rapid, bones will not be concentrated. In the extreme, the channel-lag end-member is characterized by sheet sands that result from repeated meandering, bank cutting, and reworking of sediment, and assemblages of allochthonous, abraded, unidentifiable fragments, whereas the channel-fill end-member is characterized by finer sediments and autochthonous, unabraded, complete skeletons that accumulated in abandoned channels after "channel avulsion" (Table 2.11). In typical cases, channel-lags fine upward, whereas channel-fills include thin, discontinuous, coarse beds. Channel-fill assemblages are much more variable taphonomically than channel-lags; although characterized by much better preservation, much pre-burial breakage and scratching may occur via carnivore activity and vertical reorientation via trampling around waterholes (Coombs and Coombs, 1997). The variable orientations of bones in channel-fills may also indicate accumulation in quiet water, however.

Most recently, Cutler (1998) has described the taphonomy of the land mammal fauna of the Pollack Farm Site (Delaware, U.S.A.), which has been termed "the most diverse Tertiary land mammal fauna known in eastern North America north of Florida" (Benson, 1998, p. 2). The fauna is of Early Miocene (Hemingfordian) age and consists primarily of single teeth and small disarticulated robust post-cranial elements of small to large mammals such as metatarsals. Based on weathering stages, abrasion-related features, and tooth and gnaw marks, Cutler (1998) concluded that carcasses had been exposed subaerially and later reworked and concentrated during transgression.

Table 2.11. *Modes of formation of vertebrate assemblages based on studies of Permian and Miocene assemblages. Assemblages are attritional and accumulated over intervals of $10^2–10^4$ years*

	Channel-lag	Channel-fill
Sedimentary context		
Large-scale	Lower parts of channels or erosional troughs	Above basal lags, usually in middle to upper parts of channels
Small-scale	Basal lag deposits, scour pockets, channels within channels	Discontinuous, thin, coarse beds, thicker fine-grained units
Lithology	Sands, gravels, mudclast and nodule conglomerates	Mudstones, silts, clays, fine sands, nodule conglomerates
Taphonomic attributes		
Sorting	Larger, heavier, robust elements more common (e.g., jaws, teeth); usually well sorted	Size-sorting in coarser sediments; variable to poor sorting otherwise
Abrasion	Edges often rounded, bone pebbles common	Edges fresh to rounded, usually fresh in mudstones
Fragmentation	Variable; usually broken parts	Variable; more complete in finer sediments
Associated skeletal parts	Rare	Variable; more frequent in mudstones
Orientation	Commonly aligned with paleocurrent; usually horizontal	Variable alignment with paleocurrent; random in mudstones; often at angles to a horizontal plane
Body sizes	Variable; large usually more common	A wide range usually present, including microfauna
Interpretative notes	Bones usually allochthonous; may represent large areas of the drainage basin	Bones at death site or transported short distances; most are autochthonous with respect to the local channel
	Channels represent active drainages with recurring energetic flow and reworking of banks and bedload	Channels are abandoned and have sporadic, waning flow with minor reworking of bank and bedload sediments

Behrensmeyer, 1988.

2.8.2 Reptiles

Although numerous field and experimental taphonomic analyses have been conducted on modern and ancient mammals, other vertebrate remains, including those of dinosaurs, have been studied far less intensively. This is no doubt related to the availability of mammalian remains in the field and modern bones for experimental work. Like mammals, though, disarticulation and disintegration of non-mammalian skeletons is quite variable, even within the same taxon (e.g., Meyer, 1991; and Blob, 1997, for turtles).

Dodson (1971) was among the first to study the taphonomy of dinosaur remains in detail. Based on sedimentological inferences, he concluded that the environments of the Oldman Formation (Late Cretaceous; Campanian) of Dinosaur Provincial Park (Alberta, Canada) represent fluvial meandering and braided channels formed adjacent to the highlands of the Laramide Orogeny (see Hungerbühler, 1998, for recent review of similar sites in Germany). Skeletal remains ranged from complete skeletons to isolated bones, and were most common in clean channel sands, which comprise about 70% of the rocks in the park; this relationship persisted over approximately a 70-m section and suggested that preservation of skeletons was the result of normal "background" processes, rather than catastrophic ones. Overbank (levee) deposits, along with sideritic concretions (indicative of iron-rich marginal marine or fresh water; Chapter 3), plant rootlets, and desiccation cracks were also present but relatively uncommon. Of 84 quarries examined, only one contained fossils deposited in strictly muddy sediments, all of which suggest that the setting was not analogous to many modern river settings with extensive floodplains; instead, it appears that relatively short rivers drained the highlands, much as they apparently did in the Late Permian Uralian foreland basin (Olson, 1962). Expansion of montmorillonite-rich clays may have destroyed some bones, however.

The relationship between channel sands and bones contrasts sharply with the reports of well-preserved reptilian and mammalian remains in fine-grained fluvial sediments (e.g., Olson, 1962; Clark *et al.*, 1967), including those of the Hell Creek (White *et al.*, 1998). This suggested to Dodson (1971) that many animals died in the water, possibly during floods. Dodson (1971) grouped remains into 10 preservational classes (A–J; Figure 2.21). Of 17 complete (or nearly so) skeletons (class A) 12 occurred in channel sands. Other classes of skeletons exhibited varying states of disarticulation (down to isolated bones) which, according to Dodson (1971), reflected varying degrees of post-mortem decay and transport (Figure 2.21). Among the more notable occurrences are headless, but otherwise complete, skeletons (class E), which presumably represent an early stage of decomposition (e.g., Weigelt, 1989; Schäfer, 1972), perhaps of floating carcasses. Class F contains

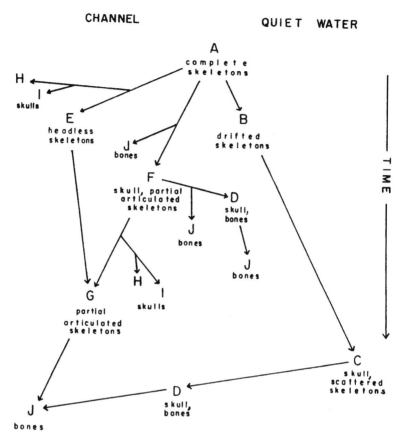

Figure 2.21. Taphonomic pathways of dinosaur remains of Oldman Formation (Late Cretaceous). (Reprinted from P. Dodson (1971) Sedimentology and taphonomy of the Oldman Formation (Campanian), Dinosaur Provincial Park, Alberta (Canada). *Palaeogeography, Palaeoclimatology, Palaeoecology* 10:21–74, with kind permission from Elsevier Science-NL, Sara Burgerhartstraat 25, 1055 KV Amsterdam, The Netherlands.)

both skulls and incomplete articulated skeletons that apparently continued to decompose as they were being transported. Other concentrations (lags) consist of pebble-size (~4–64 mm) remains, such as garpike scales, fish and ray teeth, sturgeon spines, crocodile scutes, isolated vertebrae, etc., all of which appear to represent end products of disarticulation, transport, and physicochemical alteration; interestingly, however, many teeth and scales show no signs of abrasion. Experimental studies also indicate that reptilian (dinosaur) teeth can undergo extensive transport (equivalent of 360–400 km) but exhibit little or no wear (Argast *et al.*, 1987; see also Eaton *et al.*, 1989; and Chapter 10). One of the most puzzling classes is class D, which is represented in part by skulls (with a few disarticulated postcranial bones), ribs and scapulae. Ribs and scapulae are

2.8 Vertebrates: bones as stones

Table 2.12. *Taphonomic pathways of reptilian remains of the Karroo Basin (Late Permian). Number of plus signs (+) denotes relative importance of each taphonomic pathway*

	Taphonomic pathways					
	1	2	3	4	5	6
Facies						
Channel				+	+	++
Channel bank	+	+				
Proximal floodplain	++++	++	+++			
Distal floodplain	+					

1, Disarticulated skeletons embedded at site of death, buried by vertically accreted alluvium; 2, Articulated and disarticulated skeletons preserved in underground burrows; 3, Articulated, disarticulated and transported small postcranial elements accumulated in embayments of lowstand lake margin; 4, Carcasses of animals that died in channel furrow, disarticulated during transport downstream and bones added to channel gravels; 5, Caliche encrusted bone reworked from cut bank collapse; 6, Waterhole accumulation in and around swales and chute channels.
Smith, 1993

easily moved by currents, whereas skulls are not (Voorhies, 1969; Behrensmeyer, 1975); possibly skulls projected into the water column sufficiently to be moved (section 2.2; Figure 2.18) or were dropped to the bottom from floating carcasses. Intact vertebral columns with legs attached, but no skull, were also found; according to Dodson (1971), it is unlikely that predators or scavengers (e.g., crocodiles; Weigelt, 1989) would prefer the skull only and ignore the rest of the carcass. Nevertheless, although occasionally isolated bones with long grooves or partially disarticulated skeletons were found (e.g., skull with neck still attached), there was a general paucity of evidence for the activities of carnivores and scavengers (see also section 2.8.1 regarding disarticulation of mammalian skeletons).

Like Dodson (1971), Smith (1993) found that therapsid remains were concentrated in certain environments. Using the weathering stages of Behrensmeyer (1978), Smith defined taphonomic pathways of the Karroo Basin (Late Permian), a foreland basin that formed in what is now South Africa as Pangea was being assembled (Table 2.12). A major influence on preservation was the distance of

Table 2.13. *Taphonomic characteristics of hypothetical fossil vertebrate assemblages*

	Cause of mortality		
Mode of transport	Predator	Natural trap	Sudden disaster
None	Articulation present; most specimens clustered; high % juveniles; punctures, scratches; hydraulic sorting absent	Articulation present; specimens clustered; moderate % juveniles; little bone damage; hydraulic sorting absent	Articulation common; specimens clustered; moderate % juveniles; little bone damage; hydraulic sorting absent
By predators or scavengers to a focal area	Articulation variable; some specimens clustered; moderate–high % juveniles; punctures, scratches; hydraulic sorting absent	Articulation variable; some specimens clustered; low–moderate % juveniles; punctures, scratches; hydraulic sorting absent	Articulation variable; some specimens clustered; low–moderate % juveniles; punctures, scratches; hydraulic sorting absent
By currents	Articulation absent; specimens scattered; moderate % juveniles; polish and abrasion; hydraulic sorting present	Articulation absent; specimens scattered; low % juveniles; polish and abrasion; hydraulic sorting present	Articulation absent; specimens scattered; low % juveniles; polish and abrasion; hydraulic sorting present

Badgley, 1986b.

2.8 Vertebrates: bones as stones

the burial site from the main channel, which determined the extent of disarticulation, transport, and burial by floods versus residence time at the sediment–air interface.

The concentration of hardparts in certain facies of the Oldman Formation and Karroo Group contrasts sharply with that of the Morrison (Late Jurassic; Kimmeridgian–Portlandian) of the western United States (Dodson *et al.*, 1980; see also Wood *et al.*, 1988; Eberth, 1990) and Hell Creek (White *et al.*, 1998). Dinosaur remains are not confined to particular environments in the Morrison, apparently because dinosaurs migrated in response to seasonal rainfall, a situation analogous to that of mammals in East Africa (Dodson *et al.*, 1980). The Morrison consists of sediments weathered from the ancestral Rocky Mountains that spread over a huge ($\sim$1 000 000 km^2) lowland plain after regression of Jurassic seas northward into Canada; overbank and floodplain deposits were common. Unlike the Oldman Formation, skeletons, including mass accumulations, are highly disarticulated and altered, which suggests prolonged exposure to air during dry periods before final burial. Possibly, many of these remains accumulated in soils or soil-like deposits (Chapter 3).

Like mammals, these sorts of studies laid the groundwork for the use of the taphofacies concept in the study of reptilian assemblages. In his study of the preservation of dinosaur remains in the Two Medicine and Judith River formations of north-central Wyoming, for example, Rogers (1993) adopted a taphofacies approach similar to that of Behrensmeyer (1988). These formations represent an alluvial/paralic/shallow foreshore gradient in a Late Cretaceous (Campanian) foreland basin (Two Medicine: proximal; Judith River: distal; cf. Table 2.13). The sections have provided a wealth of information on dinosaur evolution, paleoecology, and behavior since the time of Leidy (1856) and Cope (1876), and include dinosaur nests and nesting sites (e.g., Horner, 1982). The formations consist of mudstones, siltstones, and lesser amounts of sandstone; sands tend to be thicker in the Judith River Formation and often display "multi-story" lenticular or sheet-like geometries, and basal contacts are often erosional (ripup clasts, etc.)

Rogers (1993) recognized taphofacies in part by using the stages of bone weathering developed by Behrensmeyer (1978) for Recent mammal remains in Amboseli Basin (Table 2.13). Channel-lag concentrations are deposits of allochthonous, usually disassociated, bones and teeth at the base(s) of single or multi-story channels. Skeletal remains may or may not show evidence of abrasion, but often show evidence of hydraulic sorting due to differences in bone density, size, and shape (cf. Voorhies, 1969; Dodson, 1975; Behrensmeyer, 1975, 1988). Remains may originate from one of three pathways: death in the channel itself, transport from outside the channel, or reworking of older material into the channel (Behrensmeyer, 1982).

Both channel-lag and channel-fill deposits may include carbonaceous (plant) remains, including wood, and freshwater bivalves and gastropods. Channel-fill concentrations form in abandoned channels from autochthonous and parautochthonous remains. Although weathering stages are usually low (Table 2.13), remains are also typically disarticulated. Rogers (1993) found that ribs, vertebrae, and phalanges were underrepresented in the two channel-fills of his study, suggesting that the assemblages had been winnowed (Voorhies, 1969; Hanson, 1980).

Rogers (1993) also differentiated between subaerial and subaqueous floodplain concentrations, using both paleontological and sedimentological criteria. Subaqueous deposits, which contain freshwater vertebrates, invertebrates, or varves, and which are often enriched in plant debris and amber, are of two types: bone beds are usually monospecific concentrations of disarticulated hadrosaurs (*Maiasaura*, *Gryposaurus*, *Hypacrosaurus*) that exhibit minimal weathering, although vertebrae, ribs, and phalanges are underrepresented, and may include turtle, champosaurs, and crocodiles; vertebrate microfossil concentrations, on the other hand, are high-diversity assemblages of small aquatic (fish, amphibians, reptiles) and terrestrial (dinosaurs, mammals) taxa with physicochemically resistant hardparts (teeth, scales) predominating; most remains show minimal weathering, although a few bones may show advanced weathering (stages 4–5).

For biological reasons, subaerial floodplain concentrations represent nesting sites (amniotes lay eggs on dry land), including both hatched and unhatched eggs with intact embryos (Horner and Weishampel, 1988). Eggshell appears to be relatively unweathered (microstructure is often preserved) and concentrations of bones within and around nests no doubt represent hatchling cohorts (Horner, 1982), which, despite preservational biases against juvenile bone (Behrensmeyer, 1978), show little weathering (eggshell fragments may be preserved in marine sediment based on actualistic experiments of Hayward *et al.*, 1997). Subaerial concentrations of larger bones are typically monospecific assemblages of disarticulated hadrosaurs that also exhibit low amounts of weathering.

The taphofacies approach is no less important in the study of marine reptiles. Martill (1985; see also Martill, 1987*a*) recognized five preservational types in the Lower Oxford Clay (Middle Jurassic, Callovian) of central England: (1) fully articulated skeletons, with all bones in true position, are rare and are found in fissile, highly bituminous (carbon-rich) shales; (2) partly to wholly

2.8 Vertebrates: bones as stones

disarticulated skeletons (but with all bones present and in association). There is a positive correlation between grain size (environmental energy) and disarticulation. Disarticulated skeletons are common in *Gryphaea* shell beds, and are associated with storm activity, reworking, and burrowed bituminous shales, which indicate deposit feeding and scavenging; (3) isolated bones and teeth, which occur throughout the Lower Oxford Clay; bones were probably dropped from floating carcasses (Schäfer, 1972), whereas teeth were probably shed from both living animals and carcasses either floating or scavenged on the sea bottom; (4) worn bones, which are common in shell beds and sands, and which probably lay exposed to physical and bioerosive (e.g., echinoid) agents at the SWI for prolonged intervals; and (5) coprolites enriched in small bones and otoliths (which are resistant to gastric juices) and which occur throughout the Lower Oxford Clay (see also Wilson, 1987). Martill (1985) concluded that the high concentrations of vertebrates and organic carbon (up to 14%) of the Lower Oxford Clay reflects high productivity in surface waters that could have resulted from nutrient input from nearby landmasses. If such a relationship were to hold for other sites, he suggested, it could be used to develop a predictive model of marine vertebrate fossil-Lagerstätten.

Decay gradients in aquatic realms also reflect the propensity of carcasses to float during decay (Allison and Briggs, 1991b). Dispersal of carcasses reflects the strength of tissue and the rate of gas production: as tissue strength declines, expanding gases break the carcass apart, and after the gas is released the carcass sinks. The buildup of gas depends on its origin (it is easier for it to escape from coelomic cavities than from solid flesh) and carcass size: larger carcasses have smaller surface area/volume ratios, which inhibit the diffusion of oxygen within the carcass, thereby promoting methanogenesis and fermentation (cf. Chapter 3). It is likely that most large fish, reptiles, and cetaceans decay fermentatively. Once the carcass has sunk to the bottom, hydrostatic pressure may prevent its reflotation because hydrostatic pressure inhibits the production of gases from decaying flesh; the gradient in hydrostatic pressure would seem to parallel that of environmental energy and water depth and, conceivably, is discernible from the state of preservation of carcasses, although further research is required to confirm this hypothesis (Allison and Briggs, 1991b; Allison et al., 1991). But once they reach bottom, carcasses may also be stripped of flesh in a matter of days by scavengers (e.g., Smith, 1985); marine microbes also accelerate decay and disarticulation (Arnaud et al., 1978; Ascenzi and Silvestrini, 1984; Chapters 3, 6). Despite the apparent rapidity of decay, carcasses may provide "stepping stones" for faunal elements between deep sea vent habitats (Showstack, 1998).

2.8.3 Amphibians, fish, and birds

The taphonomy of amphibians, fish, and birds has received even less attention than reptiles, probably because, ironically, they are typically *not* found in the fossil record. Although this state of affairs is changing, Weigelt's (1989) and Schäfer's (1972) monographs still offer some of the most complete observations on the stages of decay and disintegration of these groups, and the remains of all three taxa are subject to the same basic processes as mammals and reptiles. Unlike the relatively tough skin of mammals and reptiles, however, that of modern amphibians is thin and fragile and decays easily, thereby accelerating disarticulation. Dodson (1973) noted that frog and toad skin disappeared after 3 weeks, but that mouse skin exhibited no obvious deterioration after 77 days.

Like other carcasses, the degree of disarticulation of fish also appears to reflect the amount of decomposition prior to burial. Schäfer (1972) noted that fish carcasses may either float or sink, depending upon body size and the extent of gas production by decay, which increased with rising water temperature. Zangerl and Richardson (1963) conducted experiments on fish decomposition in modern black muds of high-temperature (24–37 °C), low-salinity bayous (3–8.5‰) near Lake Pontchartrain (Louisiana) using carcasses enclosed in scavenger-proof, fine-mesh wire cages. They found that in some cases the soft parts of carcasses were lost and the bones entirely disarticulated in as little as 6.5 days. Such rapid decay was in conflict with the experimental results of Hecht (1933), who did not record temperature. Zangerl and Richardson (1963) cautioned that one of the primary factors in decay is water temperature, which tends to decrease with increasing water depth, especially in lakes. Zangerl and Richardson (1963) concluded that some oxygen must be present, so they attributed rapid decay to *aerobic* bacterial decomposition (see Chapter 3).

Although fossil remains were highly concentrated in black shale horizons in their study, Zangerl and Richardson (1963) concluded that fossil assemblages did not represent mass mortality; although infauna was typically absent (because of relatively low oxygen conditions in sediment and above), vertebrate specimens typically showed evidence of disarticulation via scavenging and decay gases. Concentrations of intact skeletons, however, probably represent catastrophic mortality from seasonal overturn (in lakes) and temperature changes (see Behrensmeyer and Hook, 1992, and Lyman, 1994a, for review); Martill (1991) notes that evidence of catastrophic mortality is probably destroyed in most cases by physical and biological reworking (Chapter 4). Also, fish remains are often quite small and may be overlooked when sieving with screens as coarse as 0.5–4 mm (Smith *et al.*, 1988; Lyman, 1994a).

Bird bones also tend to be relatively small and fragile because they are hollow, and may be overlooked unless careful screening is performed. Bird carcasses are often quickly removed by scavengers, and those that are left behind undergo distinctive stages of disarticulation (Bickart, 1984; Davis and Briggs, 1998). Davis and Briggs (1998) calculated that weight loss of intact modern bird carcasses (songbirds and passerines) kept in cages was exponential and could be used to estimate the duration of decay: unprotected carcasses lost more than 80% of their weight in 3–10 days, whereas the same weight loss took 10–28 days in protected specimens. Bickart (1984), on the other hand, found that avian skeletal remains were largely unweathered even after 1 year of exposure. The differing results of Bickart (1984), Davis and Briggs (1998), Oliver and Graham (1994) probably reflect differences in the degree of scavenging and ambient temperature.

2.9 Macroflora and pollen

2.9.1 Plant macrofossils

The paleobotanical record is a rich source of detailed data on past climates as long as the conditions of assemblage formation are evaluated (e.g., Wing, 1984; Wnuk and Pfefferkorn, 1987). For example, root-shoot ratios were used by Raymond (1987) to evaluate both modern (Everglades mangroves, Okefenokee blackwater swamps) and Carboniferous deposits. She found that the rank abundance of roots can be used as an indicator of the rank abundance of taxa in ancient deposits. Extremely high root-shoot ratios (>100) appear to be indicative of black(fresh)-water swamps, in which detritivores are relatively few in number (see also Labandeira *et al.*, 1997; Buatois *et al.*, 1998).

With the exceptions of most coals and peats, plant remains are typically transported to some extent (Gastaldo, 1992a). Moreover, like arthropods, macroscopic plants consist of multiple elements and organs that are shed numerous times during their lives, and that have quite different potentials for transport and preservation (see Spicer, 1989, for extensive review; Greenwood, 1991, and Spicer, 1991, present concise summaries). Tree trunks, seeds, and pollen are usually fairly durable and can be reworked a number of times before final deposition, whereas leaves tend to decay fairly rapidly, depending on the species (e.g., Liu and Gastaldo, 1992). In the case of herbaceous taxa, parts usually wither and rot on the parent plant, so that their preservation is often a function of rapid burial by floods or volcanic ashes (although preservation still varies; Burnham and Spicer, 1986).

Because of the differing durabilities of plant parts, organs of different species are often present in the same assemblage and are mistakenly assumed to belong to the same species until different associations are found. Scheihing and Pfefferkorn (1984), for example, found mangrove leaves (*Rhizophora mangle*) and palm seeds together in coastal sediments of the Orinoco Delta (eastern Venezuela). To complicate matters further, litter production by plants varies from place to place and season to season and may be biased toward litter production from disturbed, "non-equilibrium," communities (Spicer, 1989). Nevertheless, "k-strategist" (non-opportunistic or long-lived taxa) potentially contribute more plant parts to the fossil record than "r-strategist" (opportunistic or short-lived) taxa because of greater biomass production, dehiscence of vegetative parts, and occupation of long-term growth sites (Gastaldo, 1992*a*).

Plant parts vary with respect to their transportability and, in the case of fruits, seeds, and pollen, may even be adapted for transport (Spicer, 1989). The first stage in the formation of fossil plant assemblages is the shedding of parts into air, typically by *abcission*. In temperate climes, most deciduous trees shed their leaves every fall because broad, thin leaves are susceptible to freezing. On the other hand, broad-leaved angiosperms (e.g., *Rhododendron*) and conifers tend to retain their leaves (needles), which have much lower surface-to-volume ratios or thick, waxy coats, and may therefore be underrepresented in the fossil record under normal conditions. Temperate evergreens also tend to retain their leaves longer than their counterparts in tropical forests, in which low levels of leaf litter are often found and more continuous leaf loss may be related to the recycling of nutrients. In this regard, Burnham *et al.* (1992) and Burnham (1993*a*) found that samples of leaf litter from temperate forests were more likely to record the dominant source species present, whereas samples from tropical forests record much fewer numbers of species over a smaller area; spatial heterogeneity was no doubt high in ancient tropical forests as well (DiMichele and Nelson, 1989). Consequently, Burnham (1993*a*) suggested that "climatic filters" should be applied to estimates of ancient plant diversity.

Even within the same community, the amount and timing of leaf fall varies substantially, and the degree to which forest-floor autochthonous leaf assemblages reflect source vegetation depends upon the duration (time scale) of accumulation that is considered: "long-term" (5-month) assemblages may better reflect community composition than do "short-term" (10-day) accumulations (Meldahl *et al.*, 1995; cf. Gastaldo, 1997). Thus, well-preserved "snapshots" of leaf assemblages may give only a glimpse of true community dynamics (Meldahl *et al.*, 1995), whereas longer-term, time-averaged assemblages may reveal aspects of vegetational structure that were previously thought to be accessible only in well-preserved "fossil forests" with standing trunks (Burnham *et al.*, 1992; see also Chapter 5).

2.9 Macroflora and pollen

The difficulties of predicting settling behavior of plant parts such as leaves has led to empirical analysis of leaf settling. Ferguson (1985; see also Spicer, 1981) described field and laboratory experiments on the dispersal of leaves from a variety of temperate deciduous and evergreen taxa. Of overriding importance is leaf weight, which governs the velocity of leaf fall and the distance of transport; in Ferguson's (1985) study, weight ranged from 0.0057 to 0.072 g cm^{-2} for 49 taxa, and light, *chartaceous* (papery) leaves of angiosperms tended to be dispersed farther than heavy *coriaceous* or *carnose* (leathery or waxy) leaves of evergreens. Moreover, within angiosperms, leaf weight varies between species according to the extent of leaf desiccation before abcission. Large leaves are more likely to encounter obstacles and so settle closer to their source, and are also heavier than small ones, although "sun" leaves in tree crowns are smaller and heavier than leaves from the shadier portions of the same tree. Experimental studies of long narrow leaves indicate that they tend to rotate about an axis during fall and take longer to settle to the ground, but their greater potential distances of transport was not confirmed by field studies (Ferguson, 1985). If of sufficient size and weight, the *petiole* may also exert control over patterns and rates of leaf fall. Progressive loss of leaflets in compound leaves may result in greater dispersal of leaflets, whereas the remaining denuded leaves fall closer to the source (Ferguson, 1985).

Despite the variation in rates and patterns of leaf fall, most leaves probably are not dispersed in air farther than a distance equal to the height of the source plant. Fresh leaf litter on forest floors is heterogeneous, suggesting that leaves fall close to the source (Ferguson, 1985, his figure 9), and it has been reported that 90% of conifer needles fall within a 20° cone from the tip of the tree (Dunwiddie, 1987). In lakes, leaf dispersal is said to follow an exponential relation (Rau, 1976):

$$Z_x = Z_r^{(-k[r-x])} \tag{2.25}$$

where x is distance from the lake center, Z_x is deposition occurring at distance x, Z_r is deposition at shoreline, r is distance from lake center to shoreline, and $k = (r - x)^{-1} \ln(Z_x Z_r^{-1})$. In his study, Rau (1976) did not find any debris from the conifer, *Abies amabilis* (up to 80 m high), 160 m from the lake margin.

In the case of storms, however, litter, especially from the upper canopy, may be transported considerable distances and the source area may undergo changes in community structure as a result (Spicer, 1989). If leaves are not near the time of abcission, storms may also uproot trees that would otherwise contribute only leaves and other relatively small structures to litter (Spicer, 1989).

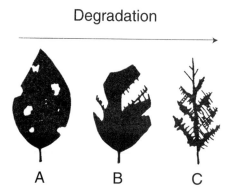

Figure 2.22. Patterns of leaf degradation caused by (A) invertebrate grazers; (B) mechanical abrasion; and (C) microbial degradation. (Redrawn from Spicer, 1989.)

Once on the ground, leaves tend to move by rolling and saltation (Spicer, 1989). Based on laboratory experiments with flat paper shapes and a fan, Spicer (1981) concluded that circular shapes tended to roll, whereas V-shapes (i.e. dry, curled leaves) were more likely to be dispersed. Wet leaves are likely to become imbricated and remain stationary, but species with a relatively high degree of lignification, such as *Fagus sylvatica* (beech) and *Quercus robur* (common oak), are water resistant and decay less rapidly, and are therefore more likely to be transported (Ferguson, 1985). Nevertheless, little ground transport occurs, as Ferguson (1985) noted for separate plots of different tree species: using spray-painted (tracer) leaves, he found that only one leaf out of 500 had moved more than 2 m after 98 days. Ferguson (1985) also conducted tumbling experiments, in which he produced leaves largely missing intervascular tissues (Figure 2.22).

Many leaves develop appearances similar to those in tumbling experiments through natural decay. Fungal, bacterial and other microbial activity degrades approximately 50% of leaf fall within 2–3 months in temperate and tropical settings (decay rates determined in the absence of mechanical destruction; see references in Burnham, 1993b). After water-soluble antifungal and antibacterial polyphenols have been leached, the leaves are attacked by bacteria, invertebrates, and especially fungi (Kaushik and Hynes, 1971; Figure 2.22). Rates and sequences of decay depend on climatic regime (faster in the tropics; see above) and composition of the litter (e.g., Nykvist, 1962). Ferguson (1985) conducted burial experiments with leaves in nylon mesh bags; he found that beech leaves (lignin-rich) decayed very slowly, but when litter consisted entirely of decay-susceptible species (e.g., sugar-rich at abcission, such as the alder, *Alnus*), litter from these species also survived (see also Spicer, 1989). Because seeds are usually more resistant to

2.9 Macroflora and pollen

mechanical degradation, their destruction by biological agents may be more important (Collinson, 1983).

Plant litter may also enter fluvial habitats, either through wind transport or flooding and erosion of bank-edge vegetation, and exhibits differential buoyancy and therefore susceptibility to downstream transport (Ferguson, 1985). Floating times range from a few hours to several weeks (Spicer, 1981; Ferguson, 1985). Leaves with thin cuticles, large numbers of stomata, little lignification, or which have undergone extensive desiccation or degradation, become rapidly waterlogged. Thus, chartaceous leaves tend to sink first and coriaceous ones last. Leaf stalks and petioles of compound and simple leaves, respectively, tend to project vertically and buoy leaves, so that relatively intact compound leaves will travel farther than their individual leaflets (Ferguson, 1985). Ferguson (1985) also noted that the "half-life" of flotation of common oak (*Quercus*) leaves increased with increasing oxygenation of the water (e.g., high-velocity mountain streams), although the *causal* relationship was not explained.

Although dry leaves tend to fragment rapidly, if fully rewetted, they fragment more slowly and settle more rapidly (Burnham, 1993b). Spicer (1989) describes experiments which indicate that angiosperms exhibit a narrow range of settling velocities (Figure 2.23). Conifer needles settle at quite different velocities, however, and so angiosperm and evergreen litter may be sorted hydraulically (Spicer and Wolfe, 1987). Fresh leaves are also quite resistant to mechanical degradation in tumblers (Ferguson, 1985), but desiccated (brittle) or decayed leaves are susceptible to fragmentation by turbulent water, including tidal flushing. Mechanically degraded leaves normally exhibit angular breaks and tears in the leaf periphery (Figure 2.22) and so may be indicative of fluvial transport, whereas in lacustrine settings, biological degradation presumably dominates (Spicer, 1989). Ferguson (1985) suggested that the taphonomic grade of leaves and fragment size could be used to infer the degree of leaf reworking; thus, fresh-appearing, intact leaves may indicate synchronous leaf fall (Spicer, 1989; but see Chapters 3, 5 regarding taphonomic grades).

More durable plant parts exhibit different floating times. Fruits and seeds typically have a greater range of buoyancy times than do leaves (weeks or longer), but their floating times are not necessarily related to size, as small seeds may float for weeks and logs may remain afloat for years (Coffin, 1983). Collinson (1983) cautioned that concentrations of hard, durable plant remains, such as seeds, should be interpreted with caution because of potential allochthony, and Scheihing and Pfefferkorn (1984) used color-coded seeds to document seed movement in response to tidal flushing in the Orinoco River. Spicer (1989) noted that on

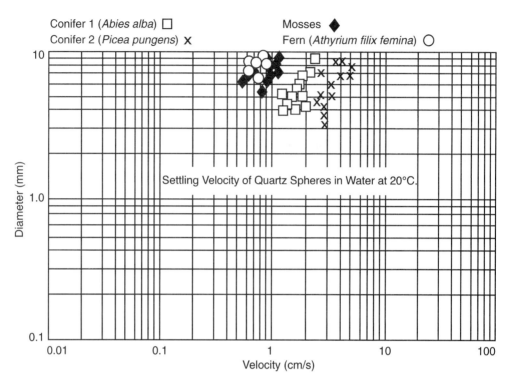

Figure 2.23. Settling velocities in still water for various saturated plant parts. Note that plant remains mostly settle within a relatively narrow range of velocities despite large differences in shape. (Redrawn from Spicer, 1989.)

beaches of the western Alaska peninsula, large concentrations of logs occur that were transported from Japan, the tree rings of which would give an erroneous regional paleoclimate signal.

While suspended in the water column or at the SWI, plant remains are subject to both entrapment by obstacles and burial, although they may undergo several cycles of entrapment (burial) and re-entrainment before final burial. Burial may occur via migrating bedforms such as ripples. Such deposits may be deceptive, however. Spicer (1989) notes that just because only conifer needles are preserved in the troughs of ripples, it does not mean that angiosperms were completely absent, as larger angiosperm litter may have been too large to settle in the troughs and were therefore bypassed downstream. Infilling of hollow stems (*pith cast*) varies with stem length and diameter, current velocity, suspended or bed load, and stem orientation with respect to current direction (Rex, 1985).

Given the metastable configurations of a point source of plant debris, the median point of the debris accumulation will tend to move downstream (cf. Aslan and Behrensmeyer, 1996). Spicer (1989) concluded that the main factor

2.9 Macroflora and pollen

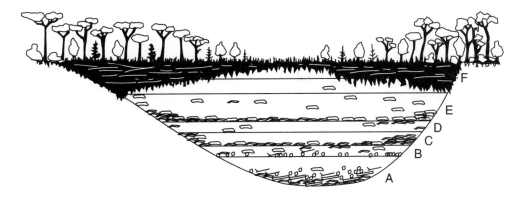

Figure 2.24. Hypothetical vertical section through an infilled abandoned channel. (A) basal channel lag containing logs, especially from bank margin erosion; (B) early infill phase dominated by riparian vegetation; (C–E) successive infilling by floods, with vegetation consisting of riparian, levee, overbank, and oxbow communities; (F) final phase of infilling, consisting of a mire community characterized by organic-rich clays and peats. (Redrawn from Spicer, 1989.)

in limiting downstream transport was water depth and stream bottom roughness in relation to the size of the plant particle (cf. Figure 2.2). These factors are in turn partly a function of stream gradient (current velocity), sediment supply, nature of the bedrock (hard or soft), and climate (e.g., temperature, rainfall), all of which influence river course and depositional facies.

These depositional facies may in turn be associated with relatively distinctive assemblages of plant remains (Spicer, 1989; Figure 2.24). As expected, channel deposits consist of lags of coarse (sand-gravel) sediment and relatively large or robust plant debris, such as logs and seeds, but may also include large amounts of riparian (streamside) debris. Gastaldo and Huc (1992) concluded that most of the degradation of wood in the Mahakam Delta (Kalimantan) results from mechanical destruction; wetting and drying or freezing and thawing, as might occur in log jams, would also be expected to accelerate wood degradation (Burnham, 1993b). Clay balls may also be reworked from underlying sediment or sourced from upstream (allochthonous). Assemblages from the channel may be relatively homogeneous (Burnham, 1989), probably as a result of post-mortem transport and mixing. Although Spicer (1989) concluded that such assemblages still contain a large local vegetation component (unless there is evidence of obstacles or rafts of debris), channel depsosits typically provide only a bulk sample of the vegetation within a particular drainage basin (Spicer and Wolfe, 1987).

In a study of the floodplain forests of the Río San Pedro (southernmost Mexico), Burnham (1989) found that assemblages in levees and back-levees are more diverse and heterogeneous than channel samples, and better reflect the local flora. Unfortunately, levees are typically poor sites for plant preservation because of oxidation via subaerial exposure and bioturbation by roots (Scheihing and Pfefferkorn, 1984; Gastaldo, 1989; Gastaldo et al., 1989).

Assemblages of crevasse splay assemblages and overbank flood deposits (such as those of Mobile Delta, Alabama) are prized because of rapid burial by fine-grained sediment when floods breach levees and because they provide an overall picture of regional floodplain vegetation, especially when depositional facies and plant assemblages are strongly related (Spicer, 1989; Davies-Vollum and Wing, 1998). These deposits consist of autochthonous rooted components, especially in swamps, where prostrate trees and erect tree bases may be preserved *in situ*, although autochthonous forest litter may be mixed with river-transported remains (from within or outside the basin) and debris from levees. Litter beds probably represent channel-bottom accumulations buried during initial flooding, whereas *detrital* peats consist of transported debris (Gastaldo et al., 1987; Spicer, 1989). Larger plant remains are sorted according to size and density. Rounded wood fragments are probably abraded during bedload transport, whereas more angular fragments are probably transported as suspension load. Scheihing and Pfefferkorn (1984) found detrital peats directly over root-bearing horizons; thus, the occurrence of root-bearing horizons directly beneath coals is not proof of accumulation *in situ* of coals. Kaye and Barghoorn (1964) concluded that in salt marshes, roots from younger horizons may penetrate in deeper (older) sediments and wood from horizons of different ages may be juxtaposed during peat autocompaction; thus, radiocarbon samples should come only from the base of peats and should only be correlated with sea-level curves using other types of data (see also Pizzuto and Schwendt, 1997, and discussion of numerical modeling of autocompaction in Chapter 4).

Floodplain sediments may become saturated by the rising water table during flood stage and trap sediment transported onto the plain, thereby enhancing burial (Gastaldo et al., 1987). Nevertheless, floodplain assemblages are typically poorly preserved. Unfortunately, these sediments are dominated by interfluves and so rooting and oxidation frequently destroy plant remains (Scheihing and Pfefferkorn, 1984). In some cases, however, marshes may form impermeable layers and perched (stagnant) water tables favorable for the preservation of fragile autochthonous remains (Wing, 1984), and oxbows may provide outstanding plant assemblages (e.g., leaf impressions) because of quiet-water sedimentation (Gastaldo et al., 1989). Indeed, oxbow deposits may record a succession from channel lag to lacustrine settings (Figure 2.24). Oxbows may be reactivated by

2.9 Macroflora and pollen

channel migration, however, so that they, as well as much of the rest of the floodplain, may be reworked. Plant assemblages from restricted environments such as oxbows or incised valleys may also differ substantially from regional climate (Demko et al., 1998). Thus, it may be necessary to look for species preserved throughout most or all fluvial facies in order to gain a perspective on regional vegetation and climate (Spicer, 1989).

Once streams or rivers debouch into lakes, current velocity diminishes rapidly and transported remains begin to settle through the water column in the form of detritus or fecal pellets (e.g., Mapes and Mapes, 1997). Under low energy settings (Silwood Lake, Berkshire, England), autochthonous components are concentrated in bottomsets, whereas allochthonous plant debris is concentrated in topset beds of prograding deltas (Spicer, 1989). Because of relatively low sedimentation rates, autochthonous debris of bottomsets is more prone to biological degradation, whereas allochthonous material of topsets is characterized by angular fragments produced by mechanical degradation (Spicer, 1989). In high energy settings (Trinity Lake, northern California), however, elements with relatively high settling velocities (e.g., seeds) were concentrated in foresets, whereas those with low fall velocities were concentrated in bottomsets (Spicer and Wolfe, 1987).

In lacustrine environments, other factors affecting plant preservation are nutrient and oxygen content of the water that in turn are associated with areal extent, depth of water, and thermal stratification and seasonal overturn. Based on modern *temperate* lakes (which are probably not good analogs for ancient lakes in different climatic settings), lacustrine environments are of two basic types: oligotrophic (low nutrient levels, low productivity) and eutrophic (abundant nutrients, high productivity; Behrensmeyer and Hook, 1992). Ferguson (1985) predicted that leaf decay is faster in oligotrophic lakes because of lower overall nutrient and food availability to bacteria, fungi, and invertebrates. But leaf degradation may be enhanced in some cases by eutrophication (Spicer, 1989; cf. section 2.4): it appears that although high productivity lakes often have diverse biotas and can potentially generate a rich fossil record, intense rates of recycling associated with high productivity often destroys most organic remains (Behrensmeyer and Hook, 1992). Conversely, eutrophication may also be correlated with increased sediment input (accompanying nutrient influx) that buries plant litter rapidly. Oligotrophic lakes, on the other hand, have relatively low diversity, but lower levels of biotic activity may decrease rates of organic destruction (Behrensmeyer and Hook, 1992).

Although broadly similar to freshwater environments, preservation in fluvio-marine settings also involves such factors as density differences between river and marine waters, tidal cycles, changes in types and rates of bioturbation, and chemistry of pore waters (Baird *et al.*, 1985, 1986; Spicer, 1989; Lin and Morse, 1991; see also Chapters 3, 4). Most notably, Scheihing and Pfefferkorn (1984) emphasized that the Orinoco Delta is a much better analog for Late Carboniferous deltas of North America because of influence of tides (as opposed to the fluvial-dominated "birds-foot" Mississippi River). Based on field experiments using painted leaves and palm seeds, Scheihing and Pfefferkorn (1984) found that tides flush large amounts of plant debris from the lower delta plain seaward. The Orinoco region is also tropical in nature, like the Late Carboniferous coal swamps, and therefore exhibits yearly alternation between wet (flood) and dry seasons that produce two completely different water tables and flow regimes. During floods, plant material is deposited in point bars of the upper delta plain but in channels and intertidal distributary banks during the low-water (dry season) stage. It is during the dry season that much plant matter, especially in levees, is exposed to air, oxidized and destroyed.

Significant changes in fossil anatomy may also occur during burial and compression. In laboratory simulations, Rex and Chaloner (1983) documented the "printing through" of anatomical features on one surface to the opposite face; thus, the presence of surface features is not an infallible indicator that one is looking at the actual surface of a specimen. The angle of fossil specimens and organic-rich layers to bedding was also found to influence the fracturing and appearance of fossil specimens of *Calamites*, *Lepidodendron*, *Sawdonia*, and *Stigmaria*.

2.9.2 Pollen

The number of pollen grains produced per individual plant varies from about 1×10^6 grains *per flower* in pine to about 10 grains per flower in beech (Erdtman, 1969). Because of its tremendous utility in biostratigraphy, paleoecological interpretation, and delineation of anthropogenic effects on the environment, the biostratinomy of pollen has received an overwhelming amount of attention. For example, Brush and DeFries (1981), Brush and Davis (1984), and Brush and Brush (1994) documented downcore changes in pollen that correlated with major anthropogenic changes in land use in the Chesapeake Bay region over an interval of approximately 200 years, beginning in about A.D. 1730. Prior to European settlement, the region was forested except for tidal wetlands and small Indian clearings; during initial European settlement (late seventeenth to mid-eighteenth centuries), approximately 20–30% of the land was cleared for tobacco farming; from the late eighteenth to mid-nineteenth centuries, expanding agriculture caused about 40% deforestation for cultivation of tobacco and grain; in the late

2.9 Macroflora and pollen

nineteenth century, small farms began to combine into larger operations (with concomitant use of heavy machinery and deep plow zones), and until the 1930s, when many farms were abandoned during the Great Depression, the amount of land under cultivation ranged from 60 to 80%; since the 1930s about 40% of the land has been reforested as a result of farm abandonment.

Because of the extensive literature, pollen transport and deposition is only briefly summarized here (see Russell, 1997; Farley, 1994, presents an annotated bibliography of modern pollen transport and deposition). Numerous studies have demonstrated the general correspondence between standing vegetation and pollen assemblages (e.g., Martin and Gray, 1962; Tauber, 1967; Davis and Webb, 1975). Nevertheless, pollen may be transported considerable distances by wind (estimates range from 10 to 150 km for "diffusion" close to the source plant up to 1500 km for minor amounts of pollen; Graham, 1957; Heusser, 1978; Birks, 1981; Bradshaw, 1994). Prentice (1985), for example, found that potential source areas of pollen grains increase with decreasing depositional velocities of pollen grains, and that lighter pollen grains are better represented with increasing basin size (cf. equation 2.25; see also Okubo and Levin, 1989). Sugita (1993) concluded that the pollen source radius for a lake is 10–30% smaller than the source radius for a point at the center (cf. Prentice, 1985), and that the difference in the source radius is greater for heavier pollen (e.g., spruce, sugar maple) than for lighter pollen grains (oak, ragweed). Pollen dispersal by wind into open ocean environments is typically unimportant, but where there is a strong directional component to winds (e.g., trade winds off the west coast of Africa), terrestrial pollen may comprise an important component of deep-sea palynomorph assemblages (Heusser, 1978), which can be used to correlate marine and terrestrial paleoclimate records (e.g., Heusser and Florer, 1973; Hooghiemstra, 1988; DuPont *et al.*, 1989). In high-energy, shallow-water environments, though, turbulence may inhibit pollen deposition (Pennington *et al.*, 1972), and in lakes, even seasonal overturn may resuspend and *focus* (sort and redeposit) pollen (Davis, 1968; Sugita, 1993).

In most cases, however, it appears that much of the pollen record in lakes and seas results from fluvial input (e.g., Heusser, 1983, 1988; Traverse, 1994), and that the pollen is hydraulically sorted during fluvial (Muller, 1959; Catto, 1985; Fall, 1987, 1992; Goodwin, 1988; Traverse, 1990) and marine transport (Cross *et al.*, 1966). Muller (1959) found that levees contained transported and reworked pollen, whereas backswamps were represented by essentially autochthonous assemblages that could be used to delineate more localized facies. Laboratory studies have also demonstrated the potential for pollen sorting (Brush and Brush, 1972;

Brush and DeFries, 1981). Ragweed (*Ambrosia*) pollen occurred in equal abundance in foreset and bottomset laminae in flume experiments involving sand mixtures, whereas *Pinus* and *Quercus* (oak) were preferentially deposited in bottomset laminae and *Betula* (birch) and *Ulmus* (elm) in topset laminae. Based on a modification of Stokes Law (equations 2.12, 2.13), grain size appears to be more important than density during settling, as pollen of different species may be preferentially deposited with certain sediment grain sizes (Fall, 1987; cf. Hall, 1989). Pollen grains that depart strongly from sphericity (e.g., bisaccate) also strongly influence settling velocity and transport (Hopkins, 1950; Holmes, 1994). Ironically higher degrees of turbulence (greater Reynolds numbers), appear to bring more pollen grains into contact with the sediment bed, where they are more likely to be trapped (Holmes, 1994). Thus, although much pollen enters lakes, estuaries, and seas via the air, and *may* be representative of the regional composition and abundance of terrestrial vegetation (Brush and Brush, 1994), large amounts of a pollen type do not necessarily imply a large amount of source vegetation in the vicinity (Brush and Brush, 1972).

2.10 Techniques of enumeration

The biostratinomic factors of dismemberment and transport not only affect the condition of individual fossils and their burial, but also the enumeration of fossil assemblages. Obviously, it is difficult to determine the ecological and taphonomic context under which an assemblage formed when recognition of that context may be dependent on the counting procedure. This subject has been dealt with extensively for certain taxa, especially vertebrates (e.g., Voorhies, 1969; Korth, 1979; Grayson, 1984; Klein and Cruz-Uribe, 1984), and the reader should be alert to potential problems with other taxa.

Taphonomic studies of mammalian assemblages have employed a variety of counting procedures, each with its own set of assumptions. Based on her work on Middle Siwalik (Late Miocene) assemblages of Pakistan, Badgley (1986*a,b*) formulated nine hypothetical fossil assemblages that produced three distinct taphonomic patterns. She concluded that the most appropriate sampling design is dependent on the natural ecological and sedimentological processes responsible for formation of the assemblage, and must be chosen on a case-by-case basis (1986*b*).

No particular method is, then, appropriate for all mammalian assemblages. The first pattern results from either a natural trap or sudden disaster without post-mortem transport (Table 2.13). Because the probability of association is

2.10 Techniques of enumeration

high in such assemblages, Badgley (1986b) recommended the *Minimum Number of Individuals* (MNI), which involves determining the minimum number of individuals necessary to account for all specimens of a taxon, as the best estimate of the true number of animals represented in the assemblage. The basic assumption of the MNI index is that skeletal elements accumulated in a more or less articulated state (although Badgley, 1986b, warns that the absence of articulated remains does not necessarily indicate a low probability of association). The MNI estimate is based on counts of the most abundant element present from one side of the body; on counts determined by consideration of all skeletal parts present; or on size and weathering state (and presumably age) of specimens. One of the primary criticisms of the MNI index is overrepresentation of rare taxa and the subjective nature of determining associations between specimens (Badgley, 1986b).

The second taphonomic pattern occurs in assemblages modified by predators and scavengers. Such assemblages may be recognizable based on body parts (preferred prey items) or size (juvenile versus adult prey). Although such assemblages are attritional and result from transport to another focal area (e.g., a cave), the probability of body part association ranges from intermediate to high. Assemblages from the channel-margin and floodplain facies of the Siwalik resemble this pattern (Table 2.14). Consequently, Badgley (1986b) recommended the MNI index as the best estimator for these assemblages as well. Nevertheless, the error in the MNI index is probably higher than for natural traps and sudden death assemblages.

The third taphonomic pattern results from post-mortem transport, regardless of the cause of mortality. Presumably, the probability of association decreases through time and increasing distance from the source (cf. Shotwell, 1955, and Voorhies, 1969). If little post-depositional breakage has occurred, the *Number of Identified Specimens per Taxon* (NISP)[2] is most appropriate, in which the number of specimens of a particular taxon is divided by the total number of specimens of all taxa in the assemblage. Badgley (1986b) used this method for Siwalik assemblages from channel and crevasse-splay deposits (Table 2.14). If the remains are highly fragmented, however, the *Minimum Number of Elements* (MNE) index

---

[2] The term *specimen* is considered as a whole or fragmentary bone or tooth, whereas the term *element* is more restrictive in that it refers to a whole body part (Voorhies, 1969; Grayson, 1984; Badgley, 1986b; Lyman, 1994b). Unfortunately, indiscriminate use of these terms has confounded comparisons of indexes published in the literature (Lyman, 1994b). I have attempted to distinguish, where appropriate, between the two terms, but in many cases I have used the term *remains* to encompass both.

Table 2.14. *Taphonomic characteristics and inferences for Siwalik fossil assemblages from four sedimentary environments*

	Sedimentary environment	
	Channel, crevasse Splay	Channel margin, floodplain
Taphonomic feature		
Articulation	None	Rare
Spatial distribution	Scattered	Most clustered, some scattered
Hydraulic equivalence	Most equivalent	Most non-equivalent
Tooth/Vertebra	3.3–3.6	1.6–1.7
Stratigraphic span	0.9–3.0 m	0.3-0.9 m
Bone damage	Polish, abrasion	Punctures, scratches, ragged edges
Taphonomic inference		
Mode of accumulation	Transport by currents	Activities of predators and scavengers
Probability of association	Low	Intermediate–high
Method of counting[a]	NISP	MNI

[a]See text for further discussion; NISP, number of identified specimens per taxon; MNI, minimum number of individuals.
Badgley, 1986*b*.

is preferred over NISP. MNE emphasizes the effects of fragmentation, and is determined by grouping together all fragments that could belong to one element even if matched breaks or overlaps between fragments are not between NISP and MNI; but, like NISP, it cannot account for association among elements.

NISP will always be greater than or equal to MNE for an assemblage (Lyman, 1994*b*). If NISP equals MNE, then either all the specimens in an assemblage are whole bones or all specimens represent the same portions of a skeletal element. If NISP is greater than MNE, then some specimens are fragments from different elements. Because an element can only be broken into a limited number of identifiable fragments, NISP and MNE are often correlated (Lyman, 1994*b*).

Although NISP and MNE measure the *extent* of fragmention (the proportion of fragmentary specimens), they do not measure its *intensity* (how small the fragments are). Lyman (1994*b*) recommended calculation of a NISP:MNE ratio as a measure of the intensity of fragmentation, in which complete bones are removed from both indices (inclusion of complete specimens – or elements –

2.10 Techniques of enumeration

depresses the ratio). Larger fragments are more likely to overlap (i.e., share a portion of the same bone) and therefore be counted as coming from the same element. As the intensity of fragmentation increases, however, fragments become smaller and they are less likely to be independent of one another. In some cases, the probability of association may be too uncertain to specify a counting method *a priori*, and several indices should be used simultaneously as estimators of high, intermediate, and low numbers of individuals (Grayson, 1984; Klein and Cruz-Uribe, 1984; Badgley, 1986*b*; and Lyman, 1994*a*, discuss other indices).

As emphasized previously, the same taphonomic patterns may result from different processes (Table 2.14; Badgley, 1986*b*), and the amount and type of surface wear may be misleading in judging age or degree of transport. Badgley (1986*b*) emphasized, however, that certain sedimentological and stratigraphic criteria may be useful in distinguishing true taphonomic histories. For example, natural traps are laterally restricted by the trap itself, whereas sudden disasters tend to occur over much larger areas and are vertically restricted because they tend to be sudden. Siwalik channel and crevasse-splay assemblages correspond closely taphonomically to assemblages produced by current-transport (Tables 2.13, 2.14); in these cases, lower concentrations of small (e.g., juvenile) bones, as well as higher concentrations of teeth, may distinguish hydraulically sorted from predator-sorted assemblages (Table 2.13). Bones and isolated teeth were also scattered through channel and crevasse-splay deposits (Table 2.14). Siwalik channel-margin and floodplain assemblages correspond closely to those produced by predators and scavengers (Tables 2.13, 2.14); although juvenile bone is not as well calcified as adult bone, and is therefore more prone to destruction by physicochemical processes, juveniles may suffer preferential predation by carnivores and scavengers so that juvenile remains are more likely to be encountered in channel-margin and floodplain assemblages. Channel-margin and floodplain units also tend to have thinner stratigraphic spans than do channels and bones and teeth are more likely to be clustered together, although scattering may also occur (Table 2.14).

Other problems arise in sample and data processing itself, sometimes in surprising and inadvertent ways (Behrensmeyer and Hook, 1992; Flessa *et al.*, 1992). In the case of leaves, for example, there may be a tendency to select those horizons displaying distinctive leaf morphology as opposed to those represented by coalified material, and to undercollect bedding planes showing no obvious organic matter (Ferguson, 1985). Specimens from the Burgess Shale are collected in such a manner (Desmond Collins, 1992, personal communication). Thus, museum collections may not be suitable for measuring variation or evolutionary rates because they are not representative of the fossil assemblage as a whole (Bell *et al.*,

1987); however, Foote (1997a) found no evidence for a tendency to describe morphologically extreme or modal species and genera of trilobites, crinoids, or blastoids during the past 150 years or so of paleontological studies.

"Analytical" time-averaging may also result from pooling samples or data sets (Behrensmeyer and Hook, 1992; Graham, 1993). Such artificial time-averaging occurs in industrial well-cuttings (ditch samples), which are normally collected over approximately 10-m intervals; the effects of mixing are not nearly as negative as intuition would seem to warrant, however, based on high-resolution biostratigraphic studies of such samples (e.g., Martin and Fletcher, 1995; Armentrout, 1996), especially if sediment accumulation rates are sufficiently high to counteract sample mixing.

Specimen size is also a consideration. Among vertebrates, large size and coloration, as well as the observer's search image, are known to bias specimen counts along field transects (e.g., Western, 1980; Behrensmeyer and Boaz, 1980). The choice of sieve size will also influence the collection of vertebrate remains (e.g., Dodson, 1973; Behrensmeyer, 1975; Lyman, 1994a), as it will microfossils. For example, the persistence of dead tests of large reef-dwelling foraminifera in sediment may obscure subtle environmental gradients (energy, light intensity) that are preserved in sediment. Martin and Liddell (1988, 1989) found that, like scleractinian corals (e.g., Boss and Liddell, 1987), foraminifera exhibited a depth-related zonation at Discovery Bay, Jamaica, but the foraminiferal zonation was only discernible when the sand-size distributions of individual foraminiferal taxa within sieve fractions were considered in cluster analyses. When the normal foraminiferal enumeration technique of counting 300 tests per entire sample (irrespective of test size) was used, no obvious foraminiferal depth zonation resulted in cluster analyses. Cluster analysis of assemblages in separate size fractions was also performed, which is probably the most statistically appropriate way to treat such samples, but also to no avail. But when cluster analysis was conducted based on *combined* size fractions ("sieve method") a zonation was obtained for two separate (i.e., *replicate*) transects that was identical or very nearly so to the coral zonation. Furthermore, one depth (escarpment at 24 m) was anomalous in its species composition and clustered with back-reef samples, which suggested that sediment was being transported to the escarpment down grooves in the forereef structure; this anomalous assemblage was not obvious in clusters based on standard procedures (300 tests per sample) or separate size fractions. Martin and Liddell (1988, 1989) suggested that larger ($\geq 1-2$ mm) taxa, such as *Amphistegina gibbosa*, which is very resistant to destruction (Kotler *et al.*, 1992) and dominates foraminiferal assemblages on the forereef, provide a more obvious search image to the

2.10 Techniques of enumeration

investigator, and are therefore more likely to be enumerated in sediment samples and obscure subtle foraminiferal (environmental) depth gradients. Although planktonic foraminifera probably exhibit the least size disparity between species, they do exhibit distinct size distributions, which, if ignored, can produce markedly different oceanographic interpretations (see Martin and Liddell, 1989, for summary). Similar considerations apply to deep-sea benthic foraminifera (Schröder et al., 1987; Sen Gupta et al., 1987).

Although some anonymous reviewers consider the "sieve method" to be "statistically invalid," they have not suggested a satisfactory alternative to avoiding size bias in specimen counts of microfossils. If these workers are as sufficiently skilled in statistics as their reviews suggest, perhaps they can devise a statistically valid alternative that would work equally as well as (or better than) the sieve method.

Despite the large numbers of foraminifera enumerated in the sieve method, in some cases relatively small samples can be used. Normally, the standard error associated with counts of less than 100 increases dramatically (Patterson and Fishbein, 1989). However, when the assemblage is dominated by one or two species (e.g., in marshes or perhaps restricted lagoons) smaller counts can be used to obtain statistically reliable results (Patterson and Fishbein, 1989; see also Fishbein and Patterson, 1993). For example, Patterson (1990) recognized six marsh foraminiferal biofacies based on a reduced data set of six (out of 17) species, with counts ranging from 46 to 2187 tests per 10 cm^3.

Methods of enumeration other than numerical abundance or size have also been used. Staff et al. (1985) found that much better agreement in rank order abundance of dominant species of living and dead molluscan assemblages when biomass (based on weight and size) was used than when numerical abundance was employed. Consequently, Staff et al. (1985) advocated using biomass as an indicator of "abundance" (Fürsich and Aberhan, 1990; see also Chapter 5).

3 Biostratinomy II: dissolution and early diagenesis

Here lies one whose name was writ in water. John Keats's epitaph

3.1 Introduction

As they reside near the sediment–water interface (SWI), biogenic remains are subject not only to physical degradation, but also chemical alteration that recycles biogeochemically important elements such as carbon that are sequestered in hardparts (Chapters 8, 9). Hardparts of different taxa typically have unique chemical properties that hinge on their mineralogy and microstructure (Table 3.1). The geological setting of chemical destruction (e.g., sedimentation rate, pore-water chemistry, bottom water oxygenation) also varies. Nevertheless, as hardparts *tend* to behave similarly with respect to dissolution, dissolution and early diagenesis can be presented in a more synthetic manner than transport and abrasion behavior (Chapter 2). As most marine invertebrate hardparts are calcareous, carbonate precipitation and dissolution are examined first.

3.2 CaCO$_3$ dissolution and precipitation

Carbon dioxide dissolves in seawater according to the following reactions (Canfield and Raiswell, 1991b):

$$CO_2 + H_2O \rightleftharpoons H_2CO_3 \tag{3.1}$$

$$H_2CO_3 \rightleftharpoons H^+ + HCO_3^- \tag{3.2}$$

$$HCO_3^- \rightleftharpoons H^+ + CO_3^{2-} \tag{3.3}$$

As pH *increases* (decreased [H$^+$]), the dominant form of dissolved carbonate shifts toward CO$_3^{2-}$: at pH < 6.0, H$_2$CO$_3$ is the dominant species; at 6.0 < pH < 9.1, HCO$_3^-$ dominates; and at pH > 9.1, CO$_3^{2-}$ is most prevalent (Canfield and Raiswell, 1991b).

3.2 CaCO₃ dissolution and precipitation

Table 3.1. *The original mineralogy of common fossil taxa*

Taxon	Aragonite	Low-Mg calcite	High-Mg calcite	Aragonite + calcite	Silica	Phosphate
Mollusca:						
Bivalves	c	c		c		x
Gastropods	c			c		
Pteropods	c					
Cephalopods	c		x			
Brachiopods		c	x			c
Corals:						
Scleractinia	c					
Rugosa + Tabulata		c	c			
Sponges	c	c	c		c	
Bryozoans	c		c	c		
Echinoderms			c			
Ostracods		c	c			
Foraminifera:						
benthic	x		c			
pelagic		c				
Algae:						
Coccolithophoridae		c				
Rhodophyta	c		c			
Chlorophyta	c					
Charophyta		c				
Annelida	c	c	c		c	c
Arthropoda	c	c	c		x	c
Diatomacea					c	
Radiolaria					c	
Vertebrata		c	c	c		c

c, common mineralogy, x, less common.
Modified from Tucker, 1991.

CaCO₃ dissolves (and precipitates) according to the equation:

$$CaCO_3 \rightleftharpoons Ca^{2+} + CO_3^{2-} \tag{3.4}$$

Variations in dissolved CO_3^{2-} are more common in seawater than changes in dissolved Ca^{2+} (*Dittmar's Law of Constant Proportions*). Thus, if $[CO_3^{2-}]$ should increase sufficiently, the equilibrium (equation 3.4) will shift to the left (LeChatelier's Principle) and CaCO₃ precipitates.

Table 3.2. *Idealized sequence of organic matter breakdown with depth in sediment*[a,b]

Aerobic decay
$(CH_2O)_{106}(NH_3)_{16}H_3PO_4 + 138O_2 \rightarrow$
$\qquad\qquad 106CO_2 + 122H_2O + 16NO_3^- + 16H^+ + H_3PO_4$

Anaerobic decay
Manganese reduction
$(CH_2O)_{106}(NH_3)_{16}H_3PO_4 + 212MnO_2 + 332CO_2 + 120H_2O \rightarrow$
$\qquad\qquad 438HCO_3^- + 16NH_4^+ + HPO_4^{2-} + 212Mn_4^{2+}$

Nitrate reduction
$(CH_2O)_{106}(NH_3)_{16}H_3PO_4 + 84.8HNO_3 \rightarrow$
$\qquad\qquad 106CO_2 + 42.4N_2 + 148.4H_2O + 16NH_3 + H_3PO_4$

Iron reduction
$(CH_2O)_{106}(NH_3)_{16}H_3PO_4 + 424Fe(OH)_3 + 756CO_2 \rightarrow$
$\qquad\qquad 862HCO_3^- + 16NH_4^+ + HPO_4^{2-} + 424Fe^{2+} + 304H_2O$

Sulfate reduction
$(CH_2O)_{106}(NH_3)_{16}H_3PO_4 + SO_4^{2-} \rightarrow 106HCO_3^- + 53H_2S + 16NH_3 + H_3PO_4$

Methanogenesis
$(CH_2O)_{106}(NH_3)_{16}H_3PO_4 + 14H_2O \rightarrow$
$\qquad\qquad 39CO_2 + 14HCO_3^- + 53CH_4 + 16NH_4^+ + HPO_4^{2-}$

Fermentation[c]
$12(CH_2O)_{106}(NH_3)_{16}H_3PO_4 \rightarrow$
$\qquad\qquad 106CH_3CH_2COOH + 106CH_3COOH + 212CH_3CH_2OH + 318CO_2$

---

[a]Reactions at the top of the column yield more free energy than those lower in the column.
[b]Stoichiometry is that of Canfield and Raiswell (1991*b*), except for Mn and Fe reduction, methanogenesis, and fermentation, which are based on Allison and Briggs (1991*a*, after Redfield, 1958). (Adapted from Allison and Briggs, 1991*a*.)
[c]Dominant below the zone of methanogenesis (carbonate reduction), but occurs throughout the sediment profile.

Canfield and Raiswell (1991*b*) reviewed the existing literature and presented a simplified model of carbonate equilibria in marine sediments (closed, steady-state system with no differential transport of solutes) that is a useful starting point for the examination of shell preservation (see also Berner, 1989; cf. Manjunatha and Shanker, 1996). The ideal order of these reactions in sediment (Table 3.2) presumably reflects free energy yield (but see Allison and Briggs, 1991*a*, p. 123), and is characterized by the presence of increasingly refractory organic matter with sediment depth because more labile components are broken down first (Allison and Briggs, 1991*a*; Lin and Morse, 1991; Smith *et al.*, 1993).

According to the Canfield–Raiswell (1991*b*) model, organic matter is oxidized via *aerobic decay* according to the equation:

$$(CH_2O)_{106}(NH_3)_{16}H_3PO_4 + 138O_2 \rightarrow$$
$$106CO_2 + 122H_2O + 16NO_3^- + 16H^+ + H_3PO_4 \qquad (3.5)$$

Marine organic matter is assumed to oxidize according to Redfield (1958) C:N:P ratios of 106:16:1, which varies according to the source of organic matter, but which is more or less typical of plankton (Canfield and Raiswell, 1991*b*). Initially, oxidation of organic matter produces slight oversaturation of aragonite and calcite, but as CO_2 is evolved and ammonia is oxidized to NO_3^- (nitric acid or HNO_3), aragonite undersaturation occurs, followed by calcite undersaturation at pH $\leq$ 8.

Once oxygen has been consumed, nitrate is used as the electron acceptor (*denitrification*) in the oxidation of organic matter:

$$(CH_2O)_{106}(NH_3)_{16}H_3PO_4 + 84.8HNO_3 \rightarrow$$
$$106CO_2 + 42.4N_2 + 148.4H_2O + 16NH_3 + H_3PO_4 \qquad (3.6)$$

This process basically maintains the carbonate saturation level present at the start of denitrification (pH 8.0). *Reduction of Mn and Fe* may also occur immediately before and after nitrate reduction, respectively (Allison and Briggs, 1991*a,b*; Table 3.2), but their role as oxidants during degradation of organic matter may be more significant in freshwater than marine sediments (Canfield and Raiswell, 1991*b*, p. 417).

Reduction of dissolved sulfate, which is abundant in seawater, then takes over once nitrate has been depleted:

$$(CH_2O)_{106}(NH_3)_{16}H_3PO_4 + SO_4^{2-} \rightarrow$$
$$106HCO_3^- + 53H_2S + 16NH_3 + H_3PO_4 \qquad (3.7)$$

The addition of bicarbonate increases *alkalinity* in porewater (mainly total dissolved HCO_3^- and CO_3^{2-} ions) and causes supersaturation of aragonite and calcite (Berner *et al.*, 1970; normal *total* seawater alkalinity, which includes minor contributions of borate and other ions, is approximately 2–2.5 mM; Chas. Culberson, 1991, personal communication). Undersaturation is, however, observed in the early stages of sulfate reduction via the oxidation of sulfide (H_2S):

$$H_2S + 2O_2 \rightarrow SO_4^{2-} + 2H^+ \qquad (3.8)$$

The degree of acid (H^+) production from sulfate oxidation depends on the extent of precipitation of iron sulfides (e.g., pyrite), which affects pH. If little or no iron is present (such as in carbonate-rich sediments), sulfide accumulates, model pH decreases to about 7, and carbonate undersaturation results until 5–7 mM of sulfate have been reduced, yielding 10–14 mM of carbonate alkalinity. Thus, carbonate dissolution can occur in shallow-water carbonate sediments and may conceivably help to buffer anthropogenic increases in atmospheric CO_2 (Mackenzie and Morse, 1990; Walter and Burton, 1990); the potential implications of anthropogenic shallow-water dissolution for actualistic studies have not been assessed, however. Once carbonate alkalinity has reached 10–14 mM, alkalinity increases and carbonate saturation and supersaturation may occur. In Fe-rich terrigenous sediments (see Raiswell and Canfield, 1998, for survey), however, iron sulfide precipitation results in carbonate supersaturation after only 0.2–0.3 mM of sulfate have been reduced.

Although sulfate reduction oxidizes approximately 50% of organic matter in nearshore sediments, it tends to decrease away from shore as sedimentation rates decrease (Canfield, 1991; Lin and Morse, 1991). In continental margin sediments with intermediate to high sediment accumulation rates (e.g., Mississippi Delta), rapid burial of large amounts of organic matter promotes sulfate reduction and pyrite formation (Lin and Morse, 1991) because with high sedimentation rates, labile organic matter undergoes a shorter period of *oxic* and *suboxic* degradation (equation 3.5) before it is incorporated into subsurface layers where it can support SO_4^{2-} reduction (Berner, 1989; Canfield, 1991; Canfield and Raiswell, 1991*a*; Lin and Morse, 1991). Nevertheless, high accumulation rates may be overcome by bioturbation and sediment resuspension and transport, so that iron oxides may also accumulate (Canfield and Raiswell, 1991*a*). Similar considerations extend to deep-sea regimes, although sediment accumulation rates are typically low (*c.* 2 cm ka^{-1}), thereby allowing iron and manganese oxides to accumulate (e.g., Dhakar and Burdige, 1996).

Rapid burial also inhibits bioturbation (Chapter 4), which promotes dissolution of $CaCO_3$. Bioturbation pumps oxygenated water downward into sediment (*bioirrigation*), thereby oxidizing organic matter and sulfides to produce acids. In his study of Long Island Sound (LIS) sediments (U.S.A.), Aller (1982*a*) found that mollusc shell preservation was substantially better at the physically disturbed shallow-water (10 m) FOAM (Friends of Anoxic Muds) site, which is bioturbated only to about 10 cm (mixed layer), than at the NWC and DEEP sites (approximately 10–15 m water depth), which harbor an extensive infauna that burrows to depths of about a meter in the sediment but which is most intense down to about 60 cm. Similar bioturbation depths have been reported by Martin *et al.* (1995, 1996; northern Gulf of California) in terrigenous tidal flat sediments and by Walter and Burton (1990) for shallow-water carbonate environments (Florida Keys). Sedimentation rates at LIS sites are approximately $0.1 \, \text{cm yr}^{-1}$, which allows for a minimum of about 300–600 years of dissolution at NWC and DEEP sites as shells pass through the surface mixed layer, but only 10–70 years at FOAM (assuming no exhumation and reburial of shells; cf. Chapter 5); thus, only about 10% of $CaCO_3$ is preserved at deeper sites, whereas at FOAM more than 50% remains (Canfield, 1991).

In anoxic marine environments, particularly those with high sediment accumulation rates and below the zone of sulfate reduction, *methanogenesis* occurs anaerobically by the net reaction:

$$(CH_2O)_{106}(NH_3)_{16}H_3PO_4 + 14H_2O \rightarrow$$
$$+ 39CO_2 + 14HCO_3^- + 53CH_4 + 16NH_4^+ + HPO_4^{2-} \quad (3.9)$$

Methane diffuses upward to become the main source of organic carbon for sulfate-reducing bacteria at the sulfate–methane interface, where peaks in sulfate-reduction rate and carbonate supersaturation are observed:

$$CH_4 + 2H^+ + SO_4^{2-} \rightarrow H_2S + 2H_2O + CO_2 \quad (3.10)$$

The lack of oxygen also inhibits bioturbation and oxic degradation of both free organic matter and that in skeletons which served in crystallite nucleation (e.g., Berger and Soutar, 1970; Glover and Kidwell, 1993; see also Risk *et al.*, 1997). Carbonate dissolution may still occur in some euxinic basins (lacking oxygenated bottom waters), however, if dissolved sulfate levels remain relatively high and sulfide accumulates (equations 3.7, 3.8; e.g., Black Sea, Curiaço Trench); the lack of oxygen not only prevents oxic decomposition of organic matter in sediment, but also allows sulfides to build up (Canfield and Raiswell, 1991*a*) and, potentially, pyrite to precipitate in the water column and settle to the SWI (Wilkin and Barnes, 1997). In some cases, this may be counteracted by input of alkaline runoff (Black Sea; Canfield and Raiswell, 1991*b*).

Fermentation, which requires neither oxidation nor reduction, may be pronounced below the level of methanogenesis (Table 3.2). This process is present throughout most of the sediment profile but is masked by the other degradative pathways (Allison and Briggs, 1991*a*).

To summarize, $CaCO_3$ (shell) preservation is maximized in anoxic marine sediments, in which (1) bioturbation is minimized because of low oxygen content in porewaters available for respiration, (2) porewater alkalinity is enhanced (via sulfate reduction), and (3) dissolved sulfides are rapidly depleted by precipitation of iron sulfides (decreasing the concentration of the weak acid H_2S). Also, rapid burial promotes quick movement of shells through the surface (mixed) layers of sediment where bioturbation and dissolution are most intense. With deeper burial (zone of methane production), carbonate precipitation may occur as a result of carbonate supersaturation.

3.3 Shell mineralogy, architecture, microstructure, and size

3.3.1 Mineralogy

Based on carbonate equilibria measured in laboratory settings, relative rates of dissolution of calcareous shells should, to a first approximation, be predictable from shell mineralogy (Table 3.1). Low-Mg calcite (1–4 mol% Mg) is much less soluble (more thermodynamically stable) in experimental studies than high-Mg calcite (11–19 mol% Mg; Mackenzie *et al.*, 1983). The concentration of Mg present in calcareous shells varies with taxon (Table 3.1) and latitude (water chemistry and temperature), however, and even within the same taxon, higher-latitude or deep(cold)-water representatives tend to have lower Mg contents (Chave, 1954; Tucker, 1991). Aragonite should be the most stable of the biogenic carbonates because of its extremely low Mg concentration (~1000 ppm; Tucker, 1991; cf. section 3.3); thus, high-Mg calcite should be chemically less stable than aragonite.

Nevertheless, in limestones, aragonite is converted to calcite, high-Mg calcite is converted to low-Mg calcite, and all three are subject to dissolution, pyritization, silicification, and dolomitization (Walter, 1985; Tucker, 1991). This sequence of alteration is not set in stone, however: Feldman (1989) found that calcareous shells in the Waldron Shale (Silurian, Indiana, U.S.A.) underwent a general sequence of alteration, but that shells of the *same* taxon reacted at different rates.

In skeletons composed of high-Mg calcite (HMC), Mg^{2+} is lost from the crystal lattice through the process of "incongruent dissolution" (Bathurst, 1975) to give diagenetic low-Mg calcite (dLMC), in which the skeletal structure is preserved at the light microscope level. If conversion of HMC to dLMC takes place in a reducing environment and iron is available, the dLMC will be ferroan (up to 6 mol% $FeCO_3$ and up to 5 mol% $MgCO_3$), which indicates that the fossils were originally HMC (Tucker, 1991); Richter and Fuchtbauer (1978) concluded that ferroan calcites in various taxa of trilobites, bryozoans, rugose corals, echinoderms, many foraminifera, ostracodes, red algae, and serpulids indicated an original HMC composition (see Smith *et al.*, 1998, for recent review of bryozoan mineralogy; see also Chapter 8).

In rocks with low water/rock ratios, the Mg that is released may form small crystals of dolomite (microdolomite) around biogenic grains (especially HMC ossicles of echinoderms). HMC fossils tend to be dolomitized by *pseudomorphic replacement*, in which dissolution of original hardparts is coupled with simultaneous, almost atom-by-atom deposition of another mineral; in the case of HMC, dolomite occurs in optical continuity with original crystallites preferentially over LMC (e.g., Blake *et al.*, 1982; Tucker, 1991; but see Bullen and Sibley, 1984), unless HMC has already been converted to dLMC, in which case dolomitization probably occurred late during burial and is very destructive (Tucker, 1991). LMC skeletons (e.g., LMC brachiopods) tend to resist dolomitization unless the surrounding sediments have been completely dolomitized, whereas the structure of aragonite is usually destroyed by dolomitization (Tucker, 1991). Bullen and Sibley (1984) found, for example, that during experimental dolomitization with $2\,M$ $CaCl_2/MgCl_2$ at $250\,°C$, that coralline algae, echinoids, *Halimeda*, and foraminifera dolomitized faster than other fossils; that echinoids and foraminifera were mimically replaced; and that the original skeletal structure of pelecypods, gastropods, and corals (the latter two taxa aragonitic) was destroyed.

In most cases, aragonite is removed by dissolution to form *external molds* that may be preserved by partial to complete lithification of sediment or micritized *internal molds* or *steinkerns*. Aragonitic shells are usually only preserved in impermeable mudstones and marls (e.g., ammonites of the Kimmeridge Clay of western Europe; some Triassic patch reefs in northern Italy), concretions, or asphalt-impregnated limestones and sandstones (Tucker, 1991). More rarely, "half-ammonites" or calcitic aptychi may be preserved (e.g., Schlager, 1974; Tanabe *et al.*, 1984). In some cases, aragonite is progressively replaced by calcite across a thin organic film so that the original shell structure is preserved (*calcitization*; Tucker, 1991). Such shells are extremely useful because they preserve not only original biogenic structures, but also trace element and stable isotope compositions (Figures 3.1, 3.2).

Subtle taphonomic alteration also points to a potential pitfall in paleoecological analysis: shells (no matter what the original mineralogy) that appear unaltered at the light microscope level may suffer from considerable alteration at the microstructural (electron microscope) scale (e.g., Glover and Kidwell, 1993; see also Brand, 1989; Mitchell and Curry, 1997). Distinguishing between altered and unaltered shells – especially through cathodoluminescence (Tucker,

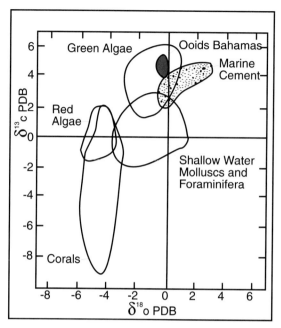

Figure 3.1. Stable isotope ratios of Recent carbonate skeletons, and abiotic cements and oöids. (Redrawn from Tucker, 1991.)

1991) – is critical to evaluating the chemistry of ancient waters (e.g., Popp *et al.*, 1986*a,b*; Veizer *et al.*, 1986; Brand, 1989, 1990; Grossman, 1992; Podlaha *et al.*, 1998), and the true microstructure of extinct taxa (e.g., Sandberg, 1975; Towe and Hemleben, 1976). In the case of the suborder Fusulinina (foraminifera), the test consists of small (~2–5 μm), equidimensional, LMC grains that were intially considered to be indicative of recrystallization but are now considered to represent the original ultrastructure (Tappan and Loeblich, 1988; Hageman and Kaesler, 1998). Other LMC biogenic particles (planktonic foraminifera, brachiopods, some bivalves) are also often well preserved (Tucker, 1991).

3.3.2 Architecture, microstructure, and size

Because the relative reactivities of low-to-high Mg calcite and aragonite differ from those predicted by purely thermodynamic considerations, other factors, such as shell architecture and microstructure (available surface area), must play substantial roles in the chemical reactivity of calcareous hardparts (Walter, 1985; see Bathurst, 1975, for review of biogenic carbonate skeletal structures). In other words, not all carbonate particles – even those of the same mineralogy – are equivalent in terms of dissolution resistance (Sorby Principle of Folk and Robles, 1964; see also Summerson, 1978).

3.3 Shell mineralogy, architecture, microstructure, and size

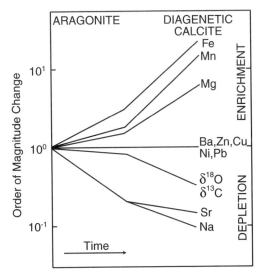

Figure 3.2. Magnitude of changes in stable isotope and trace element concentrations in aragonite skeletons as they are altered to calcite during diagenesis. (Redrawn from Tucker, 1991, based on Brand, 1989.)

Much of the work on dissolution of calcareous hardparts has been on planktonic foraminifera because of their use in biostratigraphic and paleoceanographic interpretation and their role in the long-term global carbon cycle (Chapters 8, 9). Douglas (1971) developed a dissolution-susceptible scale for planktonic foraminifera (later modified by Malmgren, 1987). Although planktonic foraminifera consist primarily of LMC, they differ substantially in their preservation potential. Planktonic species that have somewhat higher Mg test concentrations are more susceptible to dissolution (Savin and Douglas, 1973; Bender *et al.*, 1975). Still, just as for settling velocity (Chapter 2), test size, shape, architecture and structure (wall thickness, porosity, etc.) appear to be the dominant factors (e.g., Berger, 1967, 1968, 1970, 1971; Parker and Berger, 1971; see Bé, 1977, for review), as the species that are most susceptible to test dissolution also tend to settle slowest (Table 3.3; Figure 3.3). This is because most shallow-dwelling species harbor symbiotic dinoflagellates and typically have small thin-walled porous tests in order to remain afloat in the photic zone, whereas deeper-dwelling species often have dissolution-resistant features such as thick test walls and rope-like keels that decrease their buoyancy (Table 3.4; cf. Table 3.3 and Figure 3.3). Preferential settling of certain species may also affect their dissolution behavior: in the case of *Globorotalia truncatulinoides*, the exposed cone of the test is more severely affected by dissolution than the flat (spiral surface) in experiments (Bé *et al.*, 1977). Sediment infilling tends to confine dissolution to the outer surfaces

Table 3.3 *Ranks of common planktonic species of foraminifera with respect to settling velocity and dissolution resistance*

	Settling velocity A	Dissolution resistance B	
Fastest	1. *Globorotalia tumida*	*G. tumida*	Most resistant
	2. *Pulleniatina obliquiloculata*	*S. dehiscens*	
	3. *Sphaeroidinella dehiscens*	*P. obliquiloculata*	
	4. *Globoquadrina conglomerata*	*G. pachyderma*	
	5. *Globorotalia inflata*	*G. dutertrei*	
	6. *Globorotalia crassaformis*	*G. menardii*	
	7. *Globigerina pachyderma*	*G. crassaformis*	
	8. *Globigerina falconensis*	*G. inflata*	
	9. *Globoquadrina dutertrei*	*G. truncatulinoides*	
	10. *Globigerinoides conglobatus*	*G. hirsuta*	
	11. *Globorotalia menardii*	*G. conglomerata*	
	12. *Globorotalia truncatulinoides*	*G. digitata*	
	13. *Globorotalia hirsuta*	*G. hexagona*	
	14. *Globigerinoides tenellus*	*O. universa*	
	15. *Globigerinoides sacculifer*	*C. nitida*	
	16. *Globigerinoides ruber*	*G. falconensis*	
	17. *Globigerinella siphonifera*	*G. iota*	
	18. *Globigerina calida*	*G. glutinata*	
	19. *Globigerina bulloides*	*G. calida*	
	20. *Globigerina rubescens*	*G. bulloides*	
	21. *Globoquadrina hexagona*	*G. conglobatus*	
	22. *Globigerinita iota*	*G. sacculifer*	
	23. *Globigerinita glutinata*	*G. siphonifera*	
	24. *Orbulina universa*	*G. tenellus*	
	25. *Globigerina digitata*	*G. rubescens*	
	26. *Candeina nitida*	*G. ruber*	
Slowest	27. *Hastigerina pelagica*	*H. pelagica*	Least resistant

Modified from Berger and Piper, 1972.

of tests (Hecht *et al.*, 1977; Keir and Hurd, 1983) where morphological features may differ in their mineralogy and susceptibility to dissolution (Brown and Elderfield, 1996).

Although tests are typically not transported far as they settle through the water column (Chapter 2), they begin to dissolve, especially after they have reached the ocean bottom, as a result of the oxidation of organic matter and

3.3 Shell mineralogy, architecture, microstructure, and size

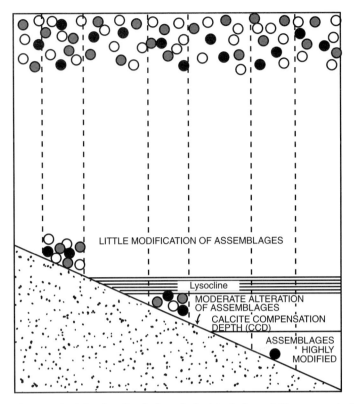

Figure 3.3. Selective dissolution of planktonic foraminifera, as ranked by Berger (1970) and Berger and Piper (1972). (Redrawn from Bë, 1977; cf. Tables 3.3, 3.4.)

sulfides (equations 3.5–3.8; Adelseck and Berger, 1977). Consequently, the number of assemblages present in the plankton may be reduced in number; Bé and Hutson (1977) found, for example, that in the tropical and subtropical Indian Ocean six "life assemblages" were reduced to two thanatocoenoses in sediment and that the number of species in the sediment assemblages was greater than in the planktonic assemblages (cf. Peterson, 1977; see Kozlova, 1986, for radiolaria). The apparent sea surface temperature (SST) reflected by the thanatocoenosis may therefore be inaccurate (e.g., Bé and Hutson, 1977; Figure 3.4). Le and Thunell (1996) developed a method which filters out the effect of differential dissolution on the calculation of SSTs.

The boundary zone between well-preserved planktonic foraminiferal assemblages and poorly preserved ones is the *lysocline* (Berger, 1968, 1977a,b), which typically lies a few hundred meters above the *calcite compensation depth* (CCD) The CCD is normally defined as the depth below which $CaCO_3$ is absent (Bé, 1977) and which grades into red clay, or in some areas, siliceous oozes. The depth to the lysocline and CCD in part reflects the input of $CaCO_3$ from land

Table 3.4. *Generalized depth habitats of planktonic foraminifera*

"Shallow-water" species living predominantly in the upper 50 m
Globigerinoides ruber
Globigerinoides sacculifer
Globigerinoides conglobatus
Globigerina quinqueloba
Globigerina rubescens

"Intermediate-water" species living in the upper 100 m, but predominantly from 50 to 100 m
Globigerina bulloides
Hastigerina pelagica
Pulleniatina obliquiloculata
Globoquadrina dutertrei
Orbulina universa
Candeina nitida
Globigerinella aequilateralis
Globigerina calida
Globigerinita glutinata

"Deep-water" species living in the upper few hundred meters and whose adult stages occur predominantly below 100 m
Globorotalia menardii
Globorotalia tumida
Globorotalia inflata
Globorotalia hirsuta
Globorotalia truncatulinoides
Globorotalia crassaformis
Globorotalia scitula
Globoquadrina pachyderma
Globoquadrina conglomerata
Globoquadrina hexagona
Globigerinella adamsi
Sphaeroidinella dehiscens
Hastigerinella digitata

Bé (1977) based on numerous sources.

and surface water production: where surface production of $CaCO_3$ is high (e.g., equatorial Pacific), the CCD tends to deepen and the distance between the lysocline and CCD is much less than where surface fertility is low and the CCD shallows. The depth to the CCD also reflects the corrosiveness (total dissolved CO_2 or alkalinity) of bottom waters (which is in turn a function of age, temperature, and pressure): the floor of the modern Pacific is bathed in "old" CO_2-rich Antarctic Bottom Water (AABW) and the CCD in most areas lies at approximately 4.5–5 km, whereas much of the bottom of the Atlantic is bathed by much less corrosive "young" North Atlantic Deep Water (NADW) and the CCD lies at approximately 5–5.5 km throughout much of the basin.

The CCD also tends to shallow near continental margins as a result of input and oxidation of organic matter from rivers and coastal upwelling (Bé, 1977; Malinky and Heckel, 1998; see also Chapter 8). Smith (1971) found that the mean production of $CaCO_3$ by foraminifera (247×10^{10} g yr^{-1}) in basins of the southern California continental borderland was approximately an order of magnitude greater than that of macrobenthos, but that much of the biogenic $CaCO_3$ dissolved. Consequently, planktonic foraminifera, which are quite stenotopic (narrow temperature and salinity tolerances), tend to decrease in number toward shore not only as a result of lack of habitat and sediment dilution, but also dissolution.

Besides the *foraminiferal lysocline*, there are also the *pteropod* (aragonite) and *coccolith* (low-Mg) lysoclines. Because aragonite tends to dissolve easily, the dissolution of most pteropods tends to occur several hundred to several thousand meters above the foraminiferal lysocline (Adelseck and Berger, 1977; Berger, 1978); consequently, pteropods can be very sensitive indicators of changes in climate and water mass structure that are not reflected by planktonic foraminifera (e.g., Almogi-Labin *et al.*, 1998). The dissolution of coccoliths also begins 1–2 km above that of foraminifera and the diversity of coccolith assemblages decreases with depth; consequently, coccoliths are sensitive indicators of dissolution above the foraminiferal lysocline, whereas foraminiferal assemblages are better indices of dissolution than coccoliths below the foraminiferal lysocline (Roth and Berger, 1977). Nevertheless, the coccolith lysocline tends to occur several hundred meters *deeper* than the foraminiferal lysocline because many coccoliths are transported to the bottom in fecal pellets (Smayda, 1971; Honjo, 1977), although they are not normally recognizable in bottom sediments (but see Thompson and Whelan, 1980; Gersonde and Wefer, 1987). The remaining coccoliths also tend to be dissolution-resistant (although they, too, may eventually dissolve; Roth and Berger, 1977; Honjo, 1977), and tend to indicate colder water than unaltered assemblages because cool-water species tend to be more robust; this tendency is more pronounced in coccolithophorids than in planktonic foraminifera (Figure 3.4; Berger and Roth, 1975).

Dissolution and recrystallization continue during burial of foraminiferal-nanofossil oozes and their transformation to chalk and limestone can further

124 Biostratinomy II: dissolution and early diagenesis

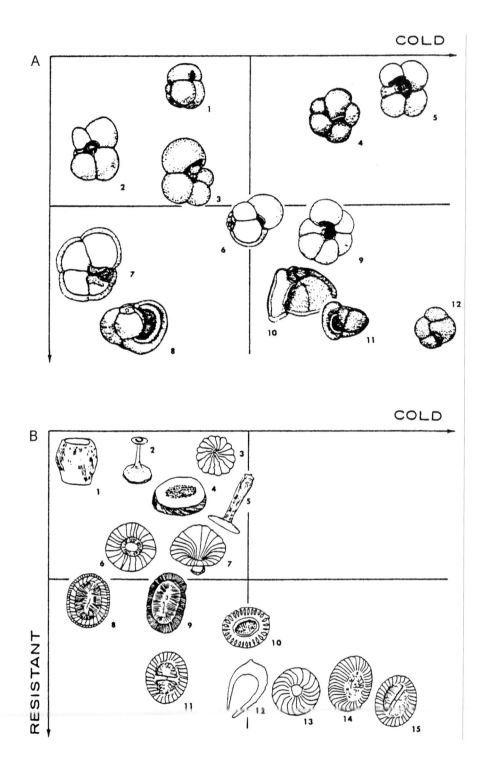

3.3 Shell mineralogy, architecture, microstructure, and size

bias paleotemperature estimates based on stable isotopes (Douglas and Savin, 1978; Garrison, 1981; Table 3.5; cf. Figures 3.1, 3.2, 3.4). During burial, compaction, and lithification, both the original ooze volume and porosity may be reduced by 50% or more (Schlanger and Douglas, 1974). Schlanger *et al.* (1973) and Schlanger and Douglas (1974) noted that the first step involves partial dissolution of foraminifera and recrystallization of nanofossils, but all of the microfossil groups are typically well preserved and abundant, especially in Neogene sediments (significant alteration may occur in sediments at least as young as Late Miocene, however; Keigwin, 1979). With the transformation to chalk, benthic miliolids and thin-walled calcareous perforate species decrease with depth: fragmention and dissolution of thin-walled planktonic foraminifera is evident, and lithification and recrystallization of nanofossils proceeds further. Nanofossils show etching, recrystallization, and secondary calcite overgrowths, all of which are species selective: certain discoasters and species with narrow or non-overlapping lathes are most strongly affected, and small species tend to dissolve at the expense of larger ones, which develop overgrowths that may alter species-specific morphological traits (e.g., Paleogene and Cretaceous sediments described by Schlanger *et al.*, 1973; see also Berger and Roth, 1975). Other species are much more robust and have been used as indicators of coccolith dissolution (e.g., discoasters and *Micula staurophora*), such as that presumably caused by impact at the Cretaceous–Tertiary boundary (Thierstein, 1980, 1981; cf. D'Hondt *et al.*, 1994).

The degree of alteration and lithification depends upon the *diagenetic potential* (Schlanger and Douglas, 1974; Garrison, 1981), which is the degree of alteration that a fossil assemblage undergoes during burial. Diagenetic potential decreases during burial (Table 3.5) and is a function of such factors as the proportion of small to large microfossils (e.g., coccoliths to foraminifera), which determines the size distribution characteristics of the sediment; amount of dissolution (which is related to water depth and proximity to the lysocline and CCD); and

Figure 3.4. Effect of preservation on foraminifera and coccolithophorids and paleotemperature interpretation. (A) Foraminifera: (1) *Globigerinoides ruber*; (2) *G. sacculifer*; (3) *Globigerinella siphonifera*; (4) *Globigerina quinqueloba*; (5) *G. bulloides*; (6) *Globigerinita glutinata*; (7) *Globorotalia menardii*; (8) *Pulleniatina obliquiloculata*; (9) *Neogloboquadrina dutertrei*; (10) *Globorotalia truncatulinoides*; (11) *G. inflata*; (12) *Globigerina pachyderma*. (B) Coccolithophorids: (1) *Scyphosphaera* sp.; (2) *Discosphaera tubifera*; (3) *Cyclococcolithina fragilis*; (4) *Pontosphaera* sp.; (5) *Rhabdosphaera clavigera*; (6) *Cyclolithella annual*; (7) *Umbellosphaera* sp.; (8) *Syracosphaera pulchra*; (9) *Helicopontosphaera kamptneri*; (10) *Emiliania huxleyi*; (11) *Gephyrocapsa oceanica*; (12) *Ceratolithus cristalus*; (13) *Cyclococcolithina leptopora*; (14) *Coccolithus pelagicus*; (15) *Gephyrocapsa caribbeanica*. (Reprinted from Berger and Roth, 1975, with permission of the American Geophysical Union.)

Table 3.5. *Diagenetic realms from the ocean surface to 10 km in subsurface sediment*

Depth	Realm	Residence time	Petrography	Porosity (ϕ) % Sonic velocity (V_c) km/s	Diagenetic potential $0 \leftarrow \rightarrow \infty$
0–200 m (surface water)	I Initial production	Weeks	Highly dispersed calcite–sea water system: $10–10^2$ forams m^{-3}, 10^4–10^6 nanoplankton m^{-3}		+++
200 m to sea floor	II Settling	Days to weeks for forams; months to years for coccoliths depending on pelletization	Pelletized coccoliths, ratio of broken to whole nanoplankton increases downward, ratio of living to empty foram tests decreasing during settling		+++++
3000–5000 m (see Diagenetic Potential)	III Deposition	Inversely proportional to sedimentation rate and dissolution rate	"Honeycombed" structure. Large foram tests supported by chains of coccoliths discs. This surface is actually part of Realm IV	$\phi \simeq 80\%+$ $V_c \simeq 1.45–1.50$	+++ ← slope at 3000 m ++ ← slope at 5000 m
0–1 m (sub-bottom)	IV Bioturbation	50 000 years (at 20 m 10^{-6} years sedimentation rate)	Remolded "honeycomb", slight compaction, burrowing, destruction by ingestion and solution	$\phi \simeq 75–80\%$ $V_c \simeq 1.45–16$	++
1–200 m (sub-bottom)	V Shallow-burial	10×10^6 years (at 20 m 10^{-6} years sedimentation rate)	Ooze affected by gravitational composition, establishment of firm grain contacts; dissolution of fossils and initiation of overgrowths	$\phi \simeq 75–60\%$ $V_c \simeq 1.6–1.8$	++
200–1000 m + (sub-bottom)	VI Deep-burial	Up to $\simeq 120 \times 10^6$ years (by then either subducted or uplifted)	Chalk with strong development of interstitial cement and overgrowth; transition down to limestone with dissolution of forams, pervasion by cement and overgrowths – grain inter-penetration and welding	$\phi \simeq 60\%$ down to 35–40% $V_c \simeq 1.8$ increasing to 3.3 km s^{-1}	++
1–10 km subsurface	VII Metamorphic	$10^6–10^7$ years	Recrystallization trending to completely interlocking crystals	$\phi \simeq 40\%$ down to <5% $V_c \simeq 3$ up to 6 km s^{-1}	+++

Schlanger and Douglas, 1974.

3.3 Shell mineralogy, architecture, microstructure, and size

surface productivity and grazing by herbivores (fecal pellet formation), which affect sedimentation rates. These factors change with changing paleoceanographic settings and affect the production of deep-sea seismic reflectors (e.g., Berger and Mayer, 1978; Kennett, 1982), which may therefore be tied to geochemical (taphonomic) signals of regional significance.

Benthic foraminifera tend to be more resistant to destruction than planktonic species during diagenesis and lithification (Schlanger *et al.*, 1973), and the relative abundance of benthic foraminifera and planktonic fragments have been used as indices of dissolution in deep-sea studies (Schlanger *et al.*, 1973; Thunell, 1976; but see Adelseck, 1977); LaMontagne *et al.* (1996) found that such dissolution indices may indicate dissolution when *total* $CaCO_3$ remains unchanged. Benthic foraminiferal assemblages are also sometimes enriched in dissolution-resistant agglutinated foraminifera not only below the deep-sea lysocline and CCD, but also off continental margins because of the decay of abundant organic matter (R. E. Martin, personal observations of industrial well cuttings taken off the Mississippi Delta). According to some workers, however, up to 95% of the initial agglutinated fauna may be destroyed at sediment depths of less than about 20 cm via predation, grazing, and bacterial (oxidative) decay of organic test cements and dissolution of ferruginous test cements in reduced sediment; fragile species – especially those belonging to the family Komokiacea – may disappear entirely with depth as a result of compaction and mechanical destruction, whereas firmly cemented tests are more likely to survive (Schröder, 1986; see also marsh foraminifera below).

Still, benthic foraminifera differ markedly in their susceptibility to dissolution. Corliss and Honjo (1981) anchored both shallow (*Amphistegina*) and deep-water tests of nine benthic species to a mooring in carbonate-undersaturated waters of the North Pacific for approximately 2 months; sample depths ranged from 306 m to 5590 m. They found that the relative susceptibility of species to carbonate undersaturation was *Amphistegina* sp. > *Pyrgo murrhina* > *Cibicidoides wuellerstorfi* > *Cibicidoides kullenbergi* = *Epistominella umbonifera* = *Hoeglundina elegans* = *Oridorsalis tener* > *Gyroidinoides orbicularis* = *G. soldanii* (see also Bremer and Lohmann, 1982). With the exception of *Pyrgo murrhina*, which is calcareous imperforate (suborder Miliolina), all of the species were trochospirally coiled and possessed pores (most belonging to the suborder Rotaliina). The typical sequence of dissolution of pore-bearing species was (1) corrosion and initial pitting of the test surface and production of shiny to dull surface textures; (2) increased surface pitting and corrosion, breakage of the last chambers, and dull surface textures; (3) breakage of additional chambers and extensive pitting and corrosion of remaining chambers; and (4) extensive breakage of chambers and highly irregular surface textures. Breakage of tests was inversely related to the order in which chambers were added, but this may have been partially the result of the orientation of tests cemented to the mooring (Metzler *et al.*, 1982). The dissolution sequence for *Pyrgo murrhina* was similar, but chambers were removed one at a time, thereby exposing smaller, earlier-formed chambers; the test walls of this species consist of thin layers of ~6 mol% Mg calcite that overlay randomly oriented crystallites (Blackmon and Todd, 1959; Corliss and Honjo, 1981); once the outer layer is removed, the crystallites are rapidly destroyed (Corliss and Honjo, 1981). *Amphistegina* sp. is a common, large reef-dwelling foraminifer characterized by a highly porous biconvex test; although it consists of low-Mg (4–5 mol%) calcite (Chave, 1954; Blackmon and Todd, 1959), this species began to dissolve at relatively shallow depths (978 m), presumably because of high test porosity (Corliss and Honjo, 1981).

Even so, the more resistant deep-sea species also possessed pores. Tests of *Cibicidoides wuellerstorfi* (and *Pyrgo murrhina*) began to break down at 2778 m, but substantial dissolution of most remaining species did not occur until 4000–5000 m. Surprisingly, *Hoeglundina elegans*, which is among the more resistant species, consists of aragonite. The exceptional behavior of *H. elegans* appears to result from its test structure, which is smoother, thicker, and less porous than those of many calcitic species. Dissolution resistance of this species is no doubt also related to the envelopment of individual crystallites by organic matter that is used as a template during test secretion (Reiss and Schneiderman, 1969; Corliss and Honjo, 1981; see also Stapleton, 1973).

The occurrence of Cd, Ba, and Sr in tests of deep-dwelling species has been used as a proxy for nutrient levels (P) in ancient (especially Pleistocene) deep-water masses (e.g., Boyle *et al.*, 1995). McCorkle *et al.* (1995) found, however, that there is preferential loss of Cd, Ba, and Sr (all calculated with respect to Ca) during dissolution of *Cibicidoides wuellerstorfi* at water depths of more than about 2.5 km (see also Dymond *et al.*, 1997). Interestingly, the Cd/Ca partitioning coefficient of *Hoeglundina elegans* is close to 1.0, and is far less depth-dependent, apparently accurately tracking Cd levels into shallow water. Also, manganese carbonate overgrowths, which spoil estimates from calcitic species, do not occur in *Hoeglundina* (Boyle *et al.*, 1995). In the case of infaunal benthic foraminifera, elevated CO_2 levels in porewaters can also cause artifactual negative shifts in δC^{13} values, which are used as indicators of surface productivity and organic matter flux to and decay at the bottom (McCorkle *et al.*, 1990).

The susceptibility of shallow-water benthic foraminifera – and other shallow-water biogenic grains – to dissolution and diagenesis also varies considerably. Kotler *et al.* (1992) exposed five species of common reef-dwelling foraminifera to Ca^{2+}-free artificial seawater in a fluidized bed reactor (Chou and Wollast, 1984) and found the following ranking of dissolution rates: *Archaias angulatus* > *Discorbis rosea* ≫ *Quinqueloculina tricarinata* > *Amphistegina gibbosa* > *Bigenerina irregularis*. Both *Archaias angulatus* and *Quinqueloculina tricarinata* are calcareous imperforate (suborder Miliolina), but in the case of *Archaias*, the test consists of 14–16 mol% Mg-calcite and has a thin outer wall for the transmission of light to algal symbionts harbored in the cytoplasm. *Bigenerina irregularis* is an agglutinated species that consists of calcareous grains cemented together with $CaCO_3$, and although the preferred habitat appears to be characterized by $CaCO_3$ saturation and precipitation in porewaters (Pigott and Land, 1986), the test grains are apparently much less soluble than the cement. By contrast, the abundance of agglutinated species in marsh and shelf assemblages may be accentuated in death assemblages via dissolution of calcareous species (e.g., Murray and Wright, 1970; Kennett, 1982; Smith, 1987; Goldstein, 1988; Murray, 1989; Jonasson and Patterson, 1992), which no doubt reflects overall undersaturation of $CaCO_3$ (e.g., Greiner, 1970). Nevertheless, agglutinated marsh foraminifera, which construct tests of sediment grains cemented with organic matter, may rapidly break down via oxidation of organic cements by oxygenation of porewaters through burrowing, bacterial oxidation, or exposure of cores to air (Scott and Medioli, 1980*b*, 1986; Jonasson and Patterson, 1992; Goldstein and Harben, 1993; Goldstein *et al.*, 1995).

Despite the results of the dissolution experiment by Kotler *et al.* (1992), the species examined persist in natural sediment assemblages (Martin and Liddell, 1988; Martin and Wright, 1988). After about 80 hours in the reactor, only two specimens of *Quinqueloculina tricarinata* and five specimens of badly corroded *Amphistegina gibbosa* remained, and the surface textures produced in the laboratory have only rarely been observed in natural assemblages (Martin, 1986; Martin and Liddell, 1988; Martin and Wright, 1988). Corliss and Honjo (1981) produced similar surface textures in specimens of *Amphistegina* sp., which testifies to the extreme corrosiveness of reactor seawater. These extreme surface textures could not, however, be produced after 2000 hours of exposure in recirculating seawater (pH 7.7) or on specimens of *Amphistegina* (and platelets of the aragonitic green alga *Halimeda*) attached to stakes and buried (up to 30 cm) for up to 15 months in back-reef sediments of Discovery Bay (Jamaica), in which bioturbation, acid production, and test dissolution should occur (Liddell and Martin, 1989; Cottey and Hallock, 1988; Kotler *et al.*, 1992).

By contrast, Walter and Burton (1990) calculated much higher rates of dissolution for other biogenic carbonate grains. For substrates implanted in nearshore back-reef sediments of the Florida Keys, they found rates of 5.6–51.2 mg yr^{-1} for *Neogoniolithon* (red algae; 18 mol% Mg-calcite); 0.5–1.4 mg yr^{-1} for echinoids (12 mol% Mg-calcite); and 0.3–1.0 mg yr^{-1} for scleractinian coral (aragonite). Their range of values is similar to that for *peak* dissolution rates of foraminifera (19.7–98.5 mg yr^{-1}) calculated by Kotler *et al.* (1992), but, as noted previously, the surface textures produced in laboratory dissolution experiments have only rarely been noted in natural assemblages.

Henrich and Wefer (1986) also noted substantial variation in dissolution of biogenic carbonates. They found that platelets of *Halimeda* and tubes of the red alga *Amphiroa fragilissima* (17.5 mol% Mg) were much less resistant to dissolution on moorings in the Drake Passage than particles (63–200 µm) of the *aragonitic* gastropod, *Strombus gigas*, whole tests of the foraminifer *Marginopora vertebralis* (18 mol% Mg), or fragments (63–200 µm) of the sea urchin *Clypeaster* (12 mol% Mg). The greater susceptibility of algal particles is apparently the result of their delicate, highly porous ultrastructure formed during biomineralization (Flajs, 1977; see also Bathurst, 1975), which disintegrates after a threshold of dissolution-induced weakness is reached (cf. Kidwell and Baumiller, 1989), and the low resistance of organic coatings to degradation. Like the pores of foraminifera, dissolution was initiated on biogenic particles along zones of weakness, including those produced in preparation procedures; Katz and Man (1980) found that cleaning of foraminiferal tests via ultrasonication in excess of 30 minutes alters the amino acid content of the organic test matrix, which may in turn affect particle reactivity.

Flessa and Brown (1983) immersed various macroinvertebrate hardparts in acid baths. They found a significant correlation ($r = 0.650$) between dissolution rate and the surface area:weight ratio of hardparts, and, like Henrich and Wefer (1986), concluded that the surface area:weight ratio is sufficient to overcome the effect of mineralogy. Solution-induced features included surface etching and chalky textures, thinning of distal margins, and the formation of holes, but it appears that bioerosion may be more significant for macroinvertebrate shell destruction in some cases than dissolution (e.g., Aller, 1995; Cutler, 1995; Chapter 2).

Particle reactivity is also a function of porewater chemistry (section 3.2). Like other workers, Walter (1985) found that aragonitic and high-Mg (18 mol%) algal

3.3 Shell mineralogy, architecture, microstructure, and size

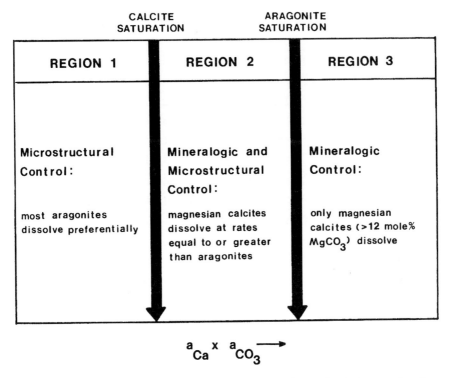

Figure 3.5. Microstructural versus mineralogical control during dissolution as a function of carbonate saturation state. (Reprinted from Walter, 1985, with permission of SEPM.)

fragments (crushed; cf. above) tended to be the most reactive skeletal remains, but that the relative reaction rates, and presumably the transformation of aragonite and high-Mg calcite to low-Mg calcite, depended on the carbonate saturation state of the medium. She recognized three regions of biogenic carbonate dissolution based on calcite–aragonite saturation, microstructure, and mineralogy (Figure 3.5). For conditions of 0.5 times *calcite* saturation, the ranking was aragonitic green alga *Halimeda* > aragonitic pelecypod (*Hippopus*) = aragonitic gastropod (*Strombus*) > 18 mol% Mg red alga (*Neogoniolithon*) > 15 mol% Mg foraminifera (*Peneroplis*) > aragonitic scleractinian corals (*Fungia*, *Acropora*) > low-Mg (<2.5 mol% Mg) barnacle (*Balanus*) > 11–12 mol% Mg echinoid (*Echinus*, *Clypeaster*, *Tripneustes*). In seawater supersaturated with respect to aragonite, only carbonates with more than 12 mol% Mg dissolved. For intermediate media of 0.8 times *aragonite* saturation (i.e., supersaturated with respect to calcite but undersaturated with respect to aragonite), reaction rates were about 5 times slower ($\sim 100\,\mu mol\,g\text{-}h^{-1}$) than for 0.5 calcite saturation, and the rankings were *Neogoniolithon* > peneroplids ≈ *Halimeda* > pelecypod = gastropod > corals > echinoids. The last ranking is very similar to the ranking observed by Purdy (1968) for grain recrystallization in natural shallow-water carbonate sediments, and which was concluded to result from both dissolution-precipitation (e.g., Alexandersson, 1972) and endolithic boring (Chapter 2). Such *micritized envelopes* may persist intact until broken by overburden pressure (Tucker, 1991).

As mentioned previously, grain size (surface area) has also been implicated in dissolution resistance and is commonly stated to be inversely related to dissolution rate (Walter and Morse, 1984; but see Adelseck, 1977). Indeed, it is almost axiomatic in taphonomy that the bigger the shell, the longer it lasts (depending on mineralogy, microstructure, etc.). One hypothesis deduced from this tenet is that big shells should be older than small shells from the same horizon. Moreover, macrofossils should *appear* older than microfossils from the same assemblage (Martin and Liddell, 1991; Martin, 1993). Martin *et al.* (1995, 1996) tested these hypotheses by comparing the surface preservation (*taphonomic grade*) and age of *Chione* (bivalve) and foraminiferal tests from Holocene tidal flats of Bahia la Choya, Mexico (northern Gulf of California). They found that hardpart size and taphonomic grade are not infallible indicators of shell age, preservability, or temporal resolution of assemblages (cf. Brett and Baird, 1986; Brandt, 1989). Disarticulated *Chione* collected from the SWI exhibited an age range of several hundred years to about 80–125 ka based on Accelerator Mass Spectrometer (AMS) ^{14}C dates and amino acid racemization (D-alloisoleucine/L-isoleucine) values, and both old and young valves ranged from highly altered to virtually pristine (see also Frey and Howard, 1986; Flessa *et al.*, 1993; Wehmiller *et al.*, 1995).

By contrast to *Chione*, almost all foraminiferal tests appeared pristine, which suggested an erroneously young age (see also Canfield and Raiswell, 1991*b*, p. 427). Nevertheless, tests were surprisingly old (up to about 2000 calendar years), which is also indicated by surface alteration at the electron microscope level (Barbieri, 1996), and which suggested that the means of test preservation differed from that of bivalves. At Choya Bay, foraminifera reproduce in discrete seasonal pulses (of approximately a few weeks), followed by periods of dissolution (of several months duration) related to low sediment accumulation rates ($\sim$0.038 cm yr^{-1}; Flessa *et al.*, 1993) and intensive bioturbation (oxidation of organic matter and sulfides to produce carbonic and sulfuric acids, respectively). Green *et al.* (1992, 1993) noted similar reproduction–dissolution cycles of foraminifera in LIS (cf. Canfield and Raiswell, 1991*b*, p. 427). Although Gobe and Fütterer (1981) claim that abrasion affects test preservation in shallow waters of Kiel Bay, this process is accompanied by dissolution (see also Smith, 1971).

In the case of Choya Bay, the depositional setting exerts a tremendous impact on the preservation of foraminiferal $CaCO_3$. After each seasonal reproductive pulse at Choya Bay, some tests are rapidly incorporated into a subsurface shell layer (up to ~1 m below the SWI) by Conveyor Belt Deposit Feeders (CDFs; Meldahl, 1987; Chapter 4) and preserved there, as the rest of the pulse rapidly dissolves. Ultimately, some of these older tests are probably exhumed by CDFs and storms and mixed with much newer tests (K. Meldahl and A. Olivera, 1995, personal communication; Martin *et al.*, 1995, 1996). Thus, different taxa may be degraded by different pathways in the same setting: depending on the setting, bivalves may be destroyed mainly by bioerosion (Aller, 1995; Cutler, 1995; Chapter 2) because of their larger size, whereas tests appear to be destroyed mainly by dissolution (Kotler *et al.*, 1992; Martin *et al.*, 1995, 1996) and secondarily by abrasion (Kotler *et al.*, 1992; Shroba, 1993) or bioerosion (Peebles and Lewis, 1988; Martin and Liddell, 1991).

In some cases, however, larger shells may be destroyed via dissolution resulting from bioturbation (cf. equations 3.5–3.8). At first glance, for example, shell hashes in the Late Miocene St Mary's Formation of the Calvert Cliffs region (Chesapeake Bay, U.S.A.) would appear to be the result of abrasion and breakage. Directionally oriented turritellids suggest occasional high-energy currents (Chapter 2), but the fine-grained sediment matrix is a contraindication of pervasive high current energy, and the habitat preferences of the fossils indicate quiet-water paralic environments (R. E. Martin, personal observations; see also Kidwell *et al.*, 1986; Kidwell, 1989). Actualistic studies in deeper shelf and slope settings indicate that abrasion tends to decrease and dissolution increase with increasing autochthony offshore; nevertheless, parautochthonous assemblages are the least altered because shelf shells are buried more rapidly than slope shells and are therefore exhumed less frequently (Callender *et al.*, 1990, 1992, 1994). Callender *et al.*'s studies also indicate that the extent of dissolution, bioerosion, etc., of bivalved shells can vary substantially within short distances in what are seemingly taphonomically homogeneous environments below wave base.

3.4 Pyritization

Pyrite is a common sedimentary mineral and is frequently associated with *permineralization* – early infiltration of cavities by fluids to give internal molds – in organic-rich, fine-grained sediments. In other cases, *petrifaction* – the addition of mineral matter to existing hardparts, especially porous ones such as bone and wood – and *replacement* has resulted in beautifully preserved structures (e.g., Grierson, 1976), although not so well preserved that "defossilization" will ever be successful (Jones, 1998)! (The term petrifaction has sometimes been used

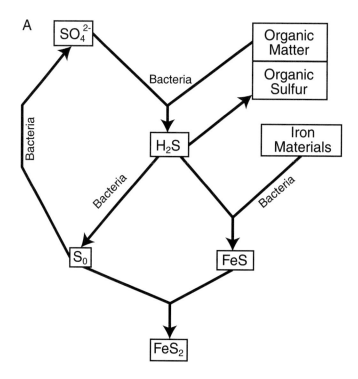

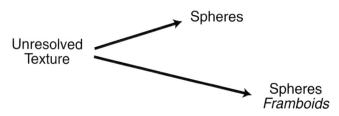

Figure 3.6. (A) Pathways and processes of sulfate reduction and pyrite formation, and (B) mineralogical phases and morphologies of iron sulfides. (Redrawn from Canfield and Raiswell, 1991a.)

synonymously with permineralization; cf. Schopf, 1975.) The quality of pyritic preservation varies substantially, however, both locally and along taphonomic gradients (e.g., Brett and Baird, 1986; Fisher, 1986; see also Chapter 6).

Canfield and Raiswell (1991a) reviewed previous work on iron sulfide formation in marine sediments and presented some general models of pyritization (Figure 3.6). The iron sulfide, pyrite (isometric FeS_2), its orthorhombic polymorph maracasite, and various precursors (mackinawite: $FeS_{0.9}$–$FeS_{0.95}$; greigite:

3.4 Pyritization

Fe_3S_4; pyrrhotite: hexagonal FeS), are the main metal sulfides found in modern marine sediments. They are formed when dissolved sulfides generated during sulfate reduction (equation 3.7) react with available dissolved iron oxides (hematite) and oxyhydroxides (mainly goethite, lepidocrocite, and ferrihydrite, based on studies of the FOAM site; Canfield and Raiswell, 1991*a*). Both marcasite and pyrrhotite are relatively rare in modern sediments, whereas greigite is more common in fresh and brackish water settings, in which dissolved sulfide is absent (low dissolved SO_4^{2-}), and in some marsh sediments in which excess iron precipitates sulfide (Canfield and Raiswell, 1991*a*). Dissolved iron is generated mainly by three reactions

(1) the reduction of iron oxides by sulfide:

$$H_2S + 2FeOOH + 4H^+ \rightarrow S^0 + 2Fe^{2+} + 4H_2O \quad (3.11)$$

(2) the reduction of iron oxides by organic compounds:

$$CH_2O + 4FeOOH + 8H^+ \rightarrow CO_2 + 4Fe^{2+} + 7H_2O \quad (3.12)$$

and (3) the oxidation of iron sulfides:

$$7O_2 + 2FeS_2 + 2H_2O \rightarrow 2Fe^{2+} + 4SO_4^{2-} + 4H^+ \quad (3.13)$$

which may occur in nearshore and marsh environments as a result of storms or seasonal changes in the redox boundary in sediment (e.g., Reaves, 1986). If sufficient iron is present or sulfate concentrations are sufficiently low (such as in fresh water), sulfide concentrations are kept low and iron may react with other ions such as bicarbonate to produce siderite ($FeCO_3$; Baird *et al.*, 1985, 1986; see also Chapter 6).

The iron sulfides that are generated vary in appearance. *Framboidal pyrite* consists of equigranular crystallites (usually cubes or pyritohedra less than 5 μm in size), often packed into spheroidal aggregates, whereas *clusters* are more variable in crystallite size and packing (Canfield and Raiswell, 1991*a*; Figure 3.6). The reactions that generate framboidal pyrite are likely

$$18CH_2O + 9SO_4^{2-} \rightarrow 18HCO_3^- + 9H_2S \quad (3.14)$$

$$6FeOOH + 9H_2S \rightarrow 6FeS \text{ (mackinawite and amorphous FeS)}$$
$$+ 3S^0 + 12H_2O \quad (3.15)$$

$$3FeS + S^0 \rightarrow Fe_3S_4 \text{ (greigite)} \quad (3.16)$$

$$Fe_3S_4 + 2S^0 \rightarrow 3FeS_2 \text{ (pyrite)} \quad (3.17)$$

Table 3.6. *Relationship of early diagenetic features to sedimentation rate and oxygenation of bottom water and sediment (cf. Figure 3.7)*

Water oxygenation	Sediment geochemistry	Sedimentation rates			
		Episodic, very rapid (1–50 cm 10^{-2} yr)	Intermediate to rapid (10–100 m 10^{-3} yr)	Low to intermediate (1–10 cm 10^{-3} yr)	
Aerobic $O_2 > 0.7$ ml l^{-1}	Oxic to depth; organic-poor	No sediment fillings; late diagenetic mineral fillings; minor pyrite	Sediment steinkerns	Partial sediment steinkerns; rare chamositic, hematitic, coatings	
Aerobic to dysaerobic $O_2 = 0.7$–0.3 ml l^{-1}	Anoxic with oxic micro-zone; organic-poor (non-sulfidic)	Pyrite steinkerns, overpyrite (euhedral); $CaCO_3$ concentrations	$CaCO_3$ concretionary mud steinkerns; minor overpyrite	Phosphatic and/or glauconitic steinkerns, often reworked; rare overpyrite	
Dysaerobic to anaerobic (euxinic) $O_2 > 0.3$ ml l^{-1}	Anoxic to surface; commonly organic-rich (commonly sulfidic)	No fillings; minor pyritic replacement; rarely, traces of soft parts	Mud steinkerns; pyrite patinas; periostracal remnants	Highly compacted mud steinkerns	

Brett and Baird, 1986.

3.4 Pyritization

Equations 3.14–3.17 may be summarized as

$$18CH_2O + 9SO_4^{2-} + 6FeOOH \rightarrow$$
$$18HCO_3^- + 3FeS + 3FeS_2 + 12H_2O \quad (3.18)$$

(Canfield and Raiswell, 1991*a*; based on Berner 1970, 1984). Thus, although pyrite is a common sedimentary mineral, it represents the terminal stage of pyritization (Figure 3.6) and is relatively unimportant to the overall *process* of pyritization. Throughout the transition from iron-dominated to sulfide-dominated porewaters, iron and sulfide typically remain at saturation with amorphous FeS, mackinawite, and greigite, as indicated by the black coloration of sediment and shells near the SWI (Canfield and Raiswell, 1991*a*). Moreover, the sum of the reactions (equation 3.18) indicates that not enough elemental sulfur (S^0) is produced to completely convert FeS to pyrite; some elemental sulfur may be produced by bioturbation and oxidation of H_2S to S^0, but below the surface mixed layer (main zone of bioturbation), the oxidant used to convert FeS to pyrite remains unknown (Canfield and Raiswell, 1991*a*).

In aerobic to dysaerobic environments (Table 3.6), pyrite formation is normally limited to local patches of metabolizable organic matter (Hecht, 1933; Zangerl and Richardson, 1963; Hudson, 1982; Berner, 1984; Brett and Baird, 1986; Fisher, 1986; Canfield and Raiswell, 1991*a*), including the shell surfaces of living organisms via reaction with excreted organic metabolites (but not periostracum; Reaves, 1984, in Canfield and Raiswell, 1991*a*; cf. Clark and Lutz, 1980). The quality of replacement depends on the initial shell structure and the access of iron and sulfate ions into the shell (Fisher, 1986). Normally, it is the more refractory organic matter left after various stages of decay that tends to be pyritized (e.g., Reaves, 1986; see also Chapters 6, 10). When the concentration of dissolved iron is high, dissolved sulfide is low, and elevated sulfide levels at the site of sulfate reduction (organic matter decomposition) will cause iron sulfide precipitation as iron diffuses toward the site (Canfield and Raiswell, 1991*a*; see also Berner, 1969). This mechanism appears to be common in Quaternary sediments (see also Fisher for the Jurassic Lower Oxford Clay) and quite rapid in continental margin sediments (a few years to decades), and includes nodular pyrite and pyrite coatings (cf. Table 3.6) of mucus linings of burrow tubes and vascular land plant debris (see Canfield and Raiswell, 1991*a*, for references). Rapid burial of organic matter may also cause isolated pockets of sulfate reduction, leading to pyritized steinkerns or fossil layers (Brett and Baird, 1986; Figure 3.7). In some cases, early replacement of the organic matrix preserves good microstructure, although the exact mechanism is apparently unclear; Fisher (1986,

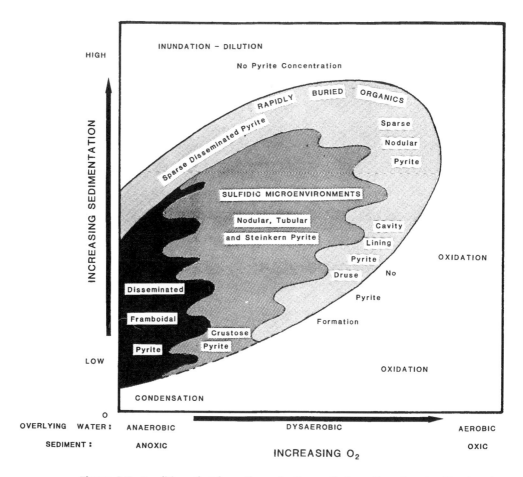

Figure 3.7. Conditions that favor the early diagenetic formation of pyrite. (Reprinted from Brett and Baird, 1986, with permission of SEPM.)

p. 583) attributed it to the limited availability of iron, which would otherwise prevent buildup of H^+ via H_2S (cf. equation 3.18).

Other mechanisms of pyritization occur and may be useful in elucidating early sediment porewater chemistry. After most iron has precipitated, dissolved sulfide diffuses toward sites of iron and may form pyrite coatings on iron oxide minerals or iron-containing sheet silicates; such coatings postdate pyritization of organic matter by hundreds to hundreds of thousands of years or more (Canfield and Raiswell, 1991a). Pyrite layers may also form in association with unconformities and migrating *reduction fronts*. Another mechanism of localized pyritization is the local dissolution of shells, which raises pH and causes the precipitation of iron sulfides:

$$CaCO_3 + Fe^{2+} + H_2S \rightarrow FeS + Ca^{2+} + CO_2 + H_2O \qquad (3.19)$$

Although this mechanism may explain many occurrences of pyritized fossils, it is unlikely to explain pyritization during early diagenesis. Calculations suggest that equation 3.19 works best when the initial porewater pH is less than about 5. This is because as $CaCO_3$ dissolves, porewater pH increases and $CaCO_3$ saturation increases, so that dissolved iron and sulfide are less available and the reaction slows (Canfield and Raiswell, 1991a). But modern environments with a porewater pH lower than 5 are not known: even the heavily irrigated sites of LIS have a pH higher than 7 (Canfield and Raiswell, 1991a,b).

In fact, ancient occurrences of pyrite exhibit a much greater range of associations with skeletal material than do modern occurrences. Although *bladed* pyrite is found in fossils, for example, it is unknown from modern sediments and may be inherited from a marcasite precursor. Other pyrite occurrences include (in approximate sequence of formation): (1) associations with internal sediment; (2) chamber linings; (3) stalactitic pyrite; and (4) overpyrite (coatings) that form during compaction as shell surfaces are cracked (Canfield and Raiswell, 1991a; cf. Table 3.6; Figure 3.7). These occurrences alone, or in conjunction with other minerals, are extremely useful in determining the diagenetic history of sediments and entombed fossils (e.g., geopetally oriented pyrite stalactites; Hudson, 1982). Pyrite formation in internal sediment probably forms as described previously, but the absence of the other occurrences in modern settings suggests that they may not form during early diagenesis (Canfield and Raiswell, 1991a; cf. Hudson, 1982). Nevertheless, following precipitation of iron sulfides, the residual sulfide-rich porewater, if it were to migrate into the zone of methanogenesis (equation 3.9) would also become carbonate-undersaturated. Iron would then be leached from surrounding sediments and precipitate with sulfides.

Thus, different occurrences of pyritization can result according to the relative timing of the passage from carbonate over- to undersaturation and iron sulfide saturation to oversaturation. A fluid that is carbonate-undersaturated and sulfide-oversaturated may cause shell replacement, whereas another fluid that is both carbonate- and sulfide-oversaturated might produce pyrite coatings. The advantage of this mechanism is that it could still occur relatively early during diagenesis (cf. Table 3.6; Figure 3.7); the only other apparent alternative is pyritization via low pH *basinal brines* by analogy to Mississippi Valley type ores (Canfield and Raiswell, 1991a).

3.5 Silicification

Silicification of fossils may, like pyritization, result in permineralization, petrifaction, and replacement. In saturated or near-saturated solutions, silicon exists in the

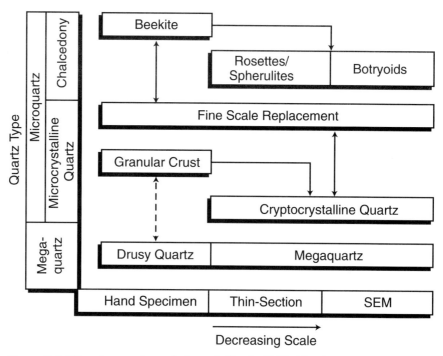

Figure 3.8. Mineralogy and morphology of silica in silicified bioclasts. Note the change in morphology with decreasing scale. SEM, scanning electron micrograph. (Redrawn from Carson, 1991.)

form of hydrated molecules of H_4SiO_4, which link to each other and eliminate water to form silica (SiO_2), and in which each tetrahedron (SiO_4) shares each of its oxygens with another tetrahedron (Carson, 1991). The two basic types of sedimentary silica are quartz and opal (Figure 3.8). Quartz exists as megaquartz (crystallites more than 20 µm) and microquartz (crystallites less than 20 µm). Microquartz in turn occurs as equant (1–4 µm) grains and chalcedony, in which the crystals have a bladed appearance. Chalcedony usually forms as a true replacement of bioclast structure, whereas mega- and microcrystalline quartz infill pore space to form molds and casts (Carson, 1991). Chalcedony is of three basic types: (1) chalcedonite (length-fast, c-axis perpendicular to the fibers), (2) quartzine (length-slow, c-axis parallel to the fibers), and (3) lutecite (length-slow, c-axis approximately 30° to fibers). Chalcedony may also alter to micro- or megaquartz (Carson, 1991).

The sources of silica are varied and include volcanic and hydrothermal, detrital, and biogenic origins (Carson, 1991). Volcanic ash and lava weather easily and probably provided the silica that petrified floras of the western United States and elsewhere, as well as microfloras of the Precambrian, when volcanic

3.5 Silicification

terrains were presumably widespread (e.g., Stein, 1982; Karowe and Jefferson, 1987). Buurman *et al.* (1973), for example, found silicified (opal-cristobalite) remnants (probably from the mangroves *Avicennia* and *Rhizophora*) at depths of 50–130 cm below the SWI in cores taken from lagoonal sediments; they concluded that silica had been weathered from adjacent volcanic soils by acids produced via oxidation of pyrite, and that silicification had occurred *within the last few centuries* (although no corroborating dates were published). Evaporation may also increase the concentration of silica in fluids when no volcanic terrains are present, including in sodium carbonate-rich lakes (Carson, 1991); silica precipitated from evaporitic solutions is likely to be length-slow chalcedony (Folk and Pittman, 1971). Interestingly, Kempe and Degens (1985) postulated that the Precambrian was characterized by a "soda ocean" enriched in sodium carbonate and bicarbonate, which may have contributed to the preservation of Precambrian microfossils (see also Chapter 9). Hydrothermal vents may have also been rich sources of silica in the Precambrian, and associated elevated geothermal gradients may have contributed to silicification in the absence of evaporation (Simonson, 1987). Other trends in silicification may be related to the diversity of siliceous sponges and land plants through time (Chapter 9).

Silicification of vascular plants must occur early to prevent decay (primarily by basidiomycetes; Leo and Barghoorn, 1976). Fortunately, organic matter appears to serve as a template for silica replacement (Carson, 1991). Although the occurrence of delicate structures argues for silicification prior to significant decay (Karowe and Jefferson, 1987), certain components, such as lignin, are relatively resistant (Leo and Barghoorn, 1976; Kenrick and Edwards, 1988; Robinson, 1990), which may introduce a preservational bias (Leo and Barghoorn, 1976). In some cases, though, delicate structures are preferentially preserved over more robust ones (Buurman *et al.*, 1973; Leo and Barghoorn, 1976). Nevertheless, a certain amount of decay appears necessary to increase the permeability of cells to silica (Leo and Barghoorn, 1976), although the amount of degradation can vary substantially between species (e.g., Hedges *et al.*, 1985). Organic decay in carbonate sediments may also be sufficient to lower pH and cause dissolution of $CaCO_3$ while silica precipitates (Knoll, 1985).

Based on experimental studies of preservation, the degradation of organic components produces functional groups that bind with silica and that reproduce the histology of the organic components in a far more faithful way than in carbonates or pyrite (Leo and Barghoorn, 1976; Karowe and Jefferson, 1987; Sigleo, 1978, 1979; Stein, 1982; Carson, 1991). The typically inferior quality of plant petrifaction in carbonates and sulfides suggests that carbonate and sulfide ions are unable to bond with ligno-cellulose components (Leo and Barghoorn, 1976; but see Grierson, 1976). In the laboratory, Leo and Barghoorn (1976) produced monosilicic acid, which is the principal soluble form of silicon in nature and is released during devitrification of volcanic glass and diagenesis of clay minerals (Murata, 1940; Karowe and Jefferson, 1987). The likely pH of the solutions infiltrating tissues is near neutral or somewhat acidic, as extremely acidic or alkaline solutions destroy plant tissues and highly alkaline solutions (pH higher than about 9) would prevent deposition of silica, which dissolves in basic solutions (Leo and Barghoorn, 1976; Buurman et al., 1973, determined pHs of 3–4). The silicic acid then presumably bonds, via hydrogen bonds, to hydroxyl groups of plant tissue, and as the concentration of silicic acid increases in plant tissue, polymerization of silicic acid takes place:

$$Si(OH)_4 + Si(OH)_4 \rightarrow Si(OH)_3OSi(OH)_3 + H_2O \qquad (3.20)$$

Opaline silica, which typifies Tertiary floras, then forms, but eventually transforms from amorphous opal (opal-A) to crystalline opal (opal-CT) to microcrystalline quartz, which is characteristic of Paleozoic floras (Leo and Barghoorn, 1976).

Leo and Barghoorn (1976) found that the structures produced in the laboratory strongly resembled those found in nature. Karowe and Jefferson (1987) argued that such a mechanism explains the preservation of trees buried by eruptions of Mt St Helens (Washington), Eocene fossil forests of Yellowstone National Park (Wyoming), and Cretaceous trees on Alexander Island (Antarctica). Also, Sigleo (1978) suggested that mild "thermal events" transformed wood of trees preserved in Petrified Forest National Park (Arizona) into a more highly stable polymer. The lack of silicified coal suggests, on the other hand, that functional groups were lost during coal formation, although silica (from, for example, volcanic ashes) may still affect the coalification process (Crowley et al., 1994).

Knoll (1985) concluded that the silicification mechanism of Leo and Barghoorn (1976) also occurs in microbes such as cyanobacteria. Microbes may be excellently preserved if silicification takes place early, perhaps within days to weeks based on experiments (Bartley, 1996). Oehler and Schopf (1971) described experiments in which modern cyanobacteria (*Lyngbya*) were embedded in silica gel that was then subjected to temperatures of approximately 150 °C and pressures of 2–4 kilobars for 2–4 weeks; the resulting chert approximated that of its Precambrian counterparts. Early decay of cell walls appears to be inhibited when functional groups bind with iron (Ferris et al., 1988), although in the case of banded iron formations, such as the Gunflint Iron Formation, silica may have originated as a primary precipitate from seawater as monosilicic and polysilicic

3.5 Silicification

acids bound to hydroxyl (OH⁻) groups on cell walls (Barghoorn and Tyler, 1965; Ferris *et al.*, 1988; Carson, 1991); surprisingly, the presence or absence of oxygen appears to make little difference (Bartley, 1996). Despite excellent preservation, however, Knoll (1985) stressed that silicified microbes give a somewhat biased view of plant life, as they often lived under unusual conditions (e.g., hypersaline lagoons; see also Allison, 1988*b*). Silicification of microbes may also result in artifactual cell walls and nuclei that mimic the structures of eukaryotes or produce new "species" from the same culture (Francis *et al.*, 1978). Sheath morphology is retained longer than cell morphology, perhaps because of cross-linkage of polysaccharides and the presence of phenols, both of which retard bacterial decay (Bartley, 1996), and different cyanobacteria species exhibit different rates of alteration and may display early taphonomic grades while *still alive* (see Bartley, 1996, and references therein).

Biogenic silica (diatoms, radiolaria, silicoflagellates) originates in the form of opal-A, which contains up to 10% water (Tucker, 1991). Opal-A is metastable and decreases in abundance with increasing age, so that it is not present in Paleozoic cherts (Tucker, 1991). Today, siliceous oozes are most common underneath high-productivity open marine settings (such as the modern eastern Pacific and Antarctica), where upwelling brings dissolved nutrients into the photic zone; Heath (1974), however, calculated that 85–90% of opaline silica is actually deposited in nearshore sediments, where most is masked by terrigenous debris. Once abundant siliceous plankton evolved (at least by the end of the Cretaceous), silica became an essential nutrient, without which metabolism and cell division are slowed or completely blocked (see Calvert, 1974, for references; see also Kilham and Kilham, 1980, and Chapters 8, 9). The rapid recycling of silica became exceedingly important, without which modern siliceous plankton would strip the oceans of silica in about 250 years; thus, many siliceous skeletons dissolve within a few hundred meters of the ocean surface, and only about 2% avoid postdepositional dissolution (Heath, 1974).

Like calcareous oozes, siliceous oozes undergo significant changes during burial (e.g., Calvert, 1974; Carson, 1991; Julson and Rack, 1992). In the ideal sequence, opal-A "matures" first to crystalline opal-CT (interlayered cristobalite and tridymite) and replaces opal-A of radiolaria, diatoms, and sponge spicules. Opal-CT then typically converts to microquartz, during which skeletal structures are normally obliterated or produce void fills (Carson, 1991; Tucker, 1991). The actual pathway of silica transformation is related to the surface area of the phase in question, which tends to decrease during the alteration sequence opal-A $\to$ opal-CT $\to$ quartz (Williams *et al.*, 1985). Since biogenic skeletons (opal-A) have very large surface areas (e.g., diatoms, radiolarians), they tend to alter relatively rapidly (Williams *et al.*, 1985).

There is ample evidence for different pathways (and timings) of silicification of calcareous fossils during early diagenesis (Carson, 1991). Silica and $CaCO_3$ tend to behave inversely: as pH increases (by, for example, evaporation), silica dissolves and $CaCO_3$ precipitates (Carson, 1991; Tucker, 1991). The occurrence of silica in intimate association with burrows and bioclasts suggests that bacterial oxidation of organic matter under *aerobic* conditions led to the build up of CO_2 in porewaters, decreased pH, and the precipitation of silica (sometimes contemporaneous with pyritization; e.g., Holdaway and Clayton, 1982; see also Jacka, 1974). If iron is present, precipitation of $CaCO_3$ and pyrite may *pre-date* silicification, but when pyrite obviously *post-dates* silicification, sulfate reduction is excluded because it would lead to increased pH and precipitation of pyrite before silicification.

Other mechanisms of silicification of calcareous fossils during diagenesis have been suggested. Since silica and $CaCO_3$ behave inversely with respect to pH, early studies of silicification commonly attributed the process to mixing of marine and meteoric waters in coastal locations so that fluids were simultaneously supersaturated with silica and undersaturated with respect to $CaCO_3$ (Knauth, 1979). Jacka (1974), who was among the first to study the geochemistry of silica replacement (Carson, 1991), concluded that such a mechanism occurred during shallow burial of the Middle Permian Getaway Limestone Member (Cherry Canyon Formation of Texas), when the sediment was only partially lithified. He found that dolomite rhombs occurred only in association with skeletal remains composed originally of high-Mg calcite (e.g., bryozoans, echinoderms). (Although Jacka also found dolomite rhombs associated with silicified fusulinids, which he stated were originally composed of HMC, fusulinids are now considered to have been originally of low-Mg calcite; cf. section 3.3.1.) To Jacka, the dolomite rhombs indicated that silica replacement occurred before HMC stabilized to LMC and that the Mg was locally derived; dolomite rhombs associated with echinoderm ossicles were aligned in linear patterns and displayed extinction simultaneous with that of unreplaced calcite (section 3.3.1).

Like dissolution of calcareous skeletons (section 3.3.2), silicification of calcareous skeletons depends on original hardpart mineralogy and microstructure, as well as the rate of supply of silica relative to rate of carbonate dissolution (Maliva and Siever, 1988). Holdaway and Clayton (1982), for example, found three distinct silica morphologies in brachiopods of the Upper Cretaceous of England: (1) fine-scale replacement of shell microstructure when silica was abundant; (2) concentric rings of silica (beekite) when silica was limited; and (3) a granular (cryptocrystalline) white silica crust that formed when $CacO_3$ dissolution was restricted and silica precipitation was fast. Brachiopods were associated with fine-scale replacement, although the degree of replacement varied with the shell layer. Rapid dissolution of small crystallites in bivalves (e.g., *Exogyra*) resulted in beekite formation, whereas in echinoids, large crystal size inhibited $CaCO_3$ dissolution and was associated with cryptocrystalline morphology.

Even within the same taxon or the same specimen, however, patterns of silicification vary substantially, and completely different silica fabrics may occur at different spatial scales (Carson, 1991; Figure 3.8). Brown *et al.* (1969) found irregular patches of chalcedony distributed over the echinoids *Micraster* and *Echinocorys* that did not preserve ambulacral and interambulacral ossicles; rhynchonellid brachiopods were typically well preserved, whereas terebratulids were normally silicified only in the umbonal regions. Schmitt and Boyd (1981) found five patterns of silicification in brachiopods and bivalves: (1) megaquartz crystals projecting inwards from the shell; (2) concentric laminae of chalcedony; (3) outer laminae of microcrystalline quartz that grades abruptly to chalcedony, which in turn grades abruptly to megaquartz; (4) megaquartz crystals aligned parallel to the shell and adjacent to beekite; and (5) megaquartz crystals aligned parallel to the shell boundary and which often exhibit relict skeletal microstructure. Aragonitic bivalves did not exhibit pattern 5. Schmitt and Boyd (1981) concluded that patterns 1–4 resulted from filling of voids created by dissolution, whereas pattern 5 resulted from concurrent carbonate dissolution and silicification.

Henderson (1984) concluded that in the Devonian stromatoporoid *Hermatoporoidea*, skeletal tissue influenced growth of megaquartz while growing quartz crystal faces were in contact with the carbonate that was being replaced; spherulitic quartzine replaced exterior surfaces in a similar manner. Megaquartz growth was concentrated near, and was simultaneous with, the formation of microscopic pressure-solution seams, and represents slow replacement associated with long diffusional pathways, whereas quartzine was associated with rapid replacement along short diffusional pathways.

3.6 Phosphatization

There are more than 300 phosphate-bearing minerals, most of which belong to the apatite group with the general formula $Ca_{10}(PO_4,CO_3)_6(F,OH,Cl)_{\geq 2}$ (Lucas and Prévôt, 1991). Cations such as Mg and Na substitute for Ca, whereas the substitution of CO_3^{2-} for PO_4^{3-} is balanced by excess F^- to produce fluorapatite ($Ca_{10}(PO_4)_6F_2$) and the rare chlorapatite ($Ca_{10}(PO_4)_6Cl_2$). Non-weathered biogenic marine grains, such as teeth, bones, or diagenetically phosphatized fossils, consist of carbonate-fluorapatite (CARFAP), which is the most common marine phosphate mineral; with the exception of tooth enamel, in bones and teeth, apatite crystals are oriented by a collagen network (Lucas and Prévôt, 1991). Original phosphatic hardparts are also known from brachiopods, arthropods, annelids, molluscs, echinoderms, and conodonts (Lowenstam and Weiner, 1983), which consist of dahllite (carbonate hydroxyapatite), francolite (carbonate fluorapatite), chitinophosphate (brachiopods and trilobites) and other minerals.

Post-mortem phosphatization of non-phosphatic skeletons can mimic original hardpart mineralogy (Bengston and Conway Morris, 1992; see also Chapter 9). In the case of originally phosphatic remains, detailed microstructure may be preserved, but is lost in non-phosphatic remains because the release of phosphoric acid during organic decay dissolves $CaCO_3$ and precipitates calcium phosphate (apatite) as microspherules; otherwise, external or internal molds form (Lucas and Prévôt, 1991). Although this criterion may be useful in discerning original from secondary phosphate in some cases, in other instances perfect preservation of bone microstructure does not exclude diagenetic recrystallization.

Phosphorus is present in low amounts in all organisms (Redfield ratios) and, like silica, is an essential nutrient, but over much longer time scales, being involved in the synthesis of nucleic acid precursors and energy transfer via ATP (Fox, 1988; nitrogen, in the form of NH_4^+ and NO_3^-, is also an essential nutrient, but over short time scales; e.g., Rau et al., 1987). The small crystal size and large surface area of biogenic hydroxyapatites results in solubilities up to approximately 10^4 those of the inorganic form (Lucas and Prévôt, 1991). Thus, biogenic phosphates such as fish bone may be quickly recycled in upwelling regimes; Suess (1981) calculated that fish debris phosphate is about 4 times more important than organically bound phosphorus in nutrient regeneration from sediments of the Peru continental margin. Microbial degradation and bioerosion may also contribute to bone destruction (Hanson and Buikstra, 1987; Piepenbrink, 1989; cf. Van Cappellen and Berner, 1988), although teeth tend to be more resistant, probably because of lower porosity (Shipman, 1981; Lucas and Prévôt, 1991). Preservation of bones and teeth in terrestrial sediments may also be augmented via burial by bioturbation (Behrensmeyer and Chapman, 1993).

The concentration of phosphorus into phosphorite deposits requires special conditions involving the juxtaposition of sea-level change and nutrient-rich waters (e.g., Riggs, 1984, for Neogene phosphorites of the U.S. Atlantic coast).

In well-oxygenated sediments, breakdown of organic matter frequently results in the dissolution of phosphatic and calcareous hardparts, whereas in organic-rich rocks in which bioturbation is inhibited, phosphatic fossils may be preserved. In addition, nutrient-rich waters are normally associated with non-calcareous plankton (diatoms, radiolarians, dinoflagellates); hence, the frequent association of phosphorites with siliceous oozes (Lucas and Prévôt, 1991). Nevertheless, the formation of phosphorites is not completely understood and the conditions under which they formed may have changed through time (see Föllmi, 1996, for review; see also Chapter 9).

3.7 Concretions

The fundamentals of porewater chemistry serve as the basis for a discussion of the formation of concretions (see Boudreau, 1997, for summary of models), which have sometimes yielded exceptionally preserved fossils belonging to a variety of taxa ranging from freshwater and marginal marine to fully marine in habitat. Baird *et al.* (1985, 1986) described a diverse fauna and flora from Mazon Creek (Pennsylvanian) assemblages of Illinois in relation to biostratinomic gradients, as did Allison (1988*b*) for the Eocene London Clay, which contains vertebrates, invertebrates, and one of the world's most diverse fossil fruit and seed assemblages (Konservat-Lagerstätten; e.g., Müller, 1985; see also Chapter 6). Although good preservation, especially of softparts, implies early cementation near the SWI, possibly in less than 50 years (Allison, 1988*b*; Allison and Pye, 1994), the fossils suggest that the genesis of concretions is quite complex, and must be evaluated in the context of the diagenetic history of the surrounding sediments (Canfield and Raiswell, 1991*b*). Unfortunately, many of the concretions used in geochemical studies have been unfossiliferous, and this may be a serious weakness in inferring the mechanism(s) of concretion formation involving fossils (Canfield and Raiswell, 1991*b*).

One of the most fundamental aspects to understanding the growth of concretions is the timing of their formation. Raiswell (1971) suggested that early concretions tend to be more spherical, and he suggested criteria for determining the relative age of concretion formation: deformed or parallel laminae, septarian structures, and cone-in-cone structures. Based on these criteria, he (1971) recognized two basic types of concretions (all unfossiliferous): type I, characterized by deformed ("bent") laminae and septarian (dehydration) structures, grew earlier than type IIa concretions, which are characterized by parallel laminae and cone-in-cone structures. Deformed laminae formed as sediment was compacted at shallow depths around the hardening concretion, whereas parallel laminae presumably formed at greater depth after much of the compaction had occurred. Cone-in-cone structures also imply crystal growth in partially-compacted sediment. These inferences are borne out by centrifugal changes in concretion porosity, which decreased from approximately 70% to 40% in type I concretions, but remained approximately the same (about 30–40%) throughout growth of type IIa concretions. Other workers have successfully used the relation between porosity and burial depth to infer relative ages of concretions (e.g., Curtis et al., 1986), but caution must be exercised because the porosity of shallow sediments varies substantially with grain size, water content, bioturbation and sedimentation rates (Canfield and Raiswell, 1991b). Type IIb concretions also have parallel laminations, but they apparently grew early without compaction, perhaps during a pause in sedimentation; such "syngenetic" concretions may be subject to encrustation by epibionts as they reside near the SWI (see Raiswell, 1971, for references). Canfield and Raiswell (1991b) suggested that most marine concretion-rich horizons or successions of horizons (e.g., Waage, 1964) form in association with changes in sedimentation rate. Concretion growth is primarily the result of carbonate supersaturation (increased alkalinity) of porewaters generated by microbial degradation of organic matter, but inferred rates of carbonate precipitation in modern anaerobic sediments with typical background sedimentation rates are too slow for carbonate precipitation to occur (Canfield and Raiswell, 1991b). Thus, the zone of $CaCO_3$ supersaturation must be immobilized stratigraphically near, *but below*, depositional breaks because if sedimentation is more or less continuous, the zone of $CaCO_3$ supersaturation will continue to move upward toward the SWI; if, however, sedimentation slows or ceases, the zone of $CaCO_3$ supersaturation will remain at a particular sediment depth long enough for cementation to occur (Raiswell, 1987). According to the model of Canfield and Raiswell (1991b; section 3.2), the zone of maximum carbonate supersaturation is reached at the interface between the zones of sulfate reduction and methanogenesis (Table 3.2), where anaerobic methane oxidation (AMO) occurs. Martin et al. (1995, 1996), for example, found a hard (but friable) gray layer of highly fragmented mollusc shells beneath the SWI at Choya Bay (Mexico) that probably resulted from both extensive sulfate reduction below the SWI (alkalinity values normally approximately 5–12 mequiv l^{-1}, but ranging up to about 50 mequiv l^{-1} in some cases) and very low sediment accumulation rates (Flessa et al., 1993; see also Pye et al., 1990; Allison and Pye, 1994).

The concentration of iron during this phase is critical in determining the pathway of concretion growth and mineralogy (Coleman, 1985; Raiswell, 1987;

3.7 Concretions

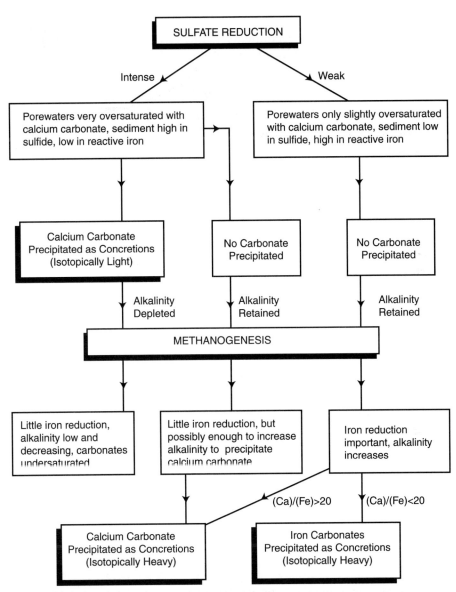

Figure 3.9. Alternative diagenetic pathways leading to the formation of carbonate precipitation and the formation of concretions. (Redrawn from Raiswell, 1987.)

Figure 3.9). In the presence of intense sulfate reduction and abundant reactive iron (i.e., marine terrigenous sediments), iron sulfides are generated and porewaters become depleted in iron. In the case of relatively low rates of sulfate reduction (e.g., freshwater or brackish sediments), however, insufficient alkalinity is generated to precipitate iron sulfides, and iron may pass through the zone of sulfate reduction to the zone of methanogenesis. However, methanogenesis produces

CO_2 (equation 3.9), thereby reducing alkalinity and $CaCO_3$ saturation. Reduction of iron in freshwater sediments (which would otherwise be consumed by iron sulfide precipitation in marine sediments), however, increases alkalinity:

$$3H_2O + 2Fe_2O_3 + CH_2O \rightarrow HCO_3^- + 4Fe^{2+} + 7OH^- \quad (3.21)$$

If iron reduction is sufficient to overcome CO_2 generation by methanogenesis, precipitation of Fe-rich siderite or Ca-rich ankerite will result, depending upon the availability of the two ions (Raiswell, 1971). Dolomite (Mg-rich) may also form, which reflects the degree of marine influence (sulfate availability; Raiswell, 1971). If sedimentation rates are sufficiently high or sulfate-reduction rates are sufficiently low, Fe may still move rapidly through the sulfate-reduction zone, thereby generating siderite, even in marine settings (e.g., Gautier, 1982; Curtis et al., 1986; Pye et al., 1990). Allison and Pye (1994) suggested that iron may also be made available in porewaters by "tidal pumping."

Despite the establishment of what appears to be a reasonable theory of concretion growth, fossil studies raise intriguing questions about the role of substrates (Canfield and Raiswell, 1991b). The occurrence of calcareous hardparts in concretions suggests that calcareous hardparts are sites of crystal nucleation, but the occurrence of both fossiliferous and unfossiliferous concretions in close lateral or stratigraphic juxtaposition (e.g., Waage, 1964) suggests that shell nucleation alone is not sufficient to initiate concretion growth, especially since unfossiliferous concretions often outnumber fossiliferous ones (Canfield and Raiswell, 1991b). This suggests that sites of localized organic decay may increase alkalinity and carbonate saturation. Since $CaCO_3$ precipitation is "poisoned" by dissolved Mg, phosphate, or organic matter, the generation of alkalinity by organic decay must apparently exceed some threshold for carbonate precipitation to occur. Much of this alkalinity presumably escapes local precipitation and raises overall porewater alkalinity of surrounding sediments. Since most concretions contain more $CaCO_3$ than could be derived from softpart decay alone, Canfield and Raiswell (1991b) suggested that if sufficient amounts of organic matter decay to promote diffusion of carbonate into the surrounding environment, overall carbonate levels are sufficient to preserve both soft- and hardparts; but, once entire carcasses have decayed, carbonate levels may become lower so that only hardparts, at best, are preserved (see also Allison and Pye, 1994).

Concretions may exhibit complex histories that also incorporate phosphatization and silicification (e.g., Müller, 1985). Carpenter et al. (1988) recognized a complex history for concretions of marginal marine sediments of the Fox Hills Formation (Upper Cretaceous, North Dakota; see also Waage, 1964; Allison,

3.7 Concretions

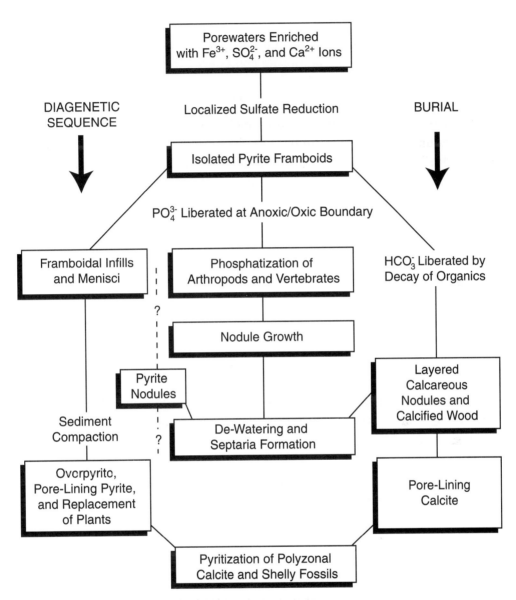

Figure 3.10. Diagenetic pathways within the London Clay. (Redrawn from Allison, 1988*b*.)

1988*b*). They concluded that oxidation (via sulfate reduction) of organic-rich sediments first released phosphate and reduced iron (as sulfides). Siderite precipitation followed in a mixed marine-phreatic zone, and then various stages of calcite precipitation related to sea-level rise and fall. Allison (1988*b*) also documented complex pathways of concretion formation in the London Clay (Eocene; Figure 3.10). As also demonstrated by Seilacher *et al.* (1976), Hudson (1982), and Carpenter *et al.* (1988), apatite (phosphate) was the first mineral to precipitate, followed by calcite and pyrite. Most fossils encountered in phosphatic concretions of the London Clay had an original phosphate component (e.g., teeth), thereby biasing the preservation of the biota (Allison, 1988*b*). Based on experiments with modern shrimp, however, the phosphate in phosphatic concretions may be derived, in part, from decaying carcasses (Briggs and Kear, 1994*b*; see also Briggs and Wilby, 1996), although the actual pathways of mineralization depended on the original composition of the skeletons (Figure 3.10). In contrast to shrimp (cf. Chapter 2), stomatopod cuticle is more robust and more heavily mineralized by calcium and phosphate and appears to be readily preserved based on laboratory experiments and fossil speciments (Hof and Briggs, 1997).

The formation of concretions in terrestrial environments is probably no less dependent on original hardpart composition and structure and depositional setting. Downing and Park (1998) discerned a number of stages in the preservation of large and small mammal bones of the Sucker Creek Formation (Miocene) of southeast Oregon. The chemical processes included partial dissolution of bone hydroxyapatite ($Ca_{10}(PO_4)_6(OH)_2$) via decay of soft tissues and groundwater, diffusion of ions into surrounding volcaniclastics, and later filling of bone voids by precipitates of calcite, quartz, hematite, and zeolites. Although some surficial bone was dissolved (producing a porous fabric), concretion formation was sufficiently fast to inhibit bone loss during later soil formation.

3.8 Soils

The formation of soils has received an enormous amount of attention for obvious agricultural and economic reasons. Unfortunately, it has often been assumed that fossil preservation in soils is *always* poor and so fossil-bearing soils have been neglected in favor of fluvial deposits, in which concentrations of fossils may be more obvious (Chapter 2). Moreover, taphonomic studies, especially those of vertebrates, have tended to emphasize the biostratinomic factors of transport, breakage, and abrasion, while overlooking the often subtle diagenetic changes that may occur in sediments (Hanson and Buikstra, 1987).

Soils differ fundamentally from most other fossiliferous strata because they form *in situ* and may more closely reflect living populations and regional and local climates and ecosystems than fossil concentrations formed by fluvial processes (e.g., Baas-Becking *et al.*, 1960; Bown and Kraus, 1981*b*). Productive, fossil-bearing horizons, including soils, commonly form in association with stratigraphic contacts that represent temporary land surfaces, and are often preserved in the

3.8 Soils

fossil record, especially in association with ancient floodplains (Behrensmeyer, 1982). Soils with exceptionally thick or well-differentiated profiles and horizons are also associated with major geological unconformities representing hiatuses of millions of years (Retallack, 1984), and despite the presumption of a strong diagenetic overprint, have been used to calculate CO_2 levels in ancient atmospheres (e.g., Mora *et al.*, 1996). Although fossils in soils are subject to some mixing with regard to habitat or age by such factors as bioturbation (trampling; Chapter 4), compaction, and predation and scavenging, they are much less affected than fossils in stream deposits (cf. Chapter 2), and many sedimentary features are often preserved (Retallack, 1984).

Fossil preservation in soils is partly a function of pH-Eh conditions (Figure 3.11). Although clay mineralogy of soils stems from source sediment and the duration of soil formation, *in general*, well-drained soils (*podsols*) are acidic (pH as low as 3.7) and have cation-poor clays such as kaolinite, whereas alkaline soils have cation-rich clays such as illite and smectite. Not surprisingly, soils of low pH become progressively depleted in $CaCO_3$ as they develop. Well-drained soils are usually oxidized (relatively high Eh), red or yellow in color, non-carbonaceous, and contain oxidized minerals such as limonite, goethite and hematite. Permanently waterlogged soils are reduced (low to negative Eh), drab-colored (bluish to greenish gray), carbonaceous, and usually contain *gley* minerals such as siderite and pyrite; pH is usually higher than 5 because lime persists indefinitely. At the other extreme are alkaline soils, which characterize arid regions. Unlike acid soils, in which the direction of groundwater movement is downward, groundwater moves upward in alkaline soils because of evaporation (Baas-Becking *et al.*, 1960). Evaporation of groundwater causes the precipitation of calcium and sodium salts (e.g., zeolites, gypsum and other evaporites), which are responsible for high pH and which may form relatively rapidly (i.e., between intervals of significant fluvial discharge; e.g., Steel, 1974, for New Red Sandstone of Scotland).

As bones are incorporated into a soil, they are subject to decay, dissolution, and sometimes encrustation, which are functions of sediment accumulation rate, soil chemistry, and bone mineralogy, microstructure, and age (Figure 3.11). Rapid sediment accumulation inhibits paleosol development but also favors low bone densities, whereas slow accumulation favors higher bone densities but also soil development and potentially bone dissolution (Asland and Behrensmeyer, 1996). Bone consists primarily of hydroxyapatite ($Ca_{10}(PO_4)_6(OH)_2$) but its alteration may begin quite rapidly, and it is best-preserved under alkaline conditions. Retallack (1984, 1988), for example, found that the most consistently fossiliferous paleosols in his study were those of the Conata Series (Late Eocene–Late

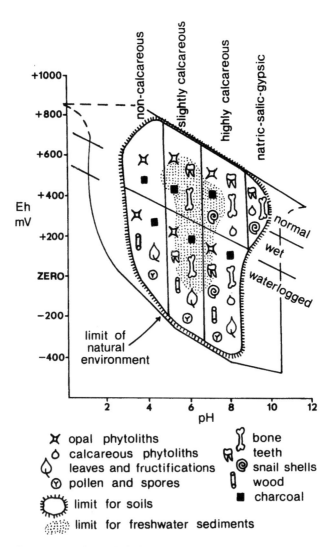

Figure 3.11. Theoretical Eh-pH stability fields for preservation of common types of terrestrial fossils in soils. (From Retallack, 1984, based on Baas-Becking et al., 1960; reprinted with permission of the editors of *Paleobiology*.)

Oligocene, Badlands National Park, South Dakota, U.S.A.), which consisted of alkaline soils of interstream savanna (see also Behrensmeyer, 1978; Behrensmeyer and Boaz, 1980); by contrast, the Interior and Yellow Mounds Series paleosols were unfossiliferous and represent well-drained, acidic, forested and wooded soils. Crystallization of salts on the lower surfaces of bones embedded in alkaline soils may accelerate bone weathering via flaking and splintering, whereas the upper exposed surfaces of the same bone may be much less weathered (Behrensmeyer, 1978); calcium carbonate may also precipitate on fossil remains in the vicinity

3.8 Soils

of karst topography (Donovan and Veltkamp, 1994). Cancellous (spongy) bone is much less dense than compact bone, and its pores provide increased surface area for physical and biological dissolution by microbes, and permineralization and petrifaction (Hanson and Buikstra, 1987; Piepenbrink, 1989). Gordon and Buikstra (1981) found significant negative correlations between adult and child bone preservation and soil pH ($r = -0.92$; $P < 0.00001$; $n = 63$ and $r = -0.48$; $P < 0.005$; $n = 32$, respectively). Soil pH explained explained 84% of the variance (r^2) in adult bones, but only 23% of the variance in preservation of children's bones, and is probably related to the greater variation in bone densities of children measured (0–14.99 years). A similar relation has been noted for mammals and dinosaurs (Behrensmeyer, 1981; Carpenter, 1982).

Rates of bone degradation can also vary dramatically within short distances. Watson (1967), for example, found that bones from near termite mounds (soil pH 4.2–5.4) exhibited obvious degradation after only 15–20 years, but bones from within mounds (pH 7.8–7.9) showed no obvious degradation, although they were 700 years old.

By contrast, teeth are made of dentine and enamel, and their weathering characteristics appear unrelated to those of bone (e.g., Behrensmeyer, 1978). Teeth may be present in acidic soils when bones are absent, and their resistance may be a function of greater density and lower permeability (Shipman, 1981). Although teeth may split when subjected to desiccation (Toots, 1965; Behrensmeyer, 1978), the lack of an unambiguous pattern of weathering suggests that the individual characteristics of each tooth (including stage of eruption, wear, ratio of enamel to dentine, overall morphology, microenvironment at or near the soil surface) are important controls on their rate of weathering (Behrensmeyer, 1978, 1981).

The diagenesis of bone and teeth is an important consideration in determining ancient dietary preferences. Different foods have different levels of Sr and Ba, for example, which may substitute for Ca in hydroxyapatite. Although Parker and Toots (1980), among other early workers, concluded that strontium is one of the few elements not affected by diagensis of bone, subsequent studies demonstrated that breakdown of the organic (proteinaceous) matrix of bone (mainly collagen) via hydrolysis is non-linear (Hare, 1974, 1980; Hanson and Buikstra, 1987), and that changes in the phosphate matrix can affect Sr levels in bone directly or indirectly through enrichment/depletion of other elements such as Ca and P (e.g., Schwarcz et al., 1989; see Price et al., 1992, for review; Whitmer et al., 1989, provide extensive tables for important elements, their sources of variation in bone, and inferences that may be drawn from them; cf. Boyle et al., 1995).

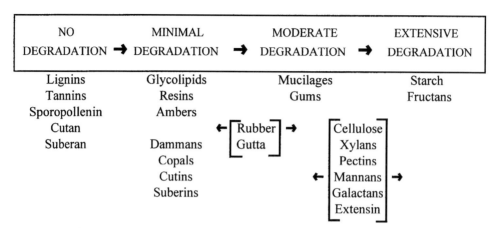

Figure 3.12. Preservation of vascular plant biomacromolecules according to general composition. (Redrawn from Gastaldo, 1992a, after Tegelaar et al., 1989.)

Land snails have also proven useful in reconstructing terrestrial climate and local habitats, and were used as evidence in early support of punctuated equilibrium (Eldredge and Gould, 1972). Most land snails consist of aragonite, although the internal shells of *Arion* and limacid slugs are composed of calcite (Evans, 1972). Not surprisingly, land snails are rapidly dissolved in acidic or neutral non-calcareous soils (especially after the protective periostracum has been destroyed), but are more common in alkaline ones, especially *rendsinas*, which commonly form on chalk and have a high $CaCO_3$ content and high pH (7.5–8.0; Evans, 1972). Nevertheless, species vary with respect to dissolution resistance, especially those with more resistant shell apices (e.g., *Pomatius*, *Cepaea*, *Arianta*, among others; Evans, 1972). Differential preservation can *usually* be detected, however, by comparing snail occurrences in unweathered parent material versus those in mature soils (Evans, 1972, pp. 212–223). Other biogenic calcareous particles (*phytoliths*), such as the endocarps of the hackberry, *Celtis occidentalis*, may also be preserved, even in oxidizied calcareous soils, because they consist of 25–64% $CaCO_3$ by dry weight.

Most other plant remains are, however, broken down to amorphous organic matter or completely destroyed, primarily through the activities of bacteria and fungi (Retallack, 1984). The preservation potential depends on the exact chemical compounds present (Figure 3.12), which may differ in their preservation potential according to the degree of molecular cross-linkage (Briggs, 1995; see also Chapter 6), and which are taxon-specific but which seem to exhibit no systematic occurrence among taxa (Gastaldo, 1992a; see also Tegelaar et al., 1989, 1991): for example, cuticles of *Beta vulgaris* and *Salicornia europaea*, both of which belong to the same superfamily (Chenopodiaceae), are composed of cutan and cutin, respectively (Figure 3.12).

Havinga (1971) conducted field and laboratory experiments and found that pollen perforation was the main cause of decay in "biologically very active" soils, river clay (pH 7.2; 3.6% $CaCO_3$; Havinga, 1967) and leaf mold in a greenhouse, but was less pronounced in horizon A of a podsol (pH 4.6); decay was least pronounced in *Sphagnum* peats (see Chapter 6). This is in general agreement with the results of Elsik (1971), who also found definite degradation scars or patterns in modern and fossil pollen that he attributed to microbial activity. The degree of degradation varied, however, within and between species (Havinga, 1971; Elsik, 1971; see Keafer *et al.*, 1992, and Zonneveld *et al.*, 1997, for similar results for dinoflagellate cysts under variable oxygen conditions). Although pollen may be present in highly alkaline (pH > 9) soils because of the inhibition of microbial activity (Potter and Rowley, 1960), pollen may also be absent from less alkaline soils (pH 6.5–8.1; Dimbleby, 1957). Hall (1981) recognized four stages of pollen preservation ("taphonomic grades"); degraded pollen grains were most common in peats and silts, but low numbers of pollen grains may also reflect dilution by sediment.

All of the above biogenic particles may be preserved in coprolites (feces), which may in turn be petrified by calcite, silica, limonite or siderite (Amstutz, 1958; Retallack, 1984). Organic-rich coprolites are produced mainly by herbivores and omnivores, and are preserved in settings in which bacterial decay is inhibited (e.g., deserts; Retallack, 1984; see Chapter 6 for permafrost, and peat bogs and other waterlogged soils). Bradley (1946), for example, found desmid (freshwater diatom)-enriched coprolites in swamp water. Moreover, because silica dissolves in waters of basic pH, diatoms can be very sensitive indicators of acidity (e.g., Birks *et al.*, 1990). Testate amoebae and ostracodes are also relatively resistant to destruction in low pH conditions (e.g., Patterson and Kumar, in press).

Coprolites may also give important clues to diet in vertebrates. Coprolites of birds and carnivores may contain large amounts of bone, although it is unlikely that they are responsible for many microvertebrate accumulations (cf. Mellett, 1974, and Bown and Kraus, 1981*b*). Amstutz (1958) briefly discussed criteria for recognizing mammalian coprolites.

Despite their appearance, soils are not static entities; they *evolve* and the preservational state of fossil assemblages changes accordingly. Soil features develop rapidly at the outset and then slow as they approach steady-state conditions (Retallack, 1984). Given the frequency and intensity of environmental disturbance such as floods (which vary according to power laws; Bak, 1996), it is unlikely that any setting will remain constant for long.

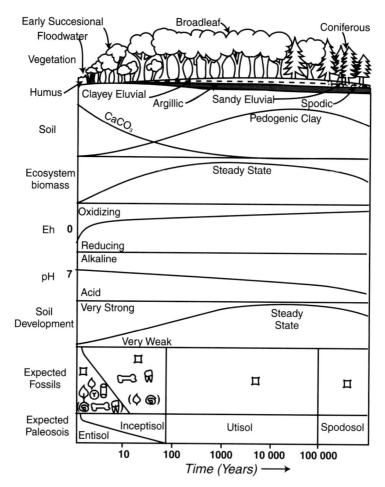

Figure 3.13. Hypothetical model illustrating physical and chemical factors during the development of soils. (Reprinted from Retallack, 1984, with permission of the editors of *Paleobiology*.)

Based on studies of Late Eocene–Oligocene soils in southwestern South Dakota, Retallack (1984) presented a model for the evolution of soils and preservation of fossils in humid temperate floodplains. He assumed a million years or more of undisturbed vegetative growth and soil development on flat-lying, moderately calcareous, alluvial, sandy siltstone in a humid climate (Figure 3.13). At the outset, conditions are briefly reducing and mildly alkaline within the moderately calcareous alluvium deposited by floodwaters. As floodwaters retreat, oxidizing conditions, which are caused by bioturbation of early successional plants and animals, set in and an *entisol* begins to form. This soil may be sufficiently calcareous to preserve snails, bones, and teeth given large enough numbers of fossils, but vegetation, such as leaves and pollen that had been deposited by floodwaters, is destroyed

3.8 Soils

(see also Davies-Vollum and Wing, 1998). As vegetational succession continues (invasion of trees and shrubs), the soils become increasingly acidic and $CaCO_3$, Mg^{2+}, Na^+, and K^+ are progressively leached from the soils to form *inceptisols*. Remaining plant fossils are typically destroyed and land snails dissolve, leaving only bones and teeth, which, if acidification proceeds far enough, also start to disappear. Given roughly a few hundred to a few thousand years, easily weathered minerals such as feldspars and micas in alluvium are weathered to clays; if clays are leached out of the upper (A) horizon and accumulate in a deeper clayey B horizon, an acidic oxidized *ultisol* results in which only siliceous skeletons (e.g., diatoms) may remain. Although this stage is *apparently* steady state (climax), further leaching of nutrients may proceed to the point where community biomass begins to decline and stunted coniferous forests become established on oxidized, acidic, nutrient-poor, quartz-rich sandy *spodosols*.

This model explains a 700-m-thick sequence of variegated floodplain and channel mudstones and sandstones in the Willwood Formation (Early Eocene, Bighorn Basin, northwest Wyoming, U.S.A.), nearly all of which display "paleopedogenic modification" such as gleying and rootlet horizons, and in which occur silicified wood, fossil hackberry fruits, and terrestrial and freshwater gastropods (Bown and Kraus, 1981a,b). Bown and Beard (1990) found that the least mature soils were closest to channels or ancient channel belts, whereas more mature soils were located farther away; less complete skeletons were more prevalent at sites of mature soils, where sediment accumulation rates were low. The mudstones are frequently tabular (2 cm–1 m) and laterally persistent (up to 12 km^2 in outcrop area, but undoubtedly larger); grade into thin, lenticular, unconnected channel sandstones; and appear to have formed during times of rapid sediment accumulation in a warm temperate to subtropical climate under alternating wet and dry conditions or fluctuating water tables, with the paleosols forming during geologically brief intervals between floods (Bown and Kraus, 1981a,b). All of the paleosols appear to represent podzolic spodosols and entisols, although some red, yellow, and purple units resembled ultisols. The spodosols frequently overlay the apparent ultisols, and appear to represent A horizons that have been overprinted by A and B horizons of younger paleosols (see also Fastovsky and McSweeney, 1991). Upsection, a decrease in gleying, an increase in number and thickness of red versus orange coloration, and an increase in abundance of calcareous aggregates, all strongly suggest better-drained soils and drier climate related to tectonism and creation of rain shadows in the Bighorn Basin. The spodosols (A horizons) themselves are associated with concentrations of disarticulated and broken vertebrate remains (Bown and Kraus, 1981b). The occurrence of vertebrates in discrete, identifiable, and widespread paleosols makes these assemblages quite valuable biostratigraphic units (zonules, faunules; Bown and Kraus, 1981*b*). Similar features have been noted in the Siwalik Series (India), Triassic Chinle Formation of Arizona, and the Late Jurassic–Early Cretaceous Morrison Formation of the Rocky Mountains (Dodson *et al.*, 1980; Bown and Kraus, 1981*a*).

The fossiliferous units accumulated gradually as litter on soil surfaces. Vertebrates represented include mainly mammals, but fish, amphibians, crocodiles and turtles (especially in association with stream channels), lacertilians, and birds (e.g., the large, flightless *Diatryma*) also occur. Of the mammals studied, no skulls, skeletons, or other articulated remains were found. Teeth are nearly 30 times more abundant (65%) than any other hardpart type, followed by jaw fragments (22%), phalanges (a distant third at 3.1%), vertebral fragments (2.2%), carpal and tarsal bones (1.2%), calcanea (0.9%), and astragali (0.7%). These elements are less damaged and were probably left behind by predators and scavengers in favor of more damaged meatier bones (e.g., femora, tibiae, humeri, radii, and ulnae). Cracked and broken teeth, bones gnawed by small mammals, and coprolites are "not uncommon" and also suggest that accumulation was associated with disarticulation and mixing by predators and scavengers; a few unbroken teeth had lost all of their enamel and may have passed through the guts of crocodiles. Most bones, however, exhibit weathering stages 0–1 (Behrensmeyer, 1978). The percentages of fossil bones is also inconsistent with fluvial transport (e.g., Voorhies, 1969; see also Dodson, 1973, and Chapter 2). Early diagenetic permineralization of bones and teeth (including gypsum and limonite) appears to have been equally common in different environments, with no obvious evidence of preferential preservation.

4 Bioturbation

Truth is the daughter of time. Aulus Gellius (AD 130–175)

4.1 Introduction

As discussed in Chapter 3, bioturbation – the mixing of sediment by organisms – has a tremendous impact on porewater chemistry and preservation. Moreover, any signal that is incorporated into the stratigraphic record, whether it be seasonal shell inputs or instantaneous volcanic ash layers, must pass through the low-pass filter of bioturbation, in which high-frequency events are damped or removed and lower-frequency events preserved. Only in rare cases do organisms impede bioturbation, such as the extensive root systems of marsh plants (e.g., Nydick *et al.*, 1995), but even in this instance, plant roots no doubt pump oxygen into the subsurface and alter porewater chemistry.

On one hand, bioturbation may be viewed favorably because it erases high-frequency "noise" and leaves behind evidence of longer-term patterns and the processes that generated them. On the other, bioturbation is a major impediment to bridging the gap between ecological (short-term) and geological or evolutionary (long-term) processes. Although methods have recently been developed for the classification and semi-quantitative estimation of the extent of bioturbation in both cross-section and on bedding planes (e.g., Droser and Bottjer, 1986; Miller and Smail, 1997; see also Bertness and Miller, 1984) and in estimating the population sizes of bioturbators of ancient sediments (Kowalewki and Demko, 1997), most earth scientists have avoided study of this process because unraveling its effects is quite complex mathematically and because it involves the integration of diverse disciplines. But if one wishes to extract high-resolution data in the ancient record – such as the type that might be useful in reconstructing pre-anthropogenic environments – this process must be confronted.

4.2 Bioturbation in terrestrial environments

Virtually all stratigraphic signals are subject to bioturbation, including those on land, the reports of which are largely anecdotal (see Graham, 1993, for review). Darwin cites numerous examples of the mixing and recycling of organic matter by earthworms (composited from Darwin, 1896, pp. 146–147):

> *Farmers in England are well aware that objects of all kinds, left on the surface of pasture-land, after a time disappear ... or work themselves downwards ... The Rev. H. C. Key had a ditch cut in a field, over which coal-ashes had been spread ... eighteen years before; and on the clean-cut perpendicular sides of the ditch, at a depth of at least seven inches, there could be seen, for a length of 60 yards, 'a distinct, very even, narrow line of coal-ashes, mixed with small coal, perfectly parallel with the top-sward' ... Secondly, Mr. Dancer states that crushed bones had been thickly strewed over a field; and some years afterwards these were found several inches below the surface, at a uniform depth ...*

Insects, such as harvester ants, have also been shown to concentrate vertebrate remains (e.g., Shipman and Walker, 1980), and similar activities also transport terrestrial gastropod shells downward (Evans, 1972). Growth and decay of plant roots and uprooting of trees may also cause mixing. Vertebrate-induced effects are most pronounced on small objects in loose sediments, and include "dinoturbation" by dinosaurs (Lockley and Conrad, 1989); trampling and subvertical–vertical reorientation of long axes of bones in swamps and lake beds or around watering holes by elephants and other mammals (e.g., Behrensmeyer and Boaz, 1980); and digging and burrowing in soils, which may be extensive (e.g., prairie dog "towns," which may extend meters deep and laterally for *miles*; Graham, 1993). Some studies have found terrestrial dispersion on the order of only centimeters, but it has been estimated that rodents may turn over 15–20% of surface soil in a single season, which results in complete mixing in 5–6 years. Physical agencies include settling of bones into desiccation cracks or karst topography ("graviturbation"), subsurface flow in subsurface springs and tar pits, liquefaction by tectonic activity, and freeze–thaw cycles or "cryoturbation" (Graham, 1993).

4.3 Diffusion models of bioturbation

4.3.1 The random walk

Mathematical models of bioturbation have been developed for marine and freshwater systems. Among the most popular bioturbation models are diffusion models, which are drawn from mathematical descriptions of thermal and chemical diffusion (see Matisoff, 1982; Cutler, 1993; Martin, 1993, for reviews). Sediment

4.3 Diffusion models of bioturbation

diffusion models assume that sediment mixing can be described mathematically as a random diffusion process: an eddy (particle) biodiffusion coefficient accounts for the redistribution of sediment particles by large numbers of organisms over a large number of individual transport events. Taken collectively, small transport events may move particles over much larger distances.

In one of the most frequently cited diffusion models, Guinasso and Schink (1975) modeled concentration profiles of microtektites in deep-sea sediments using the equation

$$D\frac{\partial^2 c}{\partial x^2} - \nu\frac{\partial c}{\partial x} - \lambda c = 0 \qquad (4.1)$$

where c is the concentration of the tracer (cm^{-3}), D is the eddy biodiffusion coefficient (cm^2 kyr^{-1}), ν is the sedimentation rate (cm kyr^{-1}), λ is the radioactive decay coefficient (kyr^{-1}) in the case of radioactive tracers, and x is the depth (increasing downward in cm). This equation states that the rate of change of tracer activity at depth x owing to mixing, minus burial of tracer and tracer decay, equals 0; in other words, $\partial c/\partial t = 0$, and the system is assumed to be in *steady state*. (In the case of conservative – non-decaying – tracers that produce *event layers*, such as volcanic ash or microtektites, the term λ of equation 4.1 may be ignored). The assumption of steady-state conditions is normally justified because most physicochemical fluctuations have a mean period much shorter than those of diffusion, advection, and decay, so that variation is time-averaged (see also Chapter 5). *Pseudo-steady-state conditions* may, however, hold in some cases: (1) if the time between fluctuations is much longer than the time scales of the other processes, the disturbance decays before the next perturbation; or (2) if the forcing is continuous over long time scales, the system can be described by a series of steady states (Boudreau, 1997). The equation also describes tracer movement and loss in only one dimension (vertical), but this description is normally justified because physicochemical gradients are typically strongly depth-related (Berner, 1980; Boudreau, 1997). Aller (1982*a*), Berg (1993) and Boudreau (1997) describe multi-dimensional models.

Based on trace fossils, the mixed layer would be expected to thin with water depth (on the order of approximately 5–10 cm in the deep sea); i.e., the depth of burrowing should become increasingly restricted to near the SWI (Figure 4.1). Ichnofacies models across continental margins presumably reflect wave energy, oxygenation, sedimentation rate, and especially food (organic carbon) availability (e.g., Trauth *et al.*, 1997; Wetzel and Uchman, 1998), all of which tend to decline away from shore (Middelburg *et al.*, 1997) and would be expected to influence

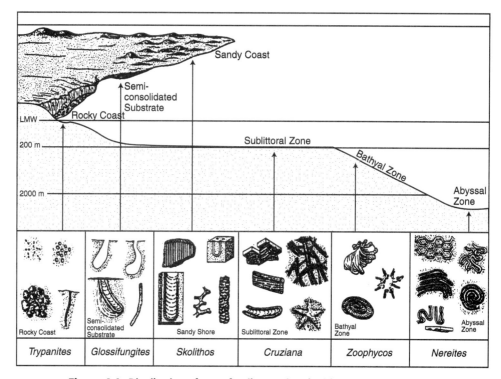

Figure 4.1. Distribution of trace fossils associated with a transect across the continental margin. Vertical traces dominate near shore, whereas more complex, horizontal traces characterize deep water. (Reprinted from Ekdale et al., 1984, with permission of SEPM.)

mixed layer thickness. Nevertheless, there is no correlation between published mixed layer thicknesses ($n = 203$) and sedimentation rate: although reported mixed layer thicknesses range up to about 1 m in shallow-water settings (e.g., Pemberton et al., 1976; Frey et al., 1978; Meldahl, 1987; Walter and Burton, 1990; Martin et al., 1995, 1996), the mean mixed layer thickness is 9.81 ± 1 cm over a wide range of environments (Boudreau, 1997; some of the data scatter may result from analytical artifacts; see review in Boudreau, 1997). The depth of bioturbation instead appears to be limited primarily by the physical difficulty and high energy costs of reworking sediment deeper than 10–15 cm (Jumars and Wheatcroft, 1989; Boudreau, 1997).

Biodiffusion coefficients (D) vary by about six orders of magnitude and tend to decrease from shallow water ($D \approx 10^{-6}$ cm^2/s) to the deep sea ($D \approx 10^{-8}$ cm^2/s; Matisoff, 1982); i.e., they tend to decrease with decreasing sedimentation rate ($r = 0.47$; $\alpha < 0.01$; $n = 203$; Boudreau, 1997; Middelburg et al., 1997). Bioturbation can obviously have a profound impact on temporal resolution over short time scales (decadal–centuries; Nittrouer and Sternberg, 1981). Myers (1977),

4.3 Diffusion models of bioturbation

for example, estimated that the turnover time of lagoonal sediment ranged between 2 days and 2 years with a mixed layer thickness of 1–10 cm. If carried to completion, bioturbation first mottles laminated sediment and ultimately homogenizes it (e.g., Moore and Scruton, 1957; see also Droser and Bottjer, 1986).

When viewed over sufficiently long time scales (more than three times the turnover time of sediment in the mixed layer; Aller and Dodge, 1974), the process of bioturbation can presumably be analogized to that of *Brownian movement*, also known as the *random walk model* (Berg, 1993). At absolute temperature T, a particle exhibiting Brownian movement has an average kinetic energy of $kT/2$ (k = Boltzmann's constant), irrespective of particle size (up to and including those seen under a microscope). This kinetic energy may be expressed as $\langle Mv^2/2 \rangle$, where M is mass, v is velocity, and $\langle\ \rangle$ denotes an average over time or the average of a group ("ensemble") of similar particles. From this relationship, the mean-square velocity and root-mean-square velocity, respectively, can be calculated:

$$\langle v^2 \rangle = \frac{kT}{M} \tag{4.2}$$

and

$$\langle v^2 \rangle^{1/2} = \left(\frac{kT}{M}\right)^{1/2} \tag{4.3}$$

Using these equations, Berg (1993) calculated a mean-square velocity of a small protein particle (lysozyme) of 1.3×10^3 cm s^{-1} ($\sim$47 km h^{-1}). In other words, an unimpeded molecule of lysozyme would traverse a room in just a few seconds. Other particles impede the progress of this molecule, however, and so its velocity is much slower. In fact, the lysozyme molecule encounters other particles so frequently that it undergoes a random walk (diffusion) unless an external force (*advection* or "drift," or in a stratigraphic context, burial) is applied.

For the development of equation 4.1, assume the following conditions (Berg, 1993): (1) an ensemble (N) of particles starts movement at time $t = 0$ and position $x = 0$ and moves along an axis (x-axis); (2) each particle moves to the right or left once every τ seconds at velocity $\pm v$ (Figure 4.2), and therefore each particle moves a distance (*step length*) $\delta = \pm v\tau$; (3) each time a particle moves, it has a 50 50 chance of moving to the right or left, and each step of a particle is statistically independent of the previous and subsequent steps (random walk model); (4) each particle does not interact with other particles (this is not true, but it simplifies the model greatly).

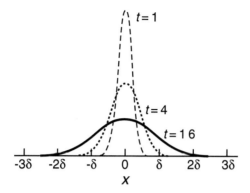

Figure 4.2. Particles undergoing a one-dimensional random walk starting at the origin ($x = 0$) and time $t = 0$, and moving in steps of length δ. Area under the curve represents the probability of finding a particle at different points along the x-axis at times $t = 1$, 4, and 16. The standard deviations (root-mean-square widths) of the distributions increase with $\sqrt{t}$, whereas peak heights decrease with $\sqrt{t}$. Compare with Figure 4.6. (Redrawn from Berg, 1993.)

Let $x_i(n)$ be the position along the x-axis of the ith particle after the nth step. The position of the particle after the nth step differs from its position at the $(n - 1)$ step by $\pm \delta$, or

$$x_i(n) = x_i(n-1) \pm \delta \tag{4.4}$$

The average movement of the ensemble is then

$$\langle x(n) \rangle = \frac{1}{N} \sum_{i=1}^{N} x_i(n) \tag{4.5}$$

Based on equation 4.4, equation 4.5 can be expressed as

$$\langle x(n) \rangle = \frac{1}{N} \sum_{i=1}^{N} [x_i(n-1) \pm \delta] \tag{4.6}$$

(The term δ in equation 4.6 is ~0, since each particle has a 50–50 chance of moving to the right or left at each step.) Therefore,

$$\langle x(n) \rangle = \frac{1}{N} \sum_{i=1}^{N} x_i(n-1) \tag{4.7}$$

which is the average particle displacement at step $n - 1$ and is expressed as $\langle x(n-1) \rangle$. Based on this set of assumptions, on average, the particles move very little. Nevertheless, the particles in the model eventually spread away from the point of origin ($x = 0$; Figure 4.2). The amount of spreading can be determined

4.3 Diffusion models of bioturbation

using the mean-square of the average particle displacement. The square of the displacement is

$$x_i^2(n) = x_i^2(n-1) \pm 2\delta x_i(n-1) + \delta^2 \qquad (4.8)$$

which is the square of the right-hand side of equation 4.4 using the quadratic equation, $a^2 + 2ab + b^2$. The *mean*-square displacement is then (cf. equation 4.6)

$$\langle x^2(n) \rangle = \frac{1}{N} \sum_{i=1}^{N} x_i^2(n-1) \pm 2\delta x_i(n-1) + \delta^2 \qquad (4.9)$$

Since $\delta \approx 0$, equation 4.9 simplifies to

$$\langle x^2(n) \rangle = \frac{1}{N} \sum_{i=1}^{N} x_i^2(n) \qquad (4.10)$$

which is the same as $\langle x_i^2(n-1) \rangle + \delta^2$ (cf. derivation of equations 4.4–4.7). Based on equation 4.10, since $x_i(0) = 0$ for $n = 1, 2, 3, \ldots$, then $\langle x^2(0) \rangle = 0$, $\langle x^2(1) \rangle = \delta^2$, $\langle x^2(2) \rangle = 2\delta^2$, $\langle x^2(3) \rangle = 3\delta^2, \ldots, \langle x^2(n) \rangle = n\delta^2$. Therefore, average particle dispersion (mean-square displacement) increases with step number (n), and mean-square displacement increases with time, since $t = n\tau$. The actual amount of spreading (the *root*-mean-square of the displacement) is thus proportional to $\sqrt{t}$. In other words, particle spreading increases with the square root of time, not time itself (Figure 4.2).

The meaning of the diffusion coefficient (D) can now be explained. Since $n = t/\tau$, $x(n) = x(t/\tau)$, and

$$\langle x^2(t) \rangle = \left(\frac{t}{\tau}\right) \delta^2 = \left(\frac{\delta^2}{\tau}\right) t \qquad (4.11)$$

(cf. preceding paragraph). Let $D = \delta^2/2\tau$; then

$$\langle x(t)^2 \rangle = (2Dt) \qquad (4.12)$$

and

$$\langle x(t)^2 \rangle^{1/2} = (2Dt)^{1/2} \qquad (4.13)$$

Thus, because the distance a particle moves is proportional to $\sqrt{t}$, in order for a particle to move twice as far, it takes four times as long, and so on. Since particles usually move only short distances, they tend to return to the same point many times before finally leaving an area (Figure 4.3). Since particles wander to new

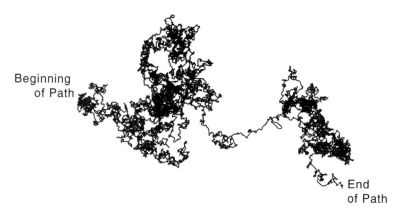

Figure 4.3. Pathway of a single particle moving in two dimensions (within the plane of the page) after 18 050 steps. Note that the particle tended to move within certain regions. Distance between beginning and end of the path represents 196 step lengths. (Modified from Berg, 1993.)

areas at random (statistically independent of previous positions), they tend to "explore" a new area fully before moving to another region and the track of the particle does not necessarily fill space uniformly. Imagine looking down on Figure 4.3 as if it were the sediment–water interface or from the side, as in a column of sediment. Then imagine coring in search of a tracer (such as a volcanic ash) whose path is indicated by the figure (e.g., Risk *et al.*, 1978). Bear in mind that for the sake of simplicity, only a one-dimensional random walk has been presented.

As hinted by Figure 4.3, particles move at different rates depending upon the time scale of observation (see Berg, 1993, pp. 10–11). The shorter the period of observation (t), the greater the particle velocity. The velocity of particles moving within a restricted area (*local mixing*) is actually much higher than the velocity of particles moving over larger distances (*non-local mixing*) because it takes much longer, on average, for a particle to move a long distance than it does a short one, even though the processes involved may be the same. Thus, the burrowing activities of animals may occur at *multiple scales* (they are patchy), even in the same species. Johnston (1995) found, for example, that pocket gophers create burrow entrance mounds at spatial scales of decimeters and temporal scales of years, and larger patches (*mima mounds*) at temporal scales of decades to centuries (approximately circular soil lenses up to about 2 m high and 25–50 m in diameter that occur at densities of 50 to more than 100 ha^{-1}). Certain shallow-water shrimps (e.g., *Callianassa*, *Upogebia*) also build mounds and subsurface lodgings, but on a much smaller scale.

4.3.2 Derivation of Fick's equations

Having derived the diffusion coefficient (D), the rest of equation 4.1 can be derived. Start with a known number of particles at a fixed point along the x-axis at time t. After the first step (τ seconds), a certain number (N) of particles will have moved to the right to point $x + \delta$. The net *flux* of particles across a surface located perpendicular to the x-axis after time τ may now be calculated. Since the particles have a 50–50 chance of moving either right or left after each step (τ), for the time τ, on average, half the particles will have moved to the right (to point $x + \delta$), and half from point $x + \delta$ to the left to point x. The net number of particles moving to the right is

$$-\tfrac{1}{2}[N(x+\delta) - N(x)] \tag{4.14}$$

To calculate the net flux (J) of particles to the right, equation 4.14 is divided by area (A) of the surface and time (τ)

$$J = -\tfrac{1}{2}[N(x+\delta) - N(x)]/A\tau \tag{4.15}$$

Multiplying by 1 (in the form of δ^2/δ^2) and rearranging terms, equation 4.15 can be rewritten as

$$J = -\left(\frac{\delta^2}{2\tau}\right)\left(\frac{1}{\delta}\right)\left[\frac{N(X+\delta)}{A\delta} - \frac{N(x)}{A\delta}\right] \tag{4.16}$$

The term $\delta^2/2\tau$ is the diffusion coefficient (D) and the two terms in the brackets are the concentrations (c) at points $x + \delta$ and x, respectively (the 2 in the denominator is present only to make the derivation of equation 4.16 easier). Equation 4.16 may be simplified to

$$J = -D\left(\frac{1}{\delta}\right)[c(x+\delta) - c(x)] \tag{4.17}$$

Since δ is very small, as $\delta \to 0$ (in the limit), equation 4.17 simplifies to

$$J = -D\left(\frac{\partial c}{\partial x}\right) \tag{4.18}$$

which is called *Fick's first equation*, after A. Fick, who, in the mid-nineteenth century, modeled diffusion by using the equations of heat conduction earlier developed by the French mathematician J. B. Fourier. Equation 4.18 is called a partial derivative (indicated by the use of ∂ rather than d) because the flux actually depends on two independent variables (position [x] and time [t]), rather than only one; i.e., J is a function of both x and t [$J(x,t)$] rather than of x or t alone. In the

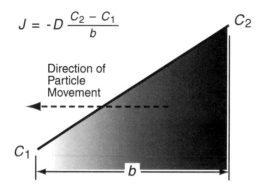

Figure 4.4. Flux of particles resulting from a concentration gradient. Particles move from right to left because there are more particles on the right than on the left. The flux J (total number of particles that move across distance b per unit time) is equal to the concentration gradient (slope of the line = rise/run = $[C_2 - C_1]/b$) times the diffusion coefficient (D) or Fick's first equation (see text for explanation). (Redrawn from Berg, 1993.)

case of a partial derivative, solutions are determined by holding all of the independent variables constant except one (see below).

Fick's first equation is a mathematical statement that the net flux (at x and t) is proportional to the slope of the concentration (at x and t), and is analogous to velocity (the rate of change of position of a particle with respect to time); $-D$ is a constant of proportionality. If particles are uniformly distributed, then the slope ($\partial c/\partial t$) and J both equal 0. If $J = 0$, then the system is at equilibrium and the particle distribution does not change with time (cf. equation 4.1). If $\partial c/\partial t$ is a non-zero constant, then J is also constant (Figure 4.4).

Fick's second equation is derived from the first (Figure 4.5). After a period of time τ, $J(x)A\tau$ particles will have entered the box (volume = Area $\times$ δ) from the left and $J(x+\delta)A\tau$ will have exited from the right. If the total number of particles in the system remains constant, then the concentration (c) of particles in the box is given by

$$\frac{1}{\tau}[c(t+\tau) - c(t)] = -\frac{1}{\tau}\frac{[J(x+\delta)A\tau - J(x)A\tau]}{A\delta} \tag{4.19A}$$

or

$$= -\frac{1}{\tau}\frac{[J(x+\delta) - J(x)]A\tau}{A\delta} \tag{4.19B}$$

which simplifies to

$$-\frac{1}{d}[J(x+\delta) - J(x)] \tag{4.20}$$

4.3 Diffusion models of bioturbation

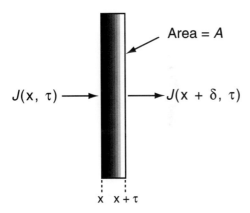

Figure 4.5. Flux (J) through a box of thickness τ used to derive Fick's second equation (see text). Area of the face perpendicular to the movement of particles = A. (Redrawn from Berg, 1993.)

As $\tau \to 0$ and $\delta \to 0$, then

$$\frac{\partial c}{\partial t} = -\frac{\partial J}{\partial x} = -\frac{\partial}{\partial x}(J) \qquad (4.21)$$

Since $J = -D\,\partial c/\partial x$ (equation 4.18),

$$\frac{\partial c}{\partial t} = -\frac{\partial}{\partial x}\left(-D\frac{\partial c}{\partial x}\right) = D\left(\frac{\partial^2 c}{\partial x^2}\right) \qquad (4.22)$$

This is Fick's second equation, which is the first term of equation 4.1. It states that the rate of change of concentration (at x and t) is equal to the rate of change of the slope (the *curvature*) of the concentration function at x and t (in other words, how much the slope is changing at each instant); again, D is a proportionality constant. This equation is analogous to that of acceleration (the rate of change of velocity with respect to time, or the second derivative of the change in particle position with respect to time). If the slope of the curve at some instant is constant, then $\partial^2 c/\partial x = 0$, $J = 0$, and the system is at equilibrium (the particle distribution does not change): just as many particles diffuse from the region of higher concentration into the box as out of the box to the region of lower concentration. Fick's second equation determines how an initially non-uniform distribution of particles redistributes itself through time.

Derivation of the second term of equation 4.1 is much shorter. If all the particles in a particular distribution move (drift) in a positive direction with (burial) velocity ν, then the flux of particles at point x must increase by $\nu c(x)$, and Fick's first equation becomes

$$J = -D\left(\frac{\partial c}{\partial x}\right) + \nu c \qquad (4.23)$$

This equation is used to rederive Fick's second equation

$$\frac{\partial c}{\partial t} = D\left(\frac{\partial^2 c}{\partial x^2}\right) - \nu\left(\frac{\partial c}{\partial x}\right) \qquad (4.24)$$

The Guinasso–Schink equation (4.1) is completed by adding the term λc for radiotracer decay.

4.3.3 Solution of the Guinasso–Schink equation

The solution of equation 4.1 is

$$C = C_0 \exp\left[\left(\frac{\nu - (\nu^2 + 4\lambda D)^{1/2}}{2D}\right)x\right] \qquad (4.25)$$

where C_0 is the original tracer concentration (Berner, 1971, p. 102). There are at least two ways to derive equation 4.25: one involves the determinants and eigenvalues of a matrix of coefficients for two equations solved simultaneously (see Causton, 1987, pp. 269–276, or Appendix B of Beltrami, 1993, for a brief treatment; Davis, 1986, Chapter 3, reviews the concepts of determinant and eigenvalue).

The other approach is simpler and is used here. Assume that $y = e^{\psi x}$, which is a common general solution of a linear first-order differential equation, is also a solution to equation 4.1, which is a linear second-order differential equation. Then $y' = \psi e^{\psi x}$ (first derivative or dy/dx) and $y'' = \psi^2 e^{\psi x}$ (second derivative or d^2y/d^2x), because the derivative of an equation involving the base of natural logarithms (e) takes the form $dy/dx = du \cdot e^u$, where $u =$ the entire exponent. For example, the first derivative (dy/dx) of $y = e^{2x}$ is $2e^{2x}$ and the second derivative is $4e^{2x}$ or $2^2 e^{2x}$). Substituting into equation 4.1,

$$D(\psi^2 e^{\psi x}) - \nu(\psi e^{\psi x}) - \lambda(e^{\psi x}) = 0 \qquad (4.26)$$

Factoring out $e^{\psi x}$,

$$e^{\psi x}[D(\psi^2) - \nu(\psi) - \lambda] = 0 \qquad (4.27)$$

4.3 Diffusion models of bioturbation

Let $D = a$, $-\nu = b$, and $-\lambda = c$. Then using the general solution for a quadratic equation

$$x = \frac{-b \pm \sqrt{b^2 - 4ac}}{2a} \tag{4.28}$$

one can solve for ψ by substitution:

$$\psi_1, \psi_2 = \frac{\nu \pm \sqrt{\nu^2 - 4\lambda D}}{2D} \tag{4.29}$$

The two values of ψ are the eigenvalues of the matrix of equations. One of these eigenvalues results in a solution that contains the square root of a negative number (an imaginary number). Fortunately, the solutions to symmetric matrices, in which the number of columns equals the number of rows, are real eigenvalues.

By making the variables in equation 4.1 dimensionless (unitless), Guinasso and Schink (1975) concluded that mixing may be described by a dimensionless parameter G, where

$$G = \frac{D}{m\nu} \tag{4.30}$$

When G is small (≤ 0.1; i.e., rapid sedimentation or slow mixing), an instantaneous or *impulse* tracer is buried below the mixed layer before it can be extensively reworked (i.e., its concentration distribution is bell-shaped or "Gaussian"; Figure 4.6). When G is large (≥ 10; slow sedimentation or rapid mixing), the tracer exhibits an exponential decrease in concentration upward but a uniform distribution in the mixed layer (Figure 4.6; cf. discussion of H in Chapter 2). For values of $G \geq \sim 0.6$, Officer and Lynch (1983) developed another method for calculating D, m, and the original depth of deposition of the tracer using

$$G = \frac{\beta_2 + \sqrt{\beta_2^2 + \pi^2 \beta_1 (\beta_2 - \beta_1)}}{2\pi^2 \beta_1} \tag{4.31}$$

and

$$m = \frac{\left(1 + \frac{1}{4G}\right)}{\beta_1} \tag{4.32}$$

where β_1 and β_2 are the slopes (fitted by linear regression) of the conservative tracer concentration profile in sediment. D is then calculated from equation 4.30 using ν obtained from radiotracers. The depth of the tracer (impulse) layer (or other

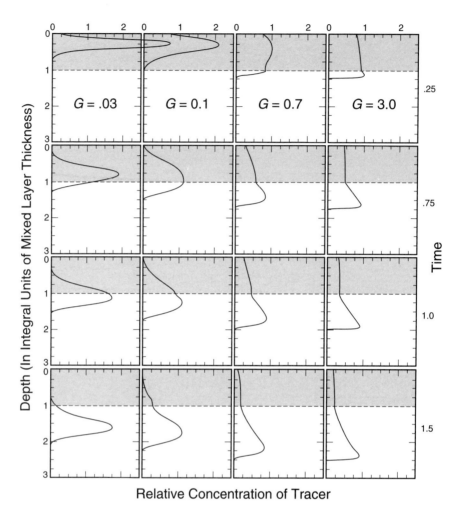

Figure 4.6. Relation between mixing parameter G and conservative impulse tracer concentration profiles (e.g., microtektites or volcanic ash) at different times. Depth in mixed layer units; time normalized to time required for one mixed layer thickness to accumulate. When G is small (rapid sedimentation, slow mixing), an impulse tracer is buried below the mixed layer before it can be extensively reworked (bell-shaped profile). If G is large (slow sedimentation, rapid reworking), the concentration of tracer decreases upward exponentially, but its concentration in the mixed layer in this case is relatively uniform. Compare with Figure 4.2 turned on its side. See text for criticisms of G. (Reprinted from Martin, 1993, with permission of the Paleontological Society.)

stratigraphic signal) that would have been observed in the core without bioturbation (x_0) can then found from the observed (or interpolated) tracer peak x_m by

$$x_0 = x_m - \left[1 - \frac{\ln\dfrac{8\pi^2 G^2 + 2}{4G + 1}}{\pi^2 G - 1}\right] m \qquad (4.33)$$

4.3 Diffusion models of bioturbation

Officer and Lynch (1983) later derived a new set of solutions to equation 4.1 using a technique called *optimization*, in which an iterative technique called the Newton–Raphson method is used to find successively better approximations to the solutions (see Causton, 1987; Boudreau, 1997; see also section 4.6).

4.3.4 Estimation of sedimentary parameters

In order to solve the Guinasso–Schink equation *analytically* (equation 4.25) or by the methods of Officer (1982) and Officer and Lynch (1982, 1983), one must estimate values for each of the sedimentary parameters m, ν, D, and λ.

Because these parameters are interrelated, determination of one affects estimation of the others (Officer, 1982). For example, given $m = 10$ cm (approximate mixed layer thickness on the middle-outer Washington, U.S.A., continental shelf) and $\nu = 0.5$ cm yr^{-1} (5 m ka^{-1}, which is quite fast in most cases), the residence time of a particle in the mixed layer is 20 years (Nittrouer and Sternberg, 1981). Therefore, multiple approaches should be used to constrain parameter estimates, as recommended by Benninger *et al.* (1979), Boudreau (1986a,b; 1997), Robbins (1986), Sharma *et al.* (1987), Anderson *et al.* (1988), and Keafer *et al.* (1992).

Estimates of sedimentary parameters can be constrained by the use of radiotracer profiles. But, the half-life of the radiotracer must be commensurate with the rate(s) of the process(es) being measured (Nittrouer and Sternberg, 1981): if the radiotracer's half-life is too small ($\lambda \gg D/m$), the isotope will decay to 0 within the mixed layer and ν cannot be measured; if, on the other hand, the half-life is too large ($\nu \ll D/m^2$), the radiotracer's activity profile will be essentially constant with depth and D will be impossible to calculate (cf. Figure 4.7). For relatively low sediment accumulation rates ($\nu^2 \ll \lambda D$), sedimentation can be ignored.

For example, based on ^{210}Pb profiles, Nittrouer and Sternberg (1981) calculated that particles reside and are reworked within the mixed layer for about 20–70 (mean $\cong$ 35) years before passing into the historical layer. Often, but by no means always, the upper portion of the ^{210}Pb profile is relatively constant within the mixed layer (mixed layer thickness can also be estimated from the depth of intensive bioturbation seen in core x-rays), but below the mixed layer ^{210}Pb is no longer supported and ^{210}Pb decays to lower levels with depth. ^{210}Pb has a half-life of 22.3 years (Figure 4.7; Benninger *et al.*, 1979; Carpenter *et al.*, 1982; Officer, 1982; Robbins, 1978, 1982; Sharma *et al.*, 1987) and is produced mainly by natural decay of the gas ^{222}Ra, which diffuses out of the Earth's crust, and is rapidly scavenged from the air by particulates settling through the atmosphere. Excess

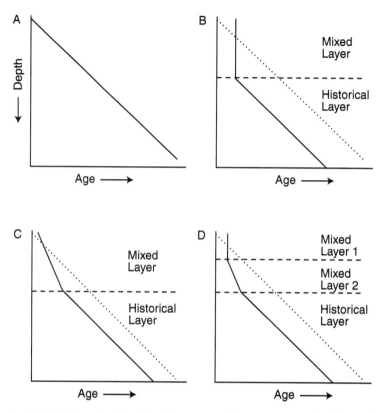

Figure 4.7. Hypothetical plots of mean age of a tracer versus depth in sediment. (A) No mixing. Slope of line is constant and determined by sediment accumulation rate only. (B) Box model (complete mixing within mixed layer; D is effectively infinite). Mean tracer age is the same at all depths within mixed layer. Age profile of tracer within historical layer is displaced toward younger mean age relative to unmixed profile (dotted line). (C) Diffusive mixing within mixed layer. Slope of age profile within mixed layer is proportional to D. Age profile is displaced, although not as much as in (A). Dotted line as in part (B). (D) Mixed layer consists of two layers with different values of D. Mixing is complete in the upper layer, but D has an intermediate value in the lower mixed layer. Dotted line as in parts (B) and (C). (Redrawn from Cutler, 1993.)

^{210}Pb (Pb$_{ex}$) levels are thus supported in the mixed layer by continued input from the atmosphere.

If Pb$_{ex}$ is assumed to be in steady state and bioturbation and sedimentation rates are assumed constant with depth, the solution to equation 4.25 may be used, in which D and ν are unknowns (Keafer et al., 1992). If ν can be ignored over the time scales under consideration (sedimentation is negligible relative to bioturbation and decay; i.e., $\nu^2 \ll 4D\lambda$), equation 4.25 simplifies to

$$\text{Pb}_{ex} = \text{Pb}_0 \exp\left(-\left[\frac{\lambda}{D}\right]^{(1/2)} x\right) \qquad (4.34)$$

4.3 Diffusion models of bioturbation

This equation has been used by many workers because on a plot of the log of Pb_{ex} versus depth, the slope of the line fit to the radiotracer profile provides an estimate of D (taking the logarithm of both sides of equation 4.34, the exponent $-\lambda/D$ or $0.031/D$ – becomes the slope of the equation for the straight line $y = mx + b$ fit to the Pb_{ex} data; Keafer et al., 1992; cf. equation 7.4). Anomalous increases in ^{210}Pb below the mixed layer probably indicate downward piping through the mixed layer via burrows (e.g., Benninger et al., 1979; Smith and Schafer, 1984). If, on the other hand, $D \cong 0$ (e.g., presumably below the surface mixed layer; cf. below), then equation 4.25 (cf. equation 4.34) reduces to

$$Pb_{ex} = Pb_0 \exp\left(-\left[\frac{\lambda}{\nu}\right]x\right) \tag{4.35}$$

Thus, the *apparent* sedimentation rate (ν') can be determined for approximately the past 100 years from the slope of the natural logarithm (ln) of ^{210}Pb concentration versus depth (x) below the mixed layer (e.g., Benninger et al., 1979; Peng et al., 1979; Officer, 1982; Aller and Cochran, 1976; Cochran and Aller, 1979; Sharma et al., 1987). ν' may confound true sedimentation and sediment resuspended by bioturbation, however, and may therefore overestimate the true value of ν depending on the magnitude of bioturbation (Officer, 1982). Instead, longer-term or "true" ν can be estimated from the downcore profile of ^{14}C (half-life ≈ 5730 years) for the past few thousand years. This is also an apparent ν, but the correction is presumably negligible for high ν rate areas like many coastal regions (Officer, 1982; cf. Keafer et al., 1992).

Even shorter-term sedimentation rates can be determined using ^{137}Cs. Unlike ^{210}Pb and ^{14}C, ^{137}Cs (half-life ≈ 30 years) is a time-dependent (impulse) tracer derived from atomic weapons testing, which peaked about 1963–1964 (Robbins, 1982), and has been used to estimate ν for about the past 20–30 years. In the case of a decaying *impulse* tracer, equation 4.34 becomes (Keafer et al., 1992):

$$\frac{Pb_{ex}}{Pb_0} = \exp\left(-\frac{x^2}{4Dt}\right) \tag{4.36}$$

^{137}Cs may be mobile because of changes in porewater redox chemistry caused by organic matter decay (T. M. Church, personal communication); nevertheless, Sharma et al. (1987) successfully modeled ^{137}Cs profiles in studies of marsh sediments. Some abnormal ^{137}Cs profiles may be the result of "levee effects" –

high sedimentation rates near tidal creeks – storm deposition or both. For example, approximately 1.5 cm of sediment was deposited at a high marsh site (Bombay Hook National Wildlife Refuge, Delaware U.S.A.), where average sedimentation rates are on the order of a few millimeters a year, near a tidal creek during a span of about 8 months (Hippensteel and Martin, work in progress); most of the sediment was probably deposited by two northeasters spaced about 1 week apart that ranked third and fourth respectively in terms of tidal height out of 30 storms (including hurricanes) measured over a half-century (Ramsey et al., 1998).

^{7}Be is not as susceptible to geochemical changes (T. M. Church, personal communication; see also Lord and Church, 1983; Luther et al., 1991), and can be used to estimate very short-term sedimentation rates (in the order of a few months; Krishnaswami et al., 1980; Sharma et al., 1987). ^{7}Be is formed by spallation reactions of cosmic ray particles with atmospheric nitrogen and oxygen (Lal and Peters, 1967) and has a half-life of approximately 28 days (T. M. Church, personal communication).

The biodiffusion coefficient (D) can be determined in several ways. First, knowing ν and λ, D may be calculated by fitting curves to radiotracer plots using equations 4.34–4.36. Other workers have used a composite layer approach to calculate D. Aller and Cochran (1976) estimated D in shallow subtidal sediments of Long Island Sound. They used the short-lived radiotracer ^{234}Th (half-life $\approx$ 24 days), which is a decay product of ^{238}U and which is most concentrated in the uppermost 5 cm or so of the sediment column beneath the SWI (it decays too rapidly to normally penetrate much deeper), to estimate D from the equation $D = m\nu'$ (where m = mixed layer thickness in centimeters; more on the derivation and limitations of this equation below). They assumed that since ^{234}Th has such a short half-life, the apparent rate of sediment accumulation ν' times m is $\approx D$ (i.e., $D \gg \nu$ for short intervals of time; see above). Later, Benninger et al. (1979) used these ^{234}Th-based calculations for D, along with D values based on ^{239,240}Pu for depths up to about 10 cm, and a ν' based on ^{210}Pb. They used a "composite-layer, mixing + sedimentation model," in which D was allowed to vary in discrete layers with depth, to determine which model best fit the data (this approach is similar to that of multiple compartment dilution models; see Simon, 1986). Indeed, the mixed layer is unlikely to be a unique surface and may be viewed more as a probabilistic concept since bioturbation decreases with depth (Sadler, 1993). Similar approaches can be used in order to meet the assumption of constant ν with depth (i.e., allow ν to differ between layers but remain constant with each layer).

Using a similar set of composite-layer models, Officer (1982) calculated the mixing parameter D from

$$D = \frac{(\nu' - \nu)\nu'}{\lambda} \quad (4.37)$$

He obtained estimates of D similar to those published by other workers for deep sea cores and the shallow subtidal of Long Island Sound using ^{210}Pb (Nozaki *et al.*, 1977; Benninger *et al.*, 1979; Benoit *et al.*, 1979; Peng *et al.*, 1979).

4.4 Caveats of diffusion-based bioturbation models

Although sedimentary chemical profiles have been successfully modeled using diffusion-based approaches (e.g., Aller and Dodge, 1974), mathematical models of bioturbation are not necessarily strictly analogous to those for thermochemical diffusion. The analogy to diffusion (*local mixing*) is justified if: (1) sediment particle motions are random in space and time; (2) the average distance traveled by a particle is much less than the scale of the process being investigated; and (3) the time interval between mixing "events" is much less than the time for reactions or the time for introduction of tracer into the sediment (Boudreau, 1986a, 1997). Because D is defined as $\delta^2/2\tau$ (equations 4.11–4.13), it depends on the square of the distance that particles are moved (step length $= \delta$) and inversely on the time (τ) between movements. What this means for *sediment particles* – as opposed to chemical molecules – is that they stay more or less in one place for relatively long periods of time before being moved to another spot where they remain for some time, and so on (Wheatcroft *et al.*, 1990; see Figure 4.3).

Also, large deposit feeders may dominate mixing (*non-local mixing*) over greater distances than "diffusive" processes (cf. Figure 4.3). "Biodiffusion" may actually be dominated by advective mixing or *apparent* burial (as opposed to sediment accumulation from the water column at the SWI) caused by bioturbating organisms such as Conveyor Belt Deposit Feeders or CDFs (e.g., various species of worms, shrimp, etc.). CDFs move sediment much more rapidly than "diffusion" alone, as these organisms are constantly transferring sediment downward and then upward to the SWI in conveyor belt-like fashion (van Straaten, 1952; Rhoads and Stanley, 1965; Meldahl, 1987). Thus, sediment and shells may move downward or *upward* (McCave, 1988), which is not accounted for by most current bioturbation models (Cutler, 1993; Boudreau, 1997, describes models and boundary conditions for non-local mixing). Bioturbators may also transfer whole parcels of sediment (not just individual sediment grains) intact into burrows in or below the mixed layer.

Rapid *horizontal* mixing coupled with vertical advection can also produce tracer profiles characteristic of diffusion models by transporting sediment grains *away* from zones of vertical transport. The net result of these activities is that the tracer profile may still *appear* to have resulted from "diffusion" alone, no matter what the actual process of sediment movement is, and can be seemingly described by a "diffusion" coefficient (D; Boudreau, 1986a,b; Boudreau and Imboden, 1987; Wheatcroft *et al.*, 1990).

To complicate matters further, organisms often sort sediment particles according to size (*size-selective feeding*). Many deposit-feeding (bioturbating) and suspension-feeding organisms have an intricate sorting apparatus and behavior patterns that are used to discriminate between sediment particles. Particles that are too large are often rejected before they are ingested because their size indicates that most likely they will provide little or no nutriment. Glass (1969), for example, found that in deep-sea cores (mainly Indo-Pacific), microtektites were dispersed over intervals of 35–90 cm (average 60 cm), representing 33–320 (average 120) thousand years. Stratigraphic distributions seemed to be controlled by the intensity of burrowing (estimated by burrow frequency), and larger microtektites were concentrated in the upper half of the bioturbated impulse layer in some cores. Extrapolating to fossils, he concluded that the apparent time of extinction of two fossil species of different average size that became extinct simultaneously might be significantly altered by bioturbation. Based on abundance curves, he also concluded that the first appearance of a species is a more reliable stratigraphic marker than its extinction.

Surprisingly, the phenomenon of size-selective feeding has been rejected out of hand by some workers as an important influence on tracer (fossil) distributions and temporal resolution at small scales. Based on studies of volcanic ash and microtektite distributions, Ruddiman *et al.* (1980) concluded that particle size selectivity by bioturbators is unimportant in the deep sea (but see dissenting view of Glass in Ruddiman *et al.*, 1980; curiously, Ruddiman and Glover, 1972, had earlier concluded that bioturbation was important in controlling ash distributions in deep-sea cores). Wheatcroft and Jumars (1987; see also Wheatcroft, 1992) statistically reanalyzed the data of Ruddiman *et al.* (1980), however, and concluded that given a particle size range of 11–500 μm (which includes the range of volcanic ash), small particles are more likely to be mixed than large ones, and that larger microfossils (e.g., foraminifera) are less likely to be advected than smaller ones (coccolithophorids).

4.4 Caveats of diffusion-based bioturbation models

Robbins (1986) also concluded (based on models of freshwater ecosystems with the oligochaete worm *Tubifex tubifex*, which is a conveyor belt feeder) that particle-selective feeding can produce non-uniform tracer profiles. Rejection of a tracer by these organisms (which are CDFs) reduces the surface concentration of the tracer and produces "biogenic graded bedding," whereas selective ingestion and transport of the tracer to the SWI increases the tracer concentration there. Thus, increasing sedimentation rate and the selectivity factor (i.e., discrimination *against* large particles) decreases the displacement and homogenization of large particles, whereas increasing feeding rate (on small particles) and depth of feeding tends to increase downward displacement and homogenization of small tracer particles (Robbins, 1986). Repeated cycles of ingestion and redeposition of a tracer at the SWI results in subsidiary tracer peaks below the SWI. Robbins's (1986, his figure 3) model-generated diagrams of tracer distributions generated by particle-selective feeding bear a strong resemblance to the distribution of ash layers published by Ruddiman and Glover (1972, their figure 8).

The simplified form of the Guinasso–Schink model (equation 4.30) has also drawn criticism (Wheatcroft, 1990; see also Sadler, 1993). First, G values for most environments lie between 0.1 and 10, not at either extreme (see also Officer and Lynch, 1983). Second, equation 4.30 is the inverse of a Péclet number, which is used in engineering to compare the relative importance of advection versus diffusion in the transfer of heat or mass away from an object. Although G, like a Péclet number, consists of a diffusion coefficient, a (burial) velocity, and a length, Wheatcroft (1990) felt that G is not *dynamically* correct. For example, if m increases, then for a given burial velocity, it takes longer for a tracer signal to move through the mixed layer; therefore, the tracer is more likely to be dispersed and the signal destroyed. But according to equation 4.30, if m increases, G should decrease and the tracer signal should be better preserved (Figure 4.6)!

Wheatcroft (1990) suggested another approach (also based on diffusion) to measure the degree of dissipation (spreading) of an impulse tracer (cf. Figures 4.2, 4.6):

$$c = \tfrac{1}{2}\mathrm{erf}\left(\frac{S}{2\sqrt{DT_m}}\right) \tag{4.38}$$

where S is signal layer thickness, T_m is transit time or the time for the signal to traverse through the mixed layer, and erf is the *error function*, which depends on the function involved and is tabulated in various engineering texts (e.g., appendix II of Carslaw and Jaeger, 1959; see chapter 2 of Crank, 1975, and Berg, 1993, for more on equation 4.38). According to Wheatcroft (1990), equation 4.38 yields more realistic values for dissipation of an impulse signal, but only if the impulse signal is relatively thin: a relatively thick layer of tens of centimeters, especially if sandy, may impede bioturbation (Hippensteel and Martin, 1998).

Despite the limitations of the G model, Hippensteel and Martin (1998) calculated mean G values of 0.032 ± 0.046 and 0.259 ± 0.384 for the high and low marsh behind barrier islands off Charleston, South Carolina, U.S.A. They estimated β_1 and β_2 of South Carolina overwash fans by measuring the slopes of obviously bioturbated (smeared upward) peaks of Oligo-Miocene foraminifera ("tracers") in vibracores that could only have been eroded from shallow offshore outcrops by storms and deposited with sands in back-barrier marshes. Their G values agree with the trend of increasing bioturbation from high to low marsh calculated by Sharma *et al.* (1987) based on equation 4.1 (see also Basan and Frey, 1977). Based on the vibracores, the shallow shelf is a major source of sediment for barrier island construction off Charleston. They calculated that the storms were of gale size or larger; such data may be of interest to insurance companies or government agencies.

4.5 Box models

In the case of slow sedimentation or rapid mixing, the Guinasso–Schink diffusion model is identical to the sediment mixing model developed by Berger and Heath (1968). This *box model* assumes that the shallow sediment mixed layer is reworked so rapidly with respect to sedimentation that the concentration of a tracer within the mixed layer is uniform before it is finally entombed in the *historical layer* beneath (Figure 4.7). The box model is equivalent to a one-compartment dilution model, in which the rate of outflow of fluid (with some predetermined concentration of a tracer) is equivalent to the rate of exit of sediment from the bottom of the mixed layer into the historical layer (Simon, 1986). On the surface, uniform mixing of the mixed layer seems a reasonable assumption for deep-sea cores (for which this model was developed) because the mixed layer is usually relatively thin (approximately 5–10 cm) and rates of deep-sea sediment accumulation are relatively slow (a few centimeters per thousand years on average). This means that, in most cases, organisms *ought* to have sufficient time to more or less homogenize the mixed layer in the deep sea. But, this is not always the case: just because sediment layers in a core do not look like they have been bioturbated does not mean that they have not been disturbed (Berger *et al.*, 1979) nor that mixing is homogeneous (e.g., photographs in Magwood and Ekdale, 1994).

4.6 Deconvolution

Assume a conservative impulse tracer, such as a microtektite or volcanic ash layer. According to the Berger–Heath model

$$\frac{dP}{P} = -\frac{ds}{m} \qquad (4.39)$$

where P is the probability of finding a particle in the mixed layer, s is the thickness of sediment deposited on top of the mixed layer, and m is the mixed layer thickness. Integrating equation 4.39 results in a decay formula:

$$P = P_0\, e^{(-s/m)} \qquad (4.40)$$

Thus, after deposition of the event layer, the stratigraphic distribution of the tracer is given by:

$$P = P_{oz}\, e^{(s_e + m)/m} \qquad (4.41)$$

where P_{oz} is the original concentration of z at a distance of m below the top of the layer, and s_e is the thickness of sediment deposited after the event layer has been deposited.

In other words, the concentration of the tracer above its true level of deposition decreases exponentially upward according to the thickness of sediment deposited after the layer has been deposited, and can be used to estimate the degree of upward or downward smearing of stratigraphic markers (see also Chapter 5).

4.6 Deconvolution

A number of attempts have been made at direct *deconvolution* ("unmixing") of marine sedimentary signals (e.g., $\delta^{18}O$) based on the Berger–Heath or Guinasso–Schink mixing models (Berger *et al.*, 1977; Schiffelbein, 1985; 1986; see also Chapter 5), but the deconvolution overestimated mixing intensity and produced artificial overshoots and offsets including artificial meltwater spikes (Jones and Ruddiman, 1982). Christensen and co-workers (Christensen 1986; Christensen and Bhunia 1986; Christensen and Goetz, 1987; Christensen and Osuna, 1989; Christensen and Klein, 1991) were more successful in reconstructing fluxes of pollutants and radiotracers in lake sediments using the Guinasso–Schink and Berger–Heath mixing models; they assumed that bioturbation and sedimentation rates are constant and they treated sediment compaction effects by assuming particular porosity profiles (see also Davis, 1974, and below). Although these assumptions appear valid for the thin mixed layer of lacustrine environments where Christensen worked, unfortunately they do not appear to hold in most – if any – marine environments.

In response to these difficulties, Bard *et al.* (1987) developed a technique based on a discretized version of the Laplace transform. They deconvolved abundance curves of "carrier" species of planktonic foraminifera and then the isotopic signals stored in the carrier shells. Other workers have noted that even at an abrupt lithological or paleoclimatic boundary, abundant species above and below the boundary that are indicative of the climatic change are worked downward and upward across the boundary (Berger and Heath, 1968; Hutson, 1980; see also Andree *et al.*, 1984; Broecker *et al.*, 1984; Peng and Broecker, 1984; Andree, 1987). In Bard *et al.*'s (1987) technique, comparison of actual data with unmixed curves enables recognition of bioturbation. Nevertheless, restored signals may be erroneous when carrier abundances approach 0; the technique also assumes that bioturbation and sedimentation rates remain constant (box model of Berger and Heath, 1968).

Peng *et al.* (1977, 1979) advocated and developed *numerical* models for *continuous* inputs, and Officer and colleagues developed them for impulse inputs (section 4.3.3). In such models, the Guinasso–Schink or Berger–Heath models (or modifications thereof) are solved in reverse by specifying various sedimentary parameters (e.g., m, v, D, λ) and solving for the unknown (e.g., test inputs). Given the spatiotemporal variation in test inputs, however (Buzas, 1968, 1970; see Murray, 1991, for brief review), it is probably best to use averaged inputs which smooth the data and prevent artificial overshoots (Hippensteel and Martin, work in progress). Test inputs can then presumably be determined by varying initial test inputs until preserved concentrations predicted by the model are similar to those sampled in cores.

Numerical models have the advantage of removing amplification of analytical noise from the data, but the disadvantage of not providing unique solutions: in theory, different values of sedimentary parameters may produce the same result (Boudreau, 1986a,b; Bard *et al.*, 1987; Boudreau, 1997, describes the numerical solution of diagenetic models). Moreover, small errors in measurement may swamp the system (Simon, 1986). Considering the difficulties of analytical solutions, however, and especially given the number of sedimentary parameters involved, if sedimentary parameters can be estimated reasonably accurately, numerical models may provide the greatest promise for reconstructing stratigraphic signals. Pizzuto and Schwendt (1997), for example, modeled 6000 years of sediment accumulation and autocompaction in a Delaware (U.S.A.) marsh using a FORTRAN program (SQUISH3) based on finite-consolidation

4.6 Deconvolution

strain theory:

$$\frac{\partial^2 e}{\partial z^2} - \lambda(\gamma_s - \gamma)\frac{\partial e}{\partial z} = \frac{1}{G}\frac{\partial e}{\partial t} \qquad (4.42)$$

where $\gamma_s - \gamma$ is the submerged unit weight of sediment, G is a finite strain consolidation coefficient, z is the volume of solids per unit area above an arbitrary datum, and

$$e = (e_i - e_f)e^{-\gamma\sigma} + e_f \qquad (4.43)$$

in which e_i and e_f are the void ratios at the beginning and end of primary consolidation, respectively, and σ is the effective stress. The model was applied to an accumulating column of sediment consisting of sediment layers of specified void ratios deposited at specified time intervals. (Material properties of the sediment were determined initially from laboratory and field studies and rates constrained by radiocarbon dates.) The model was then used to reproduce the distribution of void ratios within a vibracore, the actual thicknesses of sedimentary units, and the history of elevation changes of the SWI that was consistent with both a previously determined sea-level curve for the Delaware coast and the down-core sequence of paleoenvironments. According to the model, horizons have been lowered as much as 2.3 m by autocompaction, the rate of which has ranged from one-half to one-third of the rate of sea-level rise.

5 Time-averaging of fossil assemblages: taphonomy and temporal resolution

I can only say, there *we have been . . . And I cannot say, how long, for that is to place it in time.* T. S. Eliot, *Four Quartets*

5.1 Introduction

The primary overprint of biostratinomic processes on the fossil record is the phenomenon of *time-averaging*, in which fossils "accumulate from the local living community during the time required to deposit the containing sediment" (Walker and Bambach, 1971). In almost all cases, even without biostratinomic factors such as physical reworking, transport, dissolution, and bioturbation (Chapters 1–4), there is an inherent bias toward time-averaging because biological generation times are typically much shorter relative to net rates of sediment accumulation (Figure 5.1; Kidwell and Behrensmeyer, 1993*a*).

Thus, fossil assemblages may be viewed as resulting from multiple (often seasonal) inputs of shells to sediment. Intuitively, shell inputs would seem to be additive, so that the total (final) fossil assemblage may be viewed as the summation of multiple inputs (*principle of linear superposition*; Simon, 1986). Shell inputs decay to some extent (Chapters 3, 4), however, so that earlier inputs will, in theory, comprise less of the final assemblage than later inputs. This relationship may be expressed most simply in the form of a steady-state box model:

$$A = A_e - (A_0 - A_e) e^{-Bt} \tag{5.1}$$

in which A is the age of the assemblage, A_e is the equilibrium age of the assemblage, A_0 is the age of shell input (reproductive pulse), and B is the rate of decay that in turn depends on depositional setting (ν, D, etc.; cf. equation 4.25). A more complicated – and realistic – version of this relation is expressed by (Simon, 1986):

$$C(T) = I(0) e^{-\lambda T} + I(\Delta t) e^{-\lambda(T-\Delta t)} + I(2\Delta t) e^{-\lambda(T-2\Delta t)} + \cdots$$
$$+ I(n\Delta t) e^{-\lambda(T-n\Delta t)} \tag{5.2}$$

5.1 Introduction

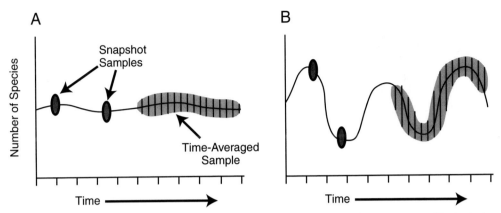

Figure 5.1. Effects of time-averaging on sampling of two different types of biological communities: (A) is relatively stable, whereas (B) fluctuates widely in number of species. "Snapshot" samples in (A) would probably not differ much from time-averaged samples ("movies"), whereas snapshots in (B) would give only glimpses of time-averaged fluctuations. (Reprinted from Behrensmeyer and Chapman, 1993, with permission of the Paleontological Society.)

where $C(T)$ is shell concentration at total time T ($\sum \Delta t$), $I(\Delta t)$ is shell input at time Δt (intervals of Δt assumed to be of equal length), and λ is the decay coefficient (assumed constant; see Chapter 4).

The *unit impulse response* (= *impulse response function* or H) is defined as the ratio of the dependent variable (in this case, C) as a function of time to the size of an input to the system at time 0; in other words, $C(t)$ is proportional to the original input or

$$\frac{C(t)}{I(0)} = H(t) \tag{5.3}$$

Assuming constancy of hardpart input and decay (which of course varies within and between sites and between hardparts of different taxa, at least on short time scales),

$$C(0) = [I(0)H(0)]$$
$$C(1) = [I(\Delta t)H(0) + I(0)H(\Delta t)]$$
$$C(2) = [I(2\Delta t)H(0) + I(\Delta t)H(\Delta t) + I(0)H(2\Delta t)]$$
$$\vdots$$
$$C(n) = [I(n\Delta t)H(0) + I[(n-1)\Delta t]H(\Delta t) + \cdots$$
$$+ I(\Delta t)H[(n-1)\Delta t] + I(0)H(n\Delta t)] \tag{5.4}$$

According to equation 5.4, $I(0)$ is convolved with H at each time increment (Δt) up to and including the nth increment, whereas the last input ($I(n\Delta t)$) is convolved only with $H(0)$. In other words, bearing in mind the above assumptions, the older the pulse of hardpart input or *cohort*, the greater the chance of its decay and the less it will comprise of the final assemblage. Ideally, then, a histogram of shell age versus frequency for most most fossil assemblages should consist predominantly of relatively young shells with a tail skewed toward older, and less frequent, shells (e.g., Flessa *et al.*, 1993; Martin *et al.*, 1995, 1996). A similar relationship has been noted in the distribution of $^{40}Ar-^{39}Ar$ dates of individual glauconies (Smith *et al.*, 1998), thereby supporting the concept of loss of older shells as new ones are added to an assemblage.

Although durations of time-averaging are typically reported as an age range (irrespective of shell frequency), based on equations 5.1–5.4, a theoretically more accurate way of reporting durations of time-averaging would be to weight shell age by frequency of age classes. Based on an exponential model of shell age distribution, Olszewski (1997) calculated that 90% of shells are added during the last half of the total interval of fossil assemblage accumulation, and that only 30 shells are required to find at least one from the oldest 10% of all shells (with 95% confidence).

Because of time-averaging, a fossil assemblage may represent a *minimal* duration of a few decades up to likely hundreds to thousands of years or more (Table 5.1), unless the assemblage is rapidly buried (i.e., Lagerstätten). Consequently, time-averaging has almost always been viewed negatively because of its effect on temporal resolution or acuity of the fossil record. Although time-averaging becomes less significant over longer scales of measurement, the incompleteness of the fossil record looms ever larger as longer time spans are considered (Chapter 7). Between the one extreme of time-averaging over short durations and stratigraphic completeness at the other, there is a "threshold" of time-averaging beyond which certain phenomena are lost.

Time-averaging is a relative term, however, and is dependent upon the scale of the phenomenon of interest (Graham, 1993; Webb, 1993; Kowalewski, 1996*a*). Short-term population (high-frequency) phenomena are in fact usually lost, but that is advantageous if one is interested in longer-term processes (similar considerations apply to the preservation of physical sedimentary structures; Nittrouer and Sternberg, 1981, p. 224). Time-averaged assemblages are more likely to be representative of long-term environmental conditions and community

5.1 Introduction

Table 5.1. *Estimated durations of time-averaging for various types of fossil assemblages*

	Years of time-averaging per assemblage (log scale)							
	−2	−1	0	1	2	3	4	5	6
	Days	Months	Years	Decades	100s	1000s	10 000s	100 000s	

Terrestrial System

Macroflora
Recent
- Fresh leaves, flowers (Days–Years)
- Cuticle, leaf fall (Months–Years)
- Cones, seeds, bark (Years–Decades)
- Tree trunks (Years–100s)
- Peat (100s)

Cretaceous–Paleogene
- Plant bed (Months–Years)
- Florule (Years–Decades)
- Biostratigraphic zone (10 000s–100 000s)
- Megafossil flora (1000s–10 000s)

Pollen
Quaternary
- Lake lamina (Years)
- 1 cm core top (Years–Decades)
- Regional pollen map (1000s)

Vertebrates
Quaternary
- Temperate/tropical (Years–100s)
- Arctic land surfaces (100s–1000s)
- Cave sites (1000s)
- Downward intrusion (Years–100s)
- Fluvial reworking (Decades–1000s)
- Bioturbation (Decades–1000s)
- Sediment flow and cryoturbation (Decades–1000s)

Cretaceous
- Nests and event beds (Months–Years)
- Attritional microsites (Decades–100s)
- Channel lags (100s–1000s)

Simulation
- Attritional land surface (Decades–1000s)

Marine System

Macroinvertebrates
Recent
- Intertidal and nearshore (Decades–100s)
- Shelf (100s)

Post-Paleozoic
- Event concentrations (Months–Years)
- Censuses (Months–Decades)
- Within-habitat t-avgd assemblgs (Years–100s)
- Composite concentrations (Decades–100s)
- Within-habitat t-avgd assemblgs (Decades–100s)
- Hiatal concentrations (1000s–10 000s)
- within-habitat t-avgd (1000s)
- environ.-cond assemblgs. (10 000s)
- biostrat.-cond. assemblgs (10 000s)

Paleozoic
- Single event beds (Months–Years)
- Composite fossil beds (Years–100s)
- Hiatal fossil beds (1000s–10 000s)
- Lag accumulations (1000s–10 000s)

Benthic Foraminifera
Recent
- Shell-poor terrigenous shelf (Years–Decades)
- Carbonate and shell-rich terrigenous shelf (Decades–1000s)
- Condensed biostratigraphic zones (10 000s–100 000s)

Within-habitat t-avgd assemblgs., within-habitat time-averaged assemblages; environ.-cond assemblgs, environmentally condensed assemblages; biostrat. cond. assemblgs, biostratigraphically condensed assemblages.
Modified from Kidwell and Behrensmeyer (1993*b*).

dynamics because the dominance of a particular set of environmental parameters will increase with time while comparatively short-term (and perhaps unrepresentative) fluctuations will be damped or completely filtered out (Fürsich and Aberhan, 1990). High loss rates in death assemblages mainly apply to the ecologically most transient parts of communities, and some death assemblages appear comparable to the results of repeated biological surveys that document changes in community species composition and diversity over several decades or more, including sudden phenomena that might be missed by short-term sampling regimes (Kidwell and Bosence, 1991; Behrensmeyer and Chapman, 1993; Kidwell and Flessa, 1995). Valentine (1989), for example, calculated that up to 85% of durably skeletonized bivalves and gastropods were preserved. Both the marine and terrestrial fossil records also document *no-analog assemblages* (Loubere, 1982) that are not observed today and which suggest that biological communities are not always the sensitive, tightly woven entities that have been portrayed in the literature (Graham, 1993).

Before sampling begins, the investigator should attempt to assess both sediment accumulation rates (which will directly affect sampling regimes) and stratigraphic completeness (Chapter 7). A fossil assemblage that has been time-averaged on one scale may not be time-averaged on another. If, for example, the duration of time-averaging, as expressed by the difference between the ages of the oldest and youngest shells in an assemblage, is approximately 1000 years, and the phenonemon occurs over a scale of approximately 10 000 years, then time-averaging is insignificant. On the other hand, if the desired scale is about 100 years, the assemblage is likely a poor choice for study (Kowalewski, 1996a). One must also be wary of *analytical* time-averaging of data sets (Chapter 1; e.g., Johnson, 1993). There is a limit to time-averaging and temporal resolution beyond which more finely spaced samples (or dates) are either unnecessary to constrain interpretations or are impossible to obtain; in fact, if absolute dates are spaced too closely, they either overlap or produce dating reversals (Webb, 1993).

Also, *the origin and magnitude of time-averaging of a fossiliferous unit should be estimated before it is used for paleobiological study, so that an appropriate sample interval is chosen* (Flessa, 1993). In order to avoid time-averaging, paleontologists have often employed fossil assemblages from bedding surfaces because the presence of a bedding surface suggests little physical or biogenic mixing. Ironically, bedding planes are often the result of non-deposition or erosion and their associated assemblages may have accumulated over significant durations of time (Schindel, 1980), so that taphonomic feedback (Kidwell and Jablonski, 1983;

Kidwell, 1986b) and early marine cementation (Walker and Diehl, 1985) may have occurred (see also Chapters 6, 8).

5.2 Consequences of time-averaging

5.2.1 Abundance

Relative or absolute abundance of species is often used in paleoecological studies. Abundance has been used, for example, to infer population dynamics of fossil assemblages. Unfortunately, the *fidelity* of fossil assemblages – "the faithfulness of the assemblage to the original biological input, in terms of taxonomic, morphometric, and age-class composition" (Kidwell and Behrensmeyer, 1993a, p. 4) – is typically altered by biostratinomic agents such as winnowing, biological mixing, and dissolution. Juveniles, especially, tend to be underrepresented in the fossil record, perhaps not so much because of their chemical reactivity (increased surface/volume ratio) but because of ontogenetic differences in the degree of mineralization: juveniles often have poorly mineralized shells and delicate architecture with high intraskeletal porosity and organic matter content (Flessa and Brown, 1983; Cummins et al., 1986; Kidwell and Bosence, 1991). Powell et al. (1984) estimated "half-lives" of 100 days for the smallest (0.8–3.1 mm) juveniles of molluscan death assemblages in Laguna Madre, Texas (see also Green et al., 1993; Martin et al., 1996).

Staff et al. (1986) demonstrated extensive alteration of lagoonal (Texas) molluscan assemblages. They found that even the rank order abundance of the dominant taxa in death assemblages rarely agreed with ranks of the same taxa in the living community or potential death assemblages. Kidwell and Bosence (1991) reached a similar conclusion based on compilations of modern molluscan death assemblages. Not surprisingly, time-averaging least affects large individuals or adults of long-lived taxa, which often dominate community trophic structure in living communities even when they are not numerically superior (Kidwell and Bosence, 1991). Consequently, Staff *et al.* (1985) advocated using biomass as an indicator of abundance because it recovered both rank order and trophic structure; unfortunately, calculation of biomass is labor-intensive (Fürsich and Aberhan, 1990).

The co-occurrence of abundant species – especially large ones since they are more likely to be preserved – has been recently used to infer long periods of faunal and ecosystem stability (*co-ordinated stasis*) in the fossil record (e.g., Brett and Baird, 1995). Co-ordinated stasis of abundant species may be real because abundant species tend to be more eurytopic and ecologically interconnected in a community framework, whereas rare species exhibit spatial and temporal flux through the framework. But abundant taxa are also the most likely to be preserved in the fossil record and sampled by paleontologists; indeed, McKinney (1996) estimates that the described fossil record is based on only the 6–7% most abundant and widespread species. As demonstrated by actualistic studies, co-ordinated stasis cannot be justified merely on the basis of the rank order of abundant species (Kidwell and Bosence, 1991, their table VI). Sampling bias toward abundant species may also mimic paleocommunity persistence because only rare species, which have greater temporal and spatial abundance variation on ecological and geological time scales, may be indicative of community (in)stability (McKinney et al., 1996).

Despite the loss of smaller shells, especially those of juveniles and rare species, time-averaged death assemblages may potentially prove quite useful in environmental studies of modern communities by providing baseline data for distinguishing between anthropogenic and pre-anthropogenic effects. In a reanalysis of the data of Cummins et al. (1986) for molluscan assemblages of Texas bays, Kidwell and Bosence (1991, pp. 144–145) concluded that "*approximately half of the species showed no significant difference between dead size-distributions and calculated mortalities, and virtually all show[ed] good qualitative agreement,*" and that the dictum that size–frequency distributions are determined primarily by taphonomy "may someday prove to be overly pessimistic" (pp. 144–145).

5.2.2 Taxonomic composition and diversity

One of the dogmas of paleoecology and taphonomy is that taxonomic composition and diversity of fossil assemblages tend to increase as the duration of time-averaging increases (Peterson, 1977; Fürsich and Aberhan, 1990). This is because, given enough time, it is unlikely that ecosystems – the physical environmental conditions and their associated biological communities – remain exactly constant. Thus, the fidelity of a death assemblage is related to the duration of study, and many species are not necessarily "exotic" but instead come from patchy or impersistent populations (Kidwell and Bosence, 1991). It is more likely that species found living at a site will also be found dead there, but not the reverse (except for the most abundant ones), and that individuals of species sampled live will dominate death assemblages at the same site (see Kidwell and Bosence, 1991, for compilation). In the marine realm, relatively constant conditions are more likely to occur in offshore shelf and deep sea settings (*autochthonous time averaging* of Fürsich and Aberhan, 1990) than in higher-energy shallow waters (*allochthonous time-averaging*) or terrestrial/aquatic environments, in which seasonal fluctuations and growth and larval (propagule) recruitment are more variable (e.g., Potter and

5.2 Consequences of time-averaging

Rowley, 1960; Thayer, 1977; Fürsich and Aberhan, 1990). Even in shallow-water environments, however, most shell accumulations are *parautochthonous* and reflect relatively little shell transport (see table IX of Kidwell and Bosence, 1991, for reports of out-of-habitat shell transport in modern marine settings; see also Chapter 2).

Ideally, the time period over which samples must be pooled (resampled) in order for the taxonomic composition of the living community to converge with the death assemblage indicates the *minimum* duration of time-averaging in actualistic studies (Kidwell and Bosence, 1991; Flessa, 1993; see also Kowalewski *et al.*, 1998). Tens to hundreds of years may be necessary for sites to experience a full range of environmental conditions and biotas in terrestrial, coastal or shelf environments, thereby also suggesting minimum durations of time-averaging (Kidwell and Bosence, 1991; Behrensmeyer and Chapman, 1993; see also Manjunatha and Shankar, 1996). But, diversity indices may be affected to different degrees by time-averaging: even small changes in the environment and in species composition may increase species richness (number of species) while evenness may remain essentially unchanged. Larger-scale environmental fluctuations, which are more likely to be associated with community replacement, will increase both indices, although species may be ecologically incompatible (e.g., in terrestrial sediments, mixtures of grazers and frugivores; Behrensmeyer and Chapman, 1993). On evolutionary time scales, immigration and extinction may also contribute to assemblage taxonomic composition and diversity (Fürsich and Aberhan, 1990).

In the *theoretical* extreme, a diverse community persisting largely unchanged through time produces a time-averaged assemblage with the same diversity pattern as several low diversity communities that replace one another through time in response to shifting environments (Fürsich and Aberhan, 1990). Fortunately, the fossil record indicates that this extreme is rarely, if ever, achieved. Despite environmental instability, low-diversity communities tend to remain low-diversity communities and the resultant fossil assemblages reflect relatively simple community structure in response to environmental stress, whereas high-diversity communities tend to reflect relatively stable conditions for long periods of time.

Recent data on coral communities challenge many of the above generalizations, however. Pandolfi and Minchin (1995) found in Madang Lagoon (Papua New Guinea) that many live coral taxa are not found in death assemblages but most taxa in death assemblages are found alive; consequently diversity (richness, Shannon–Wiener, evenness) of reef coral death assemblages is significantly less than that of corresponding life assemblages, in contrast to patterns for non-reef shelly faunas, in which most living taxa are found dead and few dead taxa are found alive. Pandolfi and Minchin (1995) attributed the differences in diversity between reef and non-reef death assemblages to two mechanisms: (1) the longevity of many reef corals may exceed the time needed to degrade their skeletons, and (2) perhaps only a subset of living assemblages is preserved because of differential preservation between coral species (Chapter 2; see also Chapter 10). The case for coral reef diversity appears to be just the opposite of that for foraminiferal diversity in settings, however, in which more dead species are found in sediment than living on vegetation or other substrates (e.g., Martin and Wright, 1988; Culver, 1990).

5.2.3 Trophic and life habits

Trophic and life habit (e.g., epifaunal/infaunal) composition of fossil assemblages appears to be less affected by time-averaging than abundance or diversity because trophic structure – at least in shallow-water communities – is determined mainly by large, long-lived individuals (Staff *et al.*, 1986). Nevertheless, some trophic groups are no doubt underrepresented in fossil assemblages (Kidwell and Bosence, 1991). If communities composed of taxa with very different trophic/life habits successively colonize the same area, the resulting assemblage will still give a good picture of long-term conditions, but not only will the diversity of the fossil assemblage artificially increase, so too will the apparent complexity of trophic relationships (niche partitioning), which will be an artifact of time-averaging and therefore meaningless because many of the species never co-existed (Fürsich and Aberhan, 1990). If successive communities composed of different taxa with *similar* trophic/life habits colonize an area, diversity will also increase artificially but the relative proportions of feeding groups will remain relatively unchanged. Shorter-term fluctuations in ecosystems will probably be damped by more persistent communities and contribute relatively little to the fossil assemblage.

5.2.4 Stratigraphic disorder

By stratigraphic disorder is meant "the degree to which fossils within a stratigraphic sequence are not in proper chronological order." Stratigraphic disorder arises through vertical mixing (e.g., bioturbation) and reworking of older fossils into younger deposits (Cutler and Flessa, 1990, p. 227).

Based on the box model of bioturbation, Berger and Heath (1968) concluded that the concentration of an extinct species above its true level of extinction decreases exponentially according to the thickness of sediment deposited after its extinction:

$$P = P_{oz} \exp{(s_e + m)/m} \tag{5.5}$$

5.2 Consequences of time-averaging

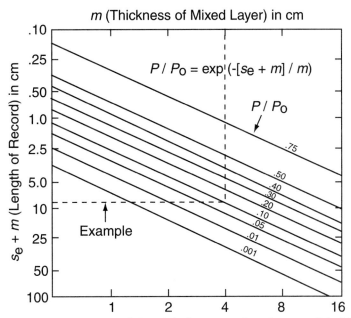

Figure 5.2. Proportion of the original concentration of a tracer (P/P_o) found at a distance (s_e) in sediment above its level of disappearance. In this example, if the thickness of the mixed layer (m) = 4 cm and upward mixing has taken place to a distance 9 cm above the original impulse layer, then the concentration of the tracer has decreased to 0.1 of its original value. (Redrawn from Berger and Heath, 1968.)

where P_{oz} is the original concentration of z at a distance of m below the top of the layer, and s_e is the thickness of sediment deposited after the event layer has been deposited.

Berger and Heath (1968) calculated the proportion of the original concentration of a tracer found in the sediment at a distance s_e above the true layer. By specifying acceptable levels of reworking (contamination), the level of stratigraphic resolution may be calculated. In their example (Figure 5.2), if $m = 4$ cm and the specified "contaminant" level (due to reworking) is 10% of the original concentration of the tracer, then the original depth of deposition may be as much as 9 cm deeper in the section. For a sedimentation rate of 1 cm ka^{-1}, this means a resolution of 9000 years; for a rate of 4 cm ka^{-1}, a resolution of 2250 years is obtained. The exact level of deposition is indicated when the tracer concentration is 1/e or 0.37 times its maximum concentration below, and is independent of m (Berger and Heath, 1968). This is because at this level, $s_e = 0$, and the exponent in equation 5.5 becomes -1; therefore, the right-hand side of equation 5.5 becomes $P_0 e^{-1}$ or $P_0(1/e)$. (The value, 1/e, is also typically used as a measure of the *return time* of a system to its original state after a disturbance). A similar model was developed by these authors for species appearances (smeared

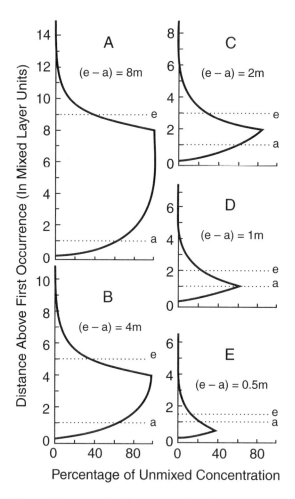

Figure 5.3. Tracer distributions resulting from bioturbation in relation to their appearance (a), extinction (e) and thickness of mixed layer (m). Distance above appearance on the x-axis is given in terms of "mixed layer units." As the stratigraphic range of a marker decreases through the sequence of diagrams, stratigraphic errors become more serious. (Redrawn from Berger and Heath, 1968.)

downward). Berger and Heath (1968) concluded that serious stratigraphic errors may occur if the stratigraphic range of a species is similar in thickness to the mixed layer (Figure 5.3); this is most likely to occur in the deep-sea during times of climatic fluctuations, when species may alternately appear and disappear in response to latitudinal movements of water masses.

Schiffelbein (1984) concluded that sedimentation rate (v) is the variable that most strongly affects the recognition of signal frequencies. Most natural processes (e.g., sedimentation rates, bioturbation; Chapter 4) are low-pass filters, which allow low-frequency signals within the *passband* region to persist in the sedimentary

5.2 Consequences of time-averaging

record from zero up to the *cutoff frequency* (f_p, which is usually defined as the frequency at which one-half of the signal amplitude – or power – is attenuated), while higher-frequency signals are lost in the *stopband* region beginning at (f_s) and extending to higher frequencies. Between f_p and f_s is a transition zone in which signals are progressively attentuated as f_s is approached. Based on a revised model of Officer and Lynch (1983; Chapter 4), Schiffelbein (1984) used linear regression to find a significant relationship between deep-sea v ($n = 16$ cores) and f_p ($r = 0.97$):

$$f_p = 0.05v \tag{5.6}$$

Rearranging in terms of period or wavelength ($\lambda = 1/f$):

$$v \text{ (in cm kyr}^{-1}) = 20/\lambda \text{ (in ka)} \tag{5.7}$$

Equation 5.7, which is based on $f_p = -3$ dB, can be used to obtain the minimum v necessary to recover a given signal period or frequency. Thus, in order to recover a signal of 4 ka in deep-sea cores, a minimum v of ~ 5 cm ka^{-1} is required (Schiffelbein, 1984). This is a fairly high sedimentation rate for most deep-sea settings, which average ~ 1–2 cm ka^{-1}. Even more than tripling f_p to -10 dB (which produces a new linear regression between v and f_p) in the hopes of resolving signals with periods of 1 ka requires a v of ~ 7 cm ka^{-1}. One way to *partially* circumvent this problem is to increase the signal-to-noise ratio (*S/N*) by *stacking* records from different sites (Imbrie *et al.*, 1984; Schiffelbein, 1985), although most stratigraphic signals (e.g., oxygen isotopes) do not have a *S/N* greater than about 10 dB (Schiffelbein, 1984).

The highest frequency that can be detected is the *Nyquist frequency*, the λ of which is exactly twice that of the distance between successive observations or samples (Davis, 1986). Irresolvable high frequencies (based on the Nyquist freqency) are incorporated into lower-frequency signals down to the Nyquist frequency (*aliasing*) and may produce significant variation in *S/N*. Cyclicity may also be induced by the sampling regime (Paul, 1992). In other words, depending on v, it may make no sense to use very closely spaced samples unless there is some way to increase *S/N* or if a core of very high v is available (Schiffelbein, 1984). Perhaps one way to resolve higher frequencies is to use outer shelf or slope records, which often have much higher sedimentation rates; the drawback of this approach is that sedimentation is often less continuous than that in the deep-sea (but see Chapter 7). Another way to overcome the effects of time-averaging (such as on presumed morphological trends in evolutionary lineages) is to take samples so widely spaced that time-averaging is unlikely to be a factor (Kidwell and Flessa, 1995).

Physical reworking – via erosion and redeposition – of substantially older (*c.* hundreds of thousands to millions of years old) *remanié* into younger sediments by upward *leaking* (or vice versa by downward *piping*) is not usually considered to be a serious problem in stratigraphy (e.g., 10 of 3000 cases for fossil mammals; Kidwell and Flessa, 1995). In most cases, microfossil-based biostratigraphic zonations are sufficiently precise at these scales that reworked specimens are typically recognizable from their anomalous stratigraphic occurrence and qualitative state of preservation (surface corrosion, breakage, infilling with foreign sediment; but see below). Similarly, Anderson and McBride (1996) distinguished *Chione* (Bivalvia) reworked from estuarine sediments upward into marine facies based on the physical appearance of shells (but see below). Berger and Heath (1968) concluded that bioturbation would not normally be responsible for such reworking in deep-sea sediments, as the concentration of the reworked marker would probably be on the order of 10^{-6} of its original value when the thickness of sediment separating Recent, for example, from older material was as little as 1 m (Figure 5.3). They suggested that erosion is the dominant process mixing older assemblages into younger pelagic sediments (but see Broecker *et al.*, 1991, below). Cutler and Flessa (1990) also concluded that bioturbation is a relatively inefficient means of producing stratigraphic disorder. Based on computer simulations with molluscs, in which stratigraphic datums were progressively shuffled (exchanged) like cards, stratigraphic order was still significant at the >95% level, even after 500 exchanges. They found that disordering was rapid at first (cf. stratigraphic maturity; Chapter 7), but that significant stratigraphic order remained in well-churned computer simulations (24 000 exchanges) in which shells had been displaced an average of two-and-a-half times the thickness of the mixed layer. They concluded that physical reworking disorders sequences much more rapidly than bioturbation. Physical reworking appears to approach a limit, however, depending on depositional setting (e.g., Anderson and McBride, 1996); Kowalewski (1996*a*, p. 318) stated that "it is hard to imagine biostratigraphic stages that have been shuffled out of their original sequence," and vastly more exchanges would be needed to shuffle small microfossils out of sequence in Cutler and Flessa's (1990) simulations (Sadler, 1993). Cutler and Flessa (1990) suggested that the effects of mixing-produced disorder can be minimized (up to a limit, depending on the amount of disorder) by increasing sample size at each stratigraphic horizon, but pooling data in reworked sequences actually worsened disorder: the disorder of the reworked shells, all of which were significantly older than *in situ* shells in computer simulations, completely overwhelmed any relict stratigraphic order (see also Kowalewski *et al.*, 1998). Increasing the spacing between samples is of limited utility in dealing with disordered sequences because the smaller number of samples makes the detection of any trends difficult (cf. Kidwell and Flessa, 1995).

At facies changes or important stratigraphic boundaries associated with significant biotic turnover, reworking by physical or biological agents may be a crucial factor in inferring sequences of events and, potentially, cause and effect. Based on the downcore abundance of microtektites in deep-sea cores, Glass (1969) concluded that stratigraphic ranges of fossils are often altered by the intensity and size-selectivity of burrowing. Extrapolating to fossils, he concluded that the apparent time of extinction of two fossil species of different average size that became extinct simultaneously might be significantly altered by bioturbation, and that the first appearance of a species is a more reliable stratigraphic marker than its extinction (Chapter 4). Similarly, early in the nineteenth century, Albert Oppel used first appearances of ammonites to construct the first detailed biostratigraphic zonation (Berry, 1987). Microfossil occurrences may also differ significantly between tops and bottoms of turbidites, which are not always easily recognizable visually (Brunner and Normark, 1985; Brunner and Ledbetter, 1987).

A more reliable method of discerning reworked specimens is the use of graphic correlation, in which anomalous stratigraphic markers (outliers) fall well off a Line of Correlation plotted through biostratigraphic and other datums (Chapter 7). Recently, using strontium isotope ratios in inoceramids and planktonic foraminifera, MacLeod and Huber (1996) concluded that at the Cretaceous–Tertiary (K–T) boundary, up to 30% of the mass of foraminifers in a sample above the K–T boundary may be derived from reworked specimens. Similarly, Trueman and Benton (1997) used rare Earth elements (REE), which are incorporated into hardparts during diagenesis, to detect reworked bones and teeth whose geochemical signatures differ from other fossil elements and geochemical signals of the assemblage (Chapter 10).

5.3 Types of time-averaged assemblages

Kidwell and Bosence (1991) recognized different magnitudes of time-averaged marine assemblages (Figure 5.4). *Census* assemblages are akin to "snapshots" since they have undergone little or no time-averaging and provide the highest possible resolution of community structure (including commensal relationships), species abundances and size–frequency distributions. Such assemblages are typically formed via sudden death by burial or *obrution* (from the Latin, *obrutus*)

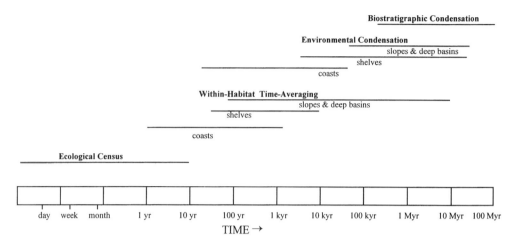

Figure 5.4. Estimates of time-averaging for various types of fossil assemblages. (Redrawn from Kidwell and Bosence, 1991.)

and may comprise Lagerstätten (Chapter 6). Census assemblages are recognizable by high proportions of articulated specimens, many in life position (e.g., echinoid tests surrounded by disarticulated spines), and evidence of rapid burial (e.g., graded bedding, early concretion formation, lack of escape burrows). Kidwell and Bosence (1991) estimated that a minimum sediment cover (about 30 cm in *Recent* sediments) is necessary to prevent the formation of extensive escape burrows. Kidwell and Bosence (1991) also estimated that in the presence of a burrowing infauna, a fossil assemblage must be at least tens of centimeters thick to sequester an assemblage below the surface mixed layer to prevent taphonomic alteration and time-averaging (cf. Hippensteel and Martin, 1998, cited in Chapter 4).

Nevertheless, census assemblages may consist of remains of both organisms previously living at the site and shells introduced from the buried layer, so that obrution deposits may in fact consist of mixtures of shells thousands of years old (Brett and Baird, 1993). Also, few mass mortalities kill all organisms or all species in a community at once. Thus, only a handful of so-called census assemblages are truly instantaneous and the duration of time-averaging for census assemblages in general ranges from zero to a few decades (Kidwell and Bosence, 1991).

Within-habitat time-averaged assemblages are more like movies, in that they represent the sum accumulation of shells through relatively continuous attrition and accumulation of more sporadic (census) assemblages (Kidwell and Bosence, 1991). Such assemblages represent the response of the biological community to environmental fluctuations over short to long time intervals and the potential effects of taphonomic feedback (Kidwell and Jablonski, 1983; Kidwell, 1986b);

5.3 Types of time-averaged assemblages

based on actualistic studies, such taphonomic feedback takes at least decades to accomplish. Resampling and the frequency and extent of environmental fluctuations in modern ecosystems suggest *minimum* durations of within-habitat time-averaging of years to decades in coastal settings and from decades to centuries on the open shelf. Areally extensive stratigraphic biofacies suggest even longer durations of time-averaging; if this were not so, then fossil assemblages would be more "mosaic" in nature and recurrent biotic associations (biofacies) would occur less frequently than observed. Also, coastal environments can shift rapidly and are unlikely to persist in one place for more than a few thousand years, whereas deeper shelf settings are more constant and respond to longer-term forcing mechanisms (sea-level fluctuations) over tens of thousands of years. The potential impact of storms on shelf assemblages also cannot be neglected. Storms (like earthquakes), vary with intensity: small storms are more frequent and affect only shallow depths, whereas strong storms are rarer but scour deeper shelves (*power law*; Kauffman, 1993, 1995; Bak, 1996). Within-habitat condensation is less likely in deep-sea sediments, however, because environmental change is often slower than rates of evolution.

Environmentally condensed assemblages form when environmental change is rapid relative to rates of sediment accumulation (Kidwell and Bosence, 1991). Under these conditions, faunas of different habitats are *telescoped* into a single assemblage or stratigraphic horizon, such as during a marine transgression of a sediment-starved shelf or salinity fluctuations in an embayment. Such assemblages may lie along facies boundaries or other discontinuities; may be shell-rich and microstratigraphically complex; and consist of shells of a wide range of preservational states (because of multiple cycles of burial and exhumation) or from ecologically incompatible species. Durations of time-averaging of environmentally condensed assemblages overlap with the upper limit of within-habitat assemblages, as indicated by the co-occurrence of modern shells with those thousands or tens of thousands of years old (Figure 5.4; Flessa *et al.*, 1993; Martin *et al.*, 1996; Anderson and McBride, 1996).

Biostratigraphically condensed assemblages consist of taxa from different biozones (i.e., mutually exclusive stratigraphic ranges; Kidwell and Bosence, 1991). Shelf and shallower assemblages that are biostratigraphically condensed will likely also be environmentally condensed because such assemblages typically accumulate over hundreds of thousands of years or more (see also section 5.2.4). Examples include lags enriched in bones, teeth, phosphatic steinkerns, and conodonts (e.g., Kidwell, 1989, 1991; Baird and Brett, 1991) or beds containing ammonites or other molluscs of different zones. Fürsich (1971) noted the condensation of fossils from three separate zones into a single highly fossiliferous unit ("calcaire à oolithes ferrugineuses") of the Bajocian (Jurassic) along the Normandy (France) coast; he concluded that reworking must have been relatively rapid, as even aragonitic ammonites and gastropods are well preserved (see also Waage, 1964, for Late Cretaceous Fox Hills Formation of north-central South Dakota; and Bandel, 1974, and Tucker, 1974, for Devonian–Carboniferous cephalopod limestones of Europe).

Biostratigraphically condensed and census assemblages are probably the easiest of the four categories to distinguish because they lie at the extremes of the classification, whereas the durations of time-averaging of within-habitat time-averaged and environmentally condensed assemblages are more variable and therefore more difficult to assess (Brett and Baird, 1993).

5.4 Recognition of time-averaging

5.4.1 Actualistic criteria

"The hallmark of time-averaged shell accumulations having complex histories is the admixture of shells in different states of preservation, from fresh to completely altered" (Kidwell and Bosence, 1991, p. 169). Although not always easily recognized or quantified, the degree of reworking and time-averaging is often recognizable from a number of qualitative paleontological and sedimentological criteria (Fürsich and Aberhan, 1990), many of which are based on studies of modern and Neogene assemblages. Artificially increased taxonomic diversity of marine fossil assemblages can often be recognized by the occurrence of ecologically incompatible species in a fossil assemblage or changes in ichnofacies or other benthos from soft-sediment to firmground (Fürsich, 1978; Kidwell, 1991). For example, abundant empty shells of opportunistic species in fossil assemblages that occur only in low abundance in living biotas is a qualitative indicator of time-averaging (Fürsich and Flessa, 1991). Sedimentological or stratigraphic criteria for time-averaging include (1) biofabric (shell-packing density and orientation; Chapter 2); (2) glauconitic, ferruginous or phosphate-rich beds that are typically associated with low rates of sediment accumulation and sometimes with unconformities (either directly or traceable in outcrop or subsurface); (3) sedimentary structures indicative of high-energy (including storm) conditions, such as ripup clasts, shell orientation and size-sorting (which may nevertheless vary substantially; Chapter 2), and laminated draft fills at angles to surrounding bedding; (4) facies relationships; (5) fossils of different preservational state or taphonomic grade (degree of articulation, fragmentation, abrasion, bioerosion,

5.4 Recognition of time-averaging

and encrustation) or with sediment infill that differs from the surrounding matrix (e.g., remanié); and (6) on evolutionary time scales, co-occurrence of biostratigraphic markers from different zones. Fürsich and Aberhan (1990) consider criterion 5 to be a sensitive indicator of time-averaging, and the condition of fossils from different zones may indeed suggest stratigraphic condensation or reworking, but sufficient (and counterintuitive) exceptions have since been identified on shorter time scales that warrant application of criterion 5 in a much more cautious manner (Kidwell, 1991; Flessa *et al.*, 1993; Martin *et al.*, 1996; see below).

Despite numerous biostratinomic studies, especially of transport and reworking (Chapter 2), it has been only recently that estimates of time-averaging have become available in terms of absolute numbers of years. Flessa (1993) briefly reviews some of the techniques that have been used to estimate durations of time-averaging, including radiocarbon dating (see Bowman, 1990; Webb, 1993; for caveats and calibrations) and amino acid racemization. As many shells are beyond the range of radiocarbon dating in actualistic studies, the identification of Pleistocene remanié is unlikely without the use of geochronological frameworks based on *ratios* of certain racemized amino acids, which is why this technique has become an invaluable tool in Holocene and Pleistocene taphonomy and chronostratigraphy (e.g., Murray-Wallace and Belperio, 1995; Wehmiller *et al.*, 1995; Martin *et al.*, 1996; Murray-Wallace *et al.*, 1996; see also Chapter 4). A related technique employs the use of *free* amino acids (FAA), but temporal calibration of the FAA method is less well documented (Powell *et al.*, 1989). Like any other geochronological tool, aminostratigraphy requires careful evaluation of the nature of the fossil assemblages involved. Disputes regarding the accuracy of aminozones (based on bivalves) versus those based on uranium-series dating (of corals) may very well be less related to the techniques themselves than to the fact that bivalves are supposedly more likely to be reworked, whereas corals are presumed to remain *in situ* and younger than the bivalves, their spat having settled on clams after their death (J. F. Wehmiller, 1997, personal communication; cf. Wehmiller and Belknap, 1982; McCartan *et al.*, 1982; Muhs, 1992; and Wehmiller *et al.*, 1995).

5.4.2 Criteria for ancient settings

It is *relatively* easy to apply "actualistic" criteria (especially when one can obtain absolute dates!) to Cenozoic, and to a lesser extent, Mesozoic assemblages, which are usually composed of taxa with extant relatives; in fact, the initial subdivision of the Cenozoic into epochs was based on the percentage of extant taxa, which decreased with increasing age. But what of Paleozoic assemblages, which are often characterized by extinct taxa (no modern analogs)? Application of the Principle of Uniformitarianism to Paleozoic assemblages is more difficult than to Meso-Cenozoic assemblages because taxonomic composition and rates of bioturbation and bioerosion during the Paleozoic undoubtedly differed from that of younger deposits (Brett and Baird, 1993; see also Chapters 8, 9). Nevertheless, comparative taphonomic analysis of Paleozoic assemblages can often distinguish between episodic and background processes that might not be readily distinguishable based on sedimentology alone. Many of these criteria are broadly similar to those for younger deposits and are substantiated by actualistic studies.

Brett and Baird (1993) distinguished several types of *single-event fossil beds*: (1) mass mortality events without rapid burial; (2) mass mortality with rapid burial; (3) simple allochthonous shell beds, in which organisms transported out of habitat are a sample of the live biota of the foreign habitat; and (4) simple autochthonous shell beds. Of these types, only obrution-related assemblages will commonly correspond to census assemblages. Taphonomic indicators of event beds are divided into those of *event-disruption* and *event-burial* (Brett and Baird, 1993). Event-disruption features include (1) evidence of turbulence (e.g., fossils incorporated into graded or migrating beds); (2) fossils rotated out of life position and into hydrodynamically unstable orientations (e.g., overturned corals and bryozoans), possibly followed by regrowth (if disruption is non-lethal) to produce contorted (geniculate) colonies or colonies containing mud layers; and (3) partial truncation of deep burrow systems by seafloor erosion. Event-burial features include (1) burial in life position and burial of articulated multi-element skeletons or soft tissues (Chapter 6), and (2) escape burrows or ichnofossils indicating attempted escape (e.g., "death marches" of Seilacher et al., 1985).

Although articulated skeletons can be transported for some distances prior to the onset of disintegration (Chapter 6), actualistic studies point to a broad spectrum of disarticulation rates in macroinvertebrates (Chapter 2) that can be used to infer durations of time-averaging in Paleozoic assemblages. Perfect articulation of multi-element skeletons, especially delicate ones of crinoids, ophiuroids, and edrioasteroids probably requires burial within hours to a few days. Among crinoids themselves, rates of disarticulation vary (Chapter 2): for example, the preservation of flexible crinoids, which tend to disintegrate more rapidly than others, is a good indicator of rapid burial. Thus, while exceptional preservation of delicate features, such as pinnulate arms of crinoids, does not *prove* autochthonous burial, rapid burial may be substantiated by, say, the presence of trilobite molt ensembles (Figure 5.5), which would otherwise not be found if substantial transport had occurred (Brett and Baird, 1993). Articulated bivalves or enrolled trilobites also

5.4 Recognition of time-averaging

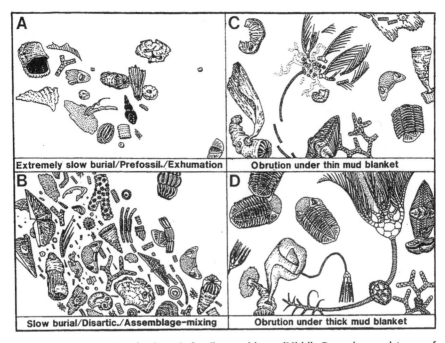

Figure 5.5. Examples of Paleozoic fossil assemblages (Middle Devonian mudstones of western New York) which record various amounts of time-averaging. (A) Condensed assemblage (phosphatized remains indicated by black); (B) within-habitat time-averaged assemblage (note variable degrees of articulation and corrosion); (C) mass mortality followed by mud deposition over the next few days (note incipient disarticulation of crinoid and trilobites); (D) census (obrution) assemblage (note excellent articulation of fossils). (Reprinted from Brett and Baird, 1993, with permission of the Paleontological Society.)

suggest rapid burial because if buried outstretched, muscles and ligaments would probably have decayed. On the other hand, slightly disrupted skeletons (*incipient disarticulation*; Figure 5.5) probably indicate somewhat slower burial, perhaps on the order of days, especially in oxygenated environments in which scavengers were present.

Paleozoic *composite* beds represent within-habitat time-averaging (Figure 5.5). Like their Meso-Cenozoic counterparts, they exhibit multiple episodes of burial, exhumation, and reburial of shells, and likely consist of two or more shell layers or lenses. In the Paleozoic, such beds are usually up to several centimeters thick and persist laterally for meters to tens of kilometers. Like their younger counterparts, taphonomic indicators include relatively high diversity compared with surrounding sediments; mixtures of ecologically incompatible species (e.g., taphonomic feedback involving taxa preferring soft or hard substrata), which may produce alternations of infaunal and epifaunal taxa (*Jeram* model of Seilacher, 1985); and shells of varying states of preservation. The criteria for recognition of *hiatal* and *lag* accumulations (Figure 5.5) are broadly similar to those for environmentally and biostratigraphically condensed assemblages, respectively (section 5.3; see also Chapter 7).

5.5 Durations of time-averaging

5.5.1 Macroinvertebrates

Before absolute dates became available for time-averaged assemblages, several workers estimated durations of time-averaging based on sedimentological, paleontological, and stratigraphic criteria. Shallow-water marine sediments and their molluscan assemblages have probably received the greatest amount of attention with regard to durations of time-averaging because of their accessibility and prominence in the fossil record. In the case of the Maryland Miocene (Chapter 1), which formed on a passive margin with relatively low sediment accumulation rates, Kidwell (1988, 1989) estimated durations of time-averaging of major shell beds on the order of a few thousands to a few tens of thousands of years and 10^5 years at most; major shell beds consist of minor shell layers condensed over ecological time scales (10^0-10^2 years). At the other extreme, major shell beds can be formed over intervals whose upper duration can be bracketed by biostratigraphic zones based on molluscs, diatoms, and planktonic foraminifera, or transgressive–regressive depositional sequences of about 1 million years' duration (Kidwell, 1988, 1989). Some of the contained fossils, especially shark's teeth, are currently being reworked into modern nearshore sediments along Chesapeake Bay and therefore comprise remanié since the duration of "time-averaging" is now on the order of 5–15 million years (Flessa, 1993). On the other hand, shell beds of the Pliocene Imperial Formation (California), which formed on a tectonically active margin and are characterized by much higher sediment accumulation rates (Chapter 7), were estimated to represent 10^2-10^3 years (Kidwell, 1988). Based on paleontological and sedimentological criteria (section 5.2), Fürsich and Aberhan (1990) estimated that time averaging in nearshore environments should be on the order of 10^2 years, whereas outer shelf and deeper environments should be characterized by durations of 10^4-10^5 years. Peterson (1976) used an indirect approach to estimate durations of 4–894 years by extrapolating from rates of shell weight loss (four molluscs and one echinoderm) over 7 months in field experiments.

The confirmation of these estimates by absolute dating techniques strongly indicates that most *qualitative* criteria of time-averaging, which are *recognizable in the field* (section 5.2), are relatively accurate (but cf. taphonomic grades

5.5 Durations of time-averaging

below). Goodfriend (1987), for example, demonstrated age mixing in land snails (*Trochoidea seetzeni*). He found that both alloisoleucine/isoleucine (A/I) amino acid ratios and ^{14}C dates on shells increased with depth, although bulk soil samples gave an *average* age that was typically older than that of snails.

Goodfriend's (1987) results parallel those of other workers who have concentrated on marine assemblages. Flessa and Kowalewski (1994) compiled 734 published radiocarbon dates (many conventional and uncalibrated) from modern nearshore (<10 m) and shelf (>10 m) habitats, assuming that the maximum age of a shell serves as an estimate of time-averaging. Nearly 70% (42 localities) of nearshore shells (66 localities total) fell into the 0–500 year class. The median age of all shells was approximately 1255 years. On the other hand, deeper shelf shells ($n = 126$ localities total) exhibited a bimodal distribution: slightly more than 25% (33 localities) fell into the 0–3000 year class, whereas about 25% (31 localities) fell into the 9000–12 000 age class, with a median of 9435 years (interestingly, Callender *et al.*, 1992, found a bimodal size–frequency distribution of bivalves in deeper shelf and slope environments). In both cases, the number of shells in greater age classes declined rapidly toward the maximum age class of 36 000–39 000 years, which is near the upper limit of radiocarbon dating. Similar age ranges have been published for other bivalved molluscs (see Flessa, 1993, for references) and indicate that shells may persist for thousands to tens of thousands of years, although taphonomic pathways and alteration would appear to vary substantially between sites and even between shells from the same sites (Callender *et al.*, 1990, 1992, 1994). Most recently, based on ^{14}C calibrated amino acid racemization dates of *Chione*, Kowalewski *et al.* (1998) concluded that processes occurring over scales of less than 10^2–10^3 years are not recorded in shell beds (cheniers).

Flessa (1993) suggested a number of explanations for the general decrease in durations of time-averaging toward shore (see also Flessa and Kowalewski, 1994). Shells may tend to be younger in nearshore settings because of: faster rates of destruction (e.g., bioerosion, abrasive reduction and fragmentation, predation; Chapter 2); greater sediment accumulation rates, which ought to leave *younger* shells nearer the SWI barring other processes; higher rates of shell production, which would bias assemblages toward younger ages (cf. Kidwell, 1986a; Broecker *et al.*, 1991; cf. principle of linear superposition; section 5.1); and the possible counteraction of increased sedimentation on bioturbation.

The Holocene rise in sea-level may also affect the availability of shells for dating: very old shells may not have been able to accumulate in many nearshore settings because sea-level reached within approximately 10 m of its present position only about 7000 years ago (Flessa, 1993; see also Manjunatha and Shankar, 1996). Thus, the relatively recent rise in sea-level may have restricted the potential duration of time-averaging in nearshore sediments, whereas shells continued to accumulate farther offshore (Anderson and McBride, 1996, came to a similar conclusion; see also Stanley and Bernasconi, 1998). Holocene sea-level rise also initiated non-steady-state diagenesis by affecting rates of sedimentation, oxic degradation of organic matter, and dissolution, all of which could potentially influence durations of time-averaging (Manjunatha and Shankar, 1996). There may also be a sampling bias toward younger shells, because many shells have been chosen to date the Holocene sea-level rise; hence the bimodal distribution of available shell ages, with the oldest mode at about 10 000 years. Nevertheless, old shells may be transported landward, whereas younger shells may be transported seaward (Flessa, 1998).

Like so many other phenomena in taphonomy and Earth history, generalizations regarding durations of time-averaging must be tempered – at least for the time being – by consideration of depositional setting. Flessa et al. (1993; see also Meldahl et al., 1997) found that disarticulated bivalve shells (*Chione*) that had been collected from the SWI of Bahia la Choya (Choya Bay) in the northern Gulf of California (Sonora, Mexico) exhibited a broad range of surface degradation and that shell condition (taphonomic grade) was not an infallible indicator of shell age (time since death; cf. Brandt, 1989): moderately old specimens (~1800 years) were sometimes relatively well preserved, whereas younger shells (several hundred years) were sometimes more highly degraded (Flessa et al., 1993). Flessa et al. (1993) suggested that a shell's surface condition is primarily indicative of the residence time of the shell at the SWI and not its age (see also Frey and Howard, 1986). Calculated sediment accumulation rates at Choya Bay for the past few thousand years are quite low (~0.038 cm/yr; Flessa et al. 1993) because sediments are derived mainly from local sources. In other words, the *taphonomic clock* (Kidwell, 1993a) that equates shell preservation with time is frequently reset at Choya Bay, and no doubt in other sedimentary regimes. Low accumulation rates at Choya Bay allow frequent resurrection and exposure of shells by bioturbators and potentially by storms (K. Meldahl and A. Olivera, 1995, personal communication) so that shell half-lives are relatively long (285–550 years); by contrast, higher accumulation rates at Bahía Concepción associated with rift basins (also northern Gulf of California) decreased shell half-lives to 90–165 years (Meldahl et al., 1997). Despite the greater potential for reworking of shells at Choya Bay, however, Springer and Flessa (1996) found that surface heterogeneity of shell distributions was accurately recorded in the subsurface shell bed (cf. section 5.2.2).

Martin et al. (1996) arrived at findings similar to those of Flessa et al. (1993) for Choya Bay, with one important exception. Amino acid racemization analyses of *Chione* indicated that the scale of time-averaging of bivalve assemblages at Choya Bay may be one to two orders of magnitude greater (~80 000–125 000 years) than that reported previously. Very old shells comprised a pocket of outliers in plots of shell age distribution, with other shells concentrated in younger age classes (Martin et al., 1996; later noted as a tail in age distributions by Meldahl et al., 1997; cf. principle of linear superposition, section 5.1). Of 17 *Chione* valves analyzed by Martin et al. (1996), nine fell into the extremely old age range, whereas the others were at most a few thousand years old (based on calibrated accelerator mass spectrometer ^{14}C dates). Significantly, six of these nine valves were indistinguishable in appearance from much younger *Chione* valves (cf. Anderson and McBride, 1996).

This scale of time-averaging is not unique to Choya Bay (cf. Meldahl et al., 1997). Based on amino acid-racemization (A/I) ratios, Wehmiller et al. (1995) found extensive age-mixing of Pleistocene and modern bivalves (mainly *Mercenaria*) at 21 beach sites between New Jersey and Florida (collection sites were concentrated in North Carolina so as to minimize anthropogenic reworking). The age distribution of the Pleistocene shells on Atlantic coast beaches appears related to the distribution of Pleistocene units in the shoreface and inner shelf and the thickness of Holocene sediment cover (see also Flessa, 1993, for other references). Murray-Wallace et al. (1996) found similar results. They used specimens of the bivalves *Pecten fumatus* (scallop) and *Placamen placidium* to identify three aminozones that corresponded to oxygen isotope stages 8, 6, and 2 (which correspond to late Pleistocene glacial intervals or sea-level lowstands) in vibracores taken off the southeast coast of Australia. Based on the lithological and chronostratigraphic framework, only a thin veneer of Holocene sediments overlies the Pleistocene section, and the degree of reworking increased toward the shoreline of each glacial maximum.

Meldahl et al. (1997) inferred from skewed shell age distributions that shells are added to the surface mixed layer at approximately constant rates and also removed from the mixed zone at random by destruction or burial. Fluctuations in productivity may, however, be detectable from lithology (e.g., phosphorite distributions) or stable (carbon) isotopes of single (ontogenetic) or multiple shell accumulations (e.g., Krantz, 1990). Fluctuations in shell input may also occur on longer time scales. Allmon et al. (1996), for example, determined that fluctuations in vertebrate and invertebrate productivity occurred off the west coast of Florida (regional scale) in response to tectonic activity (rise of the Isthmus of Panama) and its effect on circulation and upwelling regimes.

Although the above studies provide absolute durations of time-averaging, mathematical – especially numerical – models are probably necessary to better understand the actual formation of death assemblages. Powell (1992) constructed a mathematical model of the carbonate content of shallow-water terrigenous sediments (assuming maximum carbonate content at any depth of $<\sim 0.8$ g dry weight cm^{-3}). The basic equation of the model is:

$$\frac{\partial C_{ij}}{\partial t} = A_j(t) - \xi_j(t)T(t) + \frac{\partial (WC_{ij})(t)}{\partial z} \qquad (5.8)$$

where C_{ij} is the amount of carbonate present in any sediment parcel i at time t at a stratigraphic horizon z for a species j in g cm^{-3}; A_j is the rate of carbonate addition in g cm^{-3} yr^{-1} for any species; T is the rate of carbonate (taphonomic) loss in g cm^{-3} yr^{-1}; ξ_j is the fraction of total carbonate represented by species j; and W is the burial rate in cm yr^{-1} (cf. equation 5.1). In the model, the entire carbonate of each species (ξ_j) is assumed to be lost, which appears to be more or less correct for small or juvenile shells but not adults (Powell *et al.*, 1984; Green *et al.*, 1993; Martin *et al.*, 1995). Also, Powell's model assumes that all shell carbonate is gravel-size or greater (mainly bivalves in terrigenous settings), even though some studies have demonstrated that much greater quantities of carbonate are lost via dissolution of foraminifera (Smith, 1971; see also Green *et al.*, 1993; Martin *et al.*, 1995). T and W are considered to be *attributes of the environment*, whereas all shell carbonate behaves equivalently (no subscripts are used for T and W) for the sake of simplicity (cf. Chapter 3), and no taphonomic feedback is assumed to occur. Shell accumulation in the model is, then, primarily a function of production (A_j), most of which is lost (T), and, especially, burial and reworking (built into the term W).

Separate numerical models may have to be constructed for specific sites and compared within habitat (marsh, nearshore, shallow versus deep subtidal, etc.) before gross generalizations can be made. Based on his previous research, Powell (1992) considered primarily shallow Texas bays to support his model. If most shell loss occurs near the SWI, and if sediment accumulation rates are fast enough, many shells will probably be preserved (as predicted by rule 1; Chapter 1), and which Powell considered the most likely scenario. On the other hand, Powell (1992) concluded that if significant taphonomic loss occurs throughout the surface mixed layer, one way for shells to be preserved is for a temporal

5.5 Durations of time-averaging

offset to occur between shell production and loss. Thus, if bioturbation is most intense during the summer, shell input and burial during winter storms may be quite significant, a conclusion reached by Martin *et al.* (1995, 1996; see also Aller, 1982*b*; Green *et al.*, 1993). But, Powell (1992) concluded that such an offset between shell input and loss would only occur if sediment accumulation rates are high and bioturbation has little effect, a conclusion not substantiated by studies at Choya Bay (Martin *et al.*, 1996). Powell (1992) also concluded that carbonate production rates on the Texas shelf should be adequate to explain sediment shelliness there, and that shells should be roughly 1700 years old, a figure at odds with the compilations of Flessa (1993) and Flessa and Kowalewski (1994). In addition, Powell (1992) concluded that preservational processes usually occur on time scales too short to be recorded as variations in carbonate content with depth (perhaps especially after the early-middle Paleozoic, when bioturbation rates increased substantially; see below and Chapter 9), and that evidence of preservational processes is to be found mainly in taphonomic grades (but see above).

It is unlikely, however, that similar approaches can be extended to other taxa without taking into account differential preservation. Kowalewski (1996*b*), for example, noted differential loss between bivalves and lingulide brachiopods (*Glottidia palmeri*) of intertidal flats of the Colorado Delta (northern Gulf of California): bivalves persisted for at least a minimum of 100 years, whereas brachiopod valves were typically destroyed in no more than a few months. Not surprisingly, the majority of well-preserved lingulides were from larger size classes. Thus, high to catastrophic sedimentation rates are probably needed to preserve lingulides. On a positive note, extensive time-averaging or significant transport of lingulides appears unlikely, so that on those occasions when they are preserved, they would appear to record short-term aspects of their history with high fidelity.

Durations of time-averaging of Paleozoic brachiopod-rich assemblages can, however, be specified indirectly with considerable precision (Brett and Baird, 1993). In the case of disruption and burial events (section 5.4.2), graded accumulations probably developed in minutes to hours (Seilacher, 1982), whereas, based on actualistic studies, composite (within-habitat time-averaged) Paleozoic shell beds probably represent centuries to a few thousand years. Also, estimates of "standing crops" of certain taxa preserved in life position, such as brachiopods, could conceivably be used to estimate durations of shell accumulations. More rarely, rates of shell accumulation can be estimated by tracing shell-rich beds into colonies that probably accrued over several centuries. Hiatal and lag concentrations correspond to tens of thousands and hundreds of thousands to millions of years, respectively, based on actualistic studies.

5.5.2 Microfossils

Despite their relatively small size and mineralogy (Chapter 3), foraminiferal assemblages may be time-averaged for relatively long durations. Dead tests of large reef-dwelling foraminifera, such as *Archaias* and *Amphistegina*, found in sediment assemblages frequently exhibit considerable surface degradation. Based on experimental analyses of both abrasion and dissolution resistance of these and other modern reef-dwelling foraminifera from Discovery Bay, Jamaica, Kotler *et al.* (1992) concluded that once produced, foraminiferal tests in pure-carbonate environments tend to persist for long periods of time, although at that time they were uncertain exactly how long (years? decades? centuries?). Accelerator mass spectrometer (AMS) ^{14}C dates later showed that mildly to highly corroded tests of *Archaias* and *Amphistegina* ranged, respectively, between 474–2026 years and 936–1839 years. Surprisingly, pristine tests ranged from modern (as indicated by "post-bomb" ages) up to about 2000 years old, thereby exceeding the ages of mildly to highly corroded tests (Martin *et al.*, 1996). The age of fresh-appearing tests increased with depth in cores taken from Lee Stocking Island (Bahamas), but mildly and highly degraded tests showed age reversals with increasing core depth; bulk carbonate showed a similar pattern, with the oldest age (3455–3778 years) in core tops (0–2 cm). High levels of $CaCO_3$ in pure-carbonate environments appear to buffer against dissolution and promote extensive mixing of old tests and much younger shells, while inhibiting the formation of taphonomic damage on the tests themselves (cf. Chapter 3 for differences in dissolution behavior between foraminifera and other bioclasts in carbonate environments). The preservation of larger or dominant foraminifera – or other bioclasts – may mask subtle ecological gradients (Martin and Liddell, 1989).

Based primarily on size differences between foraminifera and *adult* molluscs, Martin (1993) predicted that larger (and presumably more preservable) molluscan debris ought, in general, to be significantly older than foraminiferal assemblages of the same horizon. Durations of time-averaging should therefore be greater for molluscan assemblages. Nevertheless, foraminiferal assemblages in shallow-water siliciclastic regimes may be time-averaged over intervals of duration similar to those of carbonate environments, *depending on depositional setting*. Foraminifera at Choya Bay undergo seasonal cycles of reproduction and dissolution related to periodic overturn (late March–April, late fall) of the nutrient-rich thermocline and associated phytoplankton blooms, and subsequent onset of bioturbation

5.5 Durations of time-averaging

(dissolution) as ambient air temperatures rise later in the spring and summer (Martin et al., 1995). Similar reproduction–dissolution cycles have been widely reported for foraminifera from other shallow-water environments (e.g., Buzas, 1965; Green et al., 1993; cf. Powell et al., 1984).

Most foraminifera at Choya Bay dissolve within about 3 months following reproduction, which, *intuitively*, suggests a quite young age for foraminiferal tests. Nevertheless, foraminifera from the northern margin of Choya Bay were as old as or older than the much larger *Chione*. Tests at northern flat sites are better preserved because of higher total shell content near the surface and lower rates of bioturbation (alkalinity buildup in porewaters; Chapter 3). Without high sediment accumulation rates, CDFs (primarily callianassid shrimp and polychaetes) repeatedly transport fine-grained sediment from depth up to the SWI; at the same time, this tends to concentrate coarse molluscan (mainly bivalve) debris in a relatively distinct subsurface shell layer (biogenic graded bedding; Meldahl, 1987). The depth to the shell layer decreases northward from more than 60 cm on the southern tidal flat to approximately 10 cm in some places, especially over a Pleistocene coquina that is about 125 000 years old (antecedent topography) that crops out over the northern flat. Consequently, total carbonate weight percent (shell content) of sediments near the SWI tends to increase to the north (Martin et al., 1995, 1996) and probably buffers against dissolution (Aller, 1982b). Although bioturbation by CDFs is most intense on the inner and southern flat, it decreases to the north, especially where sediment thickness is less than about 20–25 cm (Martin et al., 1995, 1996). Thus, sediment at the northern sites also tends to become anoxic and sulfate-reducing bacteria thrive, as evidenced by porewater alkalinity as high as approximately 50 mequiv l^{-1} (this situation parallels that of FOAM and NWC sites in Long Island Sound; Chapter 3).

It seems unlikely, however, that this mechanism alone can account for the surprisingly old Choya Bay test ages considering the extent of test dissolution. Instead, some tests are probably *rapidly* incorporated into the subsurface shell layer by CDFs and preserved there until they are exhumed, much later, by biological activity or storms. Thus, counterintuitively, even if a test is rapidly buried by *bioturbation* at Choya Bay, rather than by *rapid sediment influx*, it may still remain relatively pristine.

Because of low accumulation rates and extensive bioturbation and physical reworking, the *potential* for mixing or reworking of substantially older (Pleistocene) foraminifera into Holocene sediments at Choya Bay certainly exists, as demonstrated by the high A/I values of *Chione* valves mentioned previously. Foraminiferal tests that had been potentially reworked from the underlying

Pleistocene (suggested by the physical appearance of tests) were noted only rarely in samples, however.

Nevertheless, extensive mixing or reworking of microfossils with little or no surface alteration into much younger sediments may be more widespread than previously thought. Anderson *et al.* (1997) found pristine tests of large, sumbiont-bearing foraminifera ranging in age from about 1.9 to 6.4 ka in Holocene sediments of the Gulf of Mexico; based on water depth and light requirements, they inferred that tests had been transported to the site of deposition on seagrass (*Thalassia*) uprooted by storms, whereas molluscs were parautochthonous (see also Bock, 1969; Davaud and Septfontaine, 1995). Based on amino acid racemization techniques, Murray-Wallace and Belperio (1995) found that *apparently fresh*, uncemented specimens of the large, symbiont-bearing foraminifer *Marginopora vertebralis* were reworked from late Pleistocene (about 125 Ka) carbonate sediments into the overlying modern carbonate skeletal sands.

To complicate matters further, shallow-water benthic foraminifera sometimes exhibit microhabitat preferences that reflect surface productivity, organic carbon influx, and porewater oxygen content. Indeed, Buzas *et al.* (1993) argued that in soft sediments all benthic foraminifera are infaunal because a true SWI does not exist. Such occurrences represent a situation analogous to infaunal macroinvertebrates mixing into time-averaged assemblages with long-dead epifaunal species (K. Flessa, 1993, personal communication). Goldstein and Harben (1993) and Goldstein *et al.* (1995) found that the abundance of the common marsh species *Arenoparrella mexicana* changed from approximately 5% of total assemblages at the surface to approximately 50% within the top 3–5 cm of sediment because of large infaunal populations. This species, which may live to depths of about 30 cm and therefore "by-passes" the surface mixed layer, is resistant to degradation and may dominate high marsh death assemblages despite small living populations. Differential preservation of infaunal populations of this or other species could affect paleoenvironmental interpretation and the construction of Holocene sea-level curves (Ozarko *et al.* 1997; Saffert and Thomas, 1998; see also Chapter 10).

Extending across the continental shelf are probably regional preservational gradients, akin to taphofacies, that no doubt strongly influence time-averaging and stratigraphic resolution of foraminiferal assemblages (Martin, 1993). In fact, Middleburg *et al.* (1997) found that water depth – rather than sedimentation rate – was an excellent predictor of gradients in bioturbation and diagenesis. Gradients in water energy, for example, are suggested for modern shallow-shelf habitats by ichnofacies and by foraminiferal lags (Martin, 1986; Martin and

Liddell, 1988, 1991; Martin and Wright, 1988). In ancient settings, Aigner (1982*b*) concluded that the tests of the nummulite, *Nummulites gizehensis*, were hydraulically equivalent to very coarse sand and that nummulite accumulations of the Eocene of Egypt were the result of storm deposition and winnowing. Although he made no estimates of the duration of time-averaging, they would appear to be in the order of thousands of years or longer based on the time-stratigraphic framework.

Environmental (taphonomic) gradients may also change on more localized scales on continental margins. For example, in the case of large deltas, which can easily reach the shelf-edge and dump sediment onto the slope during low sea-level stands, increased shell and organic matter input (a consequence of nutrient influx) coupled with rapid burial would be expected to elevate SO_4^{2-} reduction rates, enhance preservation, and maintain a relatively normal age structure within sediment packages (cf. Chapter 7 regarding stratigraphic completeness of sections, however).

Ichnofacies models also suggest shelf-slope gradients in food (organic carbon) availability and oxygen content. Lin and Morse (1991), for example, concluded that in the northern Gulf of Mexico (Mississippi Delta and Texas–Louisiana shelf) SO_4^{2-} reduction rates and pyrite (FeS) concentrations generally decrease exponentially with increasing water depth (and decreasing sedimentation rate). Sulfate-reduction rates in the Gulf are intermediate between those of shallow nearshore organic carbon-rich sediments and organically poor deep-sea sediments (Canfield, 1991; anoxic silled basins are an exception; Chapter 3). Organic carbon concentrations in continental margin sediments are in turn a function of sedimentation rate: at high sedimentation rates (shallow water), labile organic matter undergoes a shorter period of oxic and suboxic degradation (despite typically higher rates of bioturbation; Chapter 4) before it is incorporated into subsurface layers where it can support SO_4^{2-} reduction (Chapter 3; the depth of bioturbation in pelagic settings seems, however, to be more strongly related to organic carbon flux – and nutrient supply – from the surface than to bulk sediment accumulation rates; Trauth *et al.*, 1997; see also Chapter 4). As SO_4^{2-} reduction rates decline across the continental shelf (decreased sedimentation rates), rates of bioturbation must also either decline or be counteracted by increased input of $CaCO_3$ for preservation to occur. The commonly-observed peak in foraminiferal (benthic and planktic) abundance near the shelf-slope break (e.g., Bandy, 1953) is probably a reflection of surface water productivity (shell input, as evidenced by increased planktic/benthic ratios and foraminiferal numbers) and decreased rates of terrigenous sedimentation. Differential preservation is no doubt important on the slope and rise, as well, especially if relevant biostratigraphic or paleodepth indicators are rare or susceptible to dissolution. As sedimentation rates decrease with increasing water depth, so too do SO_4^{2-}-reduction rates (despite decreased rates of bioturbation; Canfield, 1991; Lin and Morse, 1991). Also, the lysocline tends to shallow toward the continental margins in response to input and decay of marine and terrestrial organic matter (Chapter 3).

Like their nearshore counterparts, open marine (shelf to deep-sea) foraminifera also prefer certain habitats. Open-shelf species tend to be epifaunal (shallow oxic layer), whereas on the continental slope and rise, the oxidized surface layer is relatively thick and foraminifera are primarily deep infaunal species. At greater depths, deep-water epifaunal species predominate because organic carbon (food) becomes limiting. Loubere (1991) also found that certain deep-sea (East Pacific Rise) foraminifera respond to productivity gradients (see also Trauth et al., 1997). Significantly, some species appear to be highly adaptable, in terms of microhabitat preference and depth stratification, to changes in food availability and environmental conditions (Linke and Lutze, 1993).

The microhabitat preferences of foraminifera that have been demonstrated to live infaunally in shelf to deep-sea sediments, even those that deposited within the oxygen minimum zone (e.g., Bernhard, 1992), correlate to test morphology, pore size, and density (Corliss, 1985; Corliss and Chen, 1988; Corliss and Emerson, 1990; Corliss and Fois, 1991; Linke and Lutze, 1993; Jorissen et al., 1995). But it remains unclear if there is a direct link between depth-stratification and preservation. Certainly, porewater chemistry exerts subtle effects on stable isotope composition of the test: McCorkle et al. (1990) found that in deep-sea benthic foraminifera, carbon isotope composition, which is commonly used to infer past changes in oceanic productivity, can be influenced by microhabitat preferences and porewater chemistry (Chapter 3). Based on the Berger–Heath (1968) mixing model, Loubere (1989) found that significant variation in species abundances occurs in the mixed layer as a result of taxon depth stratification: infaunal taxa increase non-linearly in abundance with depth to the level of their true habitat, below which taxon abundance remains constant, whereas epifaunal assemblages may be significantly modified. He concluded that the best representation of total taxon abundance occurs at the base of the bioturbated zone (see also Loubere et al., 1993a,b; 1995). Although Douglas et al. (1980) concluded that life and death assemblages of the upper slope (100–400 m) of the southern California borderland are strongly similar (based on surface samples), Loubere and Gary (1990; see also Loubere et al., 1993a,b) concluded that there was substantial specimen loss in the upper 10 cm of boxcores from the slope (c. 1000–1100 m) off the Mississippi Delta. Denne and Sen Gupta (1989) came to similar conclusions for Texas–Louisiana slope sediments; i.e, surface grab samples, which have frequently been the basis of foraminiferal distribution studies, are not necessarily representative of subfossil assemblages of open marine environments. They found that bathyal calcareous species could be grouped into those that are well preserved (increasing downcore), moderately preserved (no trend), and poorly preserved (decreasing downcore). Agglutinated species were either disaggregation-resistant (no trend) or disaggregation-prone (decreasing downcore). Some tests are undoubtedly reworked into younger sediments from older deposits, as well (e.g., Rathburn and Miao, 1995).

Loubere (1997) later coupled the production and taphonomy of foraminiferal assemblages in outer shelf and slope sediments to organic carbon flux (productivity) and bottom water oxygen concentrations based on data from the Gulf of Mexico and deep basins of the California borderland (see also Chapter 4). He specified four different production–taphonomic regimes that correspond to changes in test assemblages across organic carbon flux–bottom water oxygen gradients reported in the literature: (1) a low organic carbon flux ($<2\,g\,C\,m^{-2}\,yr^{-1}$)–high bottom water oxygen ($>3.5\,ml\,l^{-1}$) regime, in which test production is close to the SWI, test destruction occurs in the upper 1–2 cm of sediment, and mixing reaches to depths of 5–8 cm; surface and downcore assemblages are quite similar; (2) a moderate organic carbon flux ($2-6\,g\,C\,m^{-2}\,yr^{-1}$)–high bottom water oxygen regime, in which occur vertically stratified microhabitats, a mixed layer in the upper 3–4 cm of sediment, significant infaunal test production below the mixed layer, and incomplete mixing across the mixed–historical layer boundary that may transport tests into the mixed layer from the historical layer while removing tests produced in the mixed layer to the historical layer before they are destroyed; surface and downcore assemblages may differ substantially (see also Denne and Sen Guptā, 1989; Loubere and Gary, 1990; and above); (3) a moderate organic carbon flux–moderate bottom water oxygen ($1.5-3\,ml\,l^{-1}$) setting in which there is heterogeneous (no microhabitat-specific) shell production in the sediment column, intensive bioturbation, and extensive test destruction, especially in the upper 5 cm of sediment; like regime 1, downcore assemblages are quite similar to those at the surface despite the different taphonomic setting; and (4) a high organic matter flux ($>6-7\,g\,C\,m^{-2}\,yr^{-1}$) low bottom water oxygen ($<1\,ml\,l^{-1}$) region, in which there is heterogeneous shell production and low rates of bioturbation and shell destruction, so that assemblages are quite variable on small spatial scales.

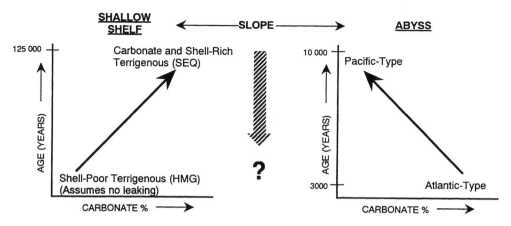

Figure 5.6. Presumed relation between apparent age and CaCO$_3$ content of foraminiferal assemblages across a passive continental margin. Slope environments are presumably gradational between shallow shelves and abyss, but rapid burial in the vicinity of deltas may counteract dissolution and time-averaging. Sequential dissolution is that of Broecker *et al.* (1991). See text for discussion of absolute ages, which can be greater than those indicated. HMG, homogeneous dissolution; SEQ, sequential dissolution. (Reprinted from Martin, 1993, with permission of the Paleontological Society.)

Despite evidence of mixing of deep-sea foraminiferal assemblages, the exact durations of time-averaging are known only very generally, and age models must be constructed taking into account each environment's taphonomic peculiarities (Figure 5.6). Adapting the Berger–Heath (1968) mixing model, Broecker *et al.* (1991) developed a steady-state age (mixing) model (Chapter 4) for deep-sea sediments based on the assumption that dissolution within the zone of bioturbation is proportional to the residence (replacement) time of grains within the mixed layer. They suggested that there are three basic forms of dissolution:

(1) *Homogeneous* dissolution, in which each grain loses a constant fraction of its mass per unit time (irrespective of grain type). Homogeneous dissolution presumably shifts the mass distribution of assemblages toward younger grains because the replacement time of grains in the mixed layer by new grains from the pelagic rain will be reduced; therefore, the greater the extent of homogeneous dissolution, the younger the age (based on ^{14}C dates) of core top assemblages

(2) *Sequential* dissolution, in which grain type A dissolves *completely* before grain type B begins to dissolve, and so on; in this case, core top ages should *increase* with the extent of dissolution

(3) *Interface* dissolution of all sediment grains at the SWI before burial. Like sequential dissolution, interface dissolution increases the replacement time

5.5 Durations of time-averaging

of grains in the mixed layer and thus increases the radiocarbon age of the core top.

Obviously, dissolution is neither purely homogeneous or sequential: benthic foraminifera tend to be more resistant than planktonics and planktonics exhibit differential resistance to dissolution (Chapter 3), and all tests that co-exist in sediment show evidence of partial dissolution (Broecker *et al.*, 1991).

For a number of reasons, Broecker *et al.* (1991) ruled out all three dissolution mechanisms for deep-sea foraminiferal assemblages (see Martin, 1993, for brief summary). Instead, these authors sought an explanation for core top assemblage ages based on changes in deep-water CO_3^{2-} ion concentration. They suggested that deep Pacific waters became more corrosive, and Atlantic waters less so, during the last glacial–interglacial transition. Under such conditions, Holocene accumulation rates would have been reduced by chemical erosion in the Pacific, thereby increasing the likelihood of reworking older (glacial) sediments into the mixing zone (see also Rathburn and Miao, 1995). In fact, Broecker *et al.* (1991) determined that radiocarbon ages for Pacific deep-sea core tops were about 8000–10 000 years greater than those for Atlantic core tops from comparable depths below the lysocline (Figure 5.6).

Differential preservation in deep-sea sediments may be more complicated, however, than Broecker *et al.* (1991) concluded. DuBois and Prell (1988) found that in the tropical Atlantic, mixed layer ages of sublysocline (>4400 m water depth) cores *decreased* (by c. 500–1000 years) relative to core tops from above the lysocline (a similar pattern was found by Anderson *et al.*, 1988, for continental shelf and slope sediments – 80 to about 2700 m water depth – off Cape Cod, Massachusetts). They found that increased weight percent fragmentation and percent resistant planktonics coincided with changes in age at this depth. Therefore, in sediments below the lysocline, the radiocarbon age structure of the mixed layer in the Atlantic is opposite to that of the Pacific (see also Boecker *et al.*, 1991). Decreased corrosiveness (relative to the Pacific) of overlying waters and higher carbonate content of sediments has apparently allowed a type of sequential dissolution to proceed in Atlantic sublysocline environments that produces *younger* (*not older*) core top ages; i.e., old skeletons are removed faster than young ones so that mixed layer ages decrease (DuBois and Prell, 1988). In terms of mixed layer age, this is just the opposite of sequential dissolution as defined by Broecker *et al.* (1991).

DuBois and Prell (1988) concluded that in the case of dissolution, the steady-state model is inappropriate for calculating temporal resolution and that ^{14}C

stratigraphies from different preservational settings cannot be strictly compared: although bulk *sediment* may have the same radiocarbon age, the proportions of the components producing that age may not be the same if the particles have different preservational histories. Dissolution-resistant benthic (or planktonic) foraminifera, for example, may erroneously increase core-top age estimates (e.g., Broecker et al., 1984). Dubois and Prell (1988) concluded that in order to use ^{14}C dates in stratigraphy, *the processes controlling hardpart input and loss must be evaluated.*

5.5.3 Vertebrates

Like inferences regarding time-averaging of marine invertebrate assemblages, durations of time-averaging of vertebrate assemblages are based primarily on actualistic studies in modern settings or on paleoecological studies of late Cenozoic sediments (Figure 5.7). And, like marine assemblages, there are advantages and drawbacks to the processes involved (Behrensmeyer and Chapman, 1993). Catastrophically produced assemblages, for example, may provide a snapshot of a community, but the picture may not be representative of the long-term dynamics (e.g., Haynes, 1985; see also Chapter 6). And while assemblages that result from predators and scavengers (Chapter 2) may give clues to their feeding habits, etc., such assemblages represent only one portion of the biotic community.

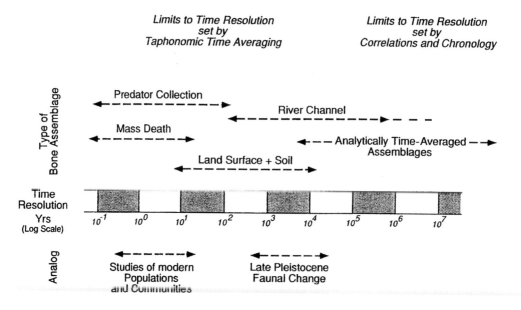

Figure 5.7. Approximate durations of time-averaging of mammalian assemblages. (Reprinted from Behrensmeyer and Chapman, 1993, with permission of the Paleontological Society.)

5.5 Durations of time-averaging

Fluvial settings are one of the most important sources of vertebrate fossils. Reworking of bones from earlier channel and overbank deposits can result in samples that have been averaged over as much as 10^3-10^5 years (Behrensmeyer, 1982; Behrensmeyer and Hook, 1992). These sorts of accumulations are likely to represent regional, rather than local, samples of the biota, given the distances that the remains have been transported. By contrast, oxbows and other abandoned channels likely fill with biota from the immediate vicinity and may even record ecological succession from aquatic to terrestrial habitats; these successions occur over 10^2-10^4 years or more (Behrensmeyer, 1982; Behrensmeyer and Hook, 1992). Graham (1993) estimated durations of time-averaging of only hundreds to thousands of years in most cases for other types of terrestrial environments, although durations of tens of thousands of years may occur in caves. Even longer durations of time-averaging may occur when one considers that it may take $4-5 \times 10^4$ years for a channel to make one complete sweep across its floodplain (Behrensmeyer, 1982).

Following up on the studies of Hanson (1980; Chapter 2), Aslan and Behrensmeyer (1996) conducted studies of temporal resolution of bone assemblages in fluvial sediments of the East Fork River (Wyoming, U.S.A.; Figure 5.8). Fossiliferous fluvial deposits that are characterized by abundant mudstones, fossiliferous ribbon sands, and poorly developed paleosols probably indicate relatively rapid sediment accumulation and limited reworking (Figure 5.8A); vertebrate assemblages in such deposits may represent durations of time-averaging as short as 10^1-10^2 years. If bones are absent from associated floodplain deposits because of rapid sediment accumulation or acidic soils (Chapter 3), then channel assemblages could be close to the lower limit (10 years), but even limited reworking of floodplain assemblages or channel assemblages themselves would produce channel assemblages that represent durations of at least 100 years. For example, the presence of the remains of *Bison*, which have not lived in the area for about a century, in channel assemblages implies that the assemblages have been time-averaged for at least 100 years or so (Behrensmeyer and Chapman, 1993). Even higher durations are suggested by the presence of mature soils (Chapter 3), which probably take on the order of 10^3-10^4 years to form (Bown and Kraus, 1981a,b; Bown and Beard, 1990); on the other hand, destruction of bones in soils (Chapter 3) could also decrease the duration of time-averaging to only the last 10-20% of the total time of soil development (Behrensmeyer and Chapman, 1993; cf. Principle of Linear Superposition). By contrast, fossiliferous multistoried sheet sands and moderately developed paleosols probably reflect slow - sediment accumulation and prolonged reworking (Figure 5.8B), thereby

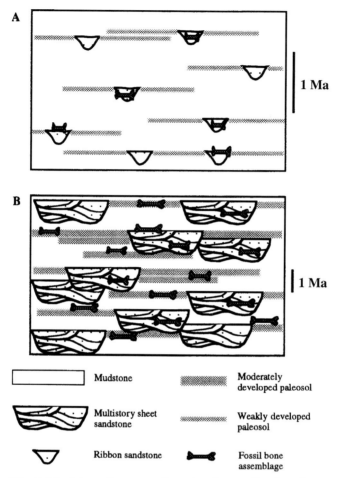

Figure 5.8. Schematic cross-sections showing predicted sandstone geometries, paleosols, and fossil distributions in fluvial channel deposits. (A) and (B) represent minimum and maximum amounts of time-averaging, respectively. Ma, million years. (Reprinted from Aslan and Behrensmeyer, 1996, with permission of SEPM)

producing assemblages of durations of up to 10^4 years. Even greater durations of time averaging may result (10^5–10^6 years) if floodplain assemblages contain substantially older fossils (East Fork River flows through Pleistocene deposits that are approximately 20 000 years old).

Using taphonomic criteria based on actualistic studies of bone accumulation (Table 5.2), the durations of modern depositional environments in which bone assemblages form, and taxonomic diversity, Rogers (1993) estimated the extent of time-averaging for each of several fossil vertebrate concentrations in the Two Medicine and Judith River formations (Late Cretaceous, Campanian) of north-central Wyoming (see also Chapter 2). Nest sites and subaerial bonebeds basically

5.5 Durations of time-averaging

represent mass mortality or very short durations ($<10^0$–10^1 years), respectively, as indicated by intact nests, intact eggshell microstructure, low dispersal and weathering of bones (all in the case of nests), and low diversity assemblages in the case of bonebeds (too little time for attritional accumulation of higher diversity assemblages). Subaqueous and channel-fill concentrations also frequently represent event beds, as they share taphonomic traits with their subaerial counterparts, but these two types of concentrations probably represent temporal end-members.

These four types of concentrations are most often found proximally (Two Medicine Formation), whereas the distal record (Judith River Formation) is characterized by vertebrate microfossil and channel-lag concentrations that have been time-averaged over about 10^2 years or more. Vertebrate microfossil concentrations formed in lacustrine and paludal settings that probably persisted for thousands of years, as the concentrations are highly diverse, dissociated, and disarticulated, and may even suffer from the effects of trampling (Lockley and Conrad, 1989). The upper limit of the duration of time-averaging in channel-lags ($\sim 10^5$ years) is based on rock accumulation rates of approximately $4.5\,\text{ka}^{-1}$ and channel thicknesses of 20 m and the reworking of microfossil concentrations by other channels.

Assuming a more or less constant input of hardparts, the change in durations of time-averaging from proximal (Two Medicine Formation) to distal (Judith River Formation) foreland basin reflects changes in sediment accumulation rate (Rogers, 1993). Accumulation rates for the Two Medicine Formation ($\sim 7\,\text{cm ka}^{-1}$) are about 1.5 times those of the Judith River Formation ($\sim 4.5\,\text{cm ka}^{-1}$). According to Rogers (1993), higher floodplain aggradation rates in the Two Medicine Formation increased the probability of preservation of event bonebeds and possibly also the dilution of bones or reworking of hardparts (i.e., channel-lags) that otherwise accumulate through slow attrition. The slower rates of accumulation in the Judith River Formation favored the formation of channel-lag assemblages (Behrensmeyer, 1987, 1988). But, accumulation rates were still too high in either formation to favor development of abandoned channels and channel-fill assemblages (Behrensmeyer 1987, 1988; Chapter 2). Thus, ironically, high rates of sedimentation may actually *increase* the durations of time-averaging of terrestrial vertebrate assemblages (Rogers, 1993).

Rogers (1993) also concluded that the prevalence of mass mortality events in the Two Medicine Formation was not just a function of high sedimentation rates. Many of the event bonebeds show evidence of brief subaerial or subaqueous exposure (e.g., disarticulation), which suggests that the event beds also represent mass

Table 5.2. *Modes of formation and durations of time-averaging of vertebrate skeletal concentrations in the Cretaceous of north-central Wyoming*

Concentration type	Sedimentary context	Taphonomic attributes	Duration of time-averaging (years)
Channel-lag concentration ($n = 10$)	Fluvial sandstone bodies, bases of single-story channels, reactivation surfaces in multi-story channels, erosive bases of cross-bed sets	Taxonomically diverse vertebrate microfossil concentrations, aquatic and terrestrial taxa, variable states of fragmentation, abrasion, and weathering, elements dissociated	$<10^2 - >10^5$
Channel-fill concentration ($n = 2$)	Abandoned segments of channels, carbonaceous siltstone, freshwater invertebrates	Mono/paucispecific concentrations of dinosaurs, bones are disarticulated, weathering stages are low and uniform	$<10^0 - 10^1$
Nest site concentration ($n = 17$)	Subaerial floodplain, silty mudstone/siltstone, associated with caliche and root traces	Clutches of hatched and unhatched eggs, occasional embryos, eggshell usually unweathered, microstructure usually intact, frequently associated with juvenile bones and bone beds (nestling cohorts), bones are disarticulated, weathering stages are typically low and uniform	$<10^0 - 10^1$

Subaerial bonebed concentration ($n = 2$)	Subaerial floodplain, silty mudstone, associated with caliche and root traces	Mono/paucispecific concentrations of dinosaurs (various age classes), bones are disarticulated, weathering stages are low and uniform	$<10^0 – 10^1$
Subaqueous bone bed concentration ($n = 7$)	Floodbasin lakes/ponds, carbonaceous silty mudstone, freshwater invertebrates	Typically mono/paucispecific concentrations of dinosaurs, bones are disarticulated, weathering stages are low and uniform	$<10^0 – 10^1$
Subaqueous microfossil concentration ($n = 11$)	Floodbasin lakes/ponds, carbonaceous silty mudstone, freshwater invertebrates	Taxonomically diverse vertebrate microfossil concentrations, aquatic and terrestrial taxa, minimal weathering (occasional elements show advanced weathering), elements dissociated	$<10^2 – >10^3$

n, number of sites.
Modified from Rogers (1993).

mortality in relation to biological or paleoclimatic phenomena that may be only indirectly related to a geological setting (e.g., drought, fire, floods, sudden changes in weather, toxic gases; cf. Weigelt, 1989).

Because the processes of time-averaging, even when geologically quite short, are often difficult to observe on human time scales, Behrensmeyer and Chapman (1993) developed a model of time-averaging in a simulated, compartmented landscape based on previous studies in Amboseli Park, Africa. They specified carcass input and body size for large (≥ 15 kg) versus small mammals (the two are usually collected differently) for different computer runs. Each bone was then run through a "destruction function," which eliminated different skeletal elements in proportion to their susceptibility to early scavenging, trampling, and weathering (Chapters 2, 3). Bones that survived were then dispersed from the landscape compartment (collecting site) where the animal died according to assessed probabilities of dispersal for each bone type. Based on field observations, larger mammals were dispersed along a gradient, whereas, since there is little actualistic data on the dispersal rates of small animals, small mammals were dispersed randomly (*not* uniformly), primarily because death of small mammals by predation is likely to be random.

"Because animals with smaller body sizes generally have faster annual turnover rates and larger populations per unit area, they should [intuitively] produce more identifiable bones than larger animals.... Working against this is ... that small bones are more susceptible to destruction, although they are more easily buried" (Behrensmeyer and Chapman, 1993, pp. 134–135; cf. Law of Numbers, Chapter 1; Chapter 2). The simulations showed that the small mammal community produced a much larger assemblage of buried bones than the large community over 1000 years. The average small mammal bone density produced was approximately 3 m^{-2}, whereas densities of only about 0.1 m^{-2} were produced for large mammals. At least 100 years would be required to produce such accumulations, which are considered to be relatively productive, barring dissolution, patterns of exposure in outcrop, and collecting techniques, all of which would reduce the number of bones found; thus, even denser accumulations might be required to produce paleontologically productive beds. The simulations also showed that for large and small mammal communities 70–80% of the input species were present in most compartments after approximately 500 years, but that occurrence of even the two most common species (cow and wildebeeste) can be quite erratic up to about 200 years. Species richness and Shannon–Weaver indices approached that of input communities in all compartments after 1000 years, although rarer species were often absent when bone densities were greater than 4 m^{-2}.

Behrensmeyer and Chapman (1993) also simulated mass mortalities by increasing by 50 times the normal annual cow and wildebeest mortality in each of two successive years after 500 years of time-averaging. They found that if mass mortality occurred randomly over the entire landscape it is swamped by background attrition, but if the mass mortality is concentrated in a compartment then the event is recognizable from higher bone densities, and dominance of cow and wildebeeste over background species in assemblages. Coupled with age structure and relative skeletal completeness, such an event ought to be recognizable in the fossil record (cf. section 2.7 for discussion of mass mortality events in reef settings).

5.5.4 Plant macrofossils and pollen

Johnson (1993) concluded that the best temporal resolution of fossil megafloras and palynofloras *assembled from correlated localities* would be no finer than approximately 1×10^5 years. Nevertheless, of all the fossil taxa that have been examined, it is the macroscopic remains of plants, generally viewed as being among the least preservable of all taxa, that probably give the best estimates of durations of time-averaging (Burnham, 1993*b*; see also Kowalewski, 1997). High temporal resolution of macroplant remains is a result of relatively low durability (short residence time of remains in sediment), warm moist climates (which accelerate decay and decrease residence times), and seasonal production of organs (which occurs in almost all climatic zones and which helps to constrain chronology). But, like multi-element skeletons of animals, the dispersal and preservation potentials of plant parts varies substantially (e.g., flowers versus trunks) and affects estimates of time-averaging (Table 5.3; cf. Chapters 2, 3).

Probably the most ephemeral of all plant parts are reproductive structures such as flowers (which are composed of sepals, petals, stamens, and ovaries) and seeds and fruits. These parts are typically produced seasonally (sometimes bimodally) and shed within a few weeks, thereby producing discrete reproductive inputs to the fossil record that decay so rapidly (1 month maximum for flower parts) in tropical forests that only a brown film is left (Burnham, 1993*b*; see also Crepet *et al.*, 1974). Similarly, regeneration of vegetative parts, such as in Carboniferous *Calamites*, could only have occurred within several weeks of burial by floods (Gastaldo, 1992*b*), which is typical of modern plants: modern plant stems rarely resprout if buried under more than 0.5 m of sediment (Burnham, 1993*b*).

Unfortunately, the complex phenologies associated with reproductive parts, especially in flowering plants, requires taxonomic assessment of the parts (e.g.,

Table 5.3. *Temporal characteristics of macroplant parts*

	Growth time[a]	Seasonal production	Decomposition time[b]	Reworking potential	Min/max time represented by parautochthonous accumulations
Trunks	<10–500 yrs	No	6 mos–150 yrs	Possible	<10–520 yrs+
Bark	0.5–30 yrs	Yes	2 mos–10 yrs	Possible	0.5–40 yrs+
Small branches	0.5–5 yrs	No/yes	3 mos–5 yrs	Possible	0.5–10 yrs+
Durable cones, fruits, seeds	2 mos–2 yrs	Yes	1 mo–5 yrs	Possible	2 mos–7 yrs
Fragile cones, fruits, seeds	2 wks–1 yr	Yes	2 wks–3 yrs	Unlikely	2 wks–4 yrs
Leaves	6 mos–3 yrs	Yes	2 wk–5 yrs	Unlikely	6 mos–8 yrs
Flowers	2 wks–2 mos	Yes	1 wk–1 yr	Unlikely	2 wks–1.2 yrs

[a]Values are meant to be typical, rather than including all extremes, including periods of growth, decay, transport and burial.
[b]Potentially longer in anoxic environments.

Burnham, 1993b.

5.5 Durations of time-averaging

Wolfe, 1971). Also, angiosperm seeds and fruits may contain lignin, which slows degradation and decreases temporal resolution to decades or centuries (e.g., Eocene Clarno Formation of Oregon, U.S.A., and London Clay). Megasporangiate cones, fruits, and seeds of evergreens are most durable and tend to remain attached to the parent plant longer, thereby producing longer reproductive pulses. Flotation of seeds and fruits for considerable distances over periods of months to years (Chapter 2) also decreases temporal resolution (e.g., Clarno Formation and London Clay, the assemblages of which may have been transported, thereby increasing the extent of time-averaging; Johnson, 1993). The decay-resistance of these plant parts also increases the likelihood of analytical time-averaging (Burnham, 1993b).

Like reproductive structures, foliage (leaves and other photosynthetic structures) input also tends to be seasonal (Burnham, 1993b), but leaves can be identified relatively easily. Seasonal foliage input is superimposed on a background input that ranges from little to none (cool temperate and dry tropical forests) to moderate (wet tropical rainforests). In temperate climes, shedding is tied to frost and rainfall and occurs over approximately a 3-month interval; conifers tend to shed more in the autumn. But even in tropical regions there may be relatively dry periods – or even drought – during which leaves may be renewed, and deciduous taxa often shed plant parts in the relatively aseasonal environments of the tropics.

Short residence times are therefore indicated by undegraded whole leaves of angiosperms because dry leaves tend to fragment rapidly and are easily degraded by microbes (Burnham, 1993b; Table 5.3; see also Chapter 2). Often only the cuticle remains, but this is still extremely useful taxonomically and environmentally because leaf shape and tooth shape and venation are present. Thus, most single-bed leaf accumulations probably represent durations of time-averaging of less than 4 years (excluding herbaceous and annual species, which tend to degrade the most rapidly), and well-preserved beds, which usually receive the greatest attention, probably represent less than 6 months. Coniferous needles, which degrade more slowly, more likely represent durations of years or longer.

The taxonomy of large woody parts (and to a much lesser extent, bark) is usually reliable, and growth rings record the response of the plant to weather and climate over relatively long durations *at one site* (Burnham, 1993b). Unfortunately, wood and bark are otherwise unsuitable for time resolution because growth tends to be continuous, and they are shed at irregular intervals by the plant (depending upon species) or are lost during storms or other catastrophes (Burnham, 1993b). In the case of storms, the regional scatter of trees should be documented in order to differentiate between small violent storms and larger, stronger cyclonic ones (Gastaldo, 1990); lithology is also useful in distinguishing other causes such as volcanic blasts or floods (Wnuk and Pfefferkorn, 1987).

Because they contain lignin and compounds that retard microbial degradation, woody plant parts degrade relatively slowly. Dense accumulations of bark probably represent decades of accumulation, although DiMichele (personal communication cited in Burnham, 1993b) suggested that accumulations of Carboniferous lycopod bark may represent less than 10 years. Gastaldo and Huc (1992), for example, found that in the Mahakam Delta (Kalimantan, Indonesia), wood fragments were substantially older than leaves. Wood fragments from a core taken in a channel bar were approximately 5950 years old, whereas leaves located at a depth of −2 m in the same core were approximately 765 years old; other leaves ranged from about 385 to 1280 years.

Like other fossil assemblages, temporal resolution of assemblages of macro-plant remains depends on sediment accumulation rates, which vary with environment (Burnham, 1993b; see also Johnson, 1993). Moreover, accumulations of plant remains can form in a variety of sedimentary settings that have different potentials for assessing durations of time-averaging (Burnham, 1993b; Johnson, 1993). Accumulations in volcanic settings represent durations of 0–1 year and have the potential to record accurately pre-eruption vegetation (e.g., Burnham and Spicer, 1986; Gastaldo, 1992a). Burnham (1993b) cites evidence of vegetation response to volcanism in the Clarno Formation, in which a 10–15 cm thick sequence consisted of a lower bed composed of angiosperms and ferns, which was overlaid less than 5 cm above by a fern-dominated bed; the lower bed appeared to represent pre-eruption vegetation, whereas the fern-dominated bed represented early colonization of the post-eruption volcanic landscape.

Fluvial sediments may also provide plant assemblages of short durations. Point bars may contain allochthonous leaf mats, seed accumulations, or log jams that formed in about 1–3 years (Burnham, 1993b). Because of the highly episodic nature of point bar deposition, however, fossil plant remains may represent more than one depositional event and mixing of plants from different communities, as suggested by age-mixing documented by Gastaldo and Huc (1992). Criteria for single (or at least short) depositional events include logs with fine rootlets still attached, absence of sediment "rinds" between logs, and the presence of randomly oriented organic matter in the sediment (Burnham, 1993b). Infilling of logs with fine-grained sediment results from multiple floods; overall sediment accumulation rates are relatively low and infilling takes

5.5 Durations of time-averaging

several hundred years (the approximate time for the log to decay; cf. model of Rex, 1985, which does not apply here because it assumes infilling by a single event; Gastaldo *et al.*, 1989). In contrast, infilling of logs by sediment clasts (bedload filling) results from one or a few events and is therefore quite rapid (Liu and Gastaldo, 1992; i.e., model of Rex, 1985, may apply). On the other hand, mixes of comminuted plant matter with whole leaves and wood fragments may indicate durations of time-averaging on the order of 6 months to 5 years; allochthonous leaf mats may be distinguished from autochthonous forest floor leaf mats based on sedimentology. In the case of seeds and fruits, the larger the modal size or the greater the durability of seeds and fruits, the greater the likelihood of dispersal over relatively long distances for long times; abrasion of propagules may become very pronounced after only 2 years, however.

Floodplains – and to a lesser extent oxbows – are very common sources of plant assemblages and probably provide the finest estimates of time-averaging (as low as several years; Burnham, 1993*b*), although carbonaceous beds may represent up to about 2000 years (Davies-Vollum and Wing, 1998). These typically fine-grained deposits are often laid down outside meander belts of reworked deposits, and contain both transported debris and evidence of local vegetation. Also, because they are episodic, plant assemblages may change from one bed to the next in a succession, so that temporal resolution becomes critical in avoiding analytical time-averaging; Burnham (1993*b*) suggests using sedimentological evidence of discontinuities to distinguish separate beds.

Clastic swamps that are associated with fluvial or estuarine deposition are often excellent sites of plant preservation because of continuous rapid deposition (e.g., Wnuk and Pfefferkorn, 1984; Burnham, 1993*b*). Nevertheless, durations of time-averaging cannot be assessed from sediment thicknesses alone because sedimentation is highly episodic. Moreover, accumulation of plant debris may exceed sediment, so that plant remains come to resemble finely-comminuted "tea grounds" (Burnham, 1993*b*).

Despite the occurrence of varves, plant assemblages from large lakes probably yield the greatest durations of time-averaging (Burnham, 1993*b*). Unlike oxbows, plant remains are less likely to be carried to deeper parts of the lake. Thus, macroplant assemblages are likely to consist of accumulations from different plant communities that colonized the lake margin over relatively long periods of time. Ironically, higher-energy settings along lake margins are more likely to produce less time-averaged deposits because of rapid deposition of different plant parts, whereas low-energy settings probably give a better estimate of the distribution of source vegetation (e.g., Spicer and Wolfe, 1987).

Mires (peats) that are associated with the above habitats, although organic-rich, are quite variable in terms of the conditions under which they form (Chapter 3) and in their rates of accumulation and compaction (thickness). Nevertheless, peats probably represent the longest durations of time-averaging of terrestrial plant assemblages because of relatively slow decay of organic matter. Moreover, Burnham (1993b) questioned the use of modern compaction ratios for Carboniferous coals – much less Late Cretaceous and Tertiary ones – because of differences in taxonomy, chemical composition, and litter production rates.

Deltaic plant assemblages usually represent mixtures of remains from different habitats (Burnham, 1993b; Chapter 2). Although intuitively this would seem highly disadvantageous, like vertebrates accumulations, such assemblages are of both regional and local nature (cf. Chapter 2). Interdistributary bays provide evidence of regional floras (Gastaldo et al., 1987), whereas swamps and abandoned delta lobes are more likely to preserve local vegetation.

As rivers drain into estuaries, the diversity of plant assemblages declines because of the stressful nature of brackish water, tidal flushing, and bioturbation (Burnham, 1993b; see, for example, Baird et al., 1985, 1986 for the Pennsylvanian; Gastaldo and Huc, 1992, for modern Mahakam Delta). Plant remains are often highly dispersed in (typically) fine-grained sediments and durations of time-averaging are relatively great because of tidal flushing and slow sediment accumulation. Probably for these reasons, estuaries have often been proposed as both sites of deposition and non-deposition of plant assemblages, even though plant preservation in modern estuaries has rarely been studied (e.g., Gastaldo et al., 1987; Raymond, 1987).

Wind-pollination, which is most prevalent in temperate to boreal climes, is notoriously patchy and inefficient (Webb, 1993; see also Chapter 2). Moreover, based on radiocarbon dates, pollen assemblages, like other fossil assemblages, may differ in age from other fossils of the same deposit. Webb (1993) suggested that pollen assemblages tend to represent only the younger portions of their time of accumulation. Brown et al. (1989), for example, found that ^{14}C ages of bulk samples were older than those of pollen assemblages (cf. section 5.3). On the other hand, pollen assemblages are more likely to be mixed and time-averaged (e.g., Davis, 1974; see also Chapters 2, 4) because pollen may persist longer than, say, leaves (even when both are from the same species; Johnson, 1993). As for marine micro- and macrofossils, different estimates of taxonomic composition may also be obtained when pollen and plant macrofossils of the same taxa are compared (Dunwiddie, 1987).

5.5 Durations of time-averaging

Despite these drawbacks, pollen has proven extremely useful in distinguishing anthropogenic from natural (pre-anthropogenic) disturbance (e.g., the frequency of natural fires and disease; Birks et al., 1976; Gajewski et al., 1985; Russell, 1997; see also Chapter 10). Although pollen from varves can provide seasonally distinct samples, most pollen samples integrate inputs over durations of 10 years or more and can be independently dated with an average precision of ±200 years for the last 12 000 years. Short cores are often taken to study short-term changes and may yield 10–20 samples in about 500 years (one sample per 25–50 years). Records of intermediate length (one sample per 250–500 years to one sample per 2×10^4 years) are also available, and a few very long records have been sampled, some stretching as far back as 3.5 myr and with temporal resolution of $\sim 3 \times 10^3$ years sample^{-1}.

Webb (1993) assessed temporal resolution of pollen assemblages based on Late Quaternary data, and his discussion provides some valuable insights into scaling and the use of paleontological data. Because the use of pollen data is dependent upon the questions to be answered, and therefore the scale required, the palynologist – or *any* paleontologist, for that matter – must choose sampling programs with respect to desired areal coverage, and spatial and temporal resolution of single samples, all of which control sample density (Webb, 1993). These factors are, in the vein of snapshots and movies of time-averaged assemblages (section 5.3), analogous to the frame of reference (areal coverage), grain size (resolution), and grain (sample) density of a photograph. The palynologist must later integrate data over spatial and temporal scales to produce three-dimensional "maps" for successive time slices. The breadth of coverage can range from local ($\sim 10^3$ km^2) to regional (10^3 km^2–3×10^5 km^2) to continental in scale, and depends on basin characteristics such as size of the basin and watershed area (which affects input and mixing of waterborne pollen) and presence or absence of a canopy over the basin. In some cases, sample densities of 1–10^{-3} km^{-2} or more have been sufficient to discriminate limits of vegetation within 20 km (see Webb, 1993, for further examples).

The spatial resolution of individual pollen samples varies with taxon and habitat (Webb, 1993). Small grains are dispersed farther than large grains (Chapter 2), which may affect distinguishing local versus regional assemblages. Pollen in soils and small hollows in forests usually accumulates within radii of 10–100 m of the source and may be useful in distinguishing forest succession (about 100 years in temperate climates), whereas pollen in lakes falls within approximately 10–50 km and may represent a regional sample of vegetation unless diluted by pollen from fringing vegetation, in which case the signal will be more local.

Gajweski et al. (1985) found, for example, that lakes greater than 4 ha ($\sim 4 \times 10^4$ m^2) in area provide regional pollen signals.

Despite the relatively thin mixed layer of lakes (which should homogenize signals; Chapter 4), spatial variation may still interfere. Davis et al. (1984) emphasized the great spatial variation of pollen assemblages within the same lake as a result of differences in pollen input, changes in sediment accumulation rates, and sediment focusing via resuspension; they (1984) recommended coring only the deepest parts of lakes (despite the propensity for sediment focusing there in modern lakes) because of the likelihood of more rapid sediment accumulation.

Bogs and mires may preserve either local pollen signals ($\sim$10 m) or regional pollen signals (1–10+ km) depending on their size. Unlike lakes, little or no post-depositional mixing occurs in bogs and mires and the regional signals are relatively uniform (Janssen, 1966).

6 Exceptional preservation

O me, why have they not buried me deep enough?
Is it kind to have made me a grave so rough,
Me, that was never a quiet sleeper?
Alfred, Lord Tennyson, *Maud*

6.1 Introduction

It has been estimated that from 66 to 79% of the biota of a marine community is not normally fossilized (e.g., Lawrence, 1968; see also Allison and Briggs, 1991*a,b*). Despite the improbability of preservation, there are numerous instances of exceptional preservation called *Lagerstätten* (singular *Lagerstätte*). The term comes from the mining industry and, loosely translated, means "fossil-motherlodes", or rocks that are unusually rich in paleontological information. Seilacher (1970) recognized two types of Lagerstätten: (1) *Konzentrat* (*concentration*) deposits, which form by sedimentological and biological processes that largely exclude the preservation of soft parts and include shell beds, bone beds, and crinoidal limestones (Allison, 1988*c*); and (2) *Konservat* (*conservation*) deposits, which are characterized by preservation of soft-bodied fossils. Thus, Konzentrat-Lagerstätten are distinguished primarily by quantity whereas Konservat deposits are distinguished by the quality of preservation (Seilacher, 1990). Both types of deposits really represent end-members of a continuum in the preservational spectrum (Allison and Briggs, 1991*a,b*). Chapters 1–5 have been concerned with the formation and occurrence of Konzentrat-Lagerstätten. This chapter is devoted to major Konservat-Lagerstätten (Figure 6.1), in which soft-bodied organisms are preserved. Long viewed like the curios in an antique shop, the tremendous value of Konservat deposits was recognized only relatively recently.

Many Konservat-Lagerstätten form as a result of mass mortality (e.g., overturn of anoxic or hypersaline waters; drought), which increases the chances of fossil preservation (Shipman, 1975). Although Lagerstätten that form through mass mortality largely escape the vagaries of time-averaging (Chapter 5), this advantage is also a potential pitfall (Seilacher, 1970). Mass mortality is only a "snapshot" of a biological community and may not be representative of the

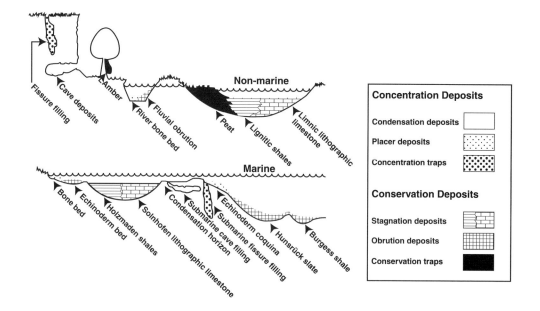

Figure 6.1. Classification of major Fossil-Lagerstätten discussed in text. (Redrawn from Seilacher *et al.*, 1985.)

long-term ecological – much less evolutionary – dynamics or "movie" of a community (cf. census assemblages, Chapter 5), but *this dictum should be evaluated carefully in each circumstance*. Haynes (1985), for example, found that mass mortality (via drought) of subadult animals of the African elephant (*Loxodonta africana*) was nearly twice that of live populations. On the other hand, based on comparison of modern and fossil stickleback fish from mass mortality assemblages, Bell *et al.* (1987, 1989) concluded that variation of fossil assemblages fell within the limits of extant populations and that in this setting the fossil assemblages were representative of short-term population fluctuations (see also Wilson, 1977; Wilson and Barton, 1996).

Despite the potential for bias toward short-term or even unrepresentative community dynamics, Lagerstätten nevertheless provide glimpses of the Earth's true biotic diversity and evolution, and, more rarely, life behaviors (e.g., molting, gregarious behavior) not necessarily preserved in time-averaged deposits (Conway Morris, 1985; Brett, 1990). Konservat-Lagerstätten also facilitate the association of disarticulated soft parts. For example, Charles D. Walcott, who initially described much of the fauna of the Burgess Shale (Middle Cambrian) of British Columbia, assigned three disarticulated remains to three separate taxa (holothurians, medusoids, and crustaceans) that, based on complete specimens discovered much later, are now known to represent the body, mouth, and appendages,

6.2 Genesis of Konservat-Lagerstätten

respectively, of the predator *Anomalocaris canadensis* (Allison and Briggs, 1991*b*). Disarticulated leaves, fruits, etc., may all be described as separate taxa until they are found preserved as they occurred in life (Spicer, 1989, 1991). Although long-recognized for their tremendously utility in biostratigraphic zonation of the Paleozoic, the affinities of conodonts remained completely unknown until their association with carbon films (Briggs *et al.*, 1983). The relatively rapid decay associated with soft-bodied biotas means that early decay can be simulated over the short time scales of the laboratory and may provide clues to the preservation and affinities of enigmatic taxa such as conodonts and graptolites (cf. Briggs and Kear, 1994*a*, and Gabbot *et al.*, 1995; Briggs, 1995; Briggs *et al.*, 1995).

6.2 Genesis of Konservat-Lagerstätten

Despite their relative scarcity, there are a finite number of situations that lead to fossil Lagerstätten (Seilacher *et al.*, 1985). Moreover, fossil Lagerstätten may be viewed as end-members of ordinary sedimentary facies and *predicting* their occurrence ("prospecting") is a realistic objective (Seilacher *et al.*, 1985; Kluessendorf, 1994). To this end, Seilacher *et al.* (1985) erected a classification of marine Konservat deposits (Figure 6.2) that recognized *stagnation* (anoxia), *obrution* (rapid smothering by sediment, including volcanic ash; Heikoop *et al.*, 1997) and *cyanobacterial sealing* as the dominant factors in the genesis of Konservat-Lagerstätten.

Obrution deposits are best-preserved in relatively low-energy environments slightly below normal storm wave base. Rapid burial (in a few hours to a few days) by turbidity currents or storms (especially in fine-grained sediments) is

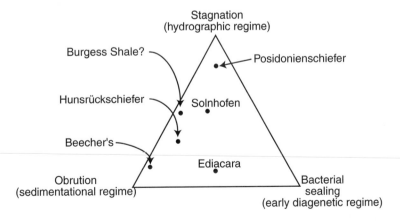

Figure 6.2. Classification of major marine *associations* of Konservat-Lagerstätten according to stagnation, obrution, and bacterial sealing. (Modified from Seilacher *et al.*, 1985.)

important because if the sediment layer is sufficiently thick, it prevents the disarticulation of carcasses by subsequent bioturbation (i.e., carcasses are buried below the surface mixed layer; Allison, 1988c; Brett, 1990). Although episodic sedimentation is most frequent in shallower waters, obrution layers are more likely to be cannibalized later by erosion in shallow settings (Brett, 1990; cf. Chapter 4). Although rapid sedimentation is conducive to the formation of obrution deposits, too high a sedimentation rate (e.g., mid to late sea-level high-stands; Chapter 7) may dilute fossil content; moreover, if a skeletal concentration of some sort is not present prior to burial, obrution deposits may be extremely cryptic (Brett, 1990, 1995). Obrution deposits may also form in condensed sedimentary sequences where they represent the only permanent accumulation (Brett, 1990).

The thickness of sediment required to smother a biota varies with the taxon. Some, such as certain bivalves (deep infaunal; pectinids), echinoids, and ophiuroids, are quite adept at escaping to the surface, although they may subsequently die at the watery SWI of the burial layer (e.g., Taylor and Brett, 1996); because of their lower mass, juvenile bivalves are more likely to escape than their adult counterparts (Schäfer, 1972; Brett, 1990). The smothering layer may be distinguished by fining-upward grain size, planar and cross-lamination indicative of deposition from a waning current, escape burrows, inverted ("flipped") epibiont-encrusted shells or concave-up shells, and possibly the occurrence of articulated carcasses; in other cases, this layer consists of unfossiliferous structureless mudstone (Brett, 1990). Hardgrounds (encrusted or bored pavements) may be exceptionally well preserved (e.g., Brett and Liddell, 1978), possibly as a result of early cementation (Walker and Diehl, 1985; Chapter 8) and along with other skeletal concentrations, may form recognizable obrution deposits. Since the buried horizon represents a former SWI, carcasses may be concentrated into windrows or preferentially aligned in other ways. In some cases, layers of enrolled or partially disarticulated fossils (e.g., trilobites) at the former SWI may be traceable for kilometers (and serve as valuable stratigraphic datums; such layers also suggest that the biota first responded to some other event, such as temperature or salinity change or asphyxiation, prior to burial (Brett, 1990).

Another factor related to softpart preservation is transport. The occurrence of soft-bodied Lagerstätten has been used to infer minimal transport of fossil remains (via turbidity currents, storms) from their living sites, the assumption of which then constrains paleoenvironmental inferences (Allison and Briggs, 1991b; cf. Brett, 1990). Nevertheless, Allison (1986) demonstrated that lightly skeletonized, soft-bodied organisms (*Nephrops, Palaemon*) can withstand disarticulation for many hours within tumblers, and Kidwell and Baumiller (1989) showed that

6.2 Genesis of Konservat-Lagerstätten

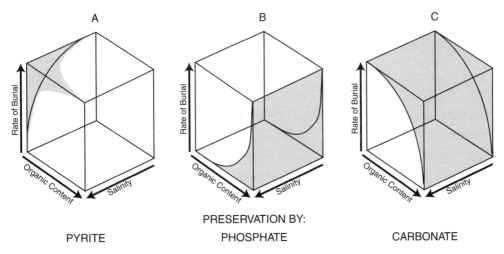

Figure 6.3. Depositional parameters required for preservation of softparts by pyrite, phosphate, or carbonate. Early diagenetic pyritization requires rapid (sometimes catastrophic) burial, low organic content, and the presence of sulfates, whereas early diagenetic phosphatization requires low burial rates and high organic content, although mineralogy varies with salinity. Preservation in carbonate requires rapid (perhaps catastrophic) burial and organic-rich sediment, although mineralogy again varies with salinity. (Redrawn from Allison, 1988c.)

the degree of echinoid fragmentation in tumblers depended on the duration of prior decay (see also Chapter 2). Thus, the degree of transport is related more closely to the amount of prior decay, and the degree of articulation of soft-bodied organisms is not necessarily an accurate indicator of transport distance.

Stagnation (anoxia) may also prevent necrolysis and disarticulation by inhibiting bioturbation, scavenging, and disarticulation, and along with obrution, promotes early diagenetic mineralization of decaying carcasses. It is now well established, however, that decay can occur in the absence of oxygen via, for example, bacterial sulfate reduction (Chapter 3; see also Allison, 1988a,c; Allison and Briggs, 1991a,b). Thus, analysis of early diagenetic (authigenic) mineralization provides important clues to the sedimentary and geochemical setting in which Lagerstätten were preserved (Allison, 1988a,b,c; Allison and Briggs, 1991a,b). Preservation of softparts is most commonly associated with carbonate minerals in the form of concretions, which may form quite rapidly (Allison and Pye, 1994; Chapter 3), or with fine-grained bedded limestone (*Plattenkalk*; Allison, 1988c). Bacterial sulfate reduction increases alkalinity and the likelihood of calcite precipitation (in marine settings) and siderite precipitation in freshwater (Allison, 1988c; Chapter 3; Figure 6.3).

Precipitation of iron sulfides may also occur if sufficient iron is present, but pyritized softparts are much less common (Allison, 1988c). As pyrite formation may be limited by carbon, iron, or, in freshwater settings, sulfate, the precise combination of conditions necessary for pyritization of soft tissues is less likely to occur. Pyrite formation is favored by rapid burial (to slow aerobic destruction of organic matter) and low-carbon settings, in which ions diffuse toward a localized carbon source (carcass). In high-carbon settings, on the other hand, sulfate reduction may be so diffuse that pyrite is disseminated throughout the sediment instead of being concentrated on carcasses (Figure 6.3; Allison, 1988c; see also Chapter 3).

Although broad high-productivity oceanic settings can yield sufficient phosphate for preservation, exceptional preservation in shallow sediments via phosphatization (e.g., apatite) appears to be restricted to localized conditions (Allison and Briggs, 1993b). Nevertheless, phosphatization of softparts exhibits the highest quality of preservation, including three-dimensional preservation of embryos (Bengston and Zhao, 1997), which has been replicated in the laboratory (Briggs and Kear, 1993b), and the retention of cellular morphology (e.g., conodonts, squid, and fish; see Allison, 1988c, for numerous references). Phosphate is liberated via bacterial decay of hard (bone) and soft organic matter (Allison, 1988b; Briggs and Kear, 1993b, 1994b; see also Chapter 3), which may take several weeks or months if decay is inhibited, the best preservation occurs when bacteria are not preserved (Briggs *et al.*, 1993). Phosphate may initially adsorb to ferric hydroxides in sediment, but the onset of anoxia reduces iron and liberates phosphate, thereby producing a phosphate concentration peak. According to this scheme, burial of larger carcasses must be sufficiently slow that decaying carcasses remain near the anoxic/oxic interface, where phosphate can be liberated to porewaters by the reduction of iron (Briggs *et al.*, 1993). In marine environments, francolite normally precipitates, but in iron-rich freshwater, vivianite is more likely to form (Figure 6.3).

Silicification of soft remains also occurs, but only rarely in animal macrofossils (Allison, 1988c). Silicification occurs primarily in plants, in which the degree of organic matter preservation varies from labile organic components (volatiles) to highly refractory lignin, cellulose, and sporopollenin of pollen grains (Figure 6.4; Table 6.1; see also van Bergen *et al.*, 1995; and Chapter 3). Among animals, arthropod cuticle is sufficiently resistant to be preserved without authigenic mineralization (Allison and Briggs, 1991a).

Seilacher *et al.* (1985) also emphasized the importance of cyanobacterial and other microbial films in soft-body preservation. These films form scums at the

6.2 Genesis of Konservat-Lagerstätten

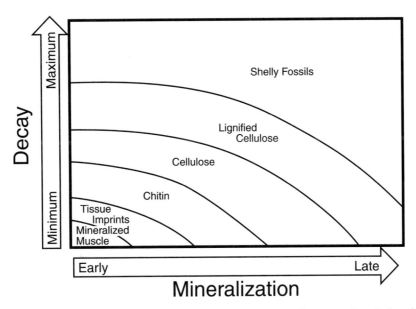

Figure 6.4. Relationship between decay, mineralization, and preservation. Reduced decay and very early diagenetic mineralization are required for soft-tissue preservation. If mineralization is impeded, decay continues, leaving more refractory components behind. (Redrawn from Allison, 1988c.)

SWI and probably facilitate the preservation of tracks and certain other ichnofossils at the surface; inhibit organic decay and erosion and bioturbation of sediment (the current velocity required to erode sediment bound by microbial mats is 2–5 times that for sediment alone; Allison and Briggs, 1991a); and protect carcasses against scavenger damage (e.g., Gall, 1990; Martill, 1988). Microbial mats may be responsible for the preservation of medusae and worms in transitional environments preserved in the Lower Triassic (Anisian) Grès-à-Voltzia shales from the Vosges of northeastern France (the Mazon Creek fauna; see below; Briggs and Gall, 1990), and appear to have pseudomorphed softparts of some vertebrates (Allison and Briggs, 1991b). Today, such films are most prominent in restricted environments such as hypersaline lagoons but were undoubtedly much more widespread in the past, especially in the Precambrian (Kaźmierczak et al., 1996). Reports of their occurrence are, however, limited to the Jurassic Solnhofen Limestone of Germany (Seilacher et al., 1985), the Jurassic Oxford Clay (Martill, 1987b, 1988), and the Eocene Messel Shale of Germany (Allison and Briggs, 1991b), although Schieber (1998) has recently reported mat-like structures from the Mid-Proterozoic Belt Supergroup (Montana, U.S.A.).

Floating vegetation, possibly as microbial mats, may also contribute to preservation. Based on the abundance of unidentifiable plant debris and the

Table 6.1. *The occurrence of refractory organic compounds in living tissues and examples of molecular and tissue preservation*

Tissue/molecule	Occurrence	Class of compound	Geological example of tissue	Geological example of molecule
Chitin	Arthropod cuticle, fungal hyphae	Polysaccharide carbohydrate	Trilobite cuticle; ceratiocaridid Crustacea	Cambrian *Hyolithellus*?
Cellulose	Plants	Polysaccharide carbohydrate	Devonian land plants from Rhynie Chert	Derivatives in Carboniferous *Calamites*
Lignin	Plants	Polyaromatic	Devonian land plant *Gosslingia*	Derivatives in Jurassic conifer
Melanin	Pigment in fungi and cephalopod ink	Polyaromatic	Jurassic ink sacs of cephalopods	Triassic ink sacs of cephalopods
Sporopollenin	Walls of plant spores	Oxidative polymer of carotenoid or carotenoid esters	Ordovician *Tasmanites* spores	Ordovician *Tasmanites* spores
Collagen	Skin and matrix of bone; graptolite periderm	Protein polymer	Cellular tissue in Ordovician graptoblast	*Possible* denatured *remnant* in Cretaceous dinosaur bone

Modified from Allison and Briggs (1991*b*).

general lack of macroscopic plant fossils and root structures, Zangerl and Richardson (1963) concluded that much of the plant debris in Pennsylvanian black shales (which resemble coals) originated from a floating mat of vegetation (*flotant*), similar to those in modern Gulf Coast (U.S.A.) environments. The vegetation produces oxygen in shallow waters that is consumed by animals, while at the same time maintaining a quiet low-oxygen bottom; such flotants may be disrupted only by hurricane-force winds (Zangerl and Richardson, 1963). Similarly, decadal-scale laminations (possibly related to cycles of solar insolation) in early Holocene sediments of Guaymas Basin (Gulf of California) have been attributed to allochthonous diatom mats that are apparently concentrated at the pycnocline and then rapidly deposited when water column stratification breaks down in early winter (Pike and Kemp, 1997).

Microbial mats may also be involved in authigenic mineralization of softparts. Some of the best evidence for the role of microbial mats in softpart phosphatization and preservation comes from Jurassic limestones near Cerin, France, which formed in a lagoonal environment (Wilby *et al.*, 1996a). Although phosphate may be released by softpart decay, much more phosphorus is present in preserved softparts than is available in living or recently dead carcasses. Thus, it appears that microbial mats concentrate phosphorus from seawater or sediment; in the case of the Cerin limestones, this amounted to approximately 2.5 times that of phosphorus levels in surrounding sediment.

Once buried, organisms are subjected to *compression* by overburden and *dewatering* of sediment. *Flattening* may produce taxonomic and morphological artifacts, and depends upon grain size, morphology and mechanical strength of the organism, orientation of the organism to bedding, nature and timing of diagenesis, and infilling of cavities (Briggs, 1990). Collapse of soft-bodied organisms usually occurs in no more than a few weeks and does not normally result in lateral expansion of the fossil (Briggs, 1990). Thus, the preservation of volatile softparts in three dimensions implies very rapid inhibition of decay and authigenic mineralization, especially by phosphate (Allison and Briggs, 1991a). Overburden compaction post-dates decay collapse and is distinguished from it by distortion and flattening of mineral infills, replacements, and coatings (Allison and Briggs, 1991a). Two-dimensional volatile tissues may be preserved as organic residues (e.g., Burgess Shale), carbonized bacterial films (Posidonienschiefer), mineral films or replacements, impressions, and concretions (e.g., Mazon Creek concretions; see below).

Once the more labile volatiles are lost, only the refractory organic components are left, which are usually structural components such as cuticles. Most

Konservat-Lagerstätten fall in this grouping, which includes most of the terrestrial record of arthropods (insects) and the bulk of the plant fossil record (Allison and Briggs, 1991*a*).

6.3 Recurrent associations of Konservat-Lagerstätten

Each Konservat-Lagerstätte has its own unique traits that prevents *rigorous* classification (Table 6.2; see Allison and Briggs, 1991*b*, for listing of major Konservat-Lagerstätten). Nevertheless, Konservat-Lagerstätten may be arranged into groups, the members of which share certain traits (*nomothetic relationships*; Chapter 1). Such classifications are useful for synthesizing information about biotic composition, evolutionary significance, environmental occurrence, and prospecting for further Lagerstätten (Seilacher *et al.*, 1985; Conway Morris, 1985; Allison and Briggs, 1991*a*).

6.3.1 Ediacaran association

The cnidarian or cnidarian-like Ediacaran biotas were first described from the late Precambrian of Australia, but have since been documented from every continent except Antarctica (Conway Morris, 1990; Crimes *et al.*, 1995; Narbonne, 1998; see McMenamin, 1998, for most recent review). Originally found in shallow-water sediments (Ediacara Hills, South Australia), they have since been discovered in deep-water environments, as well (e.g., Mistaken Point, Avalon Peninsula, Newfoundland).

These Konservat-Lagerstätten consist of impressions typically found on the base of sandstones immediately overlying finer-grained sediments; i.e., they occur on the soles of storm or turbidite beds (Narbonne, 1998). Two basic modes of preservation were originally described: fossils preserved as ridges on the soles of beds probably decayed prior to burial, whereas forms with a quilted or feathery appearance did not decompose until after burial (Wade, 1968). Those from Mistaken Point (Newfoundland) were buried by volcanic ash, however (Narbonne, 1998).

Because of the enigmatic affinities of the biota, the mode(s) of preservation are critical to determining their internal structure and whether they truly mark the dawn of animal life or instead constitute a distinct chemosynthetic kingdom or phylum ("Vendozoa" or "Vendobionta") that represents an early "failed experiment" in life (reviewed in Narbonne, 1998).

Consequently, some workers have attempted to simulate burial of the Ediacaran biota. Norris (1989) found that tentacles, mouth, and gonads were

6.3 Recurrent associations of Konservat-Lagerstätten

Table 6.2. *Questionnaire for Konservat-Lagerstätten Seilacher et al., 1985.*

(1) Basin situation
    size (km): 10^{-1} 10^0 10^2 10^3
    setting:
        oceanic
        epicontinental
        terrestrial
    origin:
        tectonic
        volcanic
        astroblemic
        suberosional
        reefal
        sedimentary
        glacial
    geographic frame:
        limestones
        clastics
        crystalline
(2) Stratigraphy
    thickness (m): 0.01 0.1 1 10 100
    duration in absolute time:
    vertical context:
        transgressive
        peak transgression
        regressive
        fining-up cycles
        coarsening-up cycles
    lateral sequence
(3) Sedimentology
    lithology:
        biograins (coccoliths, forams,
            radiolarians, spicules, etc.):
    sedimentary structures:
        lamination (varves, algal, etc.)
        slump horizons
        graded horizons
        current ripples
        wave ripples
        emersion marks (tidal channels,
            mud cracks, etc.)

(4) Geochemistry
    evaporitic precipitates (aragonite,
        calcite, dolomite, gypsum, halite):
    pyrite concretions (globular, discoid):
    isotopic deviations:
    C_{org} (particles, kerogene, bitumen):
(5) Taxonomic spectrum
    dominated by (priority):
        echinoderms
        cephalopods
        vertebrates
        crustaceans
        trilobites
        others:
(6) Ecological spectrum
    burrows (episodic, continuous):
    tracks (with or without bodies):
    endobenthos:
        hemisessile
        vagile
    epibenthos (episodic, continuous):
        sessile
        vagile
        able to swim
    pelagics:
        nekton
        floaters
        epiplankton (on wood, cephalopod
            shells, live or dead)
    terrestrial organisms:
        tracks
        skeletons
    land plants (twigs, leaves, trunks)
    flyers
(7) Necrolytic criteria
    death marches:
    landing marks (live, dead):
    softparts (impressions, films):
    organic cuticles:
    articulation (vertebrates, echinoderms,
        arthropods, bivalves, apytchi):

Table 6.2. (cont.).

(8) Stratinomic criteria	concretionary cementation (nucleus,
life positions:	pressure shadow, buckle, pedestal):
roll marks (of what?):	compactional deformation
convex up or down:	incoalation:
current orientation (azimuth):	phosphatization (of what?):
wave orientation (azimuth):	replacement (shells, bones):
(9) Diagenetic criteria	*General conclusions*:
aragonite preservation:	stagnation (thermal or halocline):
early aragonite solution (composite casts):	obrution:
pyritic steinkerns:	algal sealing:

Seilacher *et al.*, 1985.

not preserved in experimentally produced impressions of modern cnidarians, but sometimes annular rings, radial canals, and a central boss were preserved; deformed structures were more frequent in modern specimens and suggested that fossil forms were more rigid than modern ones (Briggs, 1995). Based on his results, Norris (1989) concluded that a relation between the Ediacaran biota and cnidarians could not be rejected. By contrast, field observations and laboratory experiments by Bruton (1991) demonstrated that modern *Aurelia* only form impressions if stranded subaerially, but specimens that settled into deeper water left no traces. Desiccated specimens showed structures very different from those of the Ediacaran biota and led Bruton (1991) to conclude that there was little or no affinity between the cnidarian and Ediacaran biota. More recent studies have, however, focused on offshore burial *in situ*, which may have been aided by authigenic mineralization resulting from decay and cyanobacterial mats (Briggs, 1995; Narbonne, 1998).

Although Allison and Briggs (1993b) suggested that preservation of Ediacaran biotas may be related to the absence of deep bioturbation, they had earlier (1991a) suggested that the nature of the organism is the primary factor accounting for the Ediacaran type of preservation (see also Seilacher, 1989; Fedonkin, 1992; Retallack, 1994; and Chapter 9). As diverse trace fossils indicate that bottomwaters were oxygenated and no shells have been found, the Ediacaran organisms may have had some sort of tissues that were resistant to decay. Crimes *et al.* (1995) described two forms from deep-water turbidites of the Late Cambrian of Booley Bay (southeastern Ireland). The fine detail of both *Ediacaria booleyi* n.sp. and *Nimbia occlusa* suggests that they had a rigid body wall or cuticle; also there is no evidence for the cnidarian traits of a mouth and a bilayered body wall surrounding a central cavity. Specimens of both species appear to have been transported short distances by turbidity currents (as indicated by sole marks) and then deposited, sometimes at high angles, into deep-water muds where they formed molds. As the bodies decayed, the molds were filled by sand (casts) from the overlying turbidite.

Crimes *et al.* (1995) concluded that the virtual absence of Ediacaran-type fossils from the rest of the Phanerozoic hints that the Ediacara "skeleton" may have been superseded by mineralized skeletons (cf. Chapter 8 on skeletonization). Recent evidence indicates that the Ediacaran biota survived at least to the end of the Proterozoic (Runnegar, 1996), and that members of the Ediacaran fauna possibly survived into the Cambrian by migrating into deeper water, perhaps to escape increasing competition (Crimes *et al.*, 1995). Also, molecular phylogenetic studies suggest a substantially longer period of evolution prior to the appearance of the Ediacaran biota in the fossil record, but if this is the case, the apparent lack of preservation of earlier forms remains a mystery (Runnegar, 1996; see also Runnegar, 1982, and Chapter 8).

6.3.2 Burgess Shale association

Burgess Shale associations are known from at least 34 Early–Middle Cambrian sections in North America, Spain, Poland, South Australia, China and Siberia (Conway Morris, 1990; Allison and Briggs, 1991*b*; Briggs *et al.*, 1994). Included in this association are not only the Early–Middle Cambrian Burgess Shale (see Conway Morris, 1998, for most recent review), but, among others, the Early Cambrian Emu Bay Shale, the Middle Cambrian Spence Shale (Liddell *et al.*, 1997) and Wheeler Formation, Beecher's Trilobite Bed (Late Ordovician), and the Early Devonian Hunsrückschiefer (Figure 6.2); most recently, Briggs *et al.* (1996*b*) have reported a three-dimensionally preserved Burgess Shale-type fauna (mainly arthropods and polychaetes) in calcitic concretions from the Early Silurian (Wenlock) of England. Rapid burial coupled with anoxia (in some cases) are probably the most important factors in the preservation of this association, as both rapid burial and anoxia would have inhibited bioturbation. Most examples occur in deeper water on the flanks of shallow-water carbonate belts that rimmed Laurentia (Allison and Briggs, 1991*b*), but significant soft-bodied preservation in deep-water environments is largely unknown after the first half of the Paleozoic (Allison and Briggs, 1991*b*; Figure 6.5). Increasing diversification and depth of deep burrowing infauna may account for the lack of deep-water soft-bodied Lagerstätten after the Devonian (Allison and Briggs, 1991*b*; see also Chapter 9).

248 Exceptional preservation

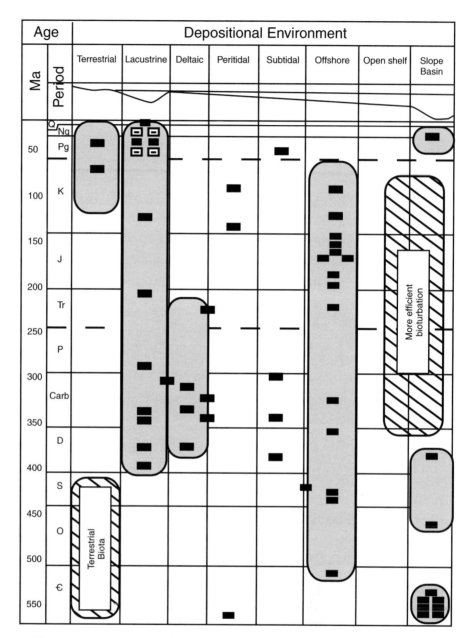

Figure 6.5. Environmental distribution of major Konservat-Lagerstätten through time. (Redrawn from Allison and Briggs, 1991b.)

Although the biotas have been intensively studied, the depositional and diagenetic conditions under which these Lagerstätten formed are still not completely understood and may actually vary (Allison and Briggs, 1991b). The Burgess Shale, which is part of the fine-grained Stephen Formation of British Columbia

(Canada), was originally inferred to have been deposited in a quiet-water lagoon, as the preservation of delicate softparts was thought to have been possible only in quiet water with little transport (Allison and Brett, 1995). The Burgess Shale-type fauna of Chengjiang, China, may also have been deposited in shallow water (Briggs et al., 1994), but soft-bodied carcasses are now known to be capable of transport with little alteration (Allison, 1986; cf. Ediacara association).

Coupled with facies relationships, the Burgess Shale biota was later inferred to have been swept into what is now the Stephen Formation from the adjacent shallower 170-m-high carbonate Cathedral Escarpment (Seilacher et al., 1985) or buried *in situ* by turbidity currents at the base of the escarpment (Allison and Briggs, 1991b), as the best-preserved organisms are found closest to the escarpment (Allison and Brett, 1995). Conway Morris (1990) concluded that the Burgess Shale biota perhaps lived at the base of the escarpment as little of the entombed fauna appears to have been derived from the shallow carbonate shoals, and that it was transported into a "hostile" (anoxic) environment, since delicate laminae of the basal Phyllopod bed indicate little or no bioturbation; also, many specimens are enrolled (which often happens in response to anoxia) and escape burrows are absent, both of which suggest that animals were "stunned" prior to delivery into deeper water. The presence of *partially* decayed delicate specimens (including trilobite molt ensembles; Brett and Baird, 1993; Chapter 5) also suggests limited transport (Conway Morris, 1990; Brett and Baird, 1993; but see above and Chapter 2). Although Ludvigsen (1989) concluded that the biota was buried by storm-generated tempestites in a *ramp* environment, based on hydrodynamic considerations, Allison and Brett (1995, p. 1081) concluded that the biota was deposited as "a high-density fluidized silt-mud flow that was never more than a few tens of centimeters thick," whereas *other* finer-grained beds may represent "low-density gravity-flow event[s] such as *distal* turbidite[s]" (my italics). Most recently, Stewart (1991; cited in Briggs et al., 1994) concluded that the Cathedral Escarpment was the result of collapse of the platform margin, and that during collapse the Burgess Shale biota could have been transported from shallower muddy sites. Each of these scenarios is compatible with the occurrence of graded beds, variable orientation of specimens relative to bedding planes (e.g., head-to-tail compaction of *Sidneyia*, which was up to 15 cm in length; Briggs et al., 1994), and seepage of sediment between appendages (Conway Morris, 1990).

Butterfield (1990) suggested that preservation in the Phyllopod bed is also the result of adsorption of clays to organisms during transport in a "sediment cloud"; the clays presumably inhibited the activity of catalytic enzymes during decay by absorbing them onto clay particle surfaces, possibly denaturing enzymes, and by stabilizing substrate (carcass) molecules. Towe (1996) objected to Butterfield's (1990, 1995) hypothesis; he (1996) suggested that Burgess Shale organisms were preserved by sediment infilling of body cavities followed by diagenesis of smectite to illite that released silica cement. Butterfield (1996) countered with further data to support the clay adsorption hypothesis; most recently, "mapping" of element distributions in clay minerals (e.g., K, Al, Si, Ca) in fossils has confirmed Butterfield's mechanism: element concentrations in fossils were controlled by tissue chemistry, whereas concentrations in the surrounding sediment matrix remained relatively constant (Orr et al., 1998).

Conway Morris (1986) concluded, based on comparison of the shelly assemblage with those from beds missing soft biota, that the diversity of the Burgess Shale is representative of the unit as a whole. Nevertheless, skeletonized taxa account for only about 20% of the genera and approximately 2% of individuals in the Burgess Shale (Conway Morris, 1986). The diversity of well-skeletonized taxa in the "pre-slide" environment suggests that the bottom was reasonably well-oxygenated, which, along with proximity to carbonate platforms (low Fe availability) *might* account for the limited pyritization that has been observed (Conway Morris, 1986).

Allison and Brett (1995) later concluded that the Burgess Shale was deposited in the vicinity of a fluctuating "oxycline." They concluded that soft-bodied organisms were preserved in anoxic sediments that lay below a low-oxygen column, as evidenced by the absence of trace fossils and the rarity of shelly benthos in beds containing softparts. Dispersed pyrite – as opposed to pyrite concentrated around carcasses – also suggests anoxic water above the SWI (Allison and Brett, 1995; see also Chapter 3), but the occurrence of interspersed low-diversity shell beds that are apparently autochthonous (based on lack of evidence for erosion beneath the shell beds), along with burrows, hints at sporadically higher oxygen levels. Allison and Brett (1995) concluded that soft parts were preserved when the oxycline was above the SWI, whereas benthic colonization by *r*-selected (opportunistic) benthos occurred when the oxycline was lowered to the SWI. Similar models have been proposed for the Posidonienschiefer (Kauffman, 1981; but see below).

On the other hand, the Early Cambrian Emu Bay Shale of South Australia may have accumulated in a shallow-water (possibly) intertidal environment, although Conway Morris (1985) suggested that the thin (3–4 m) shales with the most prolific softpart preservation may have accumulated in somewhat deeper water. These dark micaceous, finely laminated shales appear to have

6.3 Recurrent associations of Konservat-Lagerstätten

accumulated in localized anoxic "ponds" (perhaps in fault basins) with little or no bioturbation, which may account for the relatively low diversity of fossils (Conway Morris, 1985).

6.3.3 Beecher's Trilobite Bed

One of the two most significant post-Cambrian Konservat-Lagerstätten from deep-water environments is the Beecher's Trilobite Bed of the Frankfort Shale (Late Ordovician) of New York State. Beecher's Trilobite Bed is a 4-cm-thick distal "silty microturbidite," near the base of which trilobites (especially *Triarthrus*) are concentrated in a layer a few millimeters thick (Allison and Briggs, 1991*b*). Although originally considered to consist of allochthonous remains, the low-diversity fauna of the Beecher bed occurs throughout the Frankfort Shale, and suggests that the deposit is autochthonous or perhaps parautochthonous (Brett and Baird, 1993). The concentration of pyritization in the trilobite layer suggests localized pyrite formation in the vicinity of carcasses (Briggs *et al.*, 1991; Chapter 3). Pyrite occurs as framboids and cubes or pyritohedra less than 20 µm in size with occasional larger euhedral crystals (Briggs *et al.*, 1991, 1996*a*).

6.3.4 Hunsrückschiefer (Hunsrück Slate)

The younger of the two most significant post-Cambrian deep-water Konservat-Lagerstätten is the Early Devonian Hunsrückschiefer of western Germany. The Hunsrückschiefer represents a fine-grained distal turbidite and was deposited in approximately 200 m of water (Bergström, 1990). The fauna, which is most abundant in the Bundenbach area (southwest of Koblenz), is much more diverse than Beecher's Trilobite Bed, however, and includes over 100 species each of echinoderms (including beautiful ophiuroids preserved in three-dimensional relief) and molluscs, along with other shelled and soft-bodied invertebrates (including worms, ctenophores, arthropod muscles and intestines) and agnathans and placoderms (Bergström, 1990; Allison and Briggs, 1991*b*). Organisms may have been smothered or killed during transport, as many specimens are oriented at a high angle to bedding. Coupled with post-burial compaction, the high angle of fossil deposition has resulted in many deformed specimens. Although originally considered allochthonous, the presence of burrowed substrata containing fragmentary remains of the obrution fauna suggests that the background environment was not completely anoxic and that perhaps the Hunsrückschiefer may be autochthonous (Brett and Baird, 1993). Sediment surrounding carcasses is carbon-poor and iron-rich, so that pyritization of fossils is confined mainly to carcasses

(Canfield and Raiswell, 1991b; Chapter 3). The pyrite is also enriched in ^{34}S, which suggests bacterial sulfate reduction in association with decaying organic matter (Briggs *et al.*, 1996a; Chapter 9). (See note added in proof, p. 267.)

6.3.5 Orsten association

The Late Cambrian Alum Shale Formation of southern Sweden is one of the earliest formations from which Paleozoic fossils were described by Linnaeus (Müller, 1985). This formation, which consists of highly condensed organic-rich shales, contains bituminous (*anthraconitic*) nodular limestone lenses (*Orsten*). Some, but by no means all, of the nodules yield a fauna of tiny (<2 mm) three-dimensional phosphatized arthropods, including suspension-feeding branchiopods (Walossek, 1993) that may have lived in the plankton (Butterfield, 1997; see also Chapter 9) as soon as the Early Cambrian (Butterfield, 1994). Although the source of the phosphate remains unclear (Walossek, 1993), it may have been derived from the carcasses themselves, although Müller (1985) dismissed this source as inadequate; instead, he suggested that phosphate had been weathered from adjacent landmasses. He also suggested that since arthropods are the dominant representatives of this fauna, chitin may have served to nucleate phosphatic minerals; although carcasses release phosphate during decay (Briggs and Kear, 1993b, 1994b; Chapter 3), if a carcass is sufficiently calcified, it may nucleate $CaCO_3$ instead (Donovan and Veltkamp, 1994).

Faunas of similar origin but of different ages have been described from all over the globe. In some of the younger examples, the source of phosphate appears to have been vertebrate carcasses. One of the best examples is that of the Santana Formation (Early Cretaceous) of northeast Brazil, in which fishes (including striated muscle) generated during mass mortalities, were preserved in a matter of hours by phosphatization after carcasses had settled to the SWI; this phase was followed by precipitation of calcium carbonate nodules during burial (Martill, 1988, 1990). Pterosaur wing membrane and arthropod curticle are also sometimes preserved in the Santana (Martill, 1990), and phosphatized ostracodes (scavengers) and parasitic copepods have been described from vertebrate carcasses in the Santana Formation (Müller, 1985). Three-dimensional preservation of plants may also occur if early diagenetic mineralization occurs. In the three-dimensionally preserved Jurassic biota of La Voulte-sur-Rhône (France), which includes the oldest recorded octopus and other coleoid cephalopods, worms, pycnogonids, and the enigmatic thylacocephalans (mollusca), early phosphatization resulted in deposition of apatite that in turn served as a template for calcite, then pyrite, and ultimately galena (Wilby *et al.*, 1996b).

6.3.6 Posidonienschiefer (Holzmaden)

The Posidonienschiefer (Jurassic), which derives its name from an earlier name for the bivalve *Bositra*, consists of 6–8 m of thick black bituminous marls and shaly marls with intercalated bituminous allochthonous limestones that is found over large parts of central Europe (Seilacher *et al.*, 1985; Wild, 1990). The limestones represent rapid deposition, as indicated by vertically oriented fossils, and contain calcareous nodules and concretions. The preserved flora consists of coccolithophorids, but *Ginkgo*, conifers, cycads, and horsetails were washed in from land about 100 km south of Holzmaden (Wild, 1990), which is a small village located about 30 km southeast of Stuttgart. The fauna consists mostly of pelagic and epiplanktonic organisms and includes radiolarians, foraminifera, ostracodes, bivalves (some of which were attached to ammonites), gastropods, coleoids and belemnoids (including softparts), crinoids (including the giant *Seirocrinus*, which may have lived attached to floating logs much like modern gooseneck barnacles), sharks, plesiosaurs and ichthyosaurs (often with preserved stomach and intestinal contents, embryos, or giving birth to young), teleosts, crocodiles, and allochthonous pterosaurs and saurischians (Wild, 1990). The outline of vertebrates is sometimes preserved as carbonized films formed from bacteria (Wild, 1990; Allison and Briggs, 1991*a*).

The dark color of the Posidonienschiefer results from disseminated pyrite and organic material, which may reach 10% in some layers. Along with the depauperate benthic fauna, high carbon content has been used to infer stagnant bottomwaters, as opposed to *gyttja* conditions (stagnant porewaters only; Seilacher *et al.*, 1985; cf. Kauffman, 1981). The Posidonienschiefer was in fact formed during the widespread Toarcian anoxic event and is found over large parts of central Europe during a time span of the order of millions of years (Seilacher *et al.*, 1985). It appears that stagnant waters close to the bottom were overlain by oxygenated waters in which the plankton and nekton lived, and that occasionally storms oxygenated the bottom and allowed colonization by a restricted benthos (such as echinoids, ophiuroids, bivalves, and *Chondrites*) for brief periods (Seilacher *et al.*, 1985). Storms also account for current-oriented fossils at many levels (Seilacher *et al.*, 1985). Similar deposits are found elsewhere at other times, such as the Ohio Shale (Late Devonian–Early Carboniferous; Seilacher *et al.*, 1985).

6.3.7 Solnhofen Limestone

The Solnhofen Lithographic Limestone (Early Tithonian, Bavaria) is an example of the Plattenkalk lithology (micritic, even-layer limestone slabs), and was laid down primarily between algal-sponge reefs, perhaps in some sort of hypersaline lagoon (Allison and Briggs, 1991b; Davis, 1996; see also Barthel et al., 1994, for extensive updated review of Barthel's classic work). Although relatively thick (up to about 90 m), the outcrop area of the Solnhofen Limestone (~70 × 30 km) is much smaller than that of the Posidonienschiefer and the thickness much thinner (duration approximately 0.5 million years; Seilacher et al., 1985). Similar deposits of different ages are found in the French Jura and elsewhere (Seilacher et al., 1985; Allison and Briggs, 1991a,b).

The Solnhofen is famous for its pterosaurs and a possible feathered "avetheropod," the tiny delicate *Compsognathus*, which Thomas Huxley considered a link between birds and dinosaurs (Desmond, 1982). The Solnhofen is most famous, however, for the preservation of *Archaeopteryx*. *Archaeopteryx* may have been caught in storms during flight and drowned (Viohl, 1990; Davis, 1996); its carcasses probably floated (aided by hypersalinity) before final sinking and, in some cases, rapid burial by storms. More disarticulated carcasses presumably remained near the SWI for longer durations (Davis, 1996).

Kemp and Unwin (1997) assessed specimens of *Archaeopteryx* and *Compsognathus* according to skeleton completeness and degree of articulation, and found two preservational types: (A) well-articulated and almost complete, and (B) less complete and more disarticulated. They concluded that different preservational types can be present within a single basin, so that there were no general preservational patterns. Based on taphonomic studies of modern birds (Davis and Briggs, 1998), which indicated that scavenging was the primary agent of disarticulation (see Chapter 2), Kemp and Unwin (1997) concluded that in most cases the time from death to deposition ranged from a few hours or a few days (type A specimens) to approximately 27 days (type B specimens). The best-preserved specimens may have drowned during a storm, with rapid sinking facilitated by water-filled lungs and soaked plumage. Little post-mortem transport is indicated by the high degree of articulation and the preservation of feathers.

Archaeopteryx can, of course, only be identified as a bird because of the imprints of feathers that formed during early lithification (Davis and Briggs, 1995). Despite their fragility, feathers have been reported in 77% of other fossil assemblages containing avian remains (Davis and Briggs, 1995; see their table 1 for compilation). Scanning electron microscope studies of *Archaeopteryx* feathers from the Solnhofen show lithified bacteria and possible glycocalyx remains; such remains are common among carbonized traces of feathers, which are found in 69% of deposits and characterize fine-grained rocks. In the case of the

Solnhofen, bacteria colonized feather surfaces in contact with the SWI and promoted early lithification of sediments before complete decay of the feather. After decay, soft sediment above was pressed into the mold left by the decay of the feather, so that feathers are characteristically preserved as a mold of the ventral surface and the opposite surface preserves a cast of the same surface (Davis and Briggs, 1995). Feathers may also be preserved in amber and as external molds in coprolites (Davis and Briggs, 1995).

Despite its renown, however, the Solnhofen is not particularly fossiliferous and many of its approximately 600 species are, like *Archaeopteryx*, allochthonous, including crustaceans, echinoderms, and ray-like sharks, most of which died while being washed in by storms (Brett and Baird, 1993). Occasionally, a hardy horseshoe crab (mainly juvenile) survived long enough to crawl on the bottom leaving tracks to its site of death ('death march' of Seilacher *et al.*, 1985); cyanobacterial mats may have played an important role in preservation of these tracks (Seilacher *et al.*, 1985). Terrestrial plants, invertebrates (especially insects), and other vertebrates were also washed into the lagoon.

Other Solnhofen fossils are planktonic or nektonic in origin. Ammonites and belemnites were surpassed in number by the medusa-like pelagic crinoid *Saccocoma*, jellyfish, and vertebrates (Seilacher *et al.*, 1985). The lower organic carbon content of the Solnhofen suggests lowered surface productivity, however. Like the Posidonienschiefer, the Solnhofen was deposited beneath highly stratified waters (possibly represented by a halocline) that formed in a semi-arid basin restricted by reefs. The bend in the neck of preserved fish suggests, for example, a response by the carcass to dehydration as a result of hypersaline bottom-waters (Viohl, 1990; cf. Seilacher *et al.*, 1985, for *Archaeopteryx*).

Unlike the Posidonienschiefer, however, it appears that obrution was more important than stagnation during deposition of the Solnhofen (Figure 6.2; Seilacher *et al.*, 1985). The fine-grained sediment was probably derived from the margins of the basin and transported in suspension to the Solnhofen area by storms or turbidity currents. Indications of current-scour are found only toward the margins of the basins, however, whereas settling marks are found next to ammonite aptychi in more open-water deposits (Viohl, 1990).

Early diagenetic phenomena have also been documented in the Solnhofen. Cementation and sediment compaction occurred after the collapse of fossils, which are all flattened and form depressions (*collapse calderas* of Seilacher *et al.*, 1976). Aragonitic ammonite shells dissolved in shallow sediment layers, as deformed ammonites were reduced to their periostraca before slumping (Seilacher *et al.*, 1976). Softparts (e.g., fish intestines, muscle) are also preserved, possibly as a result of early bacterial activity and phosphatization, whereas the ink of cephalopods was apparently resistant to diagenesis (Viohl, 1990).

6.3.8 Mazon Creek association

The Mazon Creek biota (Middle Pennsylvanian; Westphalian D) is famous for its exceptionally well-preserved terrestrial floras and marine, freshwater, and terrestrial animals, which are found predominantly in sideritic ($FeCO_3$) concretions of the Francis Creek Shale Member of the Carbondale Formation. The biota ranges from medusae to crustaceans, insects, holothurians, fish and tetrapods, including certain problematic ones (see Shabica and Hay, 1997, for review; a freeze–thaw method for cracking concretions is also described). Similar biotas have been preserved in the coal measures of Great Britain, Montceau-les-Mines in France, and the Triassic Grès-à-Voltzia shales of northeastern France (Allison and Briggs, 1991a). The formation and preservation of Mazon Creek associations was favored by episodic sedimentation and variable salinity (inhibiting bioturbation) in widespread deltaic and estuarine settings of the Late Carboniferous (Allison and Briggs, 1991a,b); anoxia may have also been involved because estuaries have high quantities of dissolved organic matter (Simon et al., 1994). Many of the Mazon Creek taxa at the family and lower levels were adapted to euryhaline conditions; Briggs and Gall (1990) demonstrated that these same taxa survived into the Triassic Grès-à-Voltzia shales and that the similarity between the two biotas is a function primarily of environment despite the great age difference, whereas other Konservat-Lagerstätten of age closer to that of the Mazon Creek were much less similar.

Baird et al. (1985, 1986) reconstructed a taphonomic gradient for the Mazon Creek Lagerstätte of Illinois (much of this information is also presented in separate chapters in Shabica and Hay, 1997; Zangerl and Richardson, 1963, also exhaustively analyzed similar marginal marine sediments formed adjacent to a Pennsylvanian epicontinental sea in what is now west-central Indiana). The ancient depositional setting is sufficiently rare that the only modern analogs may be estuaries in the vicinity of anthropogenic deforestation and massive sediment influx that mimics erosion of the ancient Appalachians (e.g., Ganges-Brahmaputra, Mahakam, Mekong, and Orinoco deltas; Shabica and Hay, 1997; see also Scheihing and Pfefferkorn, 1984; Gastaldo and Huc, 1992). The Francis Creek Member consists of distributary channel, interdistributary bay, and proximal prodelta deposits that lie abruptly on top of the Colchester Coal Member. During its deposition, most of Illinois was inundated by an epeiric sea from the southwest, and deposition of the Francis Creek post-dated a marine transgressive event.

The initial rise of the water table in coastal areas produced conditions favorable for the formation of the Colchester Peat, and continued sea-level rise eventually inundated the swamp. Thus, the Francis Creek Member represents an estuary fill, as indicated by a change from basal laminar mudstones (low energy) to weakly rippled siltstones to climbing ripples in sandstones at the top (Baird et al., 1986). The estuarine environment is also suggested by the small size of bivalves, worms, and decapod crustaceans, among other taxa, and the total lack of normal marine taxa (e.g., corals) and the rarity of pelmatozoans.

The amount, diversity, and degree of articulation of plant detritus all decrease southwestward, probably as a result of decay, waterlogging, and buffeting by waves, as marine influence and biotas increase. The contact between nonmarine and marine biotas is abrupt, and stratigraphic evidence indicates that marine faunas were formed at the delta margin. Diverse, allochthonous clusters of leaves and fragmented foliage, bark, stems, and fruiting organs indicate transport into the estuary from diverse upstream habitats. The similar release of tremendous amounts of plant debris has been documented off the modern Orinoco Delta (eastern Venezuela). Most plants in the Mazon Creek are found on bedding surfaces, but at some localities, leaves of *Neuropteris* sp. and other species occur upended and curled within nodules, suggesting turbulence and current flow (Chapter 2). On the other hand, the occurrence of the sphenopsids *Annularia* sp. and *Sphenophyllum* sp. as vertical shafts oriented perpendicular to bedding with leaf whorls lying more or less parallel to bedding suggests burial *in situ*. In some cases, fruiting bodies have been preserved with intact interiors, waxy membranes, and spores. Three-dimensional molds of other plant remains in concretions that are partially filled with geopetal carbonaceous residue, calcite, and sphalerite suggest an early syngenetic origin for concretions, possibly via release of bicarbonate ions from decaying peat (Allison and Briggs, 1991*b*). Sometimes plant remains are permineralized with pyrite, but only rarely with siderite.

Both local and regional preservational gradients occur in the Mazon Creek fauna. The best-preserved animals occur in concretions with laminated interiors, but poor preservation is often noted at nearby sites associated with numerous burrowing taxa (bivalves, holothurians, polychaetes). Preservation also decreases along an east-to-west gradient associated with increasing bioturbation and slow sedimentation rates near the delta margin (increasing marine influence). In normal seawater, iron typically occurs only in low concentrations so that pyrite and calcite ($CaCO_3$) are precipitated in porewaters; in brackish and freshwater environments, however, dissolved iron is typically abundant in runoff and sediment and sulfate concentrations are quite low, so siderite precipitation is favored (Berner *et al.*,

1979; Walter and Burton, 1990; cf. Chapter 3). Based on actualistic studies of delta and marsh sediments (Ho and Coleman, 1969; Pye, 1984), the occurrence of abundant siderite concretions in the Mazon Creek biota suggested to Baird *et al.* (1986) that diffusion of sulfate into sediment from overlying water was inhibited by low dissolved sulfate concentrations in water; relatively low rates of bioturbation (which would otherwise pump available sulfate into sediment; Allison and Pye, 1994); and rapid sedimentation (including obrution), and loading (compaction and dewatering) of sediment that decreased sediment permeability and that may have established an upward counterflow to sulfate diffusion. Rapid burial of abundant organic matter may have also increased rates of ammonia production (including methanogenesis) that would have led to siderite precipitation (Woodland and Stenstrom, 1979; Berner, 1981; see Chapter 3). According to this reconstruction, minor amounts of pyrite typically associated with concretions may represent precipitation of sulfides released during early decay (Baird *et al.*, 1986). Depending on the concentration of organic matter, such reactions could have occurred over large areas or have been localized around buried, decaying softparts (Woodland and Stenstrom, 1979; Chapter 3).

6.3.9 Coal balls

Coal balls are associated almost exclusively with the Westphalian (Late Carboniferous) in Great Britain and Europe, but range through the Pennsylvanian in North America (Scott and Rex, 1985). Coal balls consist of peat that was permineralized by calcium and magnesium carbonates before appreciable compaction or alteration occurred, and have yielded exceptionally preserved biotas that include plant cell contents and reproductive structures (see Scott and Rex, 1985, for references). The diversity of the preserved flora is rather low, however, and may represent a "stunted" flora adapted to coal swamp environments.

Like the Mazon Creek association, the stratigraphic occurrence implies a strong relation to tectonic and climatic controls which were not uniform across the Euramerican coal province, although permineralizing fluids were basically similar. Permineralization of peat appears to have occurred in stages (zones of crystal growth) in response to fluctuations in the presence of permineralizing fluids. The dominant carbonate (typically calcite) occurs in the form of fine fibrous crystals which were initiated on cell walls surrounding the voids of plant cells. Fibrous ferroan calcite often filled central voids in fibrous calcite, which suggests that late permineralizing fluids were Fe-rich. In other cases, no distinction can be made between calcite and ferroan calcite, which suggests that both minerals were crystallizing more or less simultaneously, and that variation in the timing and nature of permineralization was commonplace. Only rarely did recrystallization and poor preservation occur, most notably in coal balls from the Netherlands (Finefrau-Nebenbank), especially in association with rootlets, which may have provided pathways for late permineralizing Mg-rich fluids. In some cases, pyrite filled late fractures in coal balls.

Coal balls are of three types: (1) *normal*, containing only plant remains (mainly in Europe); (2) *mixed*, containing animal and plant remains; and (3) *faunal*, containing only animals (common in North American coal balls, but rare to absent in European ones). Despite the relative uniformity of permineralization and preservation, the distribution of coal balls varies between and within coal seams; in some cases larger coal balls grew from smaller ones and plant structure can be traced from one coal ball to the next.

The distribution of coal balls on both large and small scales implies geochemically precise – but environmentally variable – conditions for carbonate precipitation and coal ball formation, and several mechanisms of formation have been formulated. In normal coal balls, carbonate could be precipitated directly from seawater as a marine transgression drowns the swamp; each coal-ball seam would therefore have to be overlain by a marine shale or limestone, but coal balls are sometimes absent in such stratigraphic settings, although marine beds are also rare (such as the Namurian Limestone Coal Group of Scotland). Non-marine (groundwater) sources of carbonate could also account for normal coal balls, but isotopic studies of the source of carbonate have proved inconclusive. In mixed coal ball formation, the sea breaches a barrier during a storm, bringing carbonate-rich sediment and animals into the swamp.

6.3.10 Lacustrine association

By no means do all lake sediments yield Lagerstätten, but ancient lacustrine environments – especially deep ones – are important sources of Lagerstätten, including plant macrofossils, arthropods (crustaceans such as ostracodes), molluscs, fish and terrestrial vertebrates (including skin impressions) and insects (e.g., Hanley and Flores, 1987; Allison and Briggs, 1991a,b; Behrensmeyer and Hook, 1992; Briggs *et al.*, 1998). Nutrient and oxygen content of the water appear especially important to preservation and are in turn associated with areal extent of lakes, depth of water, and thermal stratification and seasonal overturn that may produce mass mortalities (Table 6.3; see also Behrensmeyer and Hook, 1992, for brief review).

Based on modern temperate lakes (which are probably not good analogs for ancient lakes of different climatic settings), lacustrine environments are of two

Table 6.3. *Taphonomic features of lacustrine environments. See text for explanation of classification of lakes according to size, depth, and oxygenation*

Environmental context	Occurrences of terrestrial organisms			Invertebrates	Vertebrates	Ichnofossils
	Macroplant	Microplant				
Low oxygen: large lake, deep	Rare Logs, seeds	Common Pollen (allochthonous), sorted; phytoplankton		Uncommon Molluscs, zooplankton, benthos	Common Fish, articulated, disarticulated; coprolites, rare flying forms, tetrapods	Very rare
small lake, deep	Common Logs, seeds, leaves plant succession	Common Local pollen, phytoplankton		Uncommon Benthos, insects, zooplankton	Common Articulated fish, rare aquatic tetrapods, flying forms	Very rare
small lake, shallow	Common Leaves, seeds, logs, stromatolites, plant succession	Common Local pollen		Uncommon Arthropods, rare benthos	Common Aquatics, tetrapods	Uncommon Roots
Oxygenated: large lake	Uncommon Wood, debris, stromatolites	Common Pollen (autochthonous + allochthonous), benthic phytoplankton, charophytes		Common Benthos, plankton, rare insects	Common Fish (disarticulated bones, scales), tetrapods in marginal areas	Very common Burrows, roots, tracks
small lake	Uncommon Logs, leaves, debris, stromatolites	Common Benthic phytoplankton, charophytes		Very common Molluscs, crustaceans, rare insects, sponge spicules	Common Disarticulated fish, tetrapods; may be articulated in carbonates	Very common Roots, burrows, trackways

The arbitrary division between large and small lakes is set at 10 km^2, and the division between deep and shallow at 10 m. In a lacustrine context, microplants include phytoplankton such as diatoms.

Behrensmeyer and Hook, 1992.

6.3 Recurrent associations of Konservat-Lagerstätten

basic types: *oligotrophic* (low nutrient levels, low productivity) and *eutrophic* (abundant nutrients, high productivity; Behrensmeyer and Hook, 1992). Ferguson (1985) predicted that biogenic (leaf) decay is faster in oligotrophic lakes because of lower overall nutrient and food availability to bacteria, fungi, and invertebrates; on the other hand, oligotrophic lakes have relatively low diversity, but lower levels of biotic activity may decrease rates of organic destruction (Behrensmeyer and Hook, 1992). In fact, the degradation of leaves and other organic remains appears to be enhanced in some cases by eutrophication (Spicer, 1989; cf. Chapter 2): although high productivity lakes often have diverse biotas and can potentially generate a rich fossil record, intense rates of recycling associated with high productivity often destroy most organic remains (Behrensmeyer and Hook, 1992). But eutrophication may also be correlated with increased sediment input (accompanying nutrient influx) that rapidly buries biogenic remains in some cases.

Deep (>10 m) lakes tend to be large areally (>10 km^2) and stratified into an upper warm oxygenated *epilimnion* and a deeper cooler *hypolimnion* that is often anoxic and inhibits bioturbation and organic decay (Elder and Smith, 1988; Behrensmeyer and Hook, 1992). Judging from modern lakes, an overall thin sediment mixed layer may contribute to exceptional preservation in some cases (Chapter 4). The hypolimnion may be recognizable from the presence of carbon and sulfides (dark coloration) and well-developed laminations (a result of low rates of bioturbation), whereas sediments deposited in the epilimnion are of opposite nature. Soft-tissue outlines may also be preserved by siderite and vivianite in the hypolimnion (Chapter 3). Although small (<10 km^2), shallow (<10 m), oxygen-depleted lakes tend to be unstratified and oxygenated throughout, not to mention ephemeral (Behrensmeyer and Hook, 1992), preservation may be quite good in oxbow lakes (see Chapters 2, 3). Good preservation sometimes also occurs in oxygenated facies along the margins of deep lakes, such as those of rift basins (e.g., Permo-Triassic Karroo of South Africa, Middle Triassic–Early Jurassic Newark Supergroup of the eastern U.S.A., Early Cretaceous Santana of Brazil).

Among the most famous of lacustrine associations is the Green River Formation (Late Paleocene–Late Eocene), which crops out in Wyoming, Utah, and Colorado (Allison and Briggs, 1991a,b). The Green River is part of a whole series of lake deposits that existed from the Paleocene to Oligocene throughout the western United States and Canada in association with the Laramide Orogeny. Green River environments ranged from playa to large, deep, oxygen-depleted lakes in which light-colored diatomites (summer) alternated with organic-rich laminae (winter overturn) to form seasonal varves (e.g., Wilson, 1988a; cf. Wilson, 1977; there is no guarantee, however, that putative varves are really annual; see

Behrensmeyer and Hook, 1992). Vertebrates, especially fish with soft-tissue outlines, are most abundant, but the Green River also yields plants and insects. Volcanic tuffs may have also contributed to preservation in some cases and DNA has even been sequenced and amplified from *Magnolia* leaves of the Miocene (17–20 Ma) Clarkia Formation of northern Idaho (Golenberg *et al.*, 1990). Also included among large, deep, oxygen-depleted lakes are some (but by no means all) sediments of the Newark Supergroup, in which high-frequency Milankovich cycles have been identified (Olsen, 1986). And most recently, arthropod (crayfish, insect) remains have been reported from the Willershausen Lagerstätte (Germany), which formed as a small (approximately 200 m diameter), relatively shallow (about 10 m deep) lake that filled a sinkhole formed by dissolution of underlying Permian evaporites and faulting of the Triassic Middle Bunter Sandstone (Briggs *et al.*, 1998). Despite its shallowness, the lake was anoxic in its deeper portions (e.g., sediments are well laminated): the proportion of chitin in weevil cuticle from the deep portion of the paleolake is comparable in composition to that of modern beetles (40%), whereas that from more oxygenated sediments is much lower (2–5%).

The degree of preservation in lakes tends to vary along a taphonomic gradient (depth and distance from shore) and corresponds closely to the taxonomic diversity and composition of the fossil assemblages (Wilson, 1980; Behrensmeyer and Hook, 1992). In the Green River, nearshore sites are characterized by disarticulated fish, fewer complete insects, and more evergreen needles, whereas a deep-water/offshore assemblage is represented by more articulated remains and dicot leaves and wood remains that presumably floated to their site of deposition. Fishes are more common in collections made stratigraphically farther from coal seams, whereas insects and plants are more common closer to coal seams (Wilson, 1988*a*). Mass mortality of fish occurred in response to overturn of anoxic bottom-waters in winter, and differences in the degree of disarticulation of fish skeletons over many hundreds of years (as measured by counting varves) provides valuable clues to the intensity of lake overturn (Wilson and Barton, 1996; Chapter 10). Fish in the mass mortality layers exhibit a multimodal size distribution suggestive of year classes (i.e., short-term population phenomena; Wilson, 1988*a*; see also Bell *et al.*, 1987, 1989; Chapter 6).

Another exceptional case of lacustrine preservation is that of the Grube Messel, which is a former open-pit oil shale mine located about 30 km southeast of Frankfurt, Germany (Franzen, 1985, 1990). The finely laminated sediments are Middle Eocene in age and were deposited in a small, deep (but only a few tens of meters), unstratified, oxygen-depleted lake formed by rifting and surrounded by a rainforest. Despite the relatively shallow nature of the lake, the warm waters became deoxygenated as a result of algal decay following seasonal plankton blooms. Preservation is outstanding and fossils include microfossils (diatoms, pollen), plants with associated fruiting structures, thousands of insects, and a large variety of vertebrates (often with soft-body outlines preserved by bacteria), among them horses pregnant with preserved embryos and bats with preserved gut contents. Bats may have been killed by the release of CO_2 from the lake during seasonal overturn. The outlines of feathers and other soft tissues are normally preserved via bacterial autolithification as "flowing mat" structures, in which rod-shaped bacteria are preserved in siderite end-to-end; possibly, the glycocalices of the bacteria served for ion exchange with anoxic porewaters (Davis and Briggs, 1995; see also above). In fact, feathers appear to be best preserved in lacustrine sediments, probably because of relatively low rates of organic decay (Davis and Briggs, 1995).

Lacustrine associations are concentrated in the Cenozoic for two reasons: they are less likely to have been eroded and more ancient examples are more difficult to identify (Behrensmeyer and Hook, 1992; see also Chapters 7, 9). Nevertheless, impressions of dinosaur skin preserved by microbial mats have been found in Early Cretaceous lacustrine deposits of Spain (Briggs *et al.*, 1997). Paleozoic lake deposits may also contain exceptional biotas, such as insects and plants. Those from the Silurian and Devonian, however, contain less diverse biotas because they largely pre-date land plants and insects, although early fish may be preserved in rocks of the Old Red Sandstone. Still, environmental conditions and taphonomic pathways resemble those of younger lakes. The Achanarras fish bed (Middle Old Red Sandstone, Scotland), for example, resembles that of Eocene basins (Trewin, 1986): sediments consist of varved carbonate and organic-rich laminae, with the carbonates presumably resulting from algal photosynthesis and organic layers from algal blooms. Fish remains are often concentrated in laminae and apparently resulted from mass mortality caused by deoxygenation (algal blooms and organic decay) from overturn of deeper low-oxygen waters associated with storms and seasonal overturn.

6.4 Hot springs

Although not included in the ternary classification of Seilacher *et al.* (1985), hot springs are sites of exceptional preservation because they are associated with rapid silicification of biotic remains. Perhaps the most famous of such deposits is the Rhynie Chert (Early Devonian), which is part of the Lower Old Red

Sandstone exposed in northeast Scotland and which represents one of the earliest terrestrial communities preserved in the fossil record (Trewin, 1989). Although andesitic lava flows and tuffs occur in close proximity to the Rhynie Chert, it appears that silicification was associated with movement of hot waters through a fault breccia, as the chert is enriched in gold, arsenic and antimony, all of which occur in high concentrations in modern hot springs; the cherts also exhibit brecciation and resealing, which is characteristic of sinter deposits (Trewin, 1989). Desiccation cracks suggest that the community occupied an area of shallow ephemeral pools that was periodically flooded by the hot spring (see Renaut et al., 1996, for a modern analog). The famous *Rhynia major* (recently redescribed as *Aglaophyton major*) is common and occurs with well-preserved xylem cell structure, stomata, and sporangia with spores in various stages of germination. Filamentous and unicellular algae are also preserved, as are arthropods (mites, springtails, spiders and other arachnids, and shrimp-like creatures, and the earliest-known wingless insect, *Rhyniella*). Silicification must have been rapid, as plant stems up to 15 cm high are preserved in growth position. The plants are so well preserved that lesions sealed in response to sap-sucking arthropods can be seen; fungal spores and hyphae within the plant tissue also indicate early stages of decay (Trewin, 1989).

6.5 Traps

Exceptionally preserved terrestrial organisms are normally found in aquatic settings, but in some cases they are preserved in *traps* (Seilacher et al., 1985; Figure 6.1). These include (Lyman, 1994a) *freezing* in permafrost; *mummification* in caves; *permineralization* in sinkholes; entombment in sterile media such as *tar* (e.g., Pleistocene LaBrea tarpits of southern California), *salt* (Pleistocene mammals in Galicia, Poland), and *peat bogs*; and encasement in *amber*. In some cases, tissues are sufficiently well preserved that cell organelles are recognizable and DNA can be extracted (Paabo et al., 1989; Logan et al., 1991; cf. Stankiewicz et al., 1998). Other biomolecules may also be preserved, but their relative resistance to decay may vary substantially according to the degree of cross-linking, even within the same basic compounds (proteins, lipids, etc.; Briggs, 1995; Stankiewicz et al., 1997a,b; see also Chapter 3). Unfortunately, decay may be reactivated by climate change so that preservation in many traps (e.g., freezing, mummification) is largely confined to the Neogene, especially the Quaternary (Allison and Briggs, 1991a,b). Moreover, unlike Konservat-Lagerstätten, the extent of time-averaging in traps – including karst, tar, amber, and peat – may range from instantaneous to 10^4 years

6.5 Traps

(Behrensmeyer and Hook, 1992, their table 2.8). Two of the most important types of traps are emphasized here: amber and peat bogs.

6.5.1 Amber

Entrapment in plant resins is the most significant mode of preservation of soft tissues in the terrestrial realm (Allison and Briggs, 1991a,b; see Henwood, 1993, for review). Plant resins are exuded in response to tension or wounds in bark, especially in gymnosperms and angiosperms, and the molecules (terpenes) then cross-link to harden into amber. Amber is one of the most decay-resistant products of plants, and because it is nearly the same density as water, may be transported considerable distances, depending upon the density of individual specimens (Iturralde-Vinent and MacPhee, 1996). Thus, counterintuitively, amber occurrences may be allochthonous rather than autochthonous. *Copal* or subfossil resin forms mainly from legume trees (mostly belonging to the genera *Hymenaea* and *Copaifera*) and is an early stage of amber formation before full cross-linkage has taken place; the oldest copal known comes from Mizunami, Japan, and is approximately 33 000 years old (Grimaldi, 1996). Because of differences in chemical composition, ambers from different parts of the world can be differentiated using spectroscopy and other chemical techniques (Langenheim and Beck, 1965; Grimaldi, 1996). The earliest known amber is of Late Carboniferous age, but amber did not become abundant until the Early Cretaceous with the rise of the coniferous Araucariaceae, particularly in tropical and subtropical forests.

Although amber localities are known from most of the world (see Grimaldi, 1996, for maps and discussion of localities, as well as many beautiful plates), the most famous is Baltic amber, which ranges in appearance from clear brown and golden shades to deep wine-red (cherry amber) to dull milky white. Baltic amber was used as an early form of currency ("Gold of the Sea") and was worked into jewelry by the Vikings (Anonymous, 1997; see Grimaldi, 1996, for discussion of other varieties from throughout the world). Baltic amber is often stated to have been secreted by the extinct gymnosperm *Pinus succinifera* from about 50 to 35 Ma in what is now Sweden and Finland (Schlüter, 1990). Because of the distinctive presence of succinic acid, however, Baltic amber may actually have originated from *Pseudolarix* (which also belongs to the pines) and which may have been present in the region at this time; many of the plants and animals preserved in Baltic amber are closely related to species extant in Asia, where *Pseudolarix* still lives, as well as Australia and Chile (Grimaldi, 1996). In the Late Eocene–Early Oligocene, the region was inundated and the amber was washed away and reconcentrated in the glauconitic "blau Erde" (actually greenish) of Early Oligocene age along the Baltic coast of what is now Lithuania, Poland (Danzig), and Russia (Samland Peninsula). The biota of Baltic amber is dominated by Diptera (about 50% of inclusions; microbes, including slime molds, and plant remains are also found; Grimaldi, 1996), whereas in tropical amber (such as that formed in the Dominican Republic during the Oligo-Miocene), Diptera and Hymenoptera are about equal (total about 80%). Although the external morphology may be preserved, gas exchange with the atmosphere may occur (Cerling, 1989) and internal organs and tissues may decay, presumably as a result of autolysis and endogenous bacteria (Allison and Briggs, 1991a), especially in Baltic amber.

Better preservation of insects (flies) in Dominican amber versus those from the Baltic may reflect different preservative properties. Dominican amber formed from extinct species of *Hymenaea*, which produce large amounts of resin. The low molecular weight fractions (mono- and sesquiterpenes) appear to have readily diffused through body walls into carcasses, whereas Baltic amber is much more acidic (Grimaldi *et al.*, 1994; Briggs, 1995). Cellular detail is often preserved in insects and leaves in Dominican amber, whereas Baltic specimens often have a milky coating (Grimaldi *et al.*, 1994). Based on experiments, early dehydration also played a significant role (Henwood, 1992; Grimaldi *et al.*, 1994).

Dominican amber appears to have formed in a single basin during the late Early Miocene through early Middle Miocene (15–20 Ma), according to available biostratigraphic and paleogeographic data (Iturralde-Vinent and MacPhee, 1996; see also Grimaldi *et al.*, 1994). Based on microfossil and vertebrate evidence, the amber was apparently deposited in coastal lagoons fronting low, densely forested hills. There is little evidence of reworking or transport and these deposits could conceivably be used to calibrate rates of molecular evolution of preserved taxa (Iturralde-Vinent and MacPhee, 1996).

6.5.2 Peat bogs

Peat forms in a variety of environments ranging from freshwater to marine. Peat bogs are here considered a special case of waterlogged soils (Baas-Becking *et al.*, 1960; see also Chapter 3). These deposits have yielded tanned human remains (in the manner that hides are tanned to produce leather) several thousand years old that preserve cellular detail of skin, hair, and clothing. Although bones dissolve because of low pH (Glob, 1969; Shipman, 1981), tannic and fulvic acids are antibacterial and prevent softpart decay (Allison, 1988c). Geochemical and anthropogenically derived pollen signals may also be preserved in peats because bogs are very effective in trapping airborne particles (Chagué-Goff *et al.*, 1996; see also Chapter 10).

Peat bogs are of two types. *High moors*, which are typically found in the rock record, receive most of their water from rain (*ombrogenic*) and have very low salt concentrations. Sulfate reduction and iron sulfides have been reported from some high moors, but decay is typically slow in high moors and peat deposits are often sufficiently thick to build low mounds composed almost entirely of plant material. *Low moors*, on the other hand, are fed by waters that have drained through soils and rock, and therefore contain much higher salt concentrations; peat decay is enhanced so that low moors typically form depressions. *Sphagnum*, which is the major contributor to many modern peat bogs, lowers pH (despite waterlogging) by adsorbing cations on and releasing hydrogen ions from its cell walls (Baas-Becking *et al.*, 1960). Thus, high moors have a pH generally lower than 6, and sometimes as low as 2.8, whereas low moors have a pH of 7–8; low moor deposits are thus often calcareous, but, unfortunately, are rarely preserved (Baas-Becking *et al.*, 1960). Redox potential Eh in both types of bogs is about the same (mostly below +400 mV). High bogs are susceptible to drying out, however, so that Eh may reach nearly +700 mV. As sulfate reduction and iron sulfides have been reported from some high moors, though, it is likely that the lower Eh limit is negative in these cases. Low moors contain cannel coals, vitrain, clarain, and durain that probably formed under waterlogged conditions, whereas fusain, which forms under high Eh (desiccation) and low pH conditions, is more likely to be found in high moors. Despite the low probability of preservation under such conditions, extraordinary cell-by-cell preservation of flowers and other delicate reproductive structures has been reported from the Raritan Formation (Turonian, New Jersey, U.S.A.; Gandolfo *et al.*, 1998); preservation occurred via "fusainization" (charcoalification) that resulted from forest fires that preceded subsequent transportation and sedimentation (M. A. Gandolfo, personal communication)!

Note added in proof:
Bartels *et al.* (1998) *The Fossils of the Hunsrück Slate: Marine Life in the Devonian*, Cambridge University Press) concluded that the Hunsrück was part of a submarine fan complex and published many beautiful plates of the biota. Based on the epifauna and burrowing infauna, Sutcliffe *et al.* (1999, Ichnological evidence for the environmental setting of the Fossil-Lagerstätten in the Devonian Hunstrück Slate, Germany. *Geology* 27: 275–278) concluded that the water above bottom was well oxygenated, whereas the sediment was not.

7 Sedimentation and stratigraphy

Make mountains level, and the continent,
Weary of solid firmness, melt itself
Into the sea!
William Shakespeare, *Henry IV, Part 2*

7.1 Introduction

Given the growing emphasis on environmental studies (Chapter 10), stratigraphic (temporal) resolution is no less critical to reconstructing Earth history now than it has been during the past two centuries. Correlation establishes the age equivalence of rocks and fossil assemblages and is the basis for reconstructing paleogeography and facies relationships. A precise stratigraphic framework is also critical to establishing any semblance of cause and effect preserved in the stratigraphic record. Unfortunately, physical reworking (Chapters 1 and 2), dissolution and diagenesis (Chapter 3), and bioturbation (Chapter 4) typically mix (time-average) – and sometimes destroy – stratigraphic signals and temporal resolution, the exception being Lagerstätten (Chapters 5, 6).

The stratigraphic record has long been viewed as being notoriously incomplete because of non-deposition and erosion. Darwin used incompleteness to explain why there are typically no transitional fossils found between new taxa, an idea (along with a number of others) borrowed from Charles Lyell. Say that a particular spot on the Earth's surface started out with an uninterrupted sediment accumulation rate of 10 cm ka^{-1}. This is not an unusual rate for river deltas, which are sites of significant sediment influx (Enos, 1991). Then over the course of the Phanerozoic (approximately the past 540 million years), about 54 km of sediment would have accumulated, which is several times that of the average thickness of continental crust! Clearly, much of the sediment must have bypassed the site of deposition or was eroded.

Schindel (1980, 1982) evaluated the resolution necessary to document patterns claimed by seven different paleontological studies ranging from terrestrial/fluvial to deep-sea environments. He stressed that any paleontological study must first evaluate (1) *temporal scope* (the total span of geological time encompassed by a sampled sequence), which can often be determined by isotopic or biostratigraphic methods; (2) *microstratigraphic acuity* (the amount of time represented by each sample), which can be estimated from modern short-term sedimentation rates (but see Chapters 4, 5, 10); and (3) *stratigraphic completeness* (proportion of the temporal scope represented by strata), which is why many workers distinguish between rates of sedimentation and sediment *accumulation*, and why time and time-stratigraphic units have long been distinguished from one another. Such considerations are intimately related to time-averaging (Chapter 5) and the reader will no doubt notice overlap between this topic and that of stratigraphic completeness.

7.2 Stratigraphic maturity

How long does it take for a sedimentary layer to "mature" or escape physical reworking, erosion and redeposition, and bioturbation? Sadler (1993) attempted to answer this question using a mathematical model (primarily for shallow shelves) based on gambler's ruin (a type of probability model), in which the surface mixed layer is the casino; each fossil is a gambler; winning (losing) reflects increasing (decreasing) distance from the SWI; burial rate is external income irrespective of the gambler's winnings or losings; and "breaking the bank" is equivalent to escape from the mixed layer.

The maturation time of a block of sediment equals the difference between the maximum and minimum ages of the contained fossils and equals the duration of time-averaging of an assemblage. It is a *minimum* estimate of time-averaging because fossil remains do not survive indefinitely and remains that date back to the beginning of maturation may be destroyed before maturation is completed (Chapter 5). Each type of shell, of course, has its own time and depth scales with respect to maturation, so that the odds of maturation vary (e.g., Chapters 2, 3), but in this model fossils are considered equivalent for the sake of simplicity. Maturation time and the probability of being reworked may be expressed as a dimensionless reworking rate (DWR):

$$DWR = \frac{\text{S.D. of erosion} \times \text{redeposition rates}}{(\text{burial rate}) \times [\text{time scale of the problem}]^{1/2}} \quad (7.1)$$

In other words, the chances of reworking increase with the standard deviation of erosion and redeposition (e.g., large storms scour more deeply but are less frequent; cf. Bak, 1996) and decrease with depth (burial rate) and the square root of age. In the model, the amount of reworking also changes at each step but the frequency of events remains constant so the duration or time scale must

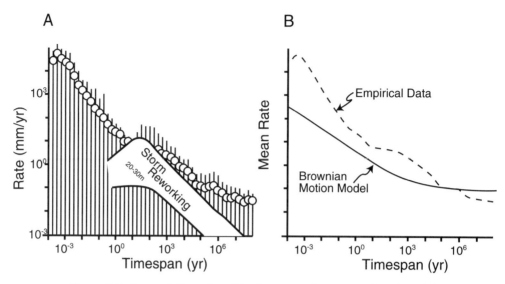

Figure 7.1. Accumulation rates of terrigenous sediments as a function of time span. (A) Empirical data (circles, mean rates; vertical bars, 1 standard deviation and are asymmetric because of logarithmic scale; field of storm reworking based on Thorne et al., 1991, for water depths of 20–30 m). (B) Mean empirical rates versus predicted (model) rates of reworking. (Redrawn from Sadler, 1993.)

be specified. Equation 7.1 may be expressed somewhat differently (Smith, 1977):

$$DWR = \frac{\text{erosion depth} \times \text{event frequency}}{\text{burial rate}} = \frac{\text{residence time}}{\text{event return time}} \quad (7.2)$$

Depth of erosion divided by burial rate yields the average residence time in the surface mixed layer and frequency is the reciprocal of return time, so DWR is now the ratio of residence time to return time. Dividing the return time by the residence time yields the expected number of reworkings before maturation.

Thorne et al. (1991) found evidence for a log-linear relationship between erosion depth and return time raised to the return time R (Figure 7.1). At low R (high-frequency reworking events, short return times), the depth of reworking remains relatively shallow and the extent of reworking and bed thickness change little (Thorne et al., 1991; Sadler, 1993). As DWR increases at low R, reworking and time-averaging increase while bed thickness remains essentially unchanged and amalgamated beds are more likely to form (Sadler, 1993). With high R values, on the other hand, there is a greater chance of deep scour; events with long return times have a greater chance of producing single event beds, and with increasing DWR there is a greater chance of reworking and bed thickness tends to increase. This is why the empirical curve for reworking is steeper and more irregular than the Brownian (random walk) model curve (Figure 7.1).

Sadler (1993) likened maturation to the curing of concrete: it never stops, but it proceeds swiftly at first and progressively slows (Cutler and Flessa, 1990, also demonstrated this with their computer models of stratigraphic disorder; see Chapter 5). The probability of reworking increases with time and reworking affects shallow sediments more often than deep ones, but to get to greater depths, shells must survive reworking in shallower sediments. Thus, the net probability of reworking is highest shortly after depositon and slows with age.

Empirical estimates of maturation time have been calculated (based on Figure 7.1) and vary substantially with environment (Figure 7.2; Sadler, 1993). Figure 7.2 was constructed using a 10-fold "aging factor" or ratio of time spans (fixed dimensionless age); this aging factor was chosen because if too small, sample artifacts would greatly affect the curves of Figure 7.2, whereas if the aging factor is too large, significant changes are lost. Based on Figure 7.2, the odds of survival for shallow shelves and platforms peak at about 30–100 years and a few thousand years (which are the longest preferable durations of time-averaging for fossil assemblages; Chapter 5); the minimum survival of sections between about 10^4–10^5 years presumably reflects low sea-level stands related to ice-cap dynamics, although the data base may be unrepresentative because Quaternary data dominate at these time spans (Flessa, 1993; Chapter 5). Nevertheless, accumulation rates for environments below wave base do not form a trough in this age interval, and survival peaks near 10^5 years for deep-water environments may reflect increased sediment influx when deltas build across exposed shelves or shelves are eroded. Note that the probability of preservation of abyssal oozes remains relatively constant (scale-invariant) although it is by no means perfect. Loess deposits undergo the most disturbance between 1 and 10^4 years, whereas floodplains are reworked the most by storms between 30 and 300 years, which may account for enhanced burial on shelves during the same interval. Beyond 10^5 years, floodplains and shelves have similar rates and it is at this age that subsidence probably begins to dominate preservation of sedimentary sections. Decreasing survival in other settings beyond about 10^5–10^6 years is probably a function of subduction in the case of active margins or sea-level change and erosion on passive ones.

7.3 Stratigraphic completeness

7.3.1 Fractals

The previous discussion introduced the concept of scale to the survival of sedimentary sections. In many circumstances the choice of a measuring scale (be it length, thickness, or time) is seemingly unimportant and is chosen for convenience.

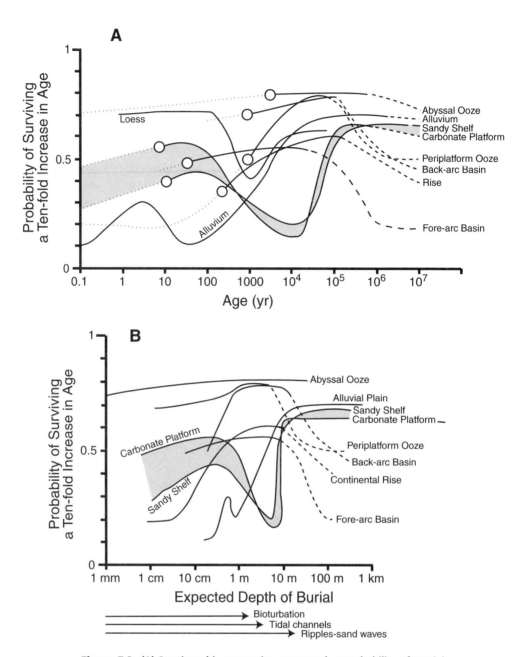

Figure 7.2. (A) Stratigraphic maturation expressed as probability of surviving a ten-fold increase in age without reworking. Open circles are estimates of mean residence times in mixed layer (mean times based on thousands to tens of thousands of estimates). Dotted curves indicate probable time scales of benthic mixing. Dashed curves indicate influence of tectonic processes. Stippled area represents shallow marine sediments subject to exposure by eustatic sea-level change. (B) Figure 7.2(A) redrawn in terms of mean thickness of sediment that would have to be eroded to exhume a fossil or keep it at the sediment–water interface. (Redrawn from Sadler, 1993.)

7.3 Stratigraphic completeness

If one were to measure the coastline of Great Britain, for example, one would undoubtedly choose the kilometer as the scale of measurement instead of centimeters. On the other hand, if the coastline of Great Britain were measured using a measuring stick of many hundreds of kilometers length, only a vaguely familiar shape would result, whereas a more familiar-looking coastline would appear using a measuring stick of about 1 km length (see Martin, 1998a). As the scale length (ℓ) decreases, then, the number of times the measuring stick must be used to measure the object increases. If N = the number of times the scale must be used to measure a particular object, then

$$F = \lim_{\ell \to 0} \left[\frac{\log N(\ell)}{\log \left(\frac{1}{\ell} \right)} \right] \quad (7.3)$$

(Kellert, 1993, p. 17). In other words, as the length of the scale decreases, the ratio of $\log N(\ell)$ over $\log(1/\ell)$ converges toward a limit called F (the *fractal dimension*) that describes the shape of the object. (Sugihara and May, 1990; Allen and Hoekstra, 1992; and Kellert, 1993, all provide relatively brief introductions to scale and fractals)

Fractal objects possess dimensions that are not whole numbers. A line has a dimension of 1 and a plane a dimension of 2. In the case of a plane, the line "fills" the entire plane ("Brownian motion" or random walk), but a fractal object with $F = 1.26$ (Koch's snowflake; see Martin, 1998a), for example, lies somewhere between a line and a plane. F, then, is a measure of the complexity of the object's shape.

Note that in equation 7.3, F is represented by the ratio of two logarithms. A log-log plot of ℓ versus the total measured perimeter (P) length of the object yields a straight line for a fractal object (Figure 7.3). The slope (m) of the line is negative (slopes downward to the right) because the longer ℓ is, the shorter the total estimated perimeter (or, conversely, the shorter is ℓ, the greater the total perimeter measured). Complex shapes have a steeper slope than simpler ones because the perimeter of a complex object will increase more as ℓ decreases than will the perimeter of a simple object.

So to determine F for an object, one can measure P as ℓ is decreased, plot $\log P$ versus $\log \ell$, and then plot (regress) a line through the points (see Sugihara and May, 1990, for other methods of calculation). The equation for this straight line is

$$\log P = (1 - F)(\log \ell) + \log K \quad (7.4)$$

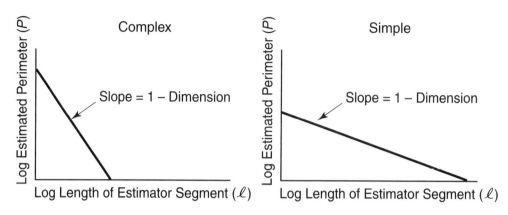

Figure 7.3. Log-log plot of the length of the measurement scale (ℓ) versus total perimeter (P) for simple and complex objects. Straight line segment indicates that the object is fractal, whereas the slope of the line (m) indicates whether it is simple or complex. See text for further explanation. (Redrawn from Allen and Hoekstra, 1992.)

where the slope (m) $= 1 - F$ and $\log K$ is a constant (y-axis intercept; $\ell = 0$); this is no more than the familiar equation for a line ($y = mx + b$). Rewriting equation 7.4 yields

$$P = K\ell^{1-F} \tag{7.5}$$

(Sugihara and May, 1990), where $\log(K\ell^{1-F}) = \log(\ell^{1-F}) + \log K = (1-F)\log\ell + \log K$. The total perimeter ($P$) of a fractal object is equal to the number of sides (N) times the length of the side (ℓ). For a fractal object, as ℓ approaches 0 (and F approaches the actual fractal dimension), the length of the perimeter approaches an asymptote, K (y-axis intercept; Figure 7.3). Each side has a fractal-dimensional length (ℓ^F), so the number of sides N equals the total true perimeter (K) divided by the fractal-dimensional length of each side (ℓ^F):

$$N = \frac{K}{\ell^F} \tag{7.6}$$

(or $N = K\ell^{-F}$). Since $P = N$ times ℓ, P may be expressed in more general form as

$$P = K\ell^{-F}(\ell) \tag{7.7}$$

where $\ell^{-F}(\ell^1) = \ell^{1-F}$. If the logarithm of both sides is taken, equation 7.4 results, in which the slope of the line is $1 - F$.

One implication of the straight line plot of Figure 7.3 is that the fractal dimension (F) is constant; the complexity of the object is scale-invariant and all fractal objects are therefore said to be self-similar. This does not mean, however, that natural objects display such strong regularity. Natural objects are much

7.3 Stratigraphic completeness

more likely to display a statistical self-similarity; the similarity at different scales is quite strong, but not exactly invariant as with man-made patterns like the Koch snowflake (Peak and Frame, 1994). In other words, for natural fractals the straight line of Figure 7.3 is likely to be plotted through a cloud of points rather than a series of points lying exactly on a straight line. Moreover, fractals are not found everywhere; in fact, given the resolution of most measuring instruments, F probably equals 1 for most natural objects (Sugihara and May, 1990). Nevertheless, natural (statistical) fractals occur all around us: coastlines (from grains of sand to coasts viewed from space), ferns, trees, rivers, landscapes (topography), surfaces of boulders, perhaps portions of stable isotope or fossil abundance curves, and, apparently, the stratigraphic record.

7.3.2 The marine record

The marine stratigraphic record reveals a pattern noted by a number of workers (e.g., Sadler and Strauss, 1990): that a log-log plot of sedimentation rate and the period of observation over which the rate was calculated results in a line with a negative slope (downward to the right; cf. Figures 7.1, 7.2). In other words, sedimentation rates are fast when measured over short time spans, but considerably slower when measured over longer intervals. More and longer intervals of erosion and non-deposition are included as the duration of observation increases. Moreover, as the absolute value of the slope of the line increases, the relative duration of the stratigraphic gaps (G) increases (Figures 7.4, 7.5).

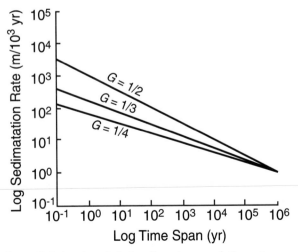

Figure 7.4. Log-log plot of sediment accumulation rate versus time span of observation (analogous to ℓ) for different gap (G) values. (Redrawn from Plotnick, 1986*b*.)

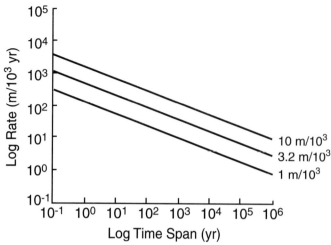

Figure 7.5. Relationship between sediment accumulation rate and time span of observation for a G (gap) value of 1/3. (Redrawn from Plotnick, 1986b.)

The time represented by the missing sediment is called a *hiatus* (typically of the order of hundreds of thousands to millions of years or more, although they can be of any duration as long as it is detectable) and the *surface* of erosion or non-deposition an *unconformity*. These gaps are analogous to "scene changes" that separate the "snapshots" or – more likely – the "movie clips" preserved in the geological record (Chapter 5). At the other extreme from unconformities are *diastems*, which are gaps in the stratigraphic record so short that they are virtually undetectable, because they are caused by very short-term changes in sedimentary regime. Sedimentation within a particular environment is not everywhere uniform and continuous throughout; storms, for example, may redistribute previously deposited sediment but leave no record of themselves. In between unconformities and diastems are stratigraphic gaps and hiatuses of all magnitudes and durations.

Based on Barrell's (1917) famous diagram (Figure 7.6), one might suspect that the distribution of stratigraphic hiatuses is fractal. Embedded within a depositional sequence are, for example, smaller-scale *parasequences* (see below), which undergo net shallowing or deepening trends (Wilgus et al., 1988; Brett, 1995, Holland, 1995; see also Brett, 1998). Using log-log plots of sediment accumulation rate and duration of observation, Plotnick (1986b) developed a fractal model for hiatus distribution based on the concept of a Cantor bar, in which successive portions (e.g., middle third) of a section are removed and the portions removed (stratigraphic gaps = G) vary from one experimental run to the next (Figures 7.4, 7.5). The rocks and the gaps are the curds (clumps) and whey (milky

7.3 Stratigraphic completeness

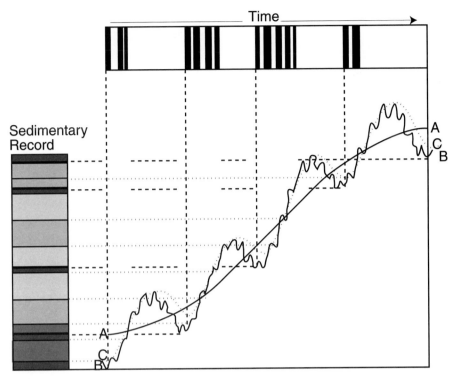

Figure 7.6. The sedimentary rock record (left) and the amount of time recorded in rocks versus depositional gaps (top) for different rates of sediment accumulation (A, B, C). (Redrawn from Dunbar and Rogers, 1957, after Barrell, 1917.)

liquid), respectively, of Mandelbrot's (1983) terminology, with periods of sedimentation being clustered in time.

The Cantor bar model provides a theoretical distribution of deposition versus hiatuses for a particular environmental setting *before* one studies it, and is consistent with compilations of sedimentation rates in different environments published by other workers (Plotnick, 1986b; see also Sadler, 1981; Sadler and Strauss, 1990; McKinney, 1991). Thus, the suitability of a sedimentary section or a particular taphofacies for, say, paleobiological or biostratigraphic studies can be assessed *beforehand*, thereby saving time, money, and confusion (Behrensmeyer and Hook, 1992).

The slopes of the lines of the log-log rate versus time plot (Figure 7.4) are calculated from

$$m = -\frac{\log(1-G)}{\log\left[\frac{(1-G)}{2}\right]} \tag{7.8}$$

Let T_n (S_n) and T_{n+1} (S_{n+1}) be successive time (T) intervals of observation represented by sediment (S) thickness. Sedimentation rates (R) are then calculated by dividing sediment thickness by time. Assume $R_{n+1} > R_n$ and let $G = \frac{1}{2}$. Then for a particular interval,

$$T_{n+1} = T_n \frac{(1-G)}{2} \tag{7.9}$$

and

$$S_{n+1} = \frac{S_n}{2} \tag{7.10}$$

Since slope (m) = rise/run, on the log-log plot of rate (R) versus time (T),

$$m = \frac{\log R_{n+1} - \log R_n}{\log T_{n+1} - \log T_n} \tag{7.11}$$

Substituting from equations 7.9 and 7.10,

$$m = \frac{\log\left[\frac{\left(\frac{S_n}{2}\right)}{\frac{T_n(1-G)}{2}}\right] - \log\left[\frac{S_n}{T_n}\right]}{\log\left[\frac{T_n(1-G)}{2}\right] - \log T_n} \tag{7.12}$$

Simplifying and cancelling terms results in equation 7.8. Based on the difference in completeness between short and long time spans, an estimate of the completeness of the section for a particular span of time can be made from

$$\text{completeness} = \left(\frac{t}{T}\right)^{-m} \tag{7.13}$$

where t is the desired short time span, T is the time span of the whole section, and m is the slope of the line. This is effectively the same approach as Sadler's (1981) measure of stratigraphic completeness, which is the ratio of the short-term accumulation rate to the long-term accumulation rate, or the probability that a time span of interest will be present in a stratigraphic section (see also Schindel, 1980, 1982).

The Cantor bar model goes beyond estimating the completeness of section to estimating the number of hiatuses and their length for the chosen period of

7.3 Stratigraphic completeness

observation (t). Rearranging equation 7.8,

$$\log(1 - G) = \frac{m}{(m+1)} \log 2 \qquad (7.14)$$

For example, from equation 7.14, for $t = 1000$ years, Plotnick (1986b) estimated a 4-million-year-long Eocene section in Wyoming to be 11.6% complete ($m = -0.26$). Assuming 1000-year intervals of depositon, 88.4% of the time is represented by hiatuses. For the same section, $G = 0.22$, which was then used to calculate iteratively that the Eocene section consisted of 256 periods of deposition, each 2141 years long, separated by 255 hiatuses ranging in length from 1208 to 880 000 years. Fortunately, things are better for deep-sea records, although they too are often incomplete (see Plotnick, 1986a, for examples).

There is a tradeoff, then, between rates of sedimentation and continuity of sedimentation, which may be highly variable between environments (Table 7.1). Depending on the desired resolution, acuity, and completeness, some environments are probably more suitable than others for studying biological (or at least paleontological) processes than others (Table 7.2). In order to achieve greater acuity (less time-averaging per sample) and completeness, one must sacrifice time span or temporal scope (Schindel, 1982). If one wanted to study, say, shallow-water fossil assemblages of relatively short temporal duration, one might pick a deltaic environment because sedimentation can be quite fast (up to about 200 m ka^{-1}; Enos, 1991). The disadvantage of picking a deltaic environment is that the sedimentary column for such an environment will tend to occur in "packages" of rapidly deposited sediment each separated by a long hiatus (because of delta lobe-switching), so that the record will likely be discontinuous (Schindel, 1980, 1982). The deep-sea record, on the other hand, is characterized by relatively slow sedimentation (about 2 cm ka^{-1} average), but sedimentation consists of a *relatively* constant rain of pelagic (primarily microfossils) or hemipelagic sediment; nevertheless, the deep-sea record may also be incomplete or time-averaged (Chapter 5).

7.3.3 The terrestrial record

Examples of stratigraphic completeness have been developed for the terrestrial record, which is notoriously incomplete. In some cases, stratigraphic gaps may be recognized from *paleosols*. Behrensmeyer (1982) proposed a hypothetical example of completeness in fluvial environments based on her field work (Figure 7.7). The hypothetical section was 15 m thick and accumulated over 3×10^4 years (composed of 30 thousand-year intervals) for an average accumulation rate of

Table 7.1. *Ranges in sedimentation rates for different environments*

	N	$\bar{x}$	CV	Min	Max	Comments
Fluvial systems	15	86 000	189	65	410 000	Rates patchy in time and space, varying widely among rivers
Deltaic systems*	29	200 000	271	400	2 740 000	Rates vary widely among deltas, within a delta and through time in an area
Deltaic systems	28	112 000	172	400	450 000	Rates more uniform for longer periods of observation
Coastal wetlands and tidal flats*	16	100 000	374	420	1 450 000	Rates vary greatly through time; instantaneous values very high
Coastal wetlands and tidal flats	15	10 000	219	420	80 000	Rates fairly uniform despite variety of coastal settings
Reefs and marine carbonate shoals*	20	95 000	364	35	1 460 000	Rates highly variable among areas; maxima are instantaneous, species-specific rates
Reefs and marine carbonate shoals	18	2500	190	35	14 000	Rates fairly uniform over longer periods of observations
Lakes	29	5800	176	150	31 700	Greatest variation is among lakes and within a lake; diverse spans of observation; small source of variation suggests fairly continuous accumulation
Bays, lagoons and estuaries	23	3600	139	460	14 300	Rates fairly uniform among areas and coastal settings; few very short-term rates available
Bathyal and abyssal	37	370	279	0.6	4500	Rates bimodal; low pelagic values and higher hemipelagic rates
Inland seas	35	290	169	1	2000	Rates fairly uniform; few extreme rates observed

Expressed in Bubnoff units $= 1\ \mu m\ yr^{-1} = 1\ mm\ ka^{-1} = 1\ m\ myr^{-1}$; not corrected for compaction. Data sets marked by asterisks represent very high, essentially instantaneous rates of sedimentation measured over short periods (<1 month) of observation. N, sample site; $\bar{x}$, average rate; CV, coefficient of variation of rates; Min, minimum rate; Max, maximum rate.

Schindel (1980); see also Enos (1991).

Table 7.2. *Scale of microstratigraphic sampling and temporal resolution required to document various biological processes in different depositional settings*

		Approximate scale of microstratigraphic sampling required for resolution of process		
Possible time span (years)	Biological process	Inland seas; bathyal and abyssal	Lakes; bays, lagoons and estuaries; reefs and marine carbonate shoals	Fluvial systems; deltaic systems; coastal wetlands and tidal flats
10^0–10^1	Colonization, population dynamics, competition and predation, speciation (??)	0.1 mm–1 mm	1 mm–10 mm	100 mm–1 m
10^1–10^2	Ecological succession, resource partitioning, local extinction and invasion, speciation (?)	1 mm–10 mm	10 mm–100 mm	1 m–10 m
10^2–10^3	Speciation, morphological adaptation, biogeographic changes, habitat destruction (?)	10 mm–100 mm	100 mm–10 m	10 m–100 m
10^3–10^4	Habitat destruction, regional extinction	100 mm–1 m	10 m–100 m	100 m–1000 m

Schindel, 1980.

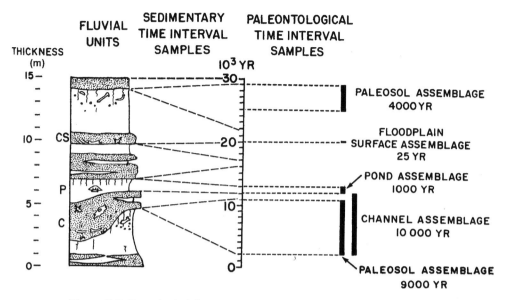

Figure 7.7. Hypothetical fluvial stratigraphic column indicating intervals of sediment accumulation versus attritional bone accumulation. C, channel; P, pond; CS, crevasse splay. (Reprinted from Behrensmeyer, 1982, with permission of the editors of *Paleobiology*.)

0.5 m 10^{-3} years. The section was 27% complete at the 10^3 year level of resolution based on Sadler's (1981) method so that eight of the 30 thousand-year intervals should be represented by sediment, as indicated in Figure 7.7. By contrast, the *paleontological* record was 53% complete: 16 of the 30 thousand intervals were fossiliferous, and only three sedimentary intervals overlapped with the fossiliferous ones (channel, pond, and crevasse splay; see also Kowalewski, 1996*a*). In the lower paleosol assemblage, all bones added over the 9000-year hiatus were preserved, whereas in the upper paleosol assemblage only the last 4000 years of the 7000-year hiatus were represented (cf. stratigraphic maturation above, and Chapter 5 on time-averaging). Although the channel assemblage spanned up to 1×10^3 years, the bones actually spanned 1×10^4 years because they were derived in part from different taphonomic pathways and older assemblages: those eroded from the paleosol assemblage represented 9×10^3 years, whereas those added during channel formation represented the last 10^3 years. Since pond and crevasse splay assemblages accumulated over much shorter durations, only the two paleosol assemblages provide more or less comparable samples in terms of taphonomic history and duration of formation.

Retallack (1984) estimated the completeness of terrestrial sections using the estimated minimum times of soil formation of paleosols based on modern soils. Minimum times were chosen for two reasons: (1) development of soil features

7.3 Stratigraphic completeness

proceeds quickly at first and then slows toward steady state, and (2) depending on climatic conditions, there may be times of soil development (e.g., sufficient moisture and temperature) sandwiched between intervals of little or no soil formation (e.g., very arid or very cold). If this is the case for the Quaternary, then estimated minimum times of soil formation may approximate those of complete development (cf. time-averaging in Chapter 5).

Retallack (1984) applied this approach to a detailed section of the White River and lower Arikaree groups (Late Eocene–Late Oligocene) of Badlands National Park, South Dakota (U.S.A.), which contains numerous paleosols. Completeness of the sections was calculated via Sadler's (1981) method using corrected radiometric and magnetostratigraphic dates. Rates of sediment accumulation and stratigraphic completeness were estimated from the minimum time of formation of particular kinds of soils by comparison with Quaternary soils. Completeness was quite low for time spans of less than 10^6 years and formations varied from 54% to only 9% for this interval, and the most complete formations were those with the fastest long-term accumulation rates. Accumulation rates based on paleosols were at least an order of magnitude faster than long-term rates based on radiometric and paleomagnetic data (cf. above). Although inaccurate in this respect, paleosol-based rates may still prove useful in estimating accumulation rates for sequences lacking absolute dates: paleosol-based rates were close to median rates for fluvial systems at their time span and they correlated well with radiometrically calibrated accumulation rates for each formation ($P < 0.05$); paleosol-based rates are therefore realistic. Also, the least complete formation exhibited the greatest visual evidence of soils (Chapter 3), whereas the most complete formation displayed much less evidence of soils. Thus, the general appearance of a sequence of paleosols may be a rough guide to its completeness.

Algeo (1993) developed a model of stratigraphic completeness using the frequency of geomagnetic reversals. Although the model was developed for marine sections, the basic approach could conceivably be used in terrestrial settings (cf. Retallack, 1984, above). According to Algeo (1993), because the timing of geomagnetic reversals is independent of sedimentation events, the distribution of reversals in a stratigraphic section can be used to evaluate completeness. According to the model, the more complete a section, the greater the probability that reversals occurred during depositional events and are recorded *within* the unit, whereas the less complete a section, the greater the probability that reversals occurred during depositional hiatuses and are recorded at contacts *between* lithological units. In Monte Carlo simulations (a kind of sampling involving probability distributions), the technique worked best when the relative frequency of reversals to stratigraphic units was between 1:8 and 2:1. Monte Carlo simulations indicated that under these conditions, completeness estimates fell within ±5% of actual completeness values approximately 70% of the time. According to Algeo (1993), given that reversal frequences in the Cenozoic are about $1-5\,\mathrm{myr}^{-1}$, the method has the potential to quantify completeness where the average time interval between reversals is between 25 ka and 2 myr, which is approximately the range of many depositional sequences and parasequences.

7.4 Sequence stratigraphy

Unconformities are the basis of the discipline of "seismic stratigraphy" (later dubbed "sequence stratigraphy") that has caused a revolution in the science of stratigraphy, especially within the petroleum industry, which immediately seized upon it when it was first published in 1977. An outgrowth of Lawrence Sloss's work on the North American craton in the 1950s and 1960s (e.g., Sloss, 1963), seismic stratigraphy was developed by Peter Vail, Robert Mitchum, and colleagues at Exxon Production Research in Houston (Vail *et al.*, 1977) in an attempt better to *predict* the location of oil and gas reservoir sands, and thereby save on exploration costs.

The basic unit of sequence stratigraphy is the *depositional sequence*, which consists of rock units bounded above and below by unconformities. Depositional sequences are thought to be cyclic in nature (related to sea-level fall and rise) and to typically represent durations of several hundred thousand to a few million years (Wilgus *et al.*, 1988). The exact cause of sea-level cycles of intermediate duration is not known because increased seafloor spreading rates (and increased mid-ocean ridge volume) are too slow, and glacial advances and retreats too fast, to account for them, but this has not prevented numerous workers from successfully applying the concepts.

In the case of relatively rapid sea-level fall, at the base of each depositional sequence is an unconformity (*sequence boundary*) formed through erosion (Figure 7.8). As sea-level continues to fall, river deltas prograde across the shelf, and if sea-level falls sufficiently, eventually the deltas may dump sediment at or beyond the shelf edge onto the continental slope to form a *lowstand systems tract* (LST; Holland, 1995, discusses the fundamentals of sequence stratigraphy without all the jargon that often obscures the basic concepts). As sea-level begins to rise and deltas retreat shoreward, a *transgressive systems tract* (TST) begins to form, and as the rate of sea-level rise slows and sea-level approaches its maximum height, a *highstand systems tract* (HST), indicative of net upward shallowing, forms. The

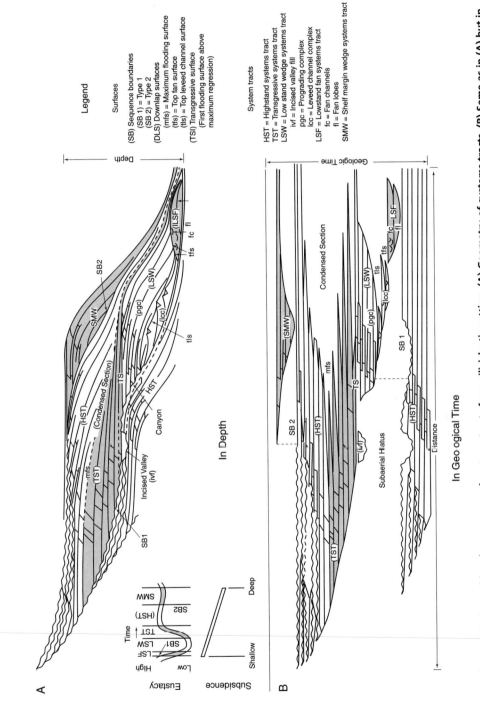

Figure 7.8. Depositional sequence and systems tracts for a siliciclastic setting. (A) Geometry of systems tracts. (B) Same as in (A) but in relation to geological time. (Redrawn from Haq et al., 1988.)

maximum height of sea-level that represents the transition from TST to HST is called a *maximum flooding surface* (MFS). Sediments bounded above and below by maximum flooding surfaces are referred to as *parasequences*; parasequences are progradational and they, or sets of them, therefore tend to shallow upward in HSTs. The MFS is commonly associated with entrapment of sediment close to shore in estuaries, bays, and lagoons, so that across the continental shelf and slope only fairly thin units, called *condensed sections*, are formed. Condensed sections are highly fossiliferous (the other systems tracts may also be fossiliferous), especially offshore, and contain deeper-water benthic faunas and diverse planktonic microfossils that have settled out of the water column and that may serve as biostratigraphic markers.

Not all systems tracts are present in any one section, however (Holland, 1995). The occurrence of systems tracts depends in part on sediment input and location with respect to the continental margin. Sequences located on the outer continental shelf are usually complete initially, but may lose the HST during the erosive phase of the subsequent LST. In continental interiors (well up on the craton), only the TST and HST are usually found because this area represents the farthest landward penetration of sea-level. Such patterns are often obvious in range charts of microfossil distributions produced by micropaleontologists for oil companies (Figure 7.9). Relatively high sea-level is represented by well-developed fossil assemblages (numerous thick lines indicating high diversity and abundance), whereas barren intervals are indicative of very shallow waters that are normally represented by low-diversity faunas and high sediment influx, which dilutes fossil assemblages. The abrupt contact below the two types of intervals represents an unconformity (rather than a fault) if it can be traced over relatively wide areas. In this example (Figure 7.9), sand is associated with the sequence boundaries and may represent deposition during an LST or subsequent reworking (basal lag sands) during the subsequent trangression (TST). The missing sediments are represented by the sequence boundary.

Holland (1995) developed a model which simulates the stratigraphic distribution of fossils in relation to systems tracts (see also Brett, 1995, 1998). First and last occurrences of fossils (which may represent *evolutionary* appearances and extinctions within a province) cluster at sequence boundaries and flooding surfaces of TSTs (Figure 7.10; cf. Figure 7.9). This is because the true upper or lower occurrences of fossils are lost through erosion (sequence boundaries) or are not present because the fossils themselves may track environmental (sea-level) change, just as rocks do. When Holland added gradients in diversity (lowest nearshore, highest offshore), taphonomy (preservation), and environmental tolerance

7.4 Sequence stratigraphy

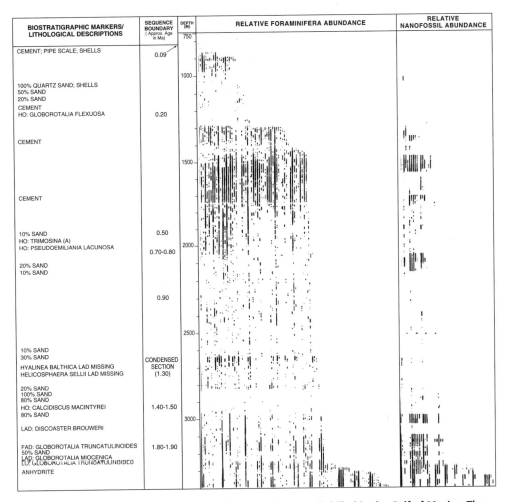

Figure 7.9. Fossil range chart for a petroleum well drilled in the Gulf of Mexico. The thickness of each line indicates relative abundance of a species, which was estimated visually by oil company paleontologists. Lithological (sand, cement) descriptions are based on visual estimates and well-log information. Sequence boundaries (based on graphic correlation) are also shown. (Redrawn from Martin and Fletcher, 1995; original data from Unocal.)

that may occur across continental shelves, the *intensity*, but not the *position*, of the stratigraphic peaks varied (cf. Signor–Lipps effect, section 7.6). Holland (1995) tested the computer model (hypothesis) against data from the Late Ordovician (Cincinnatian Series) and found that his modeled occurrences are present in the fossil record (see also section 7.7 on graphic correlation).

Sea-level change is of course intimately related to patterns of sedimentation, wave and current energy, the spatial and temporal distributions of organisms, and their preservation; thus, taphofacies, biostratigraphy, and sequence stratigraphy

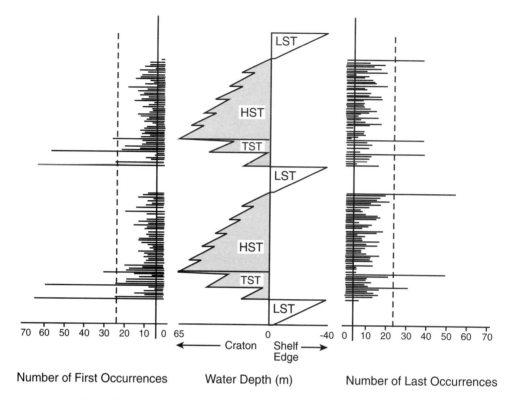

Figure 7.10. Numbers of first and last occurrences of fossil species through two depositional sequences in a single stratigraphic section deposited in a shallow seaway "upramp" of the modern continental shelf. LST, Lowstand Systems Tract; TST, Transgressive Systems Tract; HST, Highstand Systems Tract. In this example, no deposition occurs in the LST because the setting is too far onto the craton, and the time represented by the LST is contained in the sequence boundary (located between LSTs and TSTs). Any peak greater than the dashed line is statistically significant. Significant numbers of first occurrences are associated with TSTs, whereas significant last occurrences tend to occur below sequence boundaries. Compare with Figure 7.6. (Redrawn from Holland, 1995.)

would all seem to be intimately – and predictably – related, although the predictions remain largely untested (Brett, 1995). For example, coquinas are presumably most prominent at the base of parasequences and TSTs of large-scale sequences, but HSTs may display discrete skeletal hashes or beds of concretions. Bone beds signify condensed sections (see also Kidwell, 1989), and biostromes and bioherms should typically occur above MFSs (although such deposits also form at the shelf-edge during low-level stands). Konservat-Lagerstätten are presumably most likely to be found in late transgressive and early highstand systems tracts because of low-oxygen, sediment-starved conditions; sediment-starvation may also be indicated by intense burrowing, so that sequence boundaries may be recognizable from subtle-to-abrupt vertical changes in ichnofabric (e.g., Savrda, 1991, 1995). Highly corroded and fragmented remains should occur in erosive lowstand and early transgressive systems tracts, although Anderson and McBride (1996) found transgressive lags of well-preserved bioclasts (Chapter 5). Obrution deposits should, seemingly, be more common during HSTs, although increased sedimentation may dilute fossil assemblages and make them unrecognizable as such. Fossils of condensed sections may be highly corroded because of slow sedimentation and prolonged exposure at the SWI. Sediment-starved accumulations may also occur at the base and winnowed shell beds at the top of parasequences (Brett, 1995).

7.5 The stratigraphy of shell concentrations

Kidwell (1988, 1991, 1993a,b) has documented the formation of shell beds in response to sea-level change and tectonics and corresponding changes in sediment accumulation rate (cf. section 7.2; Chapters 1, 5). Both major and minor shell beds tend to be more numerous and areally more extensive in low-subsidence settings than in high-subsidence settings (Table 7.3); although shell beds in both tectonic settings form primarily in response to sediment starvation (Chapter 1), those of passive margins tend to form during intervals of sediment starvation associated with marine transgressions (Kidwell, 1988). Shell accumulations tend to be most pronounced in shallower water (presumably because of greater shell input and the greater probability of the influence of sedimentary dynamics), and may accentuate sedimentary cycles and beds that might otherwise be overlooked (e.g., Meldahl and Cutler, 1992; but see Clifton, 1989).

The relative proportions of different types of shell concentrations or possible distinctive features (marker beds) can be used to reconstruct and compare larger-scale sedimentary histories through individual depositional sequences, sets of sequences, and different basins (Figure 7.11; Kidwell, 1988, 1989, 1991, 1993a,b; see also Li and Droser, 1997; Abbott and Carter, 1997). These concentrations correspond approximately to the condensed assemblages described in Chapter 5. In *low subsidence settings* (accumulation rates up to tens of meters myr^{-1}), second and third order sequence boundaries (Chapter 8) corresponding to hiatuses of about 1 million years duration, are demarcated either by bare unconformities or erosional surfaces covered with complex concentrations of shells (e.g. Miocene Calvert Cliffs, Chesapeake Bay region; Paleogene Piney Point and Aquia formations of Virginia; Late Ordovician Cincinnatian of Cincinnati Arch region; U.S.A.; Figure 7.12). These concentrations are of two types (Table 7.3): (1) *lag*

Table 7.3. *Origins and paleontological quality of marine skeletal concentrations*

Type of concentration	Origin	Temporal resolution and spatial (habitat) fidelity of species assemblage
Event concentration	Single, ecologically brief episode of hardpart concentration (biogenic or hydraulic)	Resolution and fidelity vary depending on source of hardparts; some comprise high-resolution censuses
Composite concentration	Accretion or amalgamation of multiple event concentrations and generations; total thickness is average or expanded relative to coeval strata	Resolution and fidelity can be no better than that of component events and taxonomic fidelity will usually be poorer; potential complications from live–dead feedback and environmental condensation
Hiatal concentration	*In situ* accretion or amalgamation of events and generations during a period of slow net sedimentation; complex concentration is thin (condensed) relative to coeval strata	Expect generally lower resolution than for composite concentrations; potential complications from live–dead and diagenetic feedbacks; biostratigraphically as well as ecologically disparate faunas may be admixed
Lag concentration	Exhumation and concentration of resistant hardparts during erosion or corrosion, associated with significant stratigraphic truncation	Potentially lowest resolution and fidelity owing to mixing of hardparts from disparate facies and ages, vigorous culling and potential for recolonization

Kidwell, 1991.

concentrations, in which resistant hardparts are exhumed and concentrated from underlying (older) deposits during significant erosion and stratigraphic truncation; and (2) *hiatal concentrations*, in which numerous generations of shells accumulate in a single, relatively thin deposit while sediment accumulation remains low (but higher than for lag concentrations). Lags may grade offshore into condensed (low sediment accumulation rate) *bone beds* enriched in bones, teeth and phosphatic steinkerns (Kidwell, 1989, 1991; e.g. Maryland Miocene; Table 7.4; cf. usage of Brett and Baird, 1995). On the other hand, in *high-subsidence settings*

7.5 The stratigraphy of shell concentrations

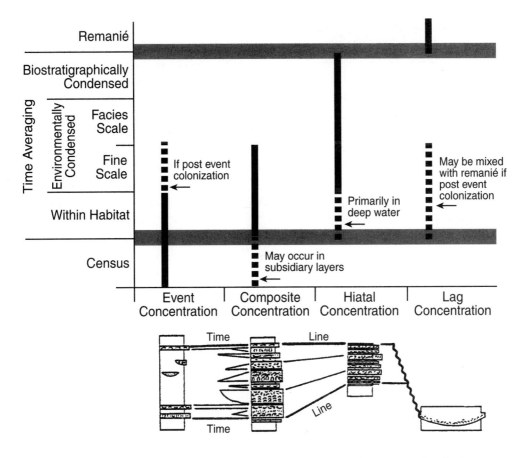

Figure 7.11 Shell concentrations in relation to degree of time-averaging. (Redrawn from Kidwell, 1993a.)

(accumulation rates to 1 km myr^{-1}; Pliocene Imperial Formation of southeastern California; Plio-Pleistocene of Pacific coast of Panama; Table 7.4), sequence boundaries of similar duration are represented by either *composite concentrations*, which are similar to hiatal concentrations but in which the total thickness of the shell bed surpasses that of coeval strata, or *event concentrations*, which represent single, ecologically brief episodes of shell accumulation. Intermediate patterns are noted in moderate subsidence settings (hundreds of meters myr^{-1}), such as the Pliocene Purisima Formation of California, Aptian-Albian Mannville Group of Alberta, and the Late Permian Cherry Canyon and San Andres Formations of the Permian Basin (west Texas; Table 7.5). In all subsidence settings, stratigraphically less significant surfaces (parasequence boundaries, bed set boundaries, and bedding planes) are bare or mantled with relatively simple event or composite shell concentrations.

292   Sedimentation and stratigraphy

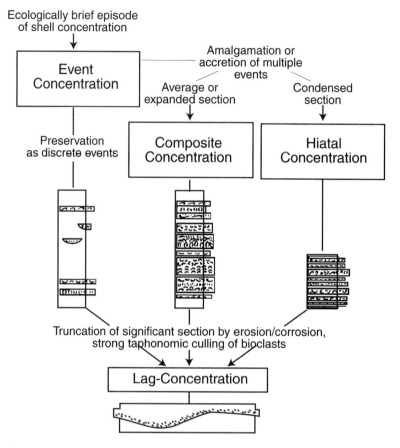

Figure 7.12. Formation of shell concentrations in relation to sediment accumulation rate. See text for further discussion. (Redrawn from Kidwell, 1991.)

As emphasized previously, certain depositional environments may be more suitable than others for paleontological studies (depending upon the questions that are being asked), and there may be tradeoffs between sample acuity, stratigraphic thickness and lateral extent (Table 7.3). Low subsidence settings along the passive margin of the eastern U.S.A. (Miocene Calvert Cliffs) are characterized by lag and hiatal concentrations, often consisting of allochthonous components to produce mixtures of α (within-habitat) and β (between-community) diversity (Kidwell, 1993b; Meldahl, 1993). Such deposits are best suited for relatively coarse biostratigraphic and paleobiological studies, as the temporal acuity of single microstratigraphic samples may be no better than that of a bulk sample because of physical and biogenic reworking (Kidwell, 1993b). Hiatal concentrations associated with *downlap* are *relatively* synchronous and may prove valuable for correlation within a basin, whereas hiatal concentrations associated with

7.5 The stratigraphy of shell concentrations

Table 7.4. *Comparison of skeletal concentrations in active and passive continental margins*

	Active California Pliocene "Imperial Fm," Latrania Mbr	Passive Maryland Miocene Calvert (Plum Point Mbr) and Choptank Fms
Geological Setting		
Average rate of sediment accumulation	4.5–5.5 mm/yr entire Fm, >0.1 mm/yr Latrania mbr	0.007–0.009 mm/yr
Depositional gradient	Few degrees to >10	<1° original dip
Major complex shell beds		
Stratigraphic frequency	1–3/~70 m marine section	4/~40 m marine record
Maximum thickness	2–7 m	1.5–10 m
Lateral extent	<10 km^2	500–3500 km^2
Geometry	Tabular to wedge-shaped over hundreds of meters; pinches out onshore in fan delta; laps out offshore against bedrock paleohighs or faulted out	Tabular to wedge-shaped over tens of kilometers; pinches out onshore; offshore, intertongues with less fossiliferous strata
Shell packing density (by volume)	10–45% calcitic shells and molds of aragonitic shells	20–70% aragonitic and calcitic shells
Time for accumulation	10^2–10^3 yr?	10^3–10^4 yr?
Origin	Sediment starved, shallow subtidal; distal edge of prograding coastal alluvial fans	Sediment starved and bypassed, shallow subtidal; condensation during marine transgression
Minor simple shell beds		
Stratigraphic frequency	<5 shell beds/5 m	4–30 shell beds/5 m
Maximum thickness	<10 cm	<10 cm
Lateral extent	<1 m^2, usually <30 cm^2	30 cm^2–1 km^2
Geometry	Clumps, lenses, thin beds	Pods, clumps, lenses, pavements, thin beds
Shell packing density	5–30% calcitic shells	20–50% calcitic and aragonitic shells
Time for accumulation	<10^2 yr	<10^2 yr
Origin	Predominantly biogenic; rare sedimentologically overprinted biogenic	Few biogenic; most are sedimentological and sedimentological overprinted biogenic

Major, equivalent or greater in dimensions than most associated facies; simple, homogeneous in features or monotonic variation.

Kidwell, 1988.

Table 7.5. *Summary of observed associations between bioclastic concentrations and stratigraphic discontinuities*

	Low subsidence (≤ 10 m myr^{-1})	Moderate subsidence (100s m myr^{-1})	High subsidence (≥ 1 km myr^{-1})
Sequence anatomy	Transgressive and highstand tracts present in third-order sequences, but very thin and commonly lack parasequence-scale subdivisions; incised valley fills uncommon, so that transgressive surface (TS) in most areas coincides with lower sequence boundary (SB); strong attenuation of sequences and component tracts at basin margin lapout rather than radical truncation	"Classic" sequence anatomy with transgressive and highstand tracts composed of discrete upwards-shallowing parasequences defined by non-depositional flooding surfaces (FS); transgressive surface in most areas well separated from lower sequence boundary	Divisible into generally transgressive and regressive phases in which parasequences may be recognized, but few or no through-going discontinuity surfaces to serve in sequence definition; abundant local erosional surfaces along basin margin; downlap surfaces very common throughout; stratigraphic buttress relation at basin margin lapout
Bioclastic concentrations			
Sequence boundary and transgressive surface	Lag concentrations only where TS coincides with lower SB. Hiatal concentrations common on TS regardless of association with SB	No examples of hiatal concentrations known	Event or composite concentrations only

Surface or interval of maximum transgression	Widespread hiatal concentrations common, may merge with TS hiatal concentration along basin margin	Widespread hiatal concentrations common	Sediment starved but uncondensed composite concentration, or event concentration; local
Parasequence boundary	Not applicable	Composite or event concentrations on lower flooding surface; Composite concentrations common in top of parasequence position	Composite or event concentrations on lower flooding surface; Composite concentrations common in top of parasequence position
Bedding planes and bed set boundaries	Event and composite concentrations common	Event and composite concentrations common	Event and composite concentrations common
Overall taphonomic pattern	Both sediment-bypassed and sediment-starved hiatal concentrations common	Taphonomically complex hiatal concentrations limited primarily to maximum transgression	Hiatal and lag concentrations rare

Kidwell, 1993a.

onlap or toplap are more likely to be diachronous (Kidwell, 1991, p. 247). Moderate subsidence settings also exhibit mixtures of α and β diversity and the temporal resolution of single samples may vary substantially. On the other hand, high-subsidence settings, such as those located along active plate margins, are characterized by composite and event concentrations (which may represent high-resolution *censuses*) largely consisting of autochthonous and parautochthonous fossils (Johnson, 1962; Meldahl, 1993) that may be extremely useful in bridging the gap between ecological and evolutionary processes; these sorts of shell concentrations are likely, however, to be less extensive areally than those of low-subsidence settings (Table 7.3). Many so-called event horizons are probably really condensed hiatal concentrations that coincided with or bracketed maximum transgressions; they are therefore of longer duration than true event horizons and of lower chronostratigraphic utility (Kidwell, 1991, p. 247).

7.6 The Signor–Lipps effect

As a biotic event or marker is approached (say an extinction or a period–system boundary, which is often marked by biotic turnover), only the most abundant species persist to the boundary, even if the event is "instantaneous" and the stratigraphic section is complete. Less abundant species will tend to disappear from the record somewhere below the boundary, never to reappear (at least in a particular section) because they are sufficiently uncommon that they are unlikely to be found. This sampling effect has been named the Signor–Lipps effect after its co-describers (Signor and Lipps, 1982; see also Dingus and Sadler, 1982).

Although frequently associated with the end-Cretaceous extinctions for obvious reasons, the Signor–Lipps effect can occur at a variety of scales. Badgley (1990), for example, found a high correlation between the square root of sample size and species richness; based on Monte Carlo simulations, she concluded that turnover of mammals in the early Eocene of Clark's Fork Basin (Wyoming, U.S.A.) was a sampling artifact.

A similar, but reversed effect, occurs following biotic turnovers as taxa rediversify. Other species may appear to die out before or during an extinction, only to be resurrected in the rock record after (above) the event has occurred (*Lazarus taxa*), whereas others that do become extinct may be replaced ("impersonated") by superficially similar, but quite different, groups called *Elvis taxa* (e.g., Erwin, 1993).

Sampling intensity and species abundances also affect the recognition of different patterns of extinction: *sudden* (*catastrophic*), *stepwise*, and *gradual* (Raup,

7.6 The Signor–Lipps effect

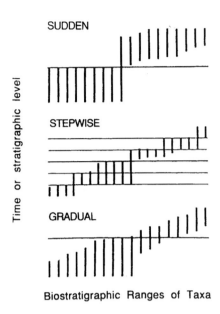

Figure 7.13. Patterns of extinction (Reprinted from Meldahl, 1990, with permission of the Geological Society of America.)

1989; Figure 7.13). These extinction patterns are significant because they suggest different mechanisms of extinction. Gradual or stepwise extinction suggests prolonged changes in, for example, climate, sea-level, or volcanism (Pospichal, 1994), whereas abrupt change suggests a more or less instantaneous mechanism (sometimes termed catastrophic).

Meldahl (1990) demonstrated the effects of taxon abundance and sampling intensity on these extinction patterns by sampling mollusc assemblages (mainly pelecypods and gastropods) at 1-cm intervals in eight cores taken through *modern* tidal flat sediments of Bahia la Choya, Mexico (northern Gulf of California; Figure 7.14; see also Flessa, 1990). For a catastrophic extinction, he assumed that the extinction event occurred at the top of the cores. For gradual and steplike extinctions he used the distribution of molluscs in the cores, and made certain species "extinct" by eliminating all biostratigraphic occurrences above certain levels in the cores. For gradual extinction, he selected species one by one for extinction at successive 1-cm intervals, beginning at 44 cm depth. For stepwise extinction, he chose 15 species at random at the 40-cm level in cores, 15 other species for extinction at the 20-cm level, and the remaining 15 species at the core tops. He ran each simulation five times, and found that "extinction" of a species was more likely to be recorded as sampling intensity increased (Figure 7.14), and that species that occurred in less than 15% of samples did not have

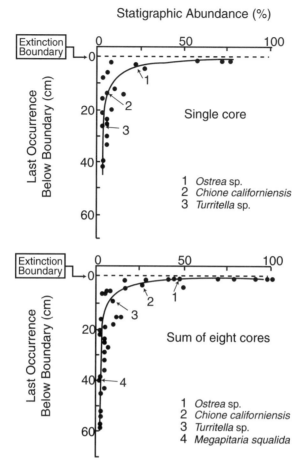

Figure 7.14. Comparison of stratigraphic ranges of mollusc species in a single core versus ranges compiled from eight cores taken through modern tidal flat sediments in the northern Gulf of California. As the number of samples increases, error in stratigraphic ranges decreases and uppermost occurrences of fossil species shift upward toward hypothetical extinction boundary (core top). New species are also found as sampling continues. (Redrawn from Meldahl, 1990.)

their uppermost occurrences accurately recorded (this amounted to 71% of the 45 species sampled). Each type of extinction produced a decline in diversity (number of species) that differed in *pattern* as the extinction boundary was approached. Sudden extinction produced a diversity decline that intensified as the boundary was approached, whereas gradual extinction produced a constant decline, and stepwise extinction resulted in a stepwise decline, although this pattern was difficult to distinguish from the gradual one.

The probability of sampling a given species depends not only on the species's original abundance, but also on the probability of the survival of the species's

7.6 The Signor–Lipps effect

remains. The more preservable a species is, then, the more likely it is to be reworked above or below a particular horizon. Moreover, in the case of extinction, there are associated changes in paleoclimate that often affect rates of sedimentation and burial and the completeness of stratigraphic sections (e.g., MacLeod, 1995; see also analysis of K–T trace fossils in northeastern Mexico by Ekdale and Stinnesbeck, 1998). In the case of Choya Bay, sediment accumulation rates are quite low (~0.038 cm/yr; Flessa et al., 1993), so that the chances of reworking or destruction of both relatively large (e.g., bivalve) and small (e.g., foraminifera) invertebrate remains by physical processes or bioturbation are considerable (Flessa et al., 1993; Martin et al., 1996; see also Chapters 2, 3).

The observed stratigraphic range of a species almost always, then, underestimates its true range. This relationship can be expressed in the form of confidence intervals (CIs),

$$\alpha = (1 - C_i)^{-1/(H-1)} - 1 \tag{7.15}$$

(Marshall, 1990; equation 7.15 is derived by Strauss and Sadler, 1989), where α is the confidence interval expressed as a fraction of the known stratigraphic range, C_i is the desired confidence level, and H is the number of known beds from which the species has been sampled. For example, if $H = 10$, then for a C_i of 95% (i.e., in 95% of the beds sampled the species will occur within the calculated range), $\alpha = 0.39$. If the oldest known bed is 60 Ma and the youngest 50 Ma (range = 10 million years), then the appearance of the fossil species most likely occurred (with 95% confidence) between 60 and 63.9 Ma and its disappearance most likely occurred between 50 and 46.1 Ma. α values at a C_i of 0.5 are much smaller than those for a C_i of 0.95. If both endpoints are to be estimated simultaneously, a slightly different version of equation 7.15 is employed.

This approach may be used on both local sections and *composite* sections developed from a number of different local sections. The main assumptions using the method are that (1) fossil horizons are *randomly* distributed (meaning that sedimentation rates and fossilization potential remained constant through time), (2) the intensity (efficiency) of collection was uniform throughout each section, and (3) fossilization events are statistically independent of one another. The assumptions are more easily fulfilled for local sections than for composite ones because of variation in sampling by investigators working on different sections or variation in preservation from outcrop to outcrop.

MacLeod (1996) pointed out some other often-overlooked, but extremely important, limitations of the use of confidence intervals. The entire local range of a species must be known and not just its range within a sample interval.

Also, a *robust* chronostratigraphic framework must be available because the calculation of confidence intervals assumes that a unit distance within the observed stratigraphic range is equivalent to a unit distance outside the range. This means that confidence intervals based on lithostratigraphic separation between upper and lower occurrences must either assume constant sediment accumulation rate throughout the observed and predicted range (with no unconformities) or that unconformities have already been recognized and the durations of the associated hiatuses can be estimated. Most stratigraphic sections conform to none of these assumptions.

Marshall (1997) addressed some of the criticisms related to non-random distributions. Preservation bias can be assessed to some extent by using a *fossil recovery potential function* based on bedding plane surfaces (cf. Chapter 5) or an *a priori* estimate of fossil preservation potential with water depth (see also Holland, 1995). This approach in turn *assumes* that the observed distribution of fossils is consistent with that predicted by the recovery function and must therefore be tested before its application; sections must also be sampled continuously or nearly so. Marshall (1997) also concluded that the probability (P) that a taxon became extinct during a hiatus may be calculated from the product of the probability that the taxon was extant at the time of onset of non-deposition ($1 - C_{top}$) and the probability that it became extinct during the hiatus:

$$P = (1 - C_{top})(1 - e^{\ln(0.5)t/T_m}) \tag{7.16}$$

where C_{top} is the probability that the taxon became extinct somewhere between the top of its range and the onset of non-deposition, and T_m is an estimate of the median longevity of the taxon (which will still require accurate biostratigraphic data from multiple sections).

Foote (1997b) also assessed the effects of preservation and stratigraphic gaps for discrete and continuous sampling regimes by comparing the assumptions of homogeneous preservation, no gaps, and exponential survivorship to actual occurrence data (Solow and Smith [1997] conducted a similar study assuming Poisson distributions). In Foote's (1997b) simulations (based on maximum likelihood estimation), preservation probability of taxa and stratigraphic completeness are increasingly overestimated as the difference in probability of preservation between two taxa increases because when sample size is finite a taxon with preservation probability near zero is unlikely to be found, hence, one taxon appears never to have existed whereas the other is likely to be recorded. Foote (1997b, p. 295) suggests that this problem can be reduced by "careful choice of taxa." Also, a species has a slightly greater chance of being found if a stratigraphic gap is centered on the stratigraphic interval in question (assuming constant diversity), but the parameters of stratigraphic completeness and preservation probability are progressively overestimated and extinction rate therefore underestimated as a centered gap increases in duration. There is little effect on parameter estimation either when a gap of constant duration is allowed to shift position or when the number of gaps increases (*sum* of gap sizes remains constant); thus, parameter estimation is more accurate when there are many small gaps than when there are a few large ones.

The results of such studies of course beg two questions: (1) how does one carefully choose taxa to avoid preservational bias when so little is known about differential preservation; and (2) how does one determine the position and extent of stratigraphic gaps and effects of changing sedimentation rates on the stratigraphic distributions of fossils? The first question would likely be answered by most workers in terms of abundance, preservational state (taphonomic grade), and perhaps depositional environment. The second question can be answered in a more straightforward manner by using graphic correlation.

7.7 Graphic correlation

With the advent of planktonic microfossil zonations in the 1950s and 1960s, the use of First and Last Appearance Datums (FADs and LADs) of certain marker species became commonplace (Berry, 1987). Planktonic foraminifera have served as the basis for subdivision of Cenozoic rocks into a series of numbered zones (P1–P22) for the Paleogene (Paleocene through Oligocene) and Neogene or Miocene to Recent (N1–N23); a similar zonation has been erected using calcareous nanoplankton. Cenozoic sediments are prominent not only on land, but also in the deep sea, and it has been the Deep Sea Drilling Program or DSDP (and its successor, the Ocean Drilling Program or ODP), along with the strong interest of petroleum companies, that have driven subsequent refinements of these zonations. Moreover, because of the need for more and more precise correlation, other types of markers, such as magnetic reversals (which are synchronous worldwide and so provide "anchor points" if they can be dated), oxygen isotope datums (which reflect changes in ice volume during the Plio-Pleistocene), and most recently strontium isotope datums, have been integrated into the biostratigraphic schemes to produce a highly calibrated *chrono*stratigraphic framework for the Cenozoic. This scale is constantly being refined and recalibrated based on new results from ODP cruises and work on outcrops.

But FADs and LADs can be diachronous, just like other datums (if FADs and LADS are in doubt, they are usually denoted as Lowest and Highest Occurrences,

LOs and HOs, respectively; see also MacLeod, 1995). The first or last appearance of a marker in one ocean basin or biogeographic province is not necessarily the same in another. Hence, the need for calibration using *independently* derived datums from other methods such as paleomagnetic reversals or oxygen isotope stages. But paleomagnetic and isotope datums *repeat themselves through time* (although each isotope stage in the last half of the Pleistocene tends to have a particular appearance that sets it apart from others), just like rocks do, and it is for this reason that biostratigraphic datums must be integrated with other types of datums to develop an accurate time scale. The refined scale, then, is dependent on both biostratigraphic and non-biostratigraphic datums, not one or the other.

Point-by-point comparisons of datums between sections is quite tedious, of course, but there is a way to evaluate the accuracy of datums in integrated data sets. Graphic correlation has existed for about 30 years, but it has not been until the last 5 to 10 years, with the widespread availability of personal computers and the availability of privately developed and commercial software (e.g., Hood, 1986), that the technique has begun to flourish (Mann *et al.*, 1995). Graphic correlation can be used to test the synchroneity of traditional (biostratigraphic) and non-traditional (oxygen isotope stage, paleomagnetic, well-log, microtektite and volcanic ash layers, ecostratigraphic) markers and to detect missing or expanded sections (e.g., Edwards, 1989*a*; Glass and Hazel, 1990; Martin *et al.*, 1990, 1993; Martin and Fletcher, 1995).

The method incorporates sections (cores, wells, outcrops, or any combination) into a composite section, against which individual sections are later replotted (*Shaw plots*, named for Alan Shaw, a former Amoco geologist who pioneered the technique while studying oil wells in the Rocky Mountain overthrust belt). Datums which plot on or close to a *Line of Correlation* (LOC), which is normally fitted based on investigator experience, are considered coeval (synchronous). *Any length of section may be used as long as there are suitable stratigraphic markers available for its subdivision.*

The LOC indicates changes in sediment accumulation rate. An LOC which is at an angle of approximately 45° from either the vertical or horizontal indicates that the sediment accumulation rates of the two cores (wells, etc.) are equal (Figure 7.15). Normally, however, the LOC consists of a series of segments ("doglegs"), the angles of which differ from 45°, indicating that the two sections differ in their respective accumulation rates. For example, high angle (from the horizontal) LOC segments indicate that the accumulation rate of the section plotted on the vertical (y) axis is higher than that of the section plotted on the horizontal (x)

7.7 Graphic correlation

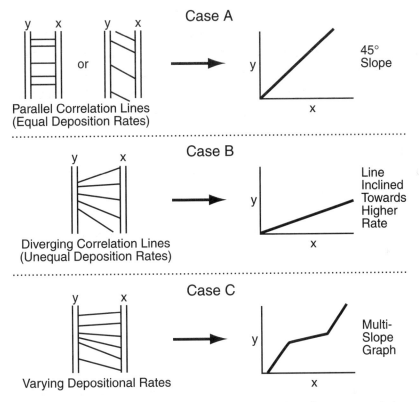

Figure 7.15. Graphic correlation (Shaw) plots. Changes in sediment accumulation rate are reflected by changes in slope of the Line of Correlation in plots of stratigraphic sections against one another. Line of Correlation is normally fitted based on investigator experience using different stratigraphic datums. See text for further discussion. (Redrawn from Martin, 1998a, after Phillips, 1986.)

axis. Conversely, low angle (from the horizontal, in this case) line segments indicate that the net sedimentation rate of the section plotted on the x-axis is higher than that of the section plotted on the y-axis (Figure 7.15). Sudden changes in slope of the LOC may indicate unconformities (sequence boundaries; e.g., Martin et al., 1993), especially if they coincide with truncated ranges of stratigraphic markers (cf. Holland, 1995). Stratigraphic markers that lie well off (above or below) the LOC are suspect because they may be reworked by erosion (or possibly bioturbation if the scale of interest is fine enough) or may represent truncated ranges or shifting environments (facies).

Graphic correlation is initiated by choosing one section as the *Standard Reference Section* (SRS). Normally, this is done by making preliminary plots of sections against each other, which are then evaluated based on investigator experience for continuity and length of section, and stratigraphic resolution (number and spacing of stratigraphic markers). It is probably also helpful to evaluate confidence intervals of appropriate taxa before choosing an SRS (Strauss and Sadler, 1989). Martin and Fletcher (1995) review some of the pitfalls in choosing an SRS (see also MacLeod and Sadler, 1995).

Once the SRS has been chosen, other sections are then plotted against the SRS to construct a *composite section* via the process of *range extension* (see Shaw, 1964; Miller, 1977; Edwards, 1989b; Mann *et al.*, 1995, for reviews). Range extension maximizes the stratigraphic range of markers. The composite section thus assimilates datums from all sections and displays their maximum stratigraphic ranges in a single, hypothetical, stratigraphic column. It is like taking all the individual sections of a fence diagram and combining them into a single stratigraphic column. The composite is normally subdivided into equal increments called *Composite Standard Units* that are equivalent to time-stratigraphic units (rocks bounded above and below by time). *Range extension is repeated for each section because the first and subsequent sections were not originally plotted against a composite based on all sections. Once completed, the sections are replotted until all datums have stabilized in position* (usually after 3–4 rounds of range extension).

Using this technique, anomalous (e.g., reworked, eroded) stratigraphic markers and stratigraphic gaps not readily apparent from biostratigraphic zonations can be detected. MacLeod (1995) analyzed a number of new sections across the Cretaceous–Tertiary (K–T) boundary that were unavailable at the time of MacLeod and Keller's (1991) graphic correlation analysis of K–T boundary sections. The new analysis confirmed the results of MacLeod and Keller's (1991) eustatic sea-level model and indicated that some sections previously *presumed* to be complete are not (Figure 7.16; see also Dingus, 1984, and Chapter 10). Inner and middle neritic sections (El Kef, Brazos River, Nye Kløv) record complete lowermost Danian sections, whereas shallow neritic (Millers Ferry) and middle bathyal (ODP sites 690 and 738) contain extensive hiatuses at or near the K–T boundary. According to the eustatic sea-level model of MacLeod and Keller (1991), these patterns result from migrations of depositional loci during eustatic sea-level rise following a late Maastrichtian lowstand. Thus, deep-sea sections are condensed and biased toward *apparent* instantaneous change at the K–T boundary, whereas sections from inner and middle neritic settings record gradual patterns of biotic, sedimentological, and geochemical change over considerable spans of section (see also section 7.4 on sequence stratigraphy). MacLeod (1995) also addresses previous – and unfounded – criticisms of the use of graphic correlation to analyze K–T sections and the graphic correlation method itself. (See also note added in proof; p. 308.)

7.7 Graphic correlation

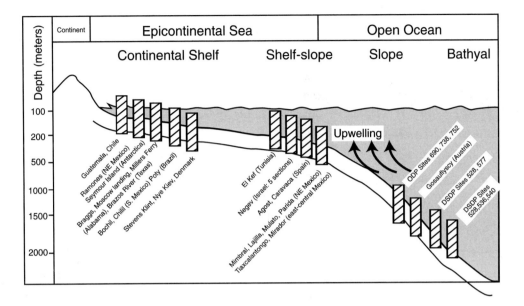

Figure 7.16. Depositional environments of Cretaceous–Tertiary (K–T) boundary sections. Deep-sea sections contain very condensed sections across the K–T boundary because of reduced sedimentation, dissolution, and erosion, whereas shallow neritic sections are eroded because of sea-level fall. ODP, Ocean Drilling Project; DSDP, Deep Sea Drilling Project. (Redrawn from Keller, 1996; see also MacLeod, 1995.)

Graphic correlation has also been used for more practical purposes. The most obvious application of paleontology and graphic correlation has been in oil and gas exploration (Mann *et al.*, 1995), which has a long history beginning about 1915–1920 with the establishment of micropaleontological laboratories in oil companies in the United States. Using "ecostratigraphic" and normal biostratigraphic datums, Martin and Fletcher (1995) used the technique to demonstrate sequence boundaries and condensed sections in what *appeared* to be a continuous section cored through the Plio-Pleistocene of the Gulf of Mexico. Based on Shaw plots they (1995) constructed a sea-level curve that closely resembled those published by other workers using extensive – and *expensive* – integrated seismic, well-log, and paleontological data bases. Despite the extensive use of sequence stratigraphy in petroleum exploration, it is still a relatively crude tool in terms of refined stratigraphic zonation and correlation, as practical resolution on seismic sections is typically no better than 10–50 m (Hallam, 1992). Most recently Armentrout (1996), among others, has used paleontology extensively to develop *predictive* models of reservoir sand occurrence that can save time, energy, and money in petroleum exploration. Armentrout's (1996) approach *integrates* facies relationships from seismic sections with "ground truth" provided by fossil data and well-logs. Without paleontology, correlating seismic lines, especially in a "wildcat" region, can be tenuous at best. Also, drilling wells may encounter subsurface "geopressured" zones, which are caused by a great thickness of overlying sediment, and which, if penetrated, can cause "blowouts" and loss of the drilling rig and crew. Denne (1994) describes the use of graphic correlation to determine the approximate depths of stratigraphic markers as the well is being drilled and the proximity of geopressured zones.

7.8 The hierarchy of taphonomic processes

Not only does the distribution of hiatuses and sedimentation rates appear to be fractal (self-similar), it also appears to be *hierarchical* (Allen and Starr, 1982; Plotnick, 1986b; McKinney, 1991; Martin, 1998a; see also Chapter 1). If the fractal dimension changes with change in scale, then there may well be a hierarchy of processes involved (Sugihara and May, 1990). In a delta, for example, relatively short-term changes in sedimentation (and hiatus frequency) are controlled by such factors as lobe-switching, whereas longer-term sedimentation rates and hiatuses reflect such processes as sea-level change. Sea-level change, sediment availability (erosion rates), and *accommodation space* (a place to put the sediment) exert "top down" control on the position and behavior of delta lobes (Plotnick, 1986b; McKinney, 1991). In turn, sea-level change reflects a variety of processes acting at different spatiotemporal scales (Chapter 8). Unconformities also correspond to changes in habitat conditions and interruptions of *within-population* processes that must have affected at least local biotic distributions (Schindel, 1982). Although unconformities are normally viewed negatively (missing record), they also provide *positive* evidence of the invasion of new local populations into an area following reestablishment of favorable conditions that are associated with renewed sedimentation (Schindel, 1982).

Like the marine record, the terrestrial record is also subject to a variety of influences on different spatiotemporal scales. Bishop (1980) emphasized the interplay of tectonics with areal extent of a fossil assemblage, its depositional setting, and quality of preservation. Continental configurations and tectonics strongly influence rainfall patterns that affect rates of erosion and sediment accumulation. Elevation (tectonics) and rainfall also affect the type of vegetation (not to mention the types of vertebrates!), which will in turn affect weathering rates. Rainfall patterns may in turn influence the frequency of floods and droughts (mass mortality), which in turn affect the preservation of fossil assemblages (e.g., Rogers, 1993; see also Chapters 2, 6).

7.8 The hierarchy of taphonomic processes

Behrensmeyer (1988) distinguished between the roles of local versus regional factors (e.g., tectonic setting) in determining the frequency of channel-lag versus channel-fill assemblages (see also Chapters 2, 5). Fluvial systems with numerous abandoned channels should provide more sites for preservation in channel-fills, whereas systems that continually rework sediments produce more channel-lag assemblages. In general, she hypothesized that sheet sands and channel-lags should predominate in areas of slow subsidence, in which there is a greater chance of reworking, whereas channel-fills should dominate when higher subsidence rates – and presumably greater rates of channel avulsion – occur (see also Aslan and Behrensmeyer, 1996; Chapter 5). Siwalik deposits (Miocene) of northern Pakistan, which accumulated adjacent to the rising Himalayas, are characterized by sheet sands 6–20 m thick that alternate with thicker fine-grained sediments that contain ribbon sands. The sheet sands are interpeted as channel deposits of a major river (on the scale of the modern Indus or Ganges), whereas ribbon sands are typically "single-story" and show little evidence of lateral reworking; paleontologically important assemblages are found mainly in fine-grained channel-fills. By contrast, the Early Permian Belle Plains and Arroyo Formations of Texas were formed on a stable craton, and should therefore be characterized by channel-lags according to Behrensmeyer's (1988) hypothesis; nevertheless, paleontologically important assemblages are also typically found in channel-fills in the Permian deposits. Thus, conditions favoring bone preservation by itself are more strongly linked to local processes associated with individual channel-fills than to overall tectonic setting and fluvial regimen. On the other hand, the frequency of channel-fills is probably more closely related to tectonic regime (see also Chapter 9).

This does not necessarily mean that the same processes produce the hierarchy of system–series–stage–age (period–epoch–age–chron). Drummond and Wilkinson (1996) found that most types of stratigraphic units (from lithofacies to depositional sequences) exhibit a lognormal distribution, in which frequency decreases with decreasing thickness, and that many supposed stratigraphic hierarchies are in reality arbitary subdivisions of a stratigraphic continuum.

The processes that form the units differ at different spatiotemporal scales. Zones are typically recognizable only within a single biogeographic province, itself the smallest unit of another hierarchy (provinces–regions–realms) used to describe the geographic distribution of organisms. Organisms are not uniformly distributed over the Earth's surface, being endemic to certain areas because of differing tolerances to environmental factors (temperature, salinity, light, etc.), pathways and barriers to dispersal and migration, and the vagaries of geological history (tectonics, sea-level change, etc.). With units of longer duration (stage, series, etc.), factors affecting the Earth's biota over larger areas (regions and realms) presumably play an ever greater role. By the time the period–system level is reached, mass extinctions of a large portion of the Earth's biota occur and account for some period–system boundaries.

Note added in proof:
Norris *et al.* (1999, Synchroneity of the K–T oceanic mass extinction and meteorite impact: Blake Nose, western North Atlantic, *Geology* 27: 419–422) indicate reworking of Cretaceous planktonic foraminifera above the K–T boundary but argued that, based on the first appearance of Danian (early Paleocene) foraminifera and calcareous nannofossils above a spherule layer, that impact eject coincide precisely with the K–T boundary; they also concluded that shallow-marine K–T boundary sections are not biostratigraphically more complete than deep-sea K–T sections.

8 Megabiases I: cycles of preservation and biomineralization

... your Father confessor trembles for you.
Charles Darwin to T. H. Huxley regarding Richard Owen
(Desmond and Moore, 1991)

8.1 Introduction

The concept of scale and its effect on measurement of stratigraphic completeness was introduced in Chapter 7. Scale is no less important to the measurement of cyclic change recorded in the stratigraphic record. As a hypothetical example: for practical purposes one would not use a meter stick to measure a sine wave with a wavelength of several kilometers. More importantly, if one used a meter stick, it would only detect wavelengths of about the same length as the meter stick, and wavelengths of much shorter or longer duration would go undetected.

It is no less true of the study of cyclicity in the geological record. What if one were attempting to document cyclic climate change through time? If each cycle were of several million years duration, one would not even see any change if the measuring scale was one of human perception (one or a few generations). Although these cycles are highly appreciated in terms of sea-level and climate change, they may produce *potential* megabiases in the fossil record that have not been fully explored and many of which require much more documentation.

This is fundamentally why the geological record is so important: there are many cycles of various – and often interrelated – phenomena in the geological record and these cycles act, and interact, at different frequencies (wavelengths) or scales that are mostly undetectable over geologically short durations of time. As more is included within the boundaries of the system being studied, more and more components, which may be thought of as behaving at different frequencies, are also included, so that the spectrum of frequencies of the entire system (hierarchy) becomes more or less continuous. Nevertheless, if the frequencies of different processes are sufficiently different (but not too different) they will interact to produce a "standing wave" or boundary recognizable to an observer (Salthe,

1985). "Higher" (more inclusive) components or *holons* will behave more slowly (have lower frequencies or cycle times of undisturbed behavior) than lower holons, which have higher frequencies. If disturbed, the *return* or *relaxation times* (*resilience*) of higher holons will also be slower than those of lower holons. Chemical weathering of continental rocks, for example, exerts a long-term (low-frequency) control on atmospheric CO_2 that occurs over tens to hundreds of millions of years; on the other hand, changes in rates of marine and terrestrial photosynthesis during the Pleistocene exerted control over atmospheric CO_2 on much shorter time scales (higher frequencies) of thousands of years or less because changes in ice cover, ocean circulation rates and land-based vegetation take only this long. These two holons may be recognized because the ratio of their two frequencies is so large: the higher holon (continental weathering) behaves so slowly, however, that it appears constant to the lower (faster) holon, which appears quite noisy to the higher holon.

In this way, constraining holons (*governors*) integrate or *filter* the signal (information) from lower holons so that the signal is averaged or *lagged* at the level of the higher holon (the signal is actually a statistical average that appears constant at the higher level). Constrained holons also filter signals from higher holons, but over such a short period that the message from the governor is read immediately and it too appears more or less constant to the constrained holon. Thus, the position of a holon in a hierarchy, which is another way of viewing its scale, is determined by its ability to process information. But the lagging (filtering) of information is not readily apparent from our human frame of reference.

8.2 Cycles of sedimentation and climate

8.2.1 Sea-level, CO_2, lithology, and biotic response

Fischer (1984) subdivided the Phanerozoic into two tectonically driven *supercycles* (each of about 300 Ma duration) of continental configuration and sea-level (Figure 8.1; Table 8.1; note usage of the term "supercycle" in Table 8.1). In this scenario, a supercontinent insulates the Earth, slowing radiogenic heat loss from the core and mantle. Eventually, heat buildup underneath continental crust becomes sufficient to cause doming, rifting, and breakup of the continent, which is followed by rapid seafloor spreading and sea-level rise. After many tens of millions of years or more, some or all of the continents ultimately collide to form yet another supercontinent and the process begins again. This cycle of supercontinent assembly–rifting–assembly has occurred at least four times in the last 2 billion years (Worsley *et al.*, 1986). In the case of the Phanerozoic, during much of the Cambrian to

8.2 Cycles of sedimentation and climate

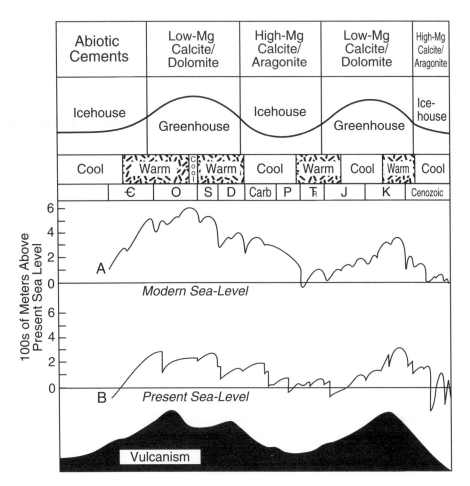

Figure 8.1. Phanerozoic cycles of continental configuration, sea-level, climate, and carbonate sedimentation. Greenhouse–icehouse cycles after Fischer (1984); carbonate cement and ooid mineralogy after Mackenzie and Agegian (1989). During greenhouses (Cambro-Devonian and Jurassic–late Eocene), continents were dispersed and seas displaced onto continents as a result of increased seafloor spreading rates and mid-ocean ridge (MOR) volume. Volcanism (Fischer, 1984) and a lack of well-developed terrestrial floras in the Cambro-Devonian (Berner, 1992, 1993a) resulted in increased atmospheric pCO_2 that presumably caused the Earth's average surface temperature to rise during these intervals. Deposition of low-Mg calcite (in cement and ooids) presumably reflects increased atmospheric pCO_2 (decreased surface ocean $CaCO_3$ saturation) and lowered oceanic Mg:Ca ratios from hydrothermal Mg–Ca exchange at MORs (Mackenzie and Agegian, 1989). Just the opposite conditions apparently prevailed during icehouses (Mississippian–Triassic; Oligocene–Recent), when atmospheric CO_2 and sea-level declined, presumably as a result of decreased spreading rates and MOR volume (Fischer, 1984) and the expansion of terrestrial floras (Berner, 1992, 1993a). During icehouses, precipitation of high-Mg calcite and aragonitic ooids and cements occurred, suggesting increased ocean $CaCO_3$ saturation. Sea-level after (A) Hallam (1992) and (B) Vail et al. (1977). Algeo and Seslavinsky (1995) recently revised peak Ordovician and mid-Cretaceous sea-level estimates downward to 100–225 m and 175–250 m, respectively, but relative magnitudes of oscillations remained largely the same. (Redrawn from Martin, 1995a.)

Table 8.1. *Stratigraphic (sea-level) cycles and their causes*

Type	Other terms	Duration (myr)	Probable cause
First order	—	200–400	Major eustatic cycles caused by formation and breakup of supercontinents
Second order	Supercycle, sequence	10–100	Eustatic cycles induced by volume changes in global midoceanic spreading ridge system
Third order	Mesothem, megacyclothem	1–10	Possibly produced by ridge changes and continental ice growth and decay
Fourth order	Cyclothem, major cycle	0.2–0.5	Milankovitch glacioeustatic cycles, astronomical forcing
Fifth order	Minor cycle	0.01–0.2	Milankovitch glacioeustatic cycles, astronomical forcing

Modified from Miall (1990), after Vail *et al.* (1977).

Devonian and Jurassic to Late Eocene, continents were dispersed and shallow, widespread epeiric seas moved onto continents as a result of increased seafloor spreading rates and mid-ocean ridge (MOR) volume, although the extent of the seas varied from continent to continent (Ronov, 1994). Conversely, sea-level underwent a broad decline during the Late Devonian to Triassic and from the Oligocene to the Recent. In the case of the Oligocene to Recent fall, the cycle has yet to be completed. As Pangea rifted apart in the Cenozoic, Antarctica and Australia moved away as a single continent. Ultimately, Antarctica and Australia separated and moved toward their present positions. By the Late Eocene, it appears that Antarctica was over the South Pole, because there was a major cooling and biotic turnover (as evidenced by benthic foraminifera) in the deep ocean or *psychrosphere* ("psychro" for cold). The waters adjacent to Antarctica apparently became sufficiently cool (and dense) to sink and set up a new pattern of deep-ocean circulation called *thermohaline* circulation: decreased temperature and increased salinity of seawater (as a result of ice formation) increases the density of seawater and causes it to sink.

At the other extreme is the last half of the Paleozoic. It appears that in the Phanerozoic, sea-level reached its nadir in the latest Permian. Based on published sea-level curves from different sources, sea-level was only slightly lower than it is today (Figure 8.1), the time of all actualistic studies that serve as the basis for so

8.2 Cycles of sedimentation and climate

much of taphonomy and the rest of the Earth sciences. This more or less monotonic sea-level decline began much earlier, however (Figure 8.1). Consequently, the *marine* record is much less complete during the last half of the Paleozoic than during the first half, which has accentuated the chronology of Late Permian marine extinctions (Erwin, 1993; see also Chapter 10).

Superimposed on the tectonically driven supercycles are smaller cycles. Changes in MOR volume may explain *second-order* changes in sea-level of 10–100 million years duration (Table 8.1), which includes the cratonic sequences (Sauk, Tippecanoe, etc.) recognized by Sloss (1963) in North America and by workers on other continents. The cause of *third-order* cycles is much more ambiguous, however. Changes in MOR volume are probably too slow to account for third-order cyclicity. Third-order cycles seem to occur on a regional basis, but can be correlated from one region to another, so that regional tectonic events caused by changes in seafloor spreading rates (regional plate rifting and convergence) and intraplate stresses (Cloetingh, 1988) may be involved (Miall, 1990; see also Hallam, 1992).

Changes in seafloor-spreading rates (a few centimeters per year on average – or about how fast one's fingernails grow in a year) and MOR volume are, however, far too slow to account for higher-frequency cycles. Cycles of solar insolation (and accompanying waxing and waning of glaciers) more likely drive sea-level fluctuations at the much shorter *Milankovich frequencies* (Table 8.1), which are the result of minor changes in the gravitational attractions between planets. These shorter cycles of sea-level may affect estimated durations of time-averaging on shallow shelves (Chapter 5).

Acting concomitantly with changes in long-term seafloor spreading rates, MOR volume, and sea-level, were long-term controls on atmospheric CO_2 (Fischer, 1984). In Fischer's cycles (Figure 8.1), seafloor spreading rates increased volcanism (as reflected by greater amounts of igneous rocks; e.g., Ronov, 1976) at spreading centers and subduction zones. According to Fischer, increased atmospheric pCO_2 from volcanism presumably caused the Earth's average surface temperature to rise during the Cambrian to Devonian and Jurassic to Late Eocene (*greenhouses*). Based on Berner's (1990, 1993b) mass balance models, atmospheric CO_2 levels during the Ordovician to Devonian were on the order of 10–15 times modern levels. Lower levels of CO_2 (about 2–5 times today's levels) prevailed during the Mesozoic, probably as a result of widespread terrestrial floras, which were largely absent during the first half of the Paleozoic (Berner, 1992, 1993a). Just the opposite conditions presumably prevailed during *icehouses* (Late Devonian–Triassic; Oligocene–Recent) as a result of decreased spreading rates and MOR volume. According to Berner's (1990, 1993b) model, CO_2 levels were very similar to today's in the Pennsylvanian (Late Carboniferous), when coal swamps were widespread, and beginning near the end of the Eocene. Coupled with the movement of Pangea near or over the South Pole, the model decrease in atmospheric CO_2 may have contributed to extensive Southern Hemisphere glaciation during much of the Carboniferous and Permian (Frakes et al., 1992). Probably also helping to cool the planet was an increase in the Earth's *albedo* (reflectivity) that resulted from increased land and ice to sea ratios.

Long-term changes in sea-level and atmospheric CO_2 had a pronounced effect on habitat and lithology, and therefore the kinds of fossil taxa likely to be encountered in the fossil record. Much of the world's petroleum comes from rocks of the Cambrian to Devonian and Mesozoic (Tissot, 1979; see also Wold and Hay, 1990). These intervals are characterized by widespread black shales, which has led some workers to suggest that ocean circulation was quite sluggish and conditions highly reducing during these times. Such conditions would have allowed preservation of large amounts of organic matter in sediments, thereby enhancing the formation of petroleum. Such a scenario is supported by high $^{34}\delta S$ values during, especially, the Cambro-Devonian. High $^{34}\delta S$ values during most of the Cambro-Devonian suggest extensive sulfate-reduction via oxidation of organic matter by oxyphobic sulfate-reducing bacteria that use SO_4^{2-} as an electron acceptor (Chapters 3, 9). The overall parallelism between sulfur isotopes and the cerium curve (which has also been suggested to indicate reducing conditions) corroborates increased anoxia of bottom-waters during the early to mid-Paleozoic (Wright et al., 1987). High atmospheric CO_2, warmer temperatures, and decreased oceanic mixing might also account for the correlation of oolitic ironstones with greenhouse phases, as the lower redox state of the oceans would presumably lead to increased ferrous iron concentrations in seawater and therefore a greater likelihood of ironstone formation (Wilkinson et al., 1985; see also Wold and Hay, 1990).

Relatively high sea-level – several hundred meters or more higher than today's, such as in the Ordovician and Cretaceous – has been the norm during most of the Phanerozoic (Table 8.1). These epeiric seaways were prime sites for the deposition of widespread limestones, such as reefs with well-developed biogenic frameworks (*bioherms*) and calcareous oozes (Ronov, 1976; Opdyke and Wilkinson, 1993). Ronov (1976) referred to this relationship as the *Law of Carbonate Deposition*, in which the amount of carbonate sediment deposited at a given time after the Precambrian was directly proportional to the intensity of volcanism and to the areal extent of epeiric seas.

8.2 Cycles of sedimentation and climate

Times of elevated atmospheric CO_2 probably accelerated the chemical weathering of continents and perhaps also the formation of limestones in shallow seaways. Chemical weathering occurs when atmospheric CO_2 dissolves in rain (acid rain) and attacks continental rocks of granitic composition to form bicarbonate ions (HCO_3^-) and release Ca^{2+} and other cations like Mg^{2+} (Mackenzie and Agegian, 1989); these and other ions are carried in dissolved form by rivers to the oceans:

$$2CO_2 + H_2O + CaSiO_3 \rightarrow Ca^{2+} + 2HCO_3^- + SiO_2 \qquad (8.1)$$

(Wollastonite – $CaSiO_3$ – has been used in place of "granite" for the sake of simplicity; Berner and Lasaga, 1989). The reaction of equation 8.1 acts as a *negative feedback* to slowly cool the Earth when atmospheric CO_2 becomes elevated (see Caldeira, 1995, for consideration of seafloor "weathering"). In marine waters, the ions from runoff are incorporated into the calcareous shells ($CaCO_3$) of marine organisms to form limestones:

$$2HCO_3^- + Ca^{2+} \rightarrow CaCO_3\downarrow + CO_2\uparrow + H_2O \qquad (8.2)$$

In the Mesozoic, calcareous plankton became a part of this cycle, with tremendous consequences for deposition of carbonates on continental shelves and in the deep-sea: calcareous oozes are eventually subducted and heated, and the CO_2 sequestered in $CaCO_3$ released back to the atmosphere through volcanoes (Chapter 9).

Negative feedback from continental weathering is supported by stable isotopes. Although the initial ratio of ^{87}Sr to ^{86}Sr is the result of radioactive decay of rubidium in igneous rocks, thereafter $^{87}Sr/^{86}Sr$ ratios behave like stable isotopes (Holser et al., 1988). The $^{87}Sr/^{86}Sr$ ratio primarily reflects a balance between weathering of high-$^{87}Sr/^{86}Sr$ continental rocks and lowering of $^{87}Sr/^{86}Sr$ ratios via hydrothermal exchange between ^{86}Sr-rich ocean crust and seawater (e.g., Raymo, 1991). Raymo et al. (1988) and Raymo (1994) attributed the overall rise of the $^{87}Sr/^{86}Sr$ curve during the late Cenozoic to mountain building and dissolved riverine fluxes to the oceans.

The relationship between tectonic activity and weathering rate seems to hold for more ancient times. Increased $^{87}Sr/^{86}Sr$ ratios in the Late Precambrian, Late Ordovician, Late Devonian, and Pennsylvanian to Permian (except for the latest Permian) correspond to intervals of Southern Hemisphere glaciation and sea-level fall that may have occurred because of increased rates of continental weathering and CO_2 drawdown as a result of orogeny in North America and Europe (Raymo, 1991; cf. Figure 9.6). These glaciations sometimes occurred when atmospheric CO_2 levels were quite high, and suggest that tectonism and resultant continental weathering (and perhaps movement of a continent over a pole) may be overriding factors in counteracting greenhouse conditions (Raymo, *et al.* 1988; Crowley and Baum, 1991). The relative proportion of exposed lithologies may also both reflect and have affected weathering rates and climate: Bluth and Kump (1991) calculated that the areal exposure of chemically resistant quartzose sandstones in the Phanerozoic is much greater than that predicted by volume because sandstones are often thin but laterally persistent. Ludwig and Probst (1998) found that sediment yields from 60 of the world's modern major river basins were not only a function of rock hardness, but also rainfall variation, runoff intensity, and basin slope; thus, erosion is greatest in young orogenic belts (because of steep morphologies and high runoff intensities) and erodibility is greatest in *arid* climates.

It appears that ocean circulation rates were stimulated by cooling during the Permo-Carboniferous: unlike the Cambro-Devonian, Parrish (1982, 1987) found a statistically significant relationship between organically rich petroleum source rocks and modeled upwelling zones during the Permo-Carboniferous. Limestones were still present during the Permo-Carboniferous, but bioherms were mostly replaced by algal and crinoidal *biostromes* (lenses) that did not have much – if any – of a biogenic framework; the prominence of algal biostromes during the Permo-Carboniferous may also reflect increased nutrient levels in the photic zone (Chapter 9). Moreover, much of the last half of the Paleozoic is characterized by non-marine or marginal marine sediments, such as coal swamps and giant cross-bedded sands that probably resulted from arid conditions and widespread interior drainage (see also Wold and Hay, 1990). Indeed, the supercontinent may have exhibited *megamonsoons*, which were much larger versions of the massive airflows found today over India and northwestern Africa (Parrish, 1993).

Such icehouse conditions of low sea-level, relatively cool climate and, presumably, relatively rapid ocean circulation rates were referred to as *oligotaxic* conditions by Fischer and Arthur (1977), because it is during these times that biotic diversity tended to be lowest (Table 8.2; see also Martin, 1998a). Conversely, *polytaxic* oceans were warm, widespread, and characterized by high diversity. Fischer and Arthur (1977) found a periodicity of about 32 million years for each of their oligotaxic–polytaxic cycles in the Meso-Cenozoic. They suggested that the mechanism was likely *endogenic* (perhaps related to sea-level change or magmatism) and not tied to solar or cosmic processes. This periodicity approximates the 26 myr periodicity of extinction for the same portion of the Phanerozoic based on improved geochronology, and which has been ascribed to swarms of impacts caused by movement of the Solar System through the central plane of

8.2 Cycles of sedimentation and climate

Table 8.2 *Phenomena associated with global icehouse and greenhouse states*

Icehouse state (example: present Earth)
Ocean highly stratified
Very stable: recycling of nutrient-rich deep water is difficult
Surface water temperatures range from <2 °C (circumpolar) to 25 °C (equatorial)
Bottom-water ranges from *c.* +2 °C (interglacial) to +1 °C (glacial)
Vigorous flow of bottom-water, rich in oxygen, hence strong oxidation, little storage of organic matter
Environments diverse; high productivity in areas of upwelling

Greenhouse state (example: Cretaceous oceans)
Much less stably stratified than icehouse state
Surface water temperatures not much higher than icehouse state at equator, but 12–15 °C in high latitudes
Deep-water temperatures from 15 °C (equatorial) to *c.* 10 °C (circumpolar)
Low-density surface-water leads to slow bottom-water flow; water at 15 °C holds half the oxygen of 2 °C water, hence little oxidative power
Therefore little recycling of organic matter, much buried in sediment; nutrient recycling much reduced
Low productivity, but good oil source beds

---

van Andel, 1994.

the Milky Way Galaxy with a periodicity of about 28–36 million years (cf. Benton, 1995; and Officer and Page, 1996, pp. 130–137, for criticisms regarding variation of periodicity and pseudoperiodicity built into data set). Other workers have suggested that there may be some minimum recovery time required by a community before it reaches sufficient diversity to suffer another round of devastation recorded as an extinction in the fossil record (*rebound effect*; Stanley, 1990). Moreover, the state of the communities themselves may determine the magnitude of their response (McKinney, 1989; Plotnick and McKinney, 1993; Chapter 10).

8.2.2 Cementation and diagenesis

The broad changes in the Earth's climate were paralleled by the mineralogy of abiotic carbonate cements and ooids (Figure 8.1; see Mackenzie and Agegian, 1989, for review). Based on a study of ooids, which are commonly accepted as being of physicochemical origin (Walker and Diehl, 1985), Sandberg (1975) proposed that variation in Mg/Ca ratios controlled the mineralogy of abiotic cements: calcite was deposited during times of lowered ratios because lower

Mg^{2+} in seawater tends to facilitate precipitation of calcite and inhibit aragonite precipitation (Walker and Diehl, 1985). Later, Mackenzie and Pigott (1981) concluded that variations in calcite and aragonite cements corresponded to variations in atmospheric CO_2 (see also Sandberg, 1983): deposition of predominantly low-Mg calcite cements and ooids during greenhouses instead reflected the lowered $CaCO_3$ saturation state of surface oceans caused by elevated atmospheric pCO_2 (Wilkinson and Given, 1986). Also, Walker and Diehl (1985) suggested that the dominance of calcitic skeletons during the Paleozoic greenhouse may have facilitated calcite precipitation by providing nucleation surfaces for incipient calcitic cement crystallization (see also section 8.2.6 and Chapter 9). Conversely, high-Mg calcite and aragonite cements and ooids dominated during icehouses, when CO_2 levels were apparently lower and dissolved Mg^{2+} levels higher (decreased seafloor spreading rates and cation exhange at MORs; Mackenzie and Pigott, 1981; Sandberg, 1983; Wilkinson and Given, 1986; Wilkinson and Algeo, 1989; Riding, 1993; see also Hardie, 1996).

Magnesium levels may have also affected the crystal habits of cements (Wilkinson and Given, 1986). In modern oceans, high-Mg calcite (>10 mol% $MgCO_3$) is *acicular* because rapid precipitation results in faster crystal growth in the c-axis direction and increased incorporation of Mg^{2+} into the crystal lattice, whereas low-Mg calcite (<10 mol% $MgCO_3$) cements are typically *equant*. Although acicular calcite cements are the predominant type of cement throughout the Phanerozoic, there was a greater abundance of equant calcite cements during greenhouses; in the case of the Paleozoic greenhouse, this appears to have been most prominent between about 500 and 400 million years ago (Figure 8.1; Wilkinson and Given, 1986). Diagenesis of sediments was also apparently enhanced during greenhouses, as suggested by increased amounts of dolomite, in which Mg^{2+} is sequestered (Walker and Diehl, 1985; Wilkinson and Algeo, 1989). Although hydrothermal alteration at ocean ridges (via Mg^{2+} for Ca^{2+} exchange) is presently the dominant sink for seawater Mg^{2+} (Wilkinson and Algeo, 1989), removal of Mg^{2+} ions to dolomite may have facilitated calcite precipitation during greenhouses (Walker and Diehl, 1985; Wilkinson and Algeo, 1989). Diagenesis during the Paleozoic greenhouse may also have been enhanced by sulfate-reducing bacteria when anoxia was widespread: Vasconcelos *et al.* (1995) found primary dolomite precipitation in an anoxic layer in a shallow lagoon in Brazil, and which was confirmed in bacterial cultures.

Bates and Brand (1990) suggested that the cyclic variation in ooid and cement mineralogy and crystal habit was *primarily* the result of diagenesis. According to them, aragonitic ooids and cements were dominant *throughout* the

Phanerozoic, but were altered to calcite when CO_2 levels were high. Although later studies have also implicated the role of alkalinity, they have largely confirmed the relationship between atmospheric pCO_2 and primary mineralogy (Morse et al., 1997). (See note added in proof, p. 329.)

8.2.3 Storms

Fischer's (1984) supercycles may have also influenced the geological record in much more subtle ways. Corresponding to the supercycles appear to be changes in storm frequency and intensity, as recorded by *tempestites* or storm beds. According to Brandt and Elias (1989), because tempestite thickness decreases with water depth, tempestite thickness should increase with more intense storms, which have longer wavelengths and therefore scour the bottom at greater depths. These workers hypothesized that tempestite thickness reflects, in part, storm intensity, which is related to atmospheric CO_2 levels: greenhouse phases should correspond to thicker tempestites and icehouse phases to relatively thin tempestites because storm intensity is presumably positively related to atmospheric pCO_2 (Barron and Moore, 1994; cf. Figure 8.1). They tested their hypothesis by examining 211 reports of tempestites in the literature; they used only data from clastic or mixed clastic–carbonate shelves and only distal storm beds that most likely represented individual events (84 storm beds from 65 literature sources; reports of amalgamated beds, which presumably represent multiple storms, were not used). They found that tempestite frequency and thickness largely paralleled greenhouse–icehouse supercycles, and that intervals with relatively large numbers of tempestites with maximum thicknesses ($n > 7$) typically included a greater range of tempestite thicknesses than intervals with few tempestites; this is to be expected if, like earthquakes, the distribution of storm intensity versus frequency follows a power law (Bak, 1996). Unfortunately, the data were insufficient to demonstrate a statistically significant pattern.

Although the relationship between apparent storm intensity and atmospheric CO_2 is a tantalizing one (Barron, 1989), there are other factors involved, as Brandt and Elias (1989) readily admitted. One important factor is paleogeography, which reflects tectonism, continent–ocean configurations, and thermal contrasts between land and epeiric seas and from equator to pole, all of which might steer hurricanes by inhibiting their poleward penetration or confine them to certain latitudes. In addition, one factor not mentioned by Brandt and Elias (1989) is that the areal extent of epeiric seas by itself might also have influenced the number of storms preserved in the rock record during greenhouse phases, when sea-level was higher and epeiric seas more widespread.

8.2.4 Fluctuations in the CCD

The entire ocean is basically a gigantic carbonate buffer system, which maintains a pH of approximately 8.1–8.3 when in equilibrium with atmospheric CO_2. If CO_2 levels rise or too much limestone is deposited on continental shelves (and therefore not enough in deep basins), the deep-sea is starved for $CaCO_3$ and the calcite compensation depth (CCD) rises to dissolve sufficient $CaCO_3$ in sediments to maintain the ocean's pH. If cratonic limestones are eroded or CO_2 falls, the CCD deepens accordingly. Thus, the CCD, which is normally defined as the depth below which $CaCO_3$ is absent (Chapter 3), has probably moved up and down in the ocean basins during the Phanerozoic (van Andel, 1975).

Fluctuations of the lysocline were frequent during the Pleistocene and occurred on Milankovich scales (e.g., Farrell and Prell, 1989). In modern oceans where surface production of $CaCO_3$ is high (e.g., equatorial Pacific), the CCD tends to deepen and the distance between the lysocline and CCD is much less than where surface fertility is low and the CCD shallows. The depth to the CCD and lysocline also reflects the corrosiveness (total dissolved CO_2) of bottom waters (which is in turn a function of water mass age, temperature, and pressure): the floor of the modern Pacific is bathed in old corrosive Antarctic Bottom Water (AABW), whereas much of the bottom of the Atlantic is bathed by much less corrosive young North Atlantic Deep Water (NADW; Chapter 3). Farrell and Prell (1989) found that during the last half of the Pleistocene (about 800 ka), the lysocline in the central equatorial Pacific shallowed by approximately 400–800 m about once every 100 ka in conjunction with Pleistocene climate cycles. Modern and interglacial sediment showed poor $CaCO_3$ preservation and greater distances (transition zones) between the lysocline and CCD, whereas glacial-aged sediment showed much better preservation and thinner transition zones as a result of deepening of the lysocline.

The deepening of the lysocline during glacials indicates an increase in bottom-water carbonate saturation (cf. Broecker et al., 1991; Chapter 5). Preservation maxima and minima tended to occur during the last half of glacial and interglacial stages and during climate transitions rather than during the middle of climatic cycles. Opdyke and Walker (1992) attributed the increased saturation to the *coral reef hypothesis*, according to which carbonate accumulation on shelves fluctuates with sea-level (but see Mylroie, 1993). Currently, up to half of the total marine carbonate accumulation may occur in reef environments, but the hypothesis indicates that a shift to deep-sea carbonate accumulation during glacials requires increased $[CO_3^{2-}]$ and lower dissolved pCO_2. Archer and Maier-Reimer (1994) implicated the degradation of sedimentary organic carbon in the dissolution of calcite inputs.

8.2 Cycles of sedimentation and climate

The depth of the CCD has fluctuated over much longer time scales, as well. Partitioning of carbonates between the cratons and oceans is determined in part by global sea-level, with accumulation of shallow-water carbonates proportional to continental freeboard and areal extent of warm, carbonate-saturated seas (Wilkinson and Walker, 1989; Law of Carbonate Deposition). Based on the $CaCO_3$ content of deep-sea cores, Van Andel (1975) was the first to note that the CCD shallowed to about 3.5 km during the Cretaceous greenhouse. The Cretaceous was a time of widespread deposition of carbonate oozes (mainly coccolithophorids and, secondarily, foraminifera) in epeiric seaways (Cretaceous comes from the root *creta* for chalk). Since the depth of the CCD is dependent on carbonate input, the CCD shallowed accordingly.

Wilkinson and Walker (1989) concluded that the CCD was also quite high during the Paleozoic greenhouse. Based on the fossil record, widespread carbonate deposition on the cratons during this time was the result of extensive calcareous benthos and not plankton (Boss and Wilkinson, 1991). Based on examination of ophiolite complexes through the Phanerozoic, Boss and Wilkinson (1991) concluded that deposition of extensive cratonic limestones likely helped keep the CCD between about 1 and 2 km depth during most of the Cambro-Devonian, thereby preventing significant deep-sea ooze formation. As sea-level fell during the last half of the Paleozoic, much of the cratonic limestones were presumably eroded, dissolved, and moved to deep basins (*calcite-push model* of Wilkinson and Walker, 1989; see also Opdyke and Wilkinson, 1988), which would have deepened the CCD. It is during this time, according to Boss and Wilkinson (1991), that preservation of calcareous oozes began and not during the Mesozoic (see also Chapter 9). The evolution of calcareous plankton could in turn have contributed to the deepening of the CCD during the late Paleozoic (*calcite-pull model* of Wilkinson and Walker, 1989), a mechanism that was originally proposed by Kuenen (1950). If the CCD was in fact deepening in the late Paleozoic, then the shallow lysocline/CCD inferred by Malinky and Heckel (1998) could be the result of organic matter input and oxidation (Bé, 1977; Chapter 3).

The composition of fossil plankton assemblages appears to reflect the $CaCO_3$ saturation of the oceans during the Paleozoic (Figure 8.2). In the case of the Cambro-Devonian, it appears that *if* (and it is a big "if") calcareous plankton existed in the Paleozoic, they must have dissolved, leaving behind dissolution-resistant non-calcareous taxa, which may partly explain the dominance of early to mid-Paleozoic plankton by non-calcareous taxa such as acritarchs, graptolites, and radiolaria (Chapter 9). Discussion of the stratigraphic distribution of the remaining plankton continues in Chapter 9.

8.2.5 The latitudinal lysocline

Modern sediments differ in their fossil content across latitude. *Heterozoan carbonates* are produced by red algae and heterotrophic benthic invertebrates, occur worldwide and are recognized as cool-water carbonates in temperate and higher latitudes (<20 °C; James, 1996; see also Hallock, 1987; Carannante *et al.*, 1995; see also Smith *et al.*, 1998). The heterozoan association is also present in warm-water (>20 °C) oligotrophic settings but is typically masked by *photozoan carbonates* that are produced by predominantly symbiont-bearing taxa (corals, foraminifera) and calcareous green algae (James, 1996). Calcite-dominated heterozoan assemblages may have flourished in ancient warm-water phototrophic settings, however, if upwelling supplied cool nutrient-rich waters. Such cool-water assemblages are calcitic and may therefore be better analogs for diagenesis of ancient limestones in upwelling settings than modern tropical carbonate aragonitic sediments (James and Bone, 1989); cool-water assemblages may, however, be mistaken for photozoan associations if facies relationships are ignored (criteria for the recognition of photosymbiosis are equivocal; James, 1996).

Thus, the mineralogy and preservation of carbonates may vary across latitude. Fluctuations in a kind of *latitudinal lysocline* may have occurred across latitude in shallow seas in concert with climate (atmospheric pCO_2), as carbonate saturation of shallow waters tends to decrease with higher (cooler) latitudes (Alexandersson, 1972; Martin and Liddell, 1991). For example, foraminifera may *tend* to be more poorly preserved in high latitudes than low because of decreased $CaCO_3$ (shell) content (Chapters 3, 5). Flessa (1995) came to similar conclusions for bivalves, although the extent of degradation is probably species-specific (see Risk *et al.*, 1997). Not all of the factors involved are, however, physicochemical in nature: at the same depth, echinoids appear to be better preserved at high latitudes because of both cooler water temperatures and decreased predation (Kidwell and Baumiller, 1990). Cool-water carbonates may also be subject to greater rates of bioerosion and erosion by physical currents (Simone and Carannante, 1988; Carranante *et al.*, 1995).

8.3 The origin and biomineralization of skeletons[1]

Why did skeletons originate in the first place? And what do they tell us about changes in climate, ocean chemistry, and, perhaps, preservation potential in the fossil record?

---

[1] This section is based on Martin (1995*a*, 1996*b*).

8.3 The origin and biomineralization of skeletons

The origin of skeletons has long perplexed paleontologists and a variety of mechanisms have been proposed to explain their appearance in the fossil record. One of the earliest proposed causes was that of increased body size and the need for support. This in turn may have resulted from an increase in atmospheric oxygen (e.g., Runnegar, 1982), a hypothesis that eventually fell into disfavor but has recently been resurrected (e.g., Canfield and Teske, 1996). Logan et al. (1995) suggested that increased production of fecal pellets in the late Precambrian sent larger amounts of organic matter to the ocean bottom, rather than allowing it to decay and use up O_2; prior to this time, bacterial degradation of organic matter was presumably common. On the other hand, Vermeij (1989), who reviewed a variety of causal mechanisms for the origin of skeletons, came down firmly in favor of predation (see also Vermeij, 1987).

One of the most intriguing causes that have been proposed is that detoxification of excess Ca^{2+} – which is toxic at high concentrations – may have occurred by secretion into hardparts (Kaźmierczak et al., 1985). By analogy to modern soda lakes associated with volcanism, early Precambrian seas had high alkalinity and high pH following early degassing of the Earth, and, as a result, Ca^{2+} and Mg^{2+} salts tended to be precipitated (Kempe and Degens, 1985; but see Grotzinger and Kasting, 1993). On the other hand, high Na^+ concentrations prevailed in this soda ocean, as sodium compounds were much more soluble. According to this scenario, subsequent leaching of chlorine from mafic crust and its accumulation in ocean waters caused the oceans to shift toward a fully halite mode by about 1 Ga (see also Ronov, 1968). With time, though, as sodium carbonate was subducted, Ca^{2+} concentrations in the oceans increased, thereby exposing the biota to calcium toxicity (Kaźmierczak et al., 1985; Kempe and Kaźmierczak, 1994). Later, during the Phanerozoic, sulfate reduction in euxinic basins would have increased alkalinity again. Slow upwelling or oceanic overturn during these phases was said to lead to increased alkalinity in the shallow oceans and high levels of $CaCO_3$ that adversely affected marine benthos (Kempe and Kaźmierczak, 1994; see also Grotzinger and Knoll, 1995; Knoll et al., 1996; Chapter 9).

In this regard, the appearance and test composition of foraminiferal suborders in the Phanerozoic presents some troublesome questions about the origin of skeletons. Cambro-Ordovician foraminifera were dominated by agglutinated species that presumably arose from the protean, organic-walled suborder Allogromiina (Figure 8.2). Paleozoic agglutinated species accreted sediment grains onto an organic lining or with organic cement. Size increase and diversification in foraminifera was aided by agglutination of sediment particles, which may have been an evolutionary outgrowth of the formation of feeding or reproductive cysts by

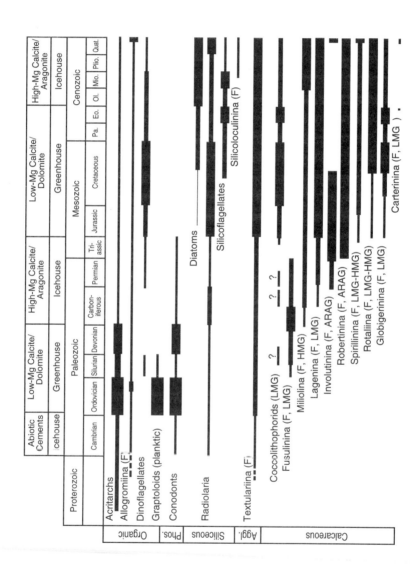

Figure 8.2. Geological distribution and composition of foraminiferal suborders. Note correspondence between mineralogy of foraminiferal suborders, abiotic cements and ooids, climate change, and ocean chemistry. Thickness of line indicates relative species diversity (period-averaged data). Geological ranges and composition of foraminiferal suborders based on Tappan and Loeblich (1988). All agglutinated foraminifera are placed in the suborder Textulariina for simplicity (cf. Loeblich and Tappan, 1989). F, foraminiferal suborder; AggL, agglutinated; LMG, low-Mg calcite; HMG, high Mg calcite; ARAG, aragonitic (cf. Figure 8.1). (Redrawn from Martin, 1995a, 1996b.)

8.3 The origin and biomineralization of skeletons

allogromiid (organic-walled) ancestors (e.g., LeCalvez, 1936). Like the macrobenthos, the test of foraminifera may have also initially provided protection from predators or indiscriminant scavengers (e.g., Lipps, 1985, 1987; Culver, 1991, 1994) that were colonizing an increasingly oxygenated and bioturbated sea-bottom in the Late Precambrian–Early Cambrian. Besides offering protection from predators and scavengers, the test would have permitted an enlarged sarcode (and pseudopodial feeding net; cf. Nicol, 1966) that allowed foraminifera more fully to exploit the meiobenthos.

The test composition of calcareous foraminiferal suborders exhibits a striking correspondence to the presumed $CaCO_3$ saturation state of oceanic surface waters when the suborders presumably originated, and which remained unchanged in composition thereafter (Figure 8.2; cf. Figure 8.1; Martin, 1995a, 1996b; see also Wilkinson, 1979; Sandberg, 1983; van de Poel et al., 1994). Carbonate-saturated waters favor extensive secretion of $CaCO_3$ in modern foraminifera, and the ability to secrete $CaCO_3$ may have allowed foraminifera further to invade benthic carbonate environments in the Paleozoic (Boersma, 1978). Based on the fossil record of foraminiferal suborders, it appears that after a calcification threshold was crossed in the Late Devonian–Early Carboniferous icehouse, most remaining suborders of calcareous foraminifera began to appear (Miliolina, Involutinina, Robertinina, Spirillinina), the mineralogy of which reflects the presumed increased $CaCO_3$ saturation state of surface waters (Figure 8.2). Low-Mg representatives of the suborders Rotaliina and Globigerinina appeared in the mid-Jurassic (Bajocian), when lower $CaCO_3$ was presumably more prevalent, and low-Mg calcite cements appeared in agglutinated foraminifera during the Cretaceous greenhouse (Loeblich and Tappan, 1989; Figure 8.2).

Only two suborders of foraminifera arose in the Cenozoic. The suborder Carterinina, which consists of only one genus, seems to have appeared in the Late Eocene, at or near the beginning of the second major Phanerozoic greenhouse to icehouse transition, when $CaCO_3$ saturation was presumably increasing. The suborder Silicoloculinina also consists of only one genus composed of opaline silica. This suborder appeared during the Neogene (mid-Miocene) icehouse, as grasslands were expanding. Grasses incorporate opaline silica into their leaves, but upon decay they release it, which presumably accelerated the flux of dissolved silica to the oceans beginning in the Neogene (Worsley et al., 1986).

Only one suborder (Lagenina), that *apparently* arose during the Late Carboniferous icehouse (mid-Pennsylvanian), has low-Mg calcite tests (Figure 8.2). Based on test structure and mineralogy, however, this group may have originated from primitive Fusulinina in the Late Silurian (Tappan and Loeblich, 1988), when

CaCO$_3$ saturation was presumably lower. The appearance of the Fusulinina appears to mark the first response of foraminifera to heightened CaCO$_3$ saturation of surface oceans, and may have resulted from decreased atmospheric pCO$_2$ during the Late Ordovician glaciation, although the mineralogy of the suborder is low-Mg calcite.

The hypothesis that foraminiferal suborder mineralogy is a response to CaCO$_3$ saturation of surface waters predicts that *test microstructure – which is secreted intracellularly in calcareous foraminifera – should parallel that of abiotic cements* (e.g., Angell, 1967, 1979). During the Ordovician and Silurian, equant (i.e., equidimensional; ~2–5 μm in size) calcite cements were present, which suggests relatively low CaCO$_3$ saturation of surface waters. Interestingly, the microgranular test of the Fusulinina consists of calcite crystals that are "closely packed," "subangular," and "*equidimensional* ... usually on the order of a few micrometers in diameter" (Lipps, 1973; my italics). In the suborder Textulariina (agglutinated foraminifera), calcareous test cement, which originated during the Cretaceous, also consists of low-Mg equidimensional crystallites (Loeblich and Tappan, 1989), and suggests a response to low surface CaCO$_3$ saturation. The test of the suborder Miliolina, which appeared during the Early Mississippian climatic transition, is secreted as intracellular acicular high-Mg calcite lathes (Berthold, 1976), and acicular calcite was the dominant cement at this time. The suborder Carterinina is also characterized by acicular calcite "spicules" (Deutsch and Lipps, 1976).

The appearance of most suborders during times of apparently increasing CaCO$_3$ saturation of surface waters could have also resulted from dissolution of earlier ancestral forms. In fact, reports of the earliest foraminifera are of agglutinated specimens from shallow-water sediments (Culver, 1991; Lipps and Rozanov, 1996). So, is the initial occurrence of a calcareous suborder in the geological record necessarily its first *evolutionary* occurrence or could early forms be missing for reasons of preservation? After all, in modern shell-poor siliciclastic settings, calcareous foraminifera may occur abundantly in shallow-water sediments, but their preservation may be quite poor because of high rates of bioturbation and dissolution (Chapters 3, 5). Arguing against a preservational bias is that most of the dominant suborders exhibit a conspicuous presence throughout their stratigraphic range after they have appeared, through both greenhouse (low CaCO$_3$ saturation) and icehouse (high CaCO$_3$ saturation) episodes (Figure 8.2). Furthermore, the large number of reports upon which the compilations of Tappan and Loeblich (1988) are based suggests that the stratigraphic ranges reported by them for different suborders will not be extended appreciably.

8.3 The origin and biomineralization of skeletons

If calcareous foraminifera had been present in abundance in the Cambro-Ordovician, for example, they would likely already have been reported, barring the discovery of exceptionally well-preserved assemblages. In fact, supposed calcareous foraminifera reported from the Cambrian have been reclassified as calcareous algae (Martin, 1995a).

In contrast to protoctists such as foraminifera, cells of metazoans live within an environment that typically buffers them from external perturbations because of the decreased surface/volume ratios of larger organisms and the presence of tissues (Bonner, 1988; Andrews, 1991). As organism size increases (which is a function of multicellularity), exposure to the external milieu correspondingly declines. Multicellularity probably also increased the potential for ion regulation – and perhaps $CaCO_3$ secretion – in metazoa because of the likelihood of secretory tissues and organs (Simkiss, 1989).

But if the geological record of appearances of foraminiferal suborders is more or less correct, why was calcification in foraminifera largely delayed until the Late Paleozoic icehouse, well after the Cambrian explosion of metazoan calcareous hardparts? The answer(s) to this question may provide important clues to the origins, diversification, biomineralization, and preservation of marine biota on Earth. Martin (1995a, 1996b) suggested that during their late Precambrian–early Phanerozoic history, foraminifera may have been constrained to secrete organic tests and cements (in agglutinated forms) because of unusual ocean chemistry. He suggested that microorganisms are much more vulnerable to changes in the external environment than larger organisms because of their small size and high surface/volume ratio. Release of high levels of dissolved PO_4^{3-} that were perhaps stored in isolated basins or deep in the oceans (*Yudomski Event*; Cook and Shergold, 1984) in the late Precambrian–Early Cambrian may have prevented calcification in early foraminifera. Phosphate ions inhibit inorganic calcite precipitation (Simkiss, 1989), and the high surface/volume ratio associated with the unicellular grade of organization in foraminifera may have facilitated entry of phosphate ions into the cell, thereby inhibiting $CaCO_3$ secretion. If the signal of oceanic eutrophication in the late Precambrian–Early Cambrian was fixed in the genome of primitive foraminifera, it may have constrained agglutinated taxa to continue to precipitate non-calcareous tests and cements through the remainder of the Cambro-Ordovician. According to this hypothesis, early metazoans were able to calcify – despite high dissolved PO_4^{3-} levels – because they were better able to regulate intracellular PO_4^{3-} concentrations, although high phosphorus levels may have stimulated phosphatic biomineralization in some early metazoa because of the presence of tissues that could become specialized for secretion

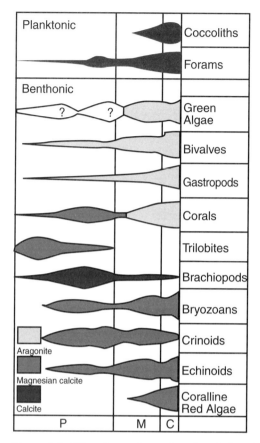

Figure 8.3. Mineralogy of macrobenthos during the Phanerozoic. P, Paleozoic; M, Mesozoic; C, Cenozoic. (Redrawn from Mackenzie and Agegian, 1989.)

(e.g., inarticulate brachiopods, conodonts; see also Bengston and Conway Morris, 1992).[2]

As phosphate levels waned, calcification in foraminifera may have been inhibited because of low $CaCO_3$ saturation levels associated with increased rates of seafloor spreading, volcanism and CO_2 production, and Mg/Ca exchange rates at MORs in the Cambro-Ordovician.

As $CaCO_3$ saturation increased in the Late Ordovician, foraminifera (suborder Fusulinina) presumably began to respond by secreting calcareous tests. Martin (1996b) suggested that this may have been a response to Ca^{2+} toxicity

[2] Although several anonymous reviewers have objected to this mechanism, all have ignored the evidence for the sensitivity of foraminifera to ocean chemistry (such as the correlation between test ultrastructure and abiotic cements), and none has proposed a satisfactory explanation for delayed calcification in foraminifera.

8.3 The origin and biomineralization of skeletons

(related to $CaCO_3$ saturation) by establishing symbiotic (mutualistic or mutually beneficial) associations with bacteria capable of secreting $CaCO_3$ (e.g., West, 1995). $CaCO_3$-secreting organelles that contain their own DNA have been found in some species of foraminifera, and suggest that these organelles were originally $CaCO_3$-secreting bacteria eaten by the ancestral foraminifer(s). According to this scenario, instead of being digested, the bacteria persisted in their hosts' cytoplasm and came to help regulate $CaCO_3$ secretion. Wallin (1927, p. 142) suggested much earlier that calcification originated through endosymbiosis. Indeed, protoctists may be viewed as highly integrated microbial communities in which symbiosis has been a major source of evolutionary innovation (Margulis, 1991, 1992; Dyer and Obar, 1994).

Still, unicellularity versus multicellularity cannot entirely explain the differences in timing of calcification in foraminifera versus metazoa. Westbroek (1997), echoing Lowenstam and Margulis (1980), suggested that biochemical pathways that evolved earlier in Precambrian metazoans were co-opted for calcification. Foraminifera, which apparently lack such sophisticated multicellular secretion pathways, ought to have been more susceptible to calcification near the Precambrian–Cambrian boundary if this mechanism is correct. Moreover, metazoans ought to have been able to *resist* increasing $CaCO_3$ during the Permo-Carboniferous to Triassic icehouse (*biocontrolled mineralization*), but many did not (*bioinduced mineralization*; Figure 8.3): the fossil record of macrobenthos indicates that most calcitic Paleozoic metazoa were often replaced by aragonitic ones following Late Permian extinctions (Wilkinson, 1979; Railsback and Anderson, 1987; Railsback, 1993).

Note added in proof:
Stanley and Hardie (1998) state that there is no temporal relationship between the mineralogy of foraminiferal tests and Mg/Ca ratios of seawater (Secular oscillations in the carbonate mineralogy of reef-building and sediment-producing organisms driven by tectonically forced shifts in seawater chemistry. *Palaeogeography, Palaeoclimatology, Palaeoecology* 144: 3–19). But, changes in rates and types of ooze deposition attributed by them to changes in seawater Mg/Ca ratios may well be related to other factors such as nutrient availability and water mass stratification (see Chapter 9). Kleypas *et al.* (1999) Geochemical consequences of increased atmospheric carbon dioxide on coral reefs, *Science* 284: 118–120) suggested that increased atmospheric pCO_2 may disrupt calcification by reef corals and other calcifying marine ecosystems.

The recent report by Pawlowski *et al.* (1999, Naked foraminiferans revealed, *Nature* 399: 27) of molecular evidence for a much earlier origin of foraminifera still leaves open the question of their skeletonization.

9 Megabiases II: secular trends in preservation

You can never step twice into the same river. Heraclitus

nature ... draws the fern's grace from the putrefaction of the forest floor, and pasturage from manure, in Latin laetamen — *and does not* laetari *mean "to rejoice"?* Primo Levi, *The Periodic Table*

9.1 Introduction

The biogeochemical cycling of elements in the Earth's crust and shallow mantle by the interaction of physical and biological entities has had a profound effect on the Earth's climate and its biota and preservation. Over relatively short time scales, the assumption of steady-state dynamics of these cycles is no doubt true. But although the cycles are described in textbooks as if they have always existed in their present steady-state form, they actually *evolved*, and gave rise to secular megabiases in preservation. And, as will quickly become apparent, many of these secular trends are intertwined with each other.

9.2 Types of secular megabiases

Kowalewski and Flessa (1996) recognized four types of secular megabiases in the fossil record. A *within-taxon* megabias is a change in the quality of a single taxon's record that results from evolutionary, environmental, or geological trends that affect the taxon's preservation potential (Figure 9.1). Such gradients may occur within taxa along: (1) depth gradients, such as in nautiloid cephalopods (e.g., Hewitt, 1988), echinoids (Kidwell and Baumiller, 1989), and articulate brachiopods (Patzkowsky, 1995); (2) with latitude (e.g., changes in temperature, rates of predation, organic matter reactivity, and $CaCO_3$ saturation), all of which affect preservation (Kidwell and Baumiller, 1989; Martin and Liddell, 1991); and (3) through time. For example, the apparent lack of large octopods (which prey on nautiloids) before the Eocene suggests that nautiloid shells settled to the bottom with softparts more or less intact before this time (but see examples in Boston and Mapes, 1991); this would have decreased the chances of post-mortem drift (which tends to remove nautiloids from the fossil record), but would have

9.2 Types of secular megabiases

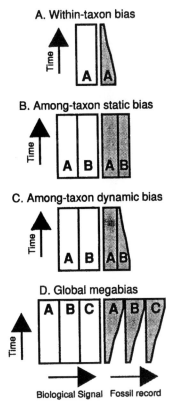

Figure 9.1. Four major types of megabias. Open boxes on left represent biological signals (e.g., diversity), whereas shaded boxes on right represent fossil record. Distortions shown are gradual, but can also be stepwise or sudden, and can happen along both temporal (shown) and spatial gradients. (Reprinted from Kowalewski and Flessa, 1996, with permission of the Geological Society of America.)

increased the chances of destruction by benthic scavengers or implosion (Hewitt, 1988). On the other hand, predation by octopods in the post-Eocene removed softparts and presumably made nautiloid shells more susceptible to transport (Hewitt, 1988), although increased encrustation of necroplanktonic conchs beginning in the Mesozoic may have decreased their buoyancy (Boston and Mapes, 1991).

An *among-taxon* megabias refers to variation in preservation potential between taxa, such as the difference in preservation potential between softbodied (e.g., worms) and shelly taxa (e.g., Lawrence, 1968; Chapter 2). Among-taxon megabiases are of two types: *static* and *dynamic distortions*. Worms, for example, have always had a poor fossil record, and some crinoids are more resistant to post-mortem transport and destruction than others (static distortion; Meyer *et al.*, 1989; see also Chapter 2). On the other hand, the fossil records of coccolithophorids (calcareous nanoplankton) and lingulide brachiopods have improved and declined, respectively, through the Phanerozoic (dynamic distortion; Boss and Wilkinson, 1991; Martin, 1995a, 1996a; Kowalewski and Flessa, 1996). The increase in calcareous plankton in the fossil record may be related to deepening of the CCD (Chapter 8) and rising nutrient levels in the photic zone of the oceans (see below). In the case of lingulides, their better preservation early in the Paleozoic may be related to less benthic scavenging during this time, whereas the decline in their preservation may be related to increased scavenging through the Phanerozoic (see also Kidwell and Brenchley, 1994; Kowalewski and Flessa, 1996; and below). The preservability of echinoids has also changed through time and is related in part to the interlocking of plates: representatives of the Paleozoic order Cidaroida are much more poorly preserved than those of the order Clypeasteroida, which is characterized by extensively interlocked plate sutures (Kidwell and Baumiller, 1989; Greenstein, 1991; Chapter 2). Differences in relative rates of decay between echinoid taxa may also partially explain patterns of taxonomic diversity: it is more likely that new species will be described from fragmentary material than more or less intact shells. For example, the number of species described in the echinoid family Cidaridae decreased from the Middle Triassic (time of apparent origination of Cidaridae) to the Pleistocene as skeletal articulation increased (Greenstein, 1992).

Global megabiases affect many or all taxa. Smith *et al.* (1988) found, for example, that diversity of fish in floodplain sediments was 5–15 times that of the Paleocene-Eocene in the Plio–Pleistocene and 5–10 times Plio-Pleistocene levels in modern habitats; they concluded that the increase was primarily a taphonomic artifact.

One of the most pronounced global megabiases involves the preservation of soft-bodied biotas. Raup (1972) suggested that Lagerstätten are probably more common in younger rocks, and in some cases this appears to be true. For example, preservation in amber becomes more commonplace in the Cretaceous with the advent of the Araucariaceae (Chapter 6). In other cases, however, preservation of soft-bodied biotas, such as the Burgess Shale-type, declines through the Phanerozoic. This would seem, intuitively, to be the result of increasing rates of bioturbation (section 9.5), oxygenation, and organic matter decay (e.g., Plotnick, 1986a; Allison and Briggs, 1993a,b).

Aronson (1992) tested the role of bioturbation by comparing observed versus expected stratigraphic distributions of Burgess Shale-type faunas and found that observed and expected distributions disagreed. He found that in order for Burgess Shale-type faunas to be burrowed away, the depth of bioturbation would have had

9.2 Types of secular megabiases

to have increased 16 times its Cambrian depth by the Ordovician–Silurian, which is at odds with the much shallower bioturbation depths reported. Therefore, Aronson (1992) concluded that other ecological or taphonomic controls must be involved in exceptional softpart preservation. In response to the criticism of Allison and Briggs (1993a) that much less of an increase in bioturbation depth would be required if all exceptionally preserved biotas had been included in Aronson's (1992) original data base, Aronson (1993) recalculated and still came to his original conclusion.

But Allison and Briggs (1993b) responded that until there are detailed data on the evolution of bioturbation intensity in low-oxygen muddy (Burgess Shale) environments, Aronson's (1992, 1993) conclusions will remain equivocal because they are based on bioturbation intensity in a variety of environments. They found that in relation to outcrop area, exceptional faunas appear to be overrepresented in the Cambrian and Jurassic, when anoxia and episodic sedimentation were apparently widespread (1993b). Times of widespread carbonate platforms (e.g., Ordovician and Permian) were not, however, favorable to soft-body preservation, presumably because siliciclastic input was limited (Chapter 8). During the Carboniferous to Triassic soft-bodied biotas (such as those of the Mazon Creek) formed mostly in "broad coastal delta plains in tropical areas characterized by high productivity, episodic sedimentation, and shifting salinity" (Allison and Briggs, 1993b, p. 529), but based on the average number of exceptional faunas per unit outcrop area, the number of exceptional faunas during the Carboniferous to Triassic was more likely to have arisen by chance. Because of increased bioturbation, Allison and Briggs (1993b) concluded that the occurrence of soft-bodied biotas after the early Paleozoic is related to conditions of paleogeography, topography, and climate that favored episodic burial, stagnation, and the formation of microbial mats (Chapter 6).

Although widespread anoxia during greenhouse episodes would seemingly contribute to soft-body preservation, it alone may be insufficient for preservation because of oxidation of organic matter via bacterial sulfate reduction (Chapters 3, 6). Soft-bodied biotas are, for example, unknown from the Cretaceous Western Interior of North America despite the presence of extensive clastic sediments that include widespread black shales (Allison and Briggs, 1993b). As discussed in Chapter 6, Butterfield (1995) suggested that animal softparts could be preserved because of the anti-enzymatic and stabilizing effects of clay minerals (see also Butterfield and Nicholas, 1996). Suites of clay minerals change through the Phanerozoic from illite (K_2O)-rich to kaolinite, which is depleted in cations and therefore silica-rich (Garrels and Mackenzie, 1971, pp. 230–237). Iron-rich smectites, which are often associated with hydrothermal activity, and their alteration products, such as illite, would be expected to be most prevalent during greenhouse episodes (Chapter 8). In fact, preservation of soft-bodied faunas declined through the Permo-Carboniferous, as illite declined and kaolinite increased, and potash (K_2O-rich) deposits became more prevalent (Hay and Wold, 1990).

9.3 Stratigraphic completeness

9.3.1 The pull of the Recent

Perhaps the most obvious of all global megabiases is that of stratigraphic completeness (Chapter 7): namely, the older the rocks, the less likely they are to be preserved. In both the terrestrial and marine records, younger rocks dominate compilations of lithological map area and volume – such as those developed by Ronov (1976) and colleagues over several decades – because there has been less time to erode them. In other words, the mass of marine sedimentary rock is a function of age and may be expressed in terms of a *survival factor*, which is the average mass of rock that has survived to the present per million years of a given time interval, and which is usually calculated at the period or epoch level (Garrels and Mackenzie, 1971; Figure 9.2). Most workers, for the sake of simplicity, have emphasized sedimentary mass over areal extent in their lithological age models. Over the years, the surviving mass of different lithologies in the Phanerozoic has been calculated and recalculated as new data have become available and the geological time scale has been revised.

There are two basic ways to view the mass/age distribution of rocks (Garrels and Mackenzie, 1971). The simpler approach is the *constant mass model*, in which a mass of sediment equal to the existing mass was formed early in the Earth's history. This model assumes early and complete degassing of the Earth, including all water, CO_2, and acidic gases (e.g., HCl) that reacted with primary igneous rocks to form sedimentary ones (e.g., Rubey, 1951). This model assumes that after degassing, there was no increase in the total mass of sediments because no new acidic gases were released to create them, although there was later *recycling (cannibalization)* by erosion and redeposition and destruction by metamorphism at a constant rate along with recycling of CO_2 and HCl (see also Ronov, 1976). According to the constant mass model, the probability of a given mass of sediment being destroyed is proportional to the ratio of the given mass to the total mass.

At the other extreme is the *linear accumulation model* (Garrels and Mackenzie, 1971). According to this view, water, CO_2, and acidic gases have been continuously degassed from the Earth's interior so that sediment mass has continually increased

9.3 Stratigraphic completeness

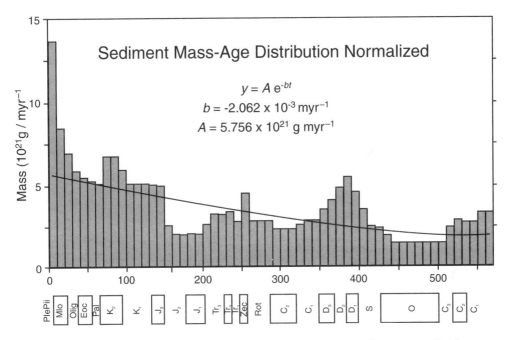

Figure 9.2. Mass/age distribution of existing Phanerozoic sediments compiled from various sources. Resolution of stratigraphic subdivisions varies. (See equation 9.1.) (Redrawn from Wold and Hay, 1993.)

with time; this model is more complex than the constant mass model. Deposition, destruction, and accumulation have all presumably occurred at constant rates, so that deposition rate minus destruction rate equals accumulation rate; like the constant mass model, the probability of a given mass of sediment being destroyed is proportional to the ratio of the given mass to total mass.

Despite the polarized assumptions of both models, the predicted mass/age distribution of either model is essentially independent of the model used if the total mass deposited is much larger than that preserved today (Garrels and Mackenzie, 1971; Figure 9.2). If, for example, the total mass of sediment deposited (and eroded and redeposited) during the Earth's history, is assumed to have been five times the mass of presently existing rocks, the two models become indistinguishable at approximately 500 Ma (Garrels and Mackenzie, 1971, figure 10.8). Moreover, in either approach, mass/age distributions of sedimentary rock can be modeled with a simple exponential relationship (cf. figures 10.5 and 10.7 of Garrels and Mackenzie, 1971).

Obviously, the constant mass model is the most easily applied. Wold and Hay (1990) assumed that the sedimentary system on a global basis is in steady state and modeled the mass of preserved sediment using a simple exponential decay curve:

$$y = Ae^{-bt} \tag{9.1}$$

where y is the remnant of original sediment at time t that would be observed today after t million years of cycling at a constant rate of erosion b and constant depositional rate A (see Wold and Hay, 1993, for later modifications). This approach reproduced some features of mass/age distributions that are relatively robust (Figure 9.2; cf. figures in Raup, 1972; Sheehan, 1977; Sepkoski et al., 1981; Signor, 1982, 1985). Minima occur in the Carboniferous (and perhaps the late Precambrian–Early Cambrian), whereas maxima occur in the middle Paleozoic (Devonian), the Paleozoic–Mesozoic boundary, and the late Mesozoic. After a decline into the Paleocene, preserved sediment mass increases through the Cenozoic to its Phanerozoic peak (cf. Garrels and MacKenzie, 1971) because at any given time a large proportion of the global sedimentary mass lies on ocean crust and so there will always be an unusually large mass of sediment with an age of less than 50 myr (Wold and Hay, 1993). The peak in preserved mass in the Pleistocene presumably occurs because of climate change that accelerated weathering, erosion, and redeposition (Wold and Hay, 1993; see also Chapter 8).

Raup (1972) pointed out the *potential* of the *pull of the Recent* in biasing the fossil record at taxonomic levels of *family and below* (see also Raup, 1976a,b). He concluded that the apparent rise in numbers of families, genera, and species after the Paleozoic (Figure 9.3) is the result of a greater likelihood of fossil preservation in younger rocks; moreover, sediment volume bias was said to be closely tied to biogeographic distribution, as it is more likely for sediment to be completely eroded from whole regions than for it to be lost via small-scale reductions in many areas. The areal extent of rocks was also thought to confound the problem because more widespread rocks are more likely to be sampled (*paleontological interest units* or PIUs of Sheehan, 1977; see also Raup, 1977). This sampling bias was said to be exacerbated by such factors as the influence of extant taxa on geological ranges and monographic effects (Raup, 1972).

Later workers concluded that the rise in global marine diversity through the Phanerozoic is real and not a reflection of preserved rock mass. Sepkoski et al. (1981; including D. M. Raup) used different estimates of marine biotic diversity (trace fossil diversity, species myr^{-1}, species richness, generic and familial diversity compiled from different sources) to assess biotic diversity trends during the Phanerozoic. They found that the correlation between these indices and geological time accounted for less than about 60% of the variance in diversity, and that when this correlation was removed, the residuals remained highly intercorrelated. To

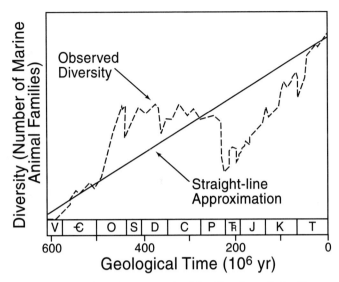

Figure 9.3. Number of marine animal families through the Phanerozoic. Note general increase through time despite extinctions. (Redrawn from Sepkoski, 1992.)

these workers, all five indices reflected the same underlying phenomenon: that a real evolutionary signal is strong enough to overcome biases in the fossil record. This signal was not, however, apparent at the ordinal level and above, perhaps because of genetic constraints that limited the origination of new bauplans (Valentine, 1995; Erwin et al., 1997); thus, higher taxa are insensitive indicators of species level trends through time (Signor, 1985).

Signor (1982) developed an iterative computer model that simulated species richness through time by using data on diversity (Raup, 1976a) and sampling intensity (rock mass/age/area relationships; PIUs of Sheehan, 1977), and by assuming a lognormal distribution of species abundances (based on ecological studies of modern communities). Signor (1982) concluded that global diversity of readily fossilizable marine invertebrates increased by 4–10 times (the higher figure is the more likely) at the end of the Mesozoic, which supported Sepkoski et al.'s (1981) conclusions, although Signor's estimates of diversity were higher than those of Sepkoski et al. (1981). Signor (1985) concluded that marine invertebrate species diversity in the Paleozoic was highest in the Devonian and Carboniferous, whereas Mesozoic levels were highest in the Jurassic and Cretaceous, and that variation in the amount of provinciality played a major role.

9.3.2 Geochemical uniformitarianism?

Coupled with the pull of the Recent are changes in lithology. Garrels and Mackenzie (1971) proposed, as a working hypothesis, the concept of *(geo)chemical uniformitarianism*: that the proportions of *primary* sedimentary lithologies has remained the same through time and that differences in lithology/age relationships reflect differential recycling of rocks. To be sure, the dominant lithologies of the Earth's crust have changed substantially through time (Garrels and Mackenzie, 1971; Veizer, 1988). This is especially pronounced in the Precambrian, when volcanogenics and graywackes gave way to arkoses and quartz sands as continental shelves became more widespread in the Proterozoic. On the other hand, the unique Banded Iron Formations (*jaspilites*) were most prevalent during the Proterozoic.

By contrast, changes in the proportions of lithologies during the Phanerozoic appear to be more subtle (see also Chapter 8). Considering the scale of the Proterozoic (about 2 billion years or roughly four "Phanerozoics" in duration), however, secular changes in lithology during the Phanerozoic might become just as pronounced as in the Precambrian given enough time. Hay *et al.* (1988), for example, suggested that because the composition of sediment on the ocean floor differs substantially from that of continental sediments, subduction (loss) of deep-sea sediment may be driving the global composition of sediment toward enrichment in silica, alumina, and potash, and toward depletion in calcium. Although carbonates are best-preserved during sea-level highstands (the areal distribution of shallow-water marine limestones has been used to determine first-order sea-level curves; cf. Law of Carbonate Deposition, Chapter 8), there appears to be a transfer underway of $CaCO_3$ from the cratons to the deep-sea (Wilkinson and Walker, 1989; Mackenzie and Morse, 1992; Chapter 8). At the same time, Mg/Ca ratios in seawater appear to have been rising (as suggested by changes in *macro*invertebrate calcareous hardpart mineralogy), perhaps in response to decreasing atmospheric CO_2 (and increasing seawater carbonate saturation) and Earth surface temperature (Mackenzie and Agegian, 1989). As this trend has presumably occurred, overall early cementation and preservation of fossil assemblages may have declined (Walker and Diehl, 1985), even as dolomitization decreased and hydrothermal alteration at ocean ridges became the dominant sink for Mg (Wilkinson and Algeo, 1989; Chapter 8; see also note added in proof at end of Chapter 8).

9.4 Pyritization

9.4.1 Modern analogs and ancient settings

Pyritization is one of the most important of all fossilization processes (Chapter 3) and is intimately tied to the biogeochemical cycles of carbon and sulfur. But the extent of pyritization and the processes involved seem to have shifted through the Phanerozoic (e.g., Feldman, 1989).

9.4 Pyritization

The fundamentals of modern biogenic pyrite and calcarenite formation seem well established (Chapter 3). In modern sediments, $CaCO_3$ (shell) content should be maximized in anoxic marine sediments, in which (1) bioturbation and dissolution (oxidation of organic matter and sulfides) is minimized because of low oxygen content in porewaters available for respiration; (2) porewater alkalinity is enhanced (via sulfate reduction); and (3) dissolved sulfides are rapidly depleted by precipitation of iron sulfides (decreasing the concentration of the weak acid H_2S). Conversely, calcareous shells deposited in oxygenated, well-bioturbated sediments should exhibit extensive dissolution (Canfield and Raiswell, 1991b).

This pattern does not, however, necessarily hold for ancient sediments (Canfield and Raiswell, 1991b). For example, calcitic fossils found in sediments of the Middle Devonian Hamilton Group (New York State; Brett et al., 1991) and Jurassic Posidonienschiefer and Solnhofen Limestone (southern Germany; Seilacher et al., 1976) exhibit extensive dissolution in black (poorly oxygenated) shales, whereas their counterparts in gray, bioturbated (presumably oxygenated) sediments display much better preservation. Previous interpretations of these units have been based on southern California borderland basins, which would appear to exhibit faunal and geochemical trends similar to the ancient settings. The deep portions of these basins are floored by black shales (suggesting anaerobic conditions above bottom) and grade into gray (dysaerobic) facies on the shallower slope. Nevertheless, in these modern basins, $CaCO_3$ preservation is best in black muds and poorest in the gray, which is exactly opposite to the preservational pattern of the ancient settings, and which raises serious questions about the use of modern deep basinal settings to infer oxygen levels of ancient taphofacies (Chapter 3).

Based on studies of intertidal mud flat and tidal creek (marsh) sediments, Reaves (1986) suggested that ancient patterns of preservation were better explained by the metabolizability of organic matter delivered to the sediments and seasonal climatic changes, rather than bottom-water oxygenation. In Connecticut (Indian Neck) sites, delivery of readily metabolizable organic matter to sediments in the summer resulted in high rates of microbial decay, establishment of anoxic–sulfidic conditions near or at the SWI, and the production of large amounts of iron sulfides. During the winter, however, when microbial activity was reduced, iron sulfides were oxidized and much $CaCO_3$ destroyed. At warmer Georgia (Sapelo Island) sites, on the other hand, destruction of reactive organic matter continued throughout the year and the resulting organic matter available for sulfate reduction became refractory; consequently, sulfidic conditions occurred farther below the SWI and $CaCO_3$ saturation was more likely to occur in surficial sediments throughout the year. Dissolution of $CaCO_3$ can, however, be extensive in shallow-water environments during the summer. McNichol *et al.* (1988) found that there was significant acid production in the spring and early summer in coastal sediments of Buzzards Bay (Massachusetts). Foraminiferal dissolution is also highest in the summer in Long Island Sound (New York; Green *et al.*, 1993) and Choya Bay (northern Gulf of California; Martin *et al.*, 1995, 1996).

There is of course the question of whether modern marsh settings are suitable analogs for modern deep basins or ancient epeiric seas. Canfield and Raiswell (1991*b*) objected to Reaves's (1986) conclusions and emphasized that the diagenetic history of sediment must be considered on a *case by case* basis. In particular, they state (p. 434) that "dissolution can only be less extensive in . . . more bioturbated sediments if less [$CaCO_3$] undersaturation results from irrigation than can result from acidity generated by sulfide-oxidation in . . . organic-rich sediments." The Jurassic Oxford Clay, for example, exhibits preservational patterns more in keeping with modern settings: the Upper Oxford Clay is a gray, bioturbated shale that is low in organic carbon and pyrite, and in which aragonite is only rarely preserved (although calcitic debris survives). The Upper Oxford Clay also contains internal pyrite molds from which aragonite must have later dissolved; in this case, bioturbation appears to have decreased alkalinity and produced weak acids (Chapter 3). In contrast, the Lower Oxford Clay is richer in carbon and pyrite (due to poorly oxygenated bottom-waters) and has a well-preserved aragonitic fauna (Canfield and Raiswell, 1991*b*), and was apparently characterized by a restricted burrowing infauna that produced a thin mixed layer; thus, alkalinity loss via bioturbation was suppressed.

In the Posidonienschiefer, however, aragonitic ammonites dissolved completely in the black shales (Chapter 6); abundant pyrite concretions also occur, and only rarely does pyrite not bear any relation to the original fossils (Hudson, 1982). Here, the production of acidic, sulfide-rich bottom-waters, which would have accumulated after the removal of iron, may have been involved (Canfield and Raiswell, 1991*b*; see also Chapter 6). Such conditions may account for the half-ammonites of Tanabe *et al.* (1984; see also Seilacher *et al.*, 1976), in which portions of ammonites buried in surficial sediment were apparently protected from dissolution by overlying bottom-waters (Canfield and Raiswell, 1991*b*), and may also explain the poor preservation of calcareous fossils in Devonian black shales of New York (Canfield and Raiswell, 1991*b*). On the other hand, more abundant pyrite in the Jet Rock of England (coeval with the Posidonienschiefer) suggests that sufficient iron was present to precipitate sulfides, decrease acidity, and preserve $CaCO_3$ (Canfield and Raiswell, 1991*b*).

9.4.2 Secular trends in pyritization

The metabolizability of organic matter has apparently changed the conditions under which pyrite formed through the Phanerozoic (Berner and Raiswell, 1983; Raiswell and Berner, 1986). Carbon and sulfur are related by the equation (Berner and Raiswell, 1983):

$$15CH_2O + 8CaSO_4 + 2Fe_2O_3 + 7MgSiO_3 \rightleftarrows$$

$$4FeS_2 + 8CaCO_3 + 7MgCO_3 + 7SiO_2 + 15H_2O \quad (9.2)$$

(The double arrows indicate that the reaction may proceed in either direction, and not equilibrium; Berner and Raiswell, 1983.) The relationship between carbon and sulfur reservoirs has probably helped to keep atmospheric pCO_2 and pO_2 *relatively* constant: according to equation 9.2, if more carbon is buried (CH_2O), much of the oxygen generated via photosynthesis is transferred to sulfur to produce $CaSO_4$, whereas less organic matter burial (lower atmospheric pO_2 and higher pCO_2) shifts O_2 to CO_2 (in the form of carbonate) and sulfur to FeS_2. Based on equation 9.2, a quantitative model was developed to explain shifts between carbon and sulfur reservoirs: from the end of the Precambrian to about the end of the Silurian, organic carbon values in sediment are relatively low and sulfur (pyrite) values are quite high, but thereafter carbon burial tended to increase as pyrite values remained relatively low (Berner and Raiswell, 1983; Figure 9.4).

From these two curves, Berner and Raiswell (1983) calculated C/S ratios for sediments that exhibit pronounced excursions through the Phanerozoic (Figure 9.5) and that apparently reflect changes in the locus of carbon deposition. During the late Precambrian to Silurian, organic matter was marine in origin. Marine organic matter degrades much more rapidly than organic matter from higher land plants; thus, marine organic matter is much more readily available for bacterial sulfate reduction and pyrite formation. Hence, lower C/S ratios prevailed in what were apparently widespread anoxic basins (Berner, 1989). Berry and Wilde (1978) suggested the oxygen minimum zone was shallowest overall during the first half of the Paleozoic, but that there was progressive ventilation and oxygenation of the oceans through the Phanerozoic caused by enhanced oceanic overturn during successive glaciations so that by the late Paleozoic the oceans were generally well oxygenated (see also Parrish, 1982, 1987; Chapter 8). Low oxygen content of early Paleozoic waters may therefore have been a relict of low-oxygen conditions of the Precambrian (Berry and Wilde, 1978; see also Tucker, 1992; Logan *et al.*, 1995).

342   Megabiases II: secular trends in preservation

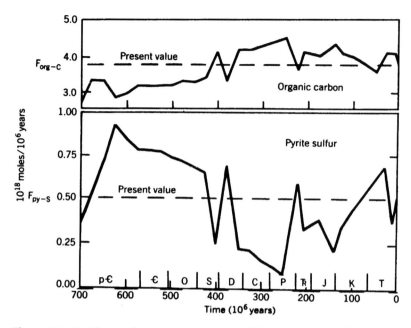

Figure 9.4. Burial rate of organic carbon and sulfide (pyrite) during the late Precambrian and Phanerozoic. (From Holser et al.,1988. Reprinted by permission of John Wiley & Sons, Inc.)

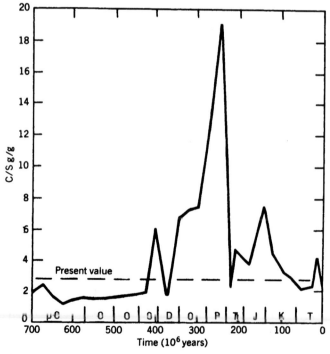

Figure 9.5. Carbon/sulfur (C/S) ratios through the Phanerozoic (From Holser et al. 1988. Reprinted by permission of John Wiley & Sons, Inc.)

9.4 Pyritization

These inferences are also supported by carbon and sulfur isotope curves, which are almost exact mirror images of each other (Figure 9.6), and which Holser *et al.* (1988, their table 4.8) implicated in cycles of seafloor spreading rates, MOR volume, CO_2 production, and sea-level (Chapter 8; see also Mackenzie and Pigott, 1981; Keith, 1982). Photosynthesis preferentially incorporates the light (^{12}C) isotope, leaving behind the heavier ^{13}C isotope in seawater, which is more likely to be incorporated into calcareous shells; thus $^{13}C/^{12}C$ ratios ($\delta^{13}C$) of shells formed during high levels of photosynthesis (whether on land or in the sea) will have heavier carbon isotope ratios. Lower $\delta^{13}C$ values prevail during times of lowered photosynthesis or when erosion and oxidation of previously existing organic matter occurs and which releases ^{12}C back to the atmosphere and oceans. Thus, low $\delta^{13}C$ values during the first half of the Paleozoic suggest relatively low marine productivity and perhaps also high atmospheric pCO_2 (Chapter 8). High $\delta^{34}S$ values during most of the Cambro-Devonian also suggest extensive sulfate reduction because of decreased deep ocean circulation rates and oxygenation (Chapter 8). Oxidation of marine organic carbon by sulfate-reducing bacteria in turn may have contributed to low $\delta^{13}C$ values (by releasing ^{12}C-rich CO_2) and to extensive anoxia during the first half of the Paleozoic (Holser *et al.*, 1988). High atmospheric CO_2 levels (Chapter 8) may have also enhanced preservation of organic carbon despite apparently low rates of marine photosynthesis. The overall parallelism between the $\delta^{34}S$ curve and the cerium curve (which has been suggested to indicate reducing conditions) also corroborates greater anoxia of bottom-waters during the early–middle Paleozoic (Wright *et al.*, 1987). The mirror-image relationship of carbon and sulfur isotopes breaks down in the Neogene, however, perhaps because sulfate reduction during this interval is favored by rapid sediment accumulation and burial of organic matter well below the SWI, where it can be attacked by sulfate-reducing bacteria (Berner, 1989; see also Chapter 3).

Carbon and sulfur isotope data for the Cambro-Devonian can also be interpreted to indicate *higher* marine productivity (van Andel, 1994, pp. 207–218). According to this scenario, warm saline deep-waters would have been unable to hold sufficient oxygen to cause unabated oxidation of organic carbon, and eventually the oxygen supply in seawater would have run out. When this happened, fewer nutrients would have been released from organic carbon back to the photic zone, and marine productivity and organic carbon formation would have slowed (see also below). With less oxygen demand (less organic carbon decay), the oxygen content of deep waters began to increase, and organic carbon decay and nutrient release would have eventually resumed, thereby stimulating productivity.

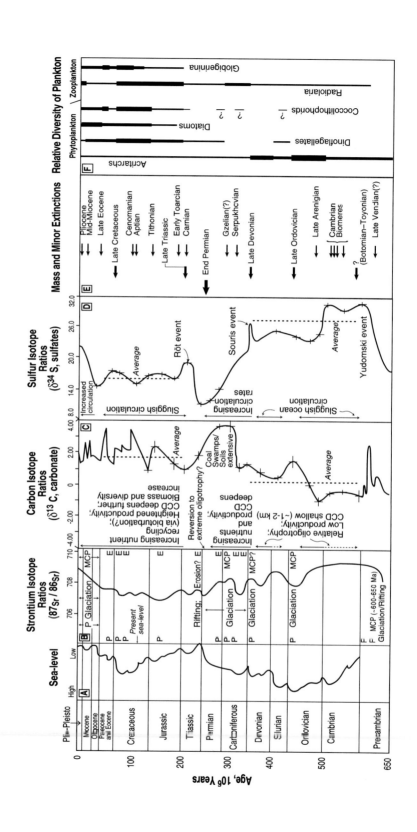

Figure 9.6. Sea-level, stable isotope, and lithological indices of nutrient fluxes and productivity, extinction episodes, and major changes in plankton assemblages through the Phanerozoic. (A) Sea-level (Hallam, 1992). (B) ^{87}Sr/^{86}Sr ratios (Holser et al., 1988; Late Precambrian portion based on Kaufman et al., 1993). Deviation from present ^{87}Sr/^{86}Sr ratios and calculated runoff rates may reflect tectonics and glaciation (Raymo, 1991) and deforestation and agriculture. P, phosphorite peak (Cook and Shergold, 1984); MCP episodes (eutrophication episodes) after Worsley et al. (1986); E, additional eutrophication discussed in text (not formally recognized as MCP episodes by Worsley et al., 1986). (C) δ^{13}C (primarily from Holser et al., 1988; late Precambrian–Early Cambrian from Magaritz et al., 1986; see also Kaufman et al., 1993; Cretaceous–Recent after Renard, 1986). Each datum point (center of cross-hair) represents single analysis aggregated at 25×10^6-yr intervals; cross-hairs indicate 1 standard error of the mean for each interval (Holser et al., 1988). Positive values indicate increased marine or terrestrial photosynthesis, whereas negative values indicate decreased productivity or oxidation of marine or terrestrial organic carbon reservoirs (release of ^{12}C-rich CO_2). General shift toward δ^{13}C positive values through the Phanerozoic suggests increased marine primary productivity and increased marine C:P burial ratios. Averages (dotted lines) fitted visually. (D) δ^{34}S (mainly after Holser et al., 1988; "events" after Claypool et al., 1980). Each datum point (center of cross-hair) represents single analysis aggregated at 25×10^6-yr intervals; cross-hairs indicate 1 standard error of the mean for each interval (Holser et al., 1988). High average values suggest extensive SO_4^{2-}-reduction in widespread anoxic basins (Cambro-Devonian, Mesozoic). Gradual shift toward lower values in late Paleozoic suggests increasing deep-water overturn rates and oxygenation. Pronounced excursions to high values (events) suggest mixing into the photic zone of nutrient-rich anoxic waters previously stored in isolated basins (Claypool et al., 1980). See text for discussion of Neogene δ^{34}S excursions. Averages fitted visually. (E) Mass and minor extinction episodes of the Phanerozoic. Thickness of arrows indicates intensity (modified from Sepkoski, 1992; late Early Cambrian extinction based on Signor, 1992). (F) Geological range and relative diversity of selected plankton groups (Martin, 1995a, 1996a). Shift from phytoplankton characteristic of presumed "superoligotrophic" conditions during the Cambro-Devonian (acritarchs) to taxa characteristic of "mesotrophic" conditions in the Mesozoic (dinoflagellates) to diatoms which prefer nutrient-rich or "eutrophic" waters in the Neogene suggests secular rise in nutrient availability and productivity through the Phanerozoic (geological ranges and diversity based on references in Martin, 1995a). (Modified from Martin, 1996a.)

Based on models of poleward heat transport, Lyle (1997) has suggested that vertical circulation – and presumably photosynthesis – in warm, early Cenozoic oceans (possibly analogous to warm Mesozoic and Paleozoic oceans) was actually *enhanced* by decreased density differences between water masses (see also Brass *et al.*, 1982). During times of *halothermal* (salinity-driven) circulation, the production of warm, highly saline, oxygen-poor water masses apparently occurred via evaporation in widespread, low-latitude shallow seaways (Brass *et al.*, 1982; Railsback *et al.*, 1990). This type of circulation was presumably much like the modern Mediterranean, where surface waters evaporate in the eastern basin, sink, and flow out through the Straits of Gibraltar, but on a much greater spatial scale. But this *lagoonal* type of circulation also depletes the photic zone of nutrients (as opposed to *estuarine* circulation, which is driven by outflow of lower-salinity surface waters and which brings nutrients to the surface). Bralower and Thierstein (1984) earlier calculated (based on modern organic carbon accumulation rates corrected for preservation) that productivity was actually lower during the mid-Cretaceous and that ocean circulation was sluggish, and Herbert and Sarmiento (1991) concluded that nutrient levels must have decreased in the photic zone in order to prevent anoxia during times of predominantly halothermal ocean circulation. Bralower and Thierstein (1984) suggested that the low oxygen demand resulting from low productivity of mid-Cretaceous waters could be balanced by a deep-water supply rate about equal to that of the present Mediterranean outflow (see also Hay, 1995, for thorough review of the unresolved relation of ocean circulation to marine productivity and anoxia).

According to the scenario invoking relatively easy overturn of salinity-stratified waters, carbon-rich black shales should be laminated, with very dark (carbon-rich) layers alternating with lighter layers; this is often the case for Mesozoic black shales (van Andel, 1994, p. 209). Early Paleozoic (especially Cambro-Silurian) black shales appear, however, to differ from their Mesozoic counterparts. This is not to deny that heightened productivity may have been important in the generation of some early Paleozoic black shales (e.g., Moore *et al.*, 1993), but early Paleozoic black shales are thicker and were deposited over much larger areas than Mesozoic black shales; this suggests that anoxia was much better developed during the first half of the Paleozoic than during later times (Thickpenny and Leggett, 1987).

Interestingly, Ingall *et al.* (1993) and Van Cappellen and Ingall (1994) suggested that nutrient regeneration is accelerated in anoxic waters (see also Passier *et al.*, 1999), but this could have resulted in precipitation and enhanced burial of phosphates rather than enhanced recycling of nutrients to the photic zone

9.4 Pyritization

(Beier and Hayes, 1989; Ruttenberg and Berner, 1993; Filippelli, 1997). Beier and Hayes (1989) concluded, for example, on the basis of the presence of trace metal-rich phosphate nodules in the euxinic New Albany Shale (Devonian–Mississippian; Indiana) that phosphate and trace metals accumulated below a chemocline because of limited vertical circulation in euxinic basins. Significant regeneration of nutrients and high marine productivity during much of the Cambro-Devonian is also not supported by the fossil record (see below).

Lower rates of bioturbation and porewater oxygenation (Chapter 3) during the early–middle Paleozoic may have enhanced the formation of black shales and pyrite (Berner and Raiswell, 1983; Raiswell and Berner, 1986; Berner, 1989; Figures 9.4, 9.5). Sepkoski et al. (1991) found that very thin (<1 cm) storm beds and flat-pebble conglomerates were much more prevalent during the Cambrian–Early Ordovician, but that by the later Paleozoic, thin tempestites and flat-pebble conglomerates were much less abundant, presumably reflecting deeper oxygenation and bioturbation of sediment (see also Brandt, 1986).

The rise of vascular plants by about the Late Silurian, and their great expansion beginning in the Devonian, resulted in increasing deposition of refractory (lignin-rich) organic matter on land (Robinson, 1990). By the Permo-Carboniferous, freshwater to marginal marine coal swamps were widespread, and higher C/S ratios are prevalent (Berner and Raiswell, 1983; Raiswell and Berner, 1986; Berner, 1989; Robinson, 1990). The productivity of terrestrial biotas was no doubt augmented by tectonic (continent–ocean) configurations conducive to coal swamp formation in the Permo-Carboniferous and the high C/P ratio of vascular plant material, which enabled greater organic carbon burial per unit phosphorus than is the case for marine organic matter (Berner, 1989).

As terrestrial floras expanded, vertebrates and invertebrates followed. The non-marine ichnological record begins as early as the Late Ordovician, with significant invasion occurring by the Siluro-Devonian (Buatois et al., 1998). Diversification of terrestrial faunas presumably affected the preservation of plant and animal remains as the capabilities of animals to catch, dismember, and digest biogenic remains evolved (e.g., changes in masticatory apparatuses, cellulose digestion in arthropods); plants responded by evolving morphological and chemical defenses to herbivores (Behrensmeyer and Hook, 1992; Labandeira et al., 1997; Farrell, 1998). Burrowing and the evolution of herding behaviors (which may have increased the probability of mass mortality events) also evolved in vertebrates (Behrensmeyer and Hook, 1992). But the exact effect of plant–animal interactions on the preservation of organic carbon (and hardparts) on land remains a fertile area of investigation (Behrensmeyer and Hook, 1992). Certainly, as biotas expanded on land, any scarcity of food or nutrient supplies presumably resulted in increased rates of recycling and destruction of fossil assemblages, just as in the sea.

9.5 Seafood through time: energy and evolution[1]

9.5.1 The Cambro-Devonian

The interaction of land, sea, and atmosphere through time may have driven one of the greatest secular megabiases of all: an increase in nutrient availability and productivity of the marine biosphere. This trend had a profound influence on the relation between the evolution and morphology and durability of hardparts and their preservation, and the relation of fossil preservation to the development of the biogeochemical cycles of carbon, phosphorus, and silica in the marine realm.

Some of the most oligotrophic waters today are in oceanic gyres (Worsley et al., 1986; Rivkin and Anderson, 1997), to which some taxa are particularly well adapted (e.g., algal symbiosis in planktonic foraminifera and radiolaria; N_2 fixation in the cyanobacterium *Trichodesmium*; Capone et al., 1997). The key points in the following argument are, however, not just that nutrient availability has increased through time, but that most nutrients have been sequestered in biomass as they became available; according to this "nutrient hypothesis," *dissolved* nutrient concentrations in open ocean surface waters have always been low, but the total quantity (*inventory*) of nutrients (and therefore biomass) in the surface oceans was far lower in the geological past ("superoligotrophic conditions").

Most nutrients may have been sequestered below the photic zone during the Cambro-Devonian because glaciers were largely absent (Fischer and Arthur, 1977; Sheldon, 1980; Berner and Raiswell, 1983, pp. 860–861; Holser et al., 1988). During the Cambro-Devonian, ice-caps occurred only in the Late Ordovician and perhaps Late Devonian (Figure 9.6; Frakes et al., 1992); thus, ocean circulation during much of this time was presumably halothermal in nature (section 9.4.2). Moreover, in modern oceans in which vigorous thermohaline circulation is driven by production of deep-water masses near both poles, dissolved PO_4^{3-} levels in the subphotic zone are several times to an order of magnitude or more higher than in the photic layer (Sverdrup et al., 1942; Hallock et al., 1991). During much of this interval, too, reduced ocean circulation and widespread anoxia may have resulted in extensive denitrification in the photic zone (by reducing NO_3^-, which is otherwise used as a nutrient; Rau et al., 1987) while

---

[1] Sections 9.5 and 9.6 are based on Martin (1995a, 1996a, 1998a); see also Bambach (1993).

high sea-levels trapped nutrients nearshore (Holser *et al.*, 1988; see also Holmden *et al.*, 1998). Also, continental weathering rates were probably too low during much of the Cambro-Devonian to deliver nutrients to the photic zone, as soils were poorly developed (Knoll and James, 1987), although root traces suggest enhanced weathering by the Early–Middle Devonian (Elick *et al.*, 1998; see Keller and Wood, 1993, for evidence of microbial weathering prior to the advent of vascular plants). Low rates of marine photosynthesis and organic carbon burial during much of the Cambro-Devonian are also suggested by relatively light (negative) $\delta^{13}C$ values (Figure 9.6). As forests were not well established much before the Devonian (Knoll and James, 1987), it is unlikely that terrestrial photosynthesis significantly influenced the $\delta^{13}C$ curve much before this time.

The record of fossil plankton supports the above interpretations, and appears consistent with changes in carbon isotope composition of crude oils through the Phanerozoic (Andrusevich *et al.*, 1998). Low overall phytoplankton densities during the Cambro-Devonian are suggested (counterintuitively) by the record of acritarch diversity (Martin, 1995a, 1996a). Acritarchs were the dominant phytoplankton of Cambro-Devonian seas (Tappan, 1980), and although of uncertain affinities, are normally considered to be cysts of marine eukaryotic unicellular algae that were resistant to inimical conditions (Tappan, 1980), presumably including low nutrient availability. Many workers have considered high acritarch diversity in ancient rocks to indicate *persistent* nutrient-rich conditions and high productivity (e.g., Tappan, 1980; Bambach, 1993, pp. 390–391), but modern plankton diversity tends to be lowest in nutrient-rich nearshore regimes and to increase basinward, where nutrient levels typically decline: because of the downward movement of water, ocean gyres are typically nutrient-poor (Hay, 1995). Like modern plankton, acritarch diversity increased in a nearshore shelf to basin (basically eutrophic to oligotrophic) direction (Tappan, 1980; see also McGowan, 1971; McGowan and Walker, 1993). Thus, *high* overall cyst diversity in the Cambro-Devonian (Figure 9.6) may reflect fairly persistent oligotrophic conditions in the photic zone rather than chronically nutrient-rich (eutrophic) ones. As nutrients became increasingly available in more open ocean settings via erosion from land, etc., larger – and presumably more diverse – populations of plankton were sustained in more open ocean waters.

Fossil zooplankton also suggest adaptation to overall low phytoplankton densities during much of the Cambro-Devonian. Radiolarians during this time apparently lived either in highly productive shallow waters, in oligotrophic oceanic gyres with symbiotic algae, or in deeper subphotic layers of the open ocean as detritivores and bacterivores (Casey, 1993). Similarly, graptolites appear to have lived in low-oxygen, nutrient-rich waters just below the photic zone, and may have migrated upward to feed on occasional phytoplankton blooms caused by intrusions of deeper nutrient-rich waters into surface waters (Berry *et al.*, 1987; see also Gutiérrez-Marco and Lenz, 1998; cf. Underwood, 1993). Recently, Finney and Berry (1997) concluded that the highest graptolite diversity during the Ordovician occurred in upwelling regimes at continental margins (see also Moore *et al.*, 1993), and they misconstrued this to contradict Martin (1995a, 1996a). On the contrary, all three authors seem to be in agreement: not only did acritarch diversity tend to increase offshore as, presumably, nutrient availability declined, acritarchs may have formed large numbers of cysts *following* opportunistic blooms in offshore upwelling regions and the subsequent exhaustion of dissolved nutrients via incorporation into living and dead biomass. Moreover, the relationship between offshore productivity and diversity appears to be a long-standing one (Rosenzweig and Abramsky, 1993). High cyst diversity and abundance in offshore settings may also be related to decreased sediment accumulation rates and increased durations of time-averaging (cf. Chapters 5, 7).

The apparently low *overall* productivity of the Cambro-Devonian was punctuated by relatively short intervals of heightened nutrient availability that may have augmented mass extinctions; "catastrophic" nutrient input need not have occurred everywhere simultaneously, however (Martin, 1998b; Miller, 1998). Positive $\delta^{13}C$ shifts in the late Precambrian through Early Cambrian (Figure 9.6), apparently in conjunction with glaciation (Kaufman *et al.*, 1993), suggest increased nutrient availability in the photic zone during this time (see also Bailey and Peters, 1998; Hoffman *et al.*, 1998).[2] A pronounced excursion of $\delta^{34}S$ to high values (Yudomski event; Figure 9.6) also suggests mixing into the photic zone of anoxic (and perhaps nutrient-rich) waters that may have been stored in isolated basins or deep in the oceans (e.g., Claypool *et al.*, 1980); increased nutrient input to the oceans may have also resulted from the Pan-African orogeny (Kaufman *et al.*, 1993). The subsequent broad Late Cambrian $\delta^{13}C$ shift occurred in conjunction with a major transgression, which may have also stimulated marine photosynthesis by releasing nutrients trapped in shelfal sediments through erosion and oxidation of sedimentary organic carbon (Broecker, 1982; Compton *et al.*, 1993). Positive shifts of the $\delta^{13}C$ curve in the Late Ordovician and Late Devonian appear to be associated with positive excursions in the $\delta^{18}O$ curve, glaciation, and

---

[2] Based on the calcification mechanism proposed in Chapter 8, marine carbon to phosphorus (MCP) episodes after the early Paleozoic must have been insufficient to affect foraminiferal calcification.

sea-level fall (Figure 9.6). Another positive δ^{34}S shift (Souris event) occurred in the Late Devonian (Claypool et al., 1980); the relatively small Late Devonian δ^{13}C shift (as opposed to the strong Late Ordovician rise) may have resulted from erosion of terrestrial carbon reservoirs during sea-level fall (Algeo et al., 1993) or poorly developed glaciers (Frakes et al., 1992) and slower deep-ocean turnover rates than in the Late Ordovician. On the other hand, Joachimski and Buggisch (1993) found positive excursions of δ^{13}C of up to approximately 2‰ just above well-documented carbon-rich shales (Kellwasser horizons) that exhibited lower (negative) δ^{13}C values (see also Caplan et al., 1996), and Racki (1998) reports localized blooms of siliceous biota in the vicinity of the Frasnian–Famennian boundary. Conceivably some of these assertions could be tested using such geochemical indicators as cadmium or barium (cf. geochemical uniformitarianism above); these elements have been used as proxies for nutrient levels primarily in Quaternary sediments but have recently been applied to more ancient sediments (Thompson and Schmitz, 1997), although they may be affected by diagenesis in some cases (Schroeder et al., 1997; see also Chapters 3, 10).

Other geochemical indicators support the hypothesis of episodic eutrophication during the Cambro-Devonian. Phosphorus is a limiting nutrient on geological time scales and is intimately involved in energy storage and transfer and other cellular pathways (Fox, 1988), and so its availability probably helped to constrain marine photosynthesis. Worsley et al. (1986; see also Sheldon, 1980) recognized a series of step-like increases in *marine carbon to phosphorus (MCP) burial ratios* that presumably reflect increased phosphorus availability and permanently enhanced marine productivity (organic carbon burial rates) during the Phanerozoic (Figure 9.6). According to Martin (1995a, 1996a), phosphogenesis during MCP episodes resulted from intensified recycling of phosphorus back to the photic zone as a result of release of waters from isolated basins; glaciation, upwelling and better oxygenation of shallow waters; enhanced rates of bioturbation (phosphorus scavenging); and the transition from high to low sediment accumulation rates during sea-level fall (Hay and Wold, 1990; see also Sheldon, 1980).

Prior to the Paleozoic, plankton may have been composed of bacteria and small eukaryotes (Signor and Vermeij, 1994; Logan et al., 1995; see also Butterfield, 1997). Diversification of epifaunal (above-bottom) suspension-feeders first took place in the Cambrian (Ausich and Bottjer, 1991) as eukaryotic plankton presumably first became abundant in the waters above bottom (Bambach, 1993; Signor and Vermeij, 1994; Martin, 1995a, 1996a; Butterfield, 1997). According to Martin's (1995a, 1996a) scenario, intervals of relatively *low* acritarch diversity during the Cambrian and Silurian reflect the elevated nutrient levels that followed the late Precambrian–Early Cambrian and Late Ordovician eutrophication (MCP) episodes, respectively (i.e., abundant nutrients precluded cyst formation; see also Vidal and Knoll, 1982; Knoll and Swett, 1987). Through the Cambrian, following the late Precambrian–Early Cambrian MCP episode, levels of *dissolved* nutrients presumably declined as they were progressively incorporated into plankton (and other) biomass along food chains; consequently, cyst diversity rose, culminating in a peak in the Ordovician (i.e., return to low levels of *dissolved* nutrients; Figure 9.6). (Similar, but smaller-scale, fluctuations in acritarch diversity have been documented by Vidal and Moczydłowska-Vidal [1997] for the late Proterozoic and Cambrian that appear to correspond to a major extinction documented by Signor [1992]). A similar decrease and increase in acritarch diversity followed in the Silurian and Devonian, respectively (Figure 9.6).

Corresponding changes took place in the macrobenthos. The Cambrian fauna was dominated by trilobites, which were vagile deposit feeders that ingested bottom sediment for organic matter and associated microbiota. The trilobite-rich fauna began to give way to a suspension-feeding brachiopod-dominated fauna in the Ordovician. Based on experiments with modern brachiopods, which today characterize oligotrophic habitats, Paleozoic brachiopods had very low energy (food) requirements and were able to survive intervals of starvation of 2 years or more (Rhodes and Thayer, 1991). Other relict Paleozoic taxa (e.g., living fossils such as the coelacanth; *Nautilus*) appear to have low metabolic rates and survive today in oligotrophic refugia such as caves or the deep sea (Vermeij, 1987, 1994; Thayer, 1992; Boutilier *et al.*, 1996); food supplies are presumably too low in such refugia to sustain the high metabolic levels of competitors and predators that arose much later (Thayer, 1992; Bambach, 1993). Diversification of the benthos occurred in part through the process of tiering, in which organisms feed at different depths above and below bottom and so more finely subdivide the available habitat, food resources, and niche space (Figure 9.7; Ausich and Bottjer, 1991; Bambach, 1993). The depth and intensity of bioturbation increased during the Early Cambrian, again during the Middle–Late Ordovician (after sea-level transgression and presumed nutrient release), and between the Ordovician and Early Devonian (following the Late Ordovician MCP episode; Larson and Rhoads, 1983; Bottjer and Droser, 1994), which suggests greater organic carbon (food) concentrations in sediment in response to an increased rain of dead organic matter to the bottom (see Kerr, 1998, for review of evidence regarding the presumed age of approximately 1 billion-year-old trace fossils from India). Although evidence of quite deep bioturbation during these times has been found,

9.5 Seafood through time: energy and evolution

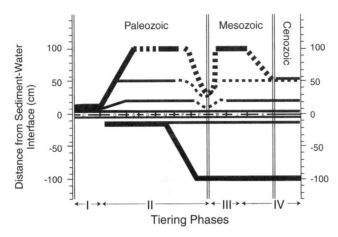

Figure 9.7. History of Phanerozoic epifaunal and infaunal tiering. Heaviest lines are based on data that indicate maximum epifaunal and infaunal tiering, whereas thinner solid lines represent tier subdivisions. Dotted lines are inferred. (Redrawn from Ausich and Bottjer, 1991.)

bioturbation was confined to shallow-water environments where nutrient levels and food were undoubtedly high (Bambach, 1993).

Signor and Vermeij (1994) suggested that the early Paleozoic plankton served as a refuge from predation and bioturbation. Nutrient and food availability may have also affected larval development. Indirect development via primary larvae is considered primitive among metazoans, whereas direct development is viewed as having evolved more than once when conditions were unfavorable for the release of large numbers of larvae into the plankton (Bengston and Zhao, 1997). Fossilized embryos from the Middle Cambrian developed directly and suggest that conditions were not always favorable for indirect development. Although these embryonic Lagerstätten may represent a biased sample (Chapter 6; Bengston and Zhao, 1997), these Lagerstätten, as well as the small size of molluscs in the early Cambrian, suggests the widespread presence of non-planktonic larvae (Signor and Vermeij, 1994).

9.5.2 The Permo-Carboniferous

Soils were well developed by the Late Devonian, which suggests increased weathering rates and nutrient flux from the continents by this time (Knoll and James, 1987; Berner, 1989, 1992). This trend continued into the Permo-Carboniferous with the spread of extensive coal swamps (Figure 9.6): illite (indicative of limited leaching) is very prominent in pre-Pennsylvanian mudrocks but tends to decline, while kaolinite (indicative of extensive weathering) tends to increase, in abundance toward the present (Weaver, 1967). Also, the production of lignin – a compound which confers structural rigidity in land plants – may have sequestered sufficient carbon in terrestrial deposits to cause a substantial increase in atmospheric pO_2 (Robinson, 1990; Graham *et al.*, 1995; see also Collinson and Scott, 1987). Conceivably, increased rates of terrestrial photosynthesis may have drawn down atmospheric CO_2, thereby contributing to southern hemisphere glaciation, sea-level fall, and enhanced ocean turnover rates during this time, and to the gradual shift toward lower $\delta^{34}S$ values during the Permo-Carboniferous (Figure 9.6).

Heightened nutrient levels (*submesotrophic conditions*) in the oceans during the Permo-Carboniferous are suggested by the broad rise in $^{87}Sr/^{86}Sr$ ratios and a fourth MCP episode in the Late Pennsylvanian (Figure 9.6). Tardy *et al.* (1989) used the $^{87}Sr/^{86}Sr$ curve as a proxy for continental runoff during the Phanerozoic; they found that the $^{87}Sr/^{86}Sr$ curve is in general agreement with calculated runoff rates, which are a function of latitudinal position of the continents and relative area of continents and oceans (Tardy *et al.*, 1989). By the Carboniferous, then, nutrient levels had apparently reached a minimum threshold value that supported relatively large, permanent populations of plankton (Pitrat, 1970; see also Knoll, 1989). After the Devonian, acritarchs are only a relatively minor component of the microfossil record (Figure 9.6), which also suggests permanently elevated nutrient levels. Coupled with a deepening CCD (Chapter 8), $CaCO_3$ pelagic rain rates to the bottom should have also increased somewhat, so that incipient calcareous oozes may have begun to accumulate at mid-ocean ridge crests by the Late Pennsylvanian at about 300 Ma (Boss and Wilkinson, 1991; see also Hay *et al.*, 1988; cf. Malinky and Heckel, 1998). In fact, modern coccolithophorids are most prominent under *mesotrophic* (intermediate) nutrient conditions.

Other biotic evidence for increased nutrient levels comes from the benthos. This includes the decline of the Fusulinina (benthic Foraminiferida) beginning about the mid-Carboniferous (Figure 8.2; Tappan and Loeblich, 1988). By analogy to modern larger reef-dwelling foraminifera, the Fusulinina were adapted to oligotrophic waters (Hallock, 1982; Tappan and Loeblich, 1988). The Fusulinina were replaced by smaller, rapidly growing foraminifera during the Middle and Late Permian (Tappan and Loeblich, 1988) and the Fusulinina themselves tended to *decrease* in size (hinting at increased rates of reproduction like those found in smaller foraminifera) and lose the shell layer known as the keriotheca after the Guadalupian (Stanley and Yang, 1994). The keriotheca was presumably an adaptation to oligotrophic conditions, as it appears to have housed algal symbionts that provided food under oligotrophic conditions. The prevalence of

9.5 Seafood through time: energy and evolution

calcareous algae in Permo-Carboniferous mounds (James, 1983; James *et al*, 1988) also suggests elevated nutrient levels after the Devonian (Hallock, 1987; Chapter 8). Increased food supplies beginning in the Carboniferous are also suggested by higher tiering above bottom (up to about 1 m) in epifaunal suspension-feeding communities and increasing depth of bioturbation (Figure 9.7; Thayer, 1983; Ausich and Bottjer, 1991; Sepkoski *et al.*, 1991), which is energetically costly (Rhodes and Thayer, 1991; Bambach, 1993). Thus, there must have been relatively reliable food supplies (plankton) high in the water column and deep in sediment (detritus), respectively (Pitrat, 1970; Valentine, 1973, p. 453; Bambach, 1993; Martin, 1995*a*, 1996*a*).

As food supplies increased, food chains presumably lengthened and metabolic rates increased (Rhodes and Thayer, 1991; Thayer, 1992; Bambach, 1993; Rhodes and Thompson, 1993; see also Vermeij, 1987, 1990; Kelley and Hansen, 1996*a,b*). In his study of the relation between body weight (nitrogen) and metabolism of modern taxa ranging from molluscs to mammals, Zeuthen (1947) showed that the metabolic rate of a taxon tends to rise with increasing evolutionary complexity. Indeed, durophagous (shell-crushing) predation increased markedly beginning in the Late Devonian (Signor and Brett, 1984; see also Ebbestad and Peel, 1997; Kowalewski *et al.*, 1998), just as acritarchs largely disappeared from the fossil record (the fossil record of predatory boring may, however, be affected by moldic and castic preservation; Harper *et al.*, 1998).

Unlike the mass extinctions in the earlier Paleozoic (and Mesozoic; see below), however, a reversion to oligotrophy may have occurred near the end of the Permian. Nutrient starvation would have essentially "pulled the rug out from under" a marine biosphere at the end of the Paleozoic that had become accustomed to greater nutrient and food availability (*submesotrophic conditions*) and higher metabolic rates. A lack of ice-caps (Frakes *et al.*, 1992), and perhaps also salinity stratification of the oceans (Holser *et al.*, 1991), may have lowered deep-water overturn rates, so that dissolved nutrients were sequestered below the photic zone as anoxia was developing.

Both δ^{13}C and ^{87}Sr/^{86}Sr ratios also declined in the Late Permian (Figure 9.6). Based on the carbon isotope analyses of an Austrian (Carnic Alps) section, Magaritz *et al.* (1992) proposed that there were two processes, each with different temporal scales, acting at the end of the Permian. The gradual negative δ^{13}C shift in the Late Permian was interpreted by these authors as resulting from erosion and oxidation of shelfal organic carbon reservoirs (Holser *et al.*, 1991; Holser and Magaritz, 1992); this process acted on a time scale of 10^6 years or more. But they contend that the sharp negative shift in the carbon isotope curve at the Permo-Triassic boundary resulted from a decline in surface productivity, which acted over a relatively short time interval (cf. Kump, 1991; see also Martin, 1995a, 1996a; Bowring et al., 1998; Kerr, 1998; Chapter 10). Interestingly, in the Austrian section, the strong negative isotope shift is associated with pyrite (anoxia) and iridium, which tends to be concentrated under anoxic conditions (Holser et al., 1991; Holser and Magaritz, 1992).

Sea-level fall (Figure 9.6) and erosion of continents should, seemingly, have added nutrients to the oceans, increased $^{87}Sr/^{86}Sr$ ratios, and stimulated marine photosynthesis. There are several possible explanations for this dilemma. First, the Permian was quite arid (Parrish, 1993): based on the low $^{87}Sr/^{86}Sr$ ratios for the Permian, Tardy et al. (1989) concluded that runoff was minimal; moreover, interior drainage may have been extensive (Wyatt, 1984; François and Walker, 1992). Second, any strontium delivered to the oceans may have been concentrated in the light isotope because of erosion of extensive basaltic lavas (Siberian Traps of about 2–3 million km^3; Renne et al., 1995) that had been extruded onto the Earth's surface at or very near the Permo-Triassic boundary (Holser and Magaritz, 1987; Holser and Magaritz, 1992, p. 3302; Bowring et al., 1997, 1998; Kerr, 1998). Third, during the Permian, gymnosperm forests were replacing the lycopod coal swamps of the Carboniferous. Evergreen leaf litter releases nutrients slowly (Tappan, 1986; Knoll and James, 1987) and may have retarded weathering of continental rocks. Charophytes ($CaCO_3$-secreting green plants of uncertain affinity), which today live in low-nutrient freshwater (Tappan, 1986), were common in the Permian (Tappan, 1986); also, the Fusulinina re-diversified somewhat at this time (Tappan and Loeblich, 1988).

Other changes in the marine benthos are consistent with lowered productivity. Extinction rates were greater for brachiopods (despite low food requirements) than for bivalves during the end-Permian extinctions, apparently because some bivalves also relied upon deposit-feeding (Rhodes and Thayer, 1991). If suspended food supplies decreased during the Permian, deposit-feeding would have become increasingly advantageous (Rhodes and Thayer, 1991; see also Ausich and Bottjer, 1991; Sheehan and Hansen, 1986; cf. Levinton, 1996). During the Late Permian there may have also been a decline in the height of epifaunal suspension-feeding communities (Ausich and Bottjer, 1991), which may have been augmented by biological *bulldozing* of suspension-feeders by deposit-feeders (Thayer, 1983).

Małkowski et al. (1989) also ascribe the end-Permian crisis to a gradual decrease in nutrient availability, but according to them the nutrient decline was caused by *increased* ocean circulation. They suggest that there was a geologically

9.5 Seafood through time: energy and evolution

rapid switch from one paleoceanographic mode to another. Prior to the end-Permian extinctions, the oceans were in an *overfed mode*, in which large amounts of carbon and nutrients accumulated in stagnant waters; nevertheless, extensive organic decay (via bacterial decomposition) deep in the oceans released nutrients back to the photic zone, thereby compensating for sluggish circulation, as evidenced, according to these authors, by a strong positive excursion in the carbon isotope curve prior to the end of the Permian (cf. Ingall *et al.*, 1993, and Van Cappellen and Ingall, 1994). As oxygen accumulated in the atmosphere (a result of carbon storage in the oceans), the Earth began to shift toward an icehouse mode, which induced more vigorous oceanic circulaton. As the Earth entered this mode, the oceans eventually switched to a *hungry mode*, and the carbon stored in them was brought to the surface and oxidized, releasing vast amounts of CO_2 into the atmosphere. According to them, recycling of nutrients to the photic zone *declined* because nutrients such as phosphorus form insoluble compounds in the presence of oxygen, and the carbon isotope curve gradually shifted toward negative values because surface productivity decreased.

The interpretations of Małkowski *et al.* (1989) are, however, largely dependent on a single section in Greenland, which does not exhibit the same characteristics as those described by Holser *et al.* (1991) for the Austrian section. The Austrian section exhibits a long, gradual negative shift in $\delta^{13}C$ before the final pronounced negative carbon shift occurs (Figure 9.6; Holser *et al.*, 1991), which suggests that the Greenland section may be stratigraphically incomplete or diagenetically altered (Hallam, 1992; Mii *et al.*, 1997).

Knoll *et al.* (1996) nevertheless followed up on the work of Małkowski *et al.* (1989) on the Greenland section. They (1996) suggested that end-Permian extinctions were caused by the release of massive quantities of CO_2 stored in the deep ocean during oceanic turnover, which may account for the sharp negative shift in $\delta^{13}C$ at this time (Kempe and Kaźmierczak, 1994). Elevated CO_2 levels would have adversely affected the metabolism of much of the Earth's biota in the Late Permian, which may account for the selectivity of extinction among benthic invertebrate taxa. Release of large amounts of CO_2 from seawater may also account for the unusually high amounts of carbonate cements in Late Permian rocks from different parts of the world (a similar scenario may also account for high amounts of carbonate cements at certain times in the Proterozoic; Grotzinger and Knoll, 1995; see also Kempe and Kaźmierczak, 1994). Knoll *et al.* (1996) contend that the oceans were only weakly stratified in the Permian and easily subject to turnover (cf. Wilde and Berry, 1984), and that CO_2 buildup in the ocean resulted from primary production in the surface layer (cf. overfed mode). Despite sluggish ocean circulation rates, Knoll *et al.* (1996) postulate that the release of phosphorus from decaying organic matter in deep anoxic waters would have been sufficient to further stimulate photosynthesis (Ingall *et al.*, 1993; Van Cappellen and Ingall, 1994) and organic decay (positive feedback). It is unclear, however, if a truly stagnant ocean could have persisted for a geologically significant length of time (Hotinski *et al.*, 1998).

Knoll *et al.* (1996) contend that their scenario is not consistent with "nutrient collapse." It is possible, however, that if ocean circulation had been sufficiently slow in the Late Permian, phytoplankton could have largely stripped the surface mixed layer of nutrients (Herbert and Sarmiento, 1991) so that some degree of nutrient collapse would have occurred (reminiscent of the early–mid Paleozoic). Moreover, based on paleoclimate models, three out of four organic-rich rocks are related to upwelling in the Early Permian, whereas only four out of 11 source rocks correspond to upwelling in the Late Permian (Parrish 1982, 1987). Also, Isozaki (1997) concluded that a highly stratified and superanoxic ocean occurred across the Permo-Triassic boundary and that it was associated with much-lowered pelagic productivity (see also exchange in 29 November 1996 issue of *Science*).

Anoxia itself is often cited as a causal agent of the end-Permian extinctions. Wilde and Berry (1984) noted that episodes of anoxia and black shale deposition appear to coincide with the Big Five mass extinctions. Wignall and Twitchett (1996) and Wignall and Hallam (1996) expanded on this relationship near the Permo-Triassic boundary. According to them, the spread of anoxic bottom-waters, which could form during sea-level *fall*, would be promoted by transgression, which would have been relatively rapid (geologically speaking) in shallow, epeiric seas. The spread of anoxic waters could also have introduced massive amounts of nutrients into shallow waters and further destabilized marine ecosystems, much like the input of sewage effluent into oligotrophic lakes that causes massive blooms of algae. The invasion of ^{12}C-rich anoxic waters onto the continents might also explain the rapid negative shift in the δ^{13}C curve at this time. Erwin (1993), who reviewed a number of extinction scenarios with respect to the end-Permian extinction, criticized this hypothesis on a number of grounds, mainly revolving around the timing of the events: many marine taxa, such as suspension-feeding blastoids, died out millions of years before the spread of anoxic waters. Also Isozaki (1997) concluded that the onset of anoxia corresponded to the end-Guadalupian decline (see also Stanley and Yang, 1994), thereby preceding the peak anoxic episodes (see also Wignall and Twitchett, 1996; Wignall and Hallam, 1996).

9.5.3 The Meso-Cenozoic

The diversification of marine plankton beginning in the Mesozoic has been attributed to sea-level rise and the resultant increase in water column stratification and habitat availability (Lipps, 1970). Carbon isotope values also started to rise again in the Triassic, suggesting enhanced marine primary production as a result of the rediversification of plankton following the Late Permian extinctions (Figure 9.6). Indeed, the strontium isotope curve exhibits a sharp rise across the Permo-Triassic boundary and closely approaches the average Phanerozoic value (Figure 9.6). Despite the overall aridity of the Permo-Triassic, Tardy *et al.* (1989) indicated a slight rise in continental runoff across the Permo-Triassic boundary, and Holser and Magaritz (1992) suggested increased erosion at this time (see also Kramm and Wedepohl, 1991). Besides erosion, the rise in the strontium isotope curve across the Permo-Triassic boundary is perhaps consistent with upwelling in incipient seaways (associated with rifting of Pangea) or overturn or release of anoxic marine waters in which terrestrially derived nutrients may have accumulated (note sharp $\delta^{34}S$ spike or Röt Event; see also Małkowski *et al.*, 1989). A slight rise in the strontium and carbon isotope curves also occurs in the Late Triassic (Figure 9.6).

During the rest of the Mesozoic, the $\delta^{13}C$ curve displays a series of sharp positive excursions that correspond to intervals of heightened carbon burial (note $\delta^{13}C$ spikes related to Oceanic Anoxic Events, OAEs; Figure 9.6). Despite apparently low productivity during the Mesozoic (Bralower and Thierstein, 1984), enhanced circulation driven by salinity differences, lessened density stratification (e.g., Lyle, 1997), wind, submarine volcanism or *caballing* (mixing of water masses of different temperature–salinity characteristics that produces a denser water mass; Hay, 1995) at times may have stimulated dramatic rises in marine productivity (see Martin, 1995a, for references). Although Worsley *et al.* (1986) did not recognize any MCP episodes in the Mesozoic (apparently because glaciers were only poorly developed at best; Frakes *et al.*, 1992), extensive phosphorites sometimes occur in the vicinity of OAEs (Figure 9.6). Föllmi *et al.* (1993) suggested that this was due to enhanced continental weathering caused by elevated atmospheric CO_2. Nutrient cycling on shelves may have also accelerated somewhat in response to rising bioturbation rates through the Mesozoic (e.g., Thayer, 1983; Sepkoski *et al.*, 1991). By the Mesozoic, then it appears that the oceans were largely well oxygenated, so that the return to anoxia at discrete times in the Mesozoic may have resulted mainly from enhanced oceanic turnover and photosynthesis and not from anoxic basins, as in the early Paleozoic (Wilde and Berry, 1982).

Other evidence suggests a gradual climate change during the latest Cretaceous. Based on a positive shift in $\delta^{13}C$ during the early–late Maastrichtian Stage transition in deep-sea cores, Barrera (1994) concluded that marine productivity was increasing in response to global cooling and enhanced oceanic overturn, and Abramovich *et al.* (1998) concluded that the pelagic ecosystem was already under considerable stress before the end of the Cretaceous (see also Barrera *et al.*, 1997). Sea-level was also falling and $^{87}Sr/^{86}Sr$ ratios (and presumably nutrient input) increasing during this time (Figure 9.6). Expansion of angiosperms in the Cretaceous perhaps also increased nutrient fluxes to shallow seas, as angiosperm leaf litter decays more rapidly than gymnosperm litter, which characterized much of the late Paleozoic and Mesozoic (Tappan, 1986; Knoll and James, 1987).

Strong positive excursions in $\delta^{13}C$ occurred again in the Cenozoic, beginning near the Eo-Oligocene boundary (Shackleton, 1986), and Hallock *et al.* (1991) and Brasier (1995) have proposed extensive turnover in symbiont-bearing foraminifera during this time because of nutrient influx. These stable isotope shifts reflect increased marine productivity as a result of formation of polar ice-caps, enhanced deep-water overturn, and enhanced continental erosion (through sea-level fall) and nutrient input from land (note rise in $^{87}Sr/^{86}Sr$ curve; Figure 9.6). Also beginning about this time was another series of MCP boosts (Figure 9.6).

Among the predominant groups of Mesozoic plankton are the dinoflagellates, which are often preserved as cysts, and the modern representatives of which tend to prefer mesotrophic nutrient levels (Kilham and Kilham, 1980). Moldowan *et al.* (1996) and Moldowan and Talyzina (1998) found high concentrations of triaromatic dinosteroids, which are indicative of dinoflagellates, during the two Phanerozoic greenhouse phases (Chapter 8; see also Butterfield and Rainbird, 1998). The occurrence of dinosteroids in the Cambro-Devonian greenhouse hints that acritarchs and dinoflagellates *may* be related; in fact, the two taxa have at times been closely allied by some micropaleontologists, although they appear to have been sustained by different nutrient levels. Also, the preservation of dinosteroids parallels the stratigraphic distribution of hydrocarbon source rocks and suggests that their pattern of preservation may be in part the result of low oxygenation in the oceans and low rates of organic matter destruction. Unlike acritarchs, then, diversification of dinoflagellate cysts in the Mesozoic – and coccolithophorids (especially by the Cretaceous) – may signal *heightened* nutrient levels (although not as high as today's; Bralower and Thierstein, 1984) that may have fueled the rise in plankton diversity.

Diatoms, which today prefer nutrient-rich *eutrophic conditions* (Kilham and Kilham, 1980), diversified explosively in the Miocene (Tappan, 1980), perhaps in response to further eutrophication driven by glaciation and oceanic overturn, the decay of silica-rich grasses (Worsley *et al.*, 1986), and increased silica input from continental weathering (Sloan *et al.*, 1997). Also, bioturbation rates continued to rise through the Cenozoic (Thayer, 1983; Sepkoski *et al.*, 1991), presumably in response to increasing surface productivity.

The rise in productivity through the Meso-Cenozoic was accompanied by significant changes in abundance and diversity of the bivalve-rich Modern Fauna through increasing reproductive output and metabolic rates and increased shell durability (Bambach, 1993; Kidwell, 1990; Kidwell and Brenchley, 1994). Modern bivalves require much higher food levels than do modern brachiopods (Rhodes and Thayer, 1991). The diversification of the Modern Fauna began in nutrient-rich Cambro-Ordovician nearshore environments (just as did the trilobite and brachiopod-dominated faunas), after which it radiated into progressively deeper habitats. The advent of abundant robust bivalve shells (as compared with brachiopods) may have also imparted a taphonomic bias to the fossil record by increasing durations of time-averaging and thickening shell beds (Kidwell and Brenchley, 1994).

Rising food levels in the Mesozoic may have also permitted reef-building taxa (Scleractinia) to establish extensive mutualistic relationships with photosynthetic algae. The algae live in the tissues of the corals and receive nutrients (in the form of waste products from the corals), and through photosynthesis they in turn provide food for their animal hosts. By using CO_2 available in the corals' tissues, the algae presumably accelerate calcification in corals by driving equation 9.3 to the right:

$$Ca^{2+} + HCO_3^- \rightarrow CaCO_3\downarrow + CO_2\uparrow + H_2O \tag{9.3}$$

Wood (1993) suggested that this relationship was not established until the Mesozoic, when branching skeletons among coral taxa (Scleractinia) became common (see also Copper, 1974). Prior to this time, many reef-building taxa secreted massive skeletons (e.g., stromatoporoids). Enhanced nutrient availability in the Mesozoic may have sustained coral growth rates sufficiently to build the branching skeletons necessary for extensive photosymbiosis, as well as the diversification of anti-fouling predators and grazers, which keep the corals free of other organisms that might otherwise overgrow them (Vermeij, 1987, 1994; Wood, 1993; Martin, 1995*a*, 1996*a*). This "Mesozoic marine revolution" may have also contributed to the shift in occurrence of bryozoan-rich carbonates from the tropics of the Paleozoic to temperate latitudes beginning in the Mesozoic (Taylor and Allison, 1998). The onset of the revolution may have been earlier, however, as evidence of boreholes is unlikely to be found in Paleozoic aragonitic shells that are typically preserved as molds or casts (Harper *et al.*, 1998).

9.5.4 Secular increase in biomass and diversity

The records of micro- and macrofossils, stable isotopes, and lithology (MCP episodes) all suggest that metabolic rates (level of the individual), biomass (approximately the level of the ecosystem), and diversity (level of the biosphere) were all being ratcheted upward through the Phanerozoic by a step-like rise in nutrient (food) availability (see also Brooks and Wiley, 1988). Indeed, perhaps the Paleozoic, Mesozoic, and Cenozoic Eras should be referred to respectively as the *Oligozoic* (for oligotrophic), *Mesozoic* (mesotrophic), and *Euzoic* (eutrophic). Other workers have previously suggested these secular trends based on either the stable isotope record (Benton, 1979) or the fossil record alone (Vermeij, 1987, 1994; Bambach, 1993; Kidwell and Brenchley, 1994).

All three Phanerozoic faunas (Cambrian trilobite-rich; Paleozoic brachiopod-rich; Meso-Cenozoic mollusc-rich) shifted their preferred habitats from onshore to offshore through the Phanerozoic. The typical explanation for these patterns revolves around biological specialization during the exploitation of new habitats and niches. This is no doubt part of the explanation, but could these patterns have also been related to the availability of phosphorus and other nutrients that in turn constrain rates of marine photosynthesis and the availability of food (energy) to higher levels of food chains? If so, it would appear that the macrobenthos was tracking the distribution of plankton (food) as the primary locus of calcareous and siliceous ooze deposition also moved away from shore during the Phanerozoic (section 9.6). But how could increased food (energy) translate into increased biomass and diversity?

Closed systems increase in entropy (disorganization) and are the basis of the Second Law of Thermodynamics (energy is conserved). Organisms and ecosystems, on the other hand, are open, dissipative systems that undergo self-organization because they exchange matter and energy with the surrounding environment, thereby increasing their negentropy or internal organization. If a system is driven away from equilibrium by changing the parameters of the system, it can shift to a new stable state or attractor (see also Chapter 10). For example, adding heat (energy) to a layer of fluid sandwiched between two hard surfaces at room temperature causes the fluid to self-organize into convection (Bénard) cells (just as in the Earth's mantle). Changing the concentrations of reactants in a highly controlled chemical reaction, such as the Belousov–Zhabotinsky reaction, causes homogeneously distributed reactants in solution to display highly organized patterns through time (Nicolis and Prigogine, 1989). In responding to the fluctuations in the physical system, the reaction's patterns change and the system is said to undergo a series of bifurcations or branchings through time (Föllmi et al., 1993; see also May and Oster, 1976), which produces a pathway or *history*.

Similarly, if nutrient (phosphorus) supply is increased to an open system, the amount of carbon fixation (productivity) by the system, and therefore the amount of food (energy) at the base of a food web available to higher levels of herbivores and carnivores, also increases. Most of the energy transferred from one level to the next is lost as heat (about 90%) and becomes unavailable to make new tissue (biomass). But if more food is available, then presumably more energy can be transferred through the food web before it is lost as heat. Ulanowicz (1980) and Wicken (1980) both proposed largely theoretical mechanisms (Ulanowicz also offered an actual example) of how newly available energy sources can be converted into new links in a food web (see also Wright et al., 1993; Ulanowicz, 1997). Larger populations of different species can also be sustained because more energy is available to sustain maintenance and reproduction. Because of their greater numbers, populations are more likely to disperse over larger areas, become isolated, and give rise to new species (see also Van Valen, 1976; Vermeij, 1987, pp. 400–401; Bambach, 1993).

In this way, the marine biosphere could have branched repeatedly in response to each episode of catastrophic eutrophication immediately following most mass – and many minor – extinctions (Figure 9.6). Nutrients cycled to the photic zone were incorporated into larger amounts of biomass (increased C/P ratios) and larger populations of organisms. If total trophic resources increase as niche overlap remains the same or decreases, then species diversity ought to increase (Krassilov, 1996). Because of their larger size, populations (demes) were better able to exploit available niches and diversify into new taxa. Higher metabolic rates of taxa were also sustained, thereby contributing to rising diversity via predation and other biotic relationships (Bambach 1993; Martin 1995*a*, 1996*a*). It may not be entirely coincidental that the diversification of rapidly growing branching corals (*Acropora*, *Porites*, *Pocillopora*) that dominate modern undisturbed reef crests, began in the late Pliocene and Pleistocene: rapid growth in *Acropora palmata* may be an adaptation to environmental fluctuations (McNeill et al., 1997).

These properties may have been immanent in certain species prior to massive eutrophication and served as a kind of preadaptation (see also Benton, 1995; Courtillot and Gaudemer, 1996). Some workers have recently reported the activation of mutation-prone "evolutionary tuning knobs" in response to stress in both prokaryotes and eukaryotes that affect early developmental pathways; such simple sequence repeats (SSRs) may drive species to evolve continuously just to stay in place ("Red Queen Hypothesis"; e.g., King et al., 1997; Pennisi, 1998; Rutherford and Lindquist, 1998; Silverman and Williams, 1999). Moreover, opportunistic species tend to reproduce rapidly, and it is these taxa that tend to survive extinction and re-diversify in the post-recovery period. In constrast to stenotopic species, eurytopic (opportunistic) species have a much higher degree of "anticipation" (Salthe, 1985, p. 147), in which they "match and therefore ... dissipate environmental perturbations." Ironically, anthropogenic eutrophication of the oceans (e.g., Anderson, 1994; Smil, 1997) may be accelerating a trend started by nature hundreds of millions of years ago (see also popularized account of Barker, 1997).

The above assertions can be corroborated indirectly. If the nutrient hypothesis is correct, MCP ratios should have increased through the Phanerozoic (Martin, 1997). Based on analyses of Jackson and Moore (1976), the percent total carbon in marine rocks increased with younger age for approximately the past 1.25–1.5 billion years (Spearman's $\rho = -0.827$; $\alpha < 0.01$; $n = 26$) as did MCP burial ratios (Spearman's $\rho = -0.748$; $\alpha < 0.01$; $n = 26$). During progressive removal of data points that represent euxinic basins and cherts (in both of which the percent total carbon burial is exceptionally high), and diagenetically altered samples or samples contaminated by terrestrial input (based on paleoenvironmental interpretations of Jackson and Moore, 1976), the correlation between age and percent total carbon and MCP burial ratios remained robust. After all suspect samples had been removed, the significance of the correlation between MCP burial ratios and age weakened somewhat (because of the decreased number of data points) but remained quite high for the past 0.9–1.0 billion years (Spearman's $\rho = -0.857$; $\alpha < 0.05$; $n = 7$). Final samples were all from nearshore settings, which argues against preservational bias via bioturbation, which ought to have *decreased* MCP burial ratios through the Phanerozoic via destruction of organic matter (see below). Confirmation of the nutrient hypothesis suggests that marine productivity may have contributed more to the positive $\delta^{13}C$ trend through the Phanerozoic than previously thought. The initiation of the rise in MCP ratios roughly coincides with the advent of eukaryotes at about 1.5–2 billion years BP; moreover, MCP ratios may have crossed a threshold in the late Precambrian that is associated with the rise of metazoans (e.g., Brasier, 1992; Tucker, 1992; Logan et al., 1995; Butterfield, 1997) but this important step may be only one step in a much longer trend (see also Bambach, 1993; Martin, 1995a, 1996a).

A trend similar to the marine record appears to have occurred on land (McMenamin and McMenamin, 1994). Based on bone structure, Chinsamy and Dodson (1995) concluded that dinosaur metabolism ranged between the low rates of modern reptiles on the one hand and modern birds and mammals on the other, a relation predicted by Zeuthen's (1947) work. As land plants evolved they would have been faced with obtaining water and conducting it to tissues. At the same time, dissolved nutrients would have been carried along. Ultimately a plumbing system formed from the plant tissues xylem and phloem, which also conferred structural rigidity to land plants (via lignin) and allowed them to grow to larger size, thus establishing trees and forests. As terrestrial plants evolved, they also became intimately involved in mutualistic and parasitic relationships with fungi and other organisms, and terrestrial biodiversity and productivity, which vastly exceeds that of marine ecosystems (Tappan, 1982, 1986), rose accordingly. Rising plant productivity and sequestration of carbon on land may have contributed to the cooling trend of the Earth during the Cenozoic (Volk, 1989*b*).

9.5.5 Alternative interpretations

The previous scenario suggests that nutrient input to the oceans occurred repeatedly but in a catastrophic manner. Filippelli and Delaney (1992; see also Filippelli and Delaney, 1994), on the other hand, calculated that phosphorus accumulation rates in ancient phosphorite deposits and modern environments are comparable. They concluded that ancient rates fall within the range of fluxes off the modern Peru margin, although Cretaceous rates fell at the low end of the range for modern fluxes (see also Bralower and Thierstein, 1984). If modern and ancient phosphorus fluxes are in fact comparable, then nutrient fluxes to the photic zone, and corresponding increases in productivity, biomass and diversity of the marine biosphere, must have occurred much more frequently through the Phanerozoic than is indicated by MCP episodes alone. In other words, MCP episodes are preservable in the geological record because of suitable tectonic and paleoceanographic circumstances (e.g., Compton *et al.* 1993; Ruttenberg and Berner, 1993), and are the most extreme versions of much more numerous phosphogenic intervals. If this is the case, nutrient input to the oceans may have been much more continuous than described here.

Others have denied that such trends in productivity and biomass have occurred (e.g., Van Valen, 1976). Schidlowski (1991), for example, sees no trend in the carbon isotope curve or carbon burial rates that is indicative of rising productivity. This may be the result of looking at data over long periods of time: the variation exhibited by a population (in this case, carbon isotope values) increases the longer one counts (samples) the population (Pimm, 1991; McKinney and Frederick, 1992). Moreover, it is possible that as productivity increased through the Phanerozoic, increasing rates of bioturbation and atmospheric oxygen levels destroyed much of the carbon produced (Garrels et al., 1976), so that the fraction of organic carbon buried through time may have remained more or less constant; thus, net carbon burial rates would *appear* to remain relatively constant.

Garrels et al. (1976) described a scenario involving numerical simulation of increased rates of *anthropogenic* erosion. In their model, they increased the global erosion rate (which would include nutrient input to the oceans) threefold, which tripled organic carbon flux to the seafloor. But a threefold increase in oxygen demand also occurred through the erosion of organic carbon and iron sulfides buried on land (cf. equation 9.2). Atmospheric oxygen reached a new steady-state about 15% below present levels in about 2 million years, while CO_2 rose to approximately 2.5 times current values (to about 800 ppm), which accentuated burial of carbon and sulfide in the oceans even as they were being eroded on land. Thus, the total amount of organic carbon and $\delta^{13}C$ values remained relatively unchanged.

Vermeij (1995) made many of the same points as Martin (1995a, 1996a) based on the Earth's biota alone (he did not discuss stable isotope data), but he came to different conclusions regarding the causal agents (cf. Vermeij, 1987). He hypothesized that heightened global temperature (a result of increased volcanism) promoted marine biological revolutions during the Cambro-Ordovician and Mesozoic. He argued that greater per capita availability of "energy" (in terms of ambient temperature, not nutrients) was primarily responsible for speeding up biological metabolism and resource–competitor interactions and therefore evolution. But increased nutrient levels (through increased weathering rates, climate change and increased rates of ocean circulation, for example) would have increased nutrient levels in the marine realm, which would have, in turn, promoted increased food levels (e.g., plankton) needed to fuel evolution and expansion of the marine biosphere. Also, Vermeij argues that extinction is associated only with *decreased* primary productivity, but isotope data also argue for prolonged intervals of enhanced marine productivity during some extinctions, and which may have helped to destabilize the marine biosphere (cf. Figure 9.6).

9.5.6 Evolution of the biogeochemical cycles of carbon and silica

Based on the fossil record, calcareous plankton (planktonic foraminifera and calcareous nanofossils such as coccolithophorids) did not evolve until the Mesozoic (Figure 9.6), and so their integration into the long-term global carbon cycle could presumably not have begun until the Mesozoic. This appears to be true

9.5 Seafood through time: energy and evolution

for planktonic foraminifera, but enigmatic calcareous nanofossils have been reported from cratonic sediments of the Paleozoic (Figure 9.6). These reports have typically been dismissed either as contaminants from younger rocks or for lack of adequate description (e.g., Lord and Hamilton, 1982). If taken seriously, though, these microfossil Lagerstätten suggest a much longer history for the group. If calcareous plankton were present in the Paleozoic, most of them must have dissolved because of greater rates of diagenesis and dissolution.

As discussed in Chapter 8, elevated atmospheric pCO_2, mainly from increased volcanism, weathers continental rock, which acts as a negative feedback and draws down CO_2. In marine waters, the ions from runoff are incorporated into the calcareous shells ($CaCO_3$) of marine plankton. Eventually the plankton die and a significant proportion settle to the bottom before dissolution. Over intervals of millions to tens of millions of years, the calcareous ooze that is formed is eventually carried by spreading ocean crust to oceanic trenches, where it is subducted, heated, and the CO_2 present in the skeletons is released back to the atmosphere, thereby completing the cycle (Chapter 8). Without the advent of this portion of the cycle, today's climate might be substantially cooler (Volk, 1989a; Beck et al., 1995).

But why did calcareous plankton apparently originate in shallow seaways and only later begin to populate the open ocean in the Mesozoic? Perhaps because in modern oceans, coccolithophorids prefer mesotrophic waters, which would have developed closer to land and progressed offshore through time. Other microfossil evidence supports this contention. Siliceous oozes preceded calcareous ones by several hundred million years (Figure 9.6), possibly because the use of SiO_2 for shell construction is energetically more efficient than $CaCO_3$ (Brasier, 1986, p. 244). Interestingly, the locus of siliceous ooze deposition shifted across continental shelves during the Phanerozoic. Maliva et al. (1989) concluded that late Proterozoic cherts were mainly peritidal and of abiological origin (based on lack of petrographic evidence for skeletal sources of silica). By the Cambrian, biogenic cherts were predominantly shallow subtidal in occurrence and the byproduct of evolutionary diversification of demosponges and problematic siliceous scale-producing protists (Maliva et al., 1989). After the rise of radiolarians in the Cambro-Ordovician, the locus of siliceous ooze deposition shifted across the shelf to the open ocean (Maliva et al., 1989; Casey, 1993); Schubert et al. (1997) attributed the decline in shallow-water siliceous fossils after the Paleozoic to the offshore shift of the locus of silica deposition. In the Mesozoic, radiolarians began to give way to silicoflagellates (which prefer moderate to high nutrient levels; Lipps and McCartney, 1993), and in the Miocene, to diatoms (Figure 9.6; Tappan, 1980). Today, diatoms account for approximately 20–25% of total global primary production (Werner, 1977; P. Roessler, 1995, personal communication, derived similar figures calculated from Tréguer *et al.*, 1995); in fact, diatoms secrete so much silica (approximately 80% of which is originally supplied by rivers) that only about 3% accumulates in siliceous oozes (via fecal pellets; Schrader, 1971; Chapter 3), the rest being dissolved in the upper water column and recycled back to the surface where it is used again (Tréguer *et al.*, 1995; see also Bareille *et al.*, 1998). Heath (1974) estimated that if rapid recycling of silica did not occur, diatoms (along with other siliceous plankton) would strip the oceans of dissolved silica in about 250 years, but Tréguer *et al.* (1995) estimated a slightly longer steady-state (assumed) residence time (total mass dissolved in the oceans/rate of supply or removal) relative to biological uptake of about 400 years. Still, for either estimate, diatoms use silica so rapaciously that Harper and Knoll (1975) suggested that this is the primary reason that radiolarians began to produce lattice-like skeletons in the Cenozoic. In contrast to diatoms, Paleozoic radiolarians had relatively robust shells (Casey, 1993), and were more likely to be preserved in shallow environments of the Cambro-Ordovician.

The oldest reliable record of diatoms is the Jurassic (Toarcian; Barron, 1993). Of potential significance in this respect is the report of spindle-shaped (pennate?) diatom-like structures from late Proterozoic stromatolitic carbonates (Licari, 1978, their plate 3, figure 9). If diatoms existed during the early Paleozoic or perhaps even Precambrian (Round and Crawford, 1981, 1984), their apparent absence from open oceans during the Paleozoic and early Mesozoic may indicate their restriction to nearshore nutrient-rich habitats, in which fragile siliceous eukaryotes – like diatoms – are unlikely to be preserved (Margulis *et al.*, 1980). Another possibility is that naked ancestors existed prior to the Mesozoic but that diatoms did not begin to silicify until the Jurassic (Harwood and Gersonde, 1990); Sorhannus (1997) suggested that diatoms originated only about 330–400 million years ago (Devonian to Carboniferous) based on ribosomal RNA evidence (cf. Schopf, 1996, regarding a similar controversy about the origins of the first cells).

Note added in proof:
Bidle and Azam (1999, Accelerated dissolution of diatom silica by bacterial assemblages, *Nature* 397: 508–512) implicated bacterial decay in the dissolution of diatom frustules and the recycling of silica. Benitez-Nelson and Buesseler (1999, Variability of inorganic and organic phosphorus turnover rates in the coastal ocean, *Nature* 398: 502–505) found that turnover rates of phosphorus in coastal waters are quite high, and that relatively low phosphorus concentrations can support relatively high primary production. The entire February (1999) issue of *Palaios* (volume 14, number 1) is devoted to the recognition and evolution of microbial mats and their role in biogeochemical cycles.

Marzoli *et al.* (1999, Extensive 200-million-year-old continental flood basalts of the central Atlantic magmatic province, *Science* 284: 616–618) concluded that the extrusion of extensive flood basalts during the rifting of Pangaea coincided with the Triassic/Jurassic extinction. There is also a strong negative shift in $^{87}Sr/^{86}Sr$ ratios about this time (Figure 9.6), as would be expected based on the explanation of a similar shift near the end of the Permian (pp. 355–356).

10 Applied taphonomy

History is more or less bunk ... We want to live in the present, and the only history that is worth a tinker's damn is the history we make today.
Henry Ford

The greatest objects of Nature are, methinks, the most pleasing to behold; and next to the great Concave of the Heavens, and those boundless Regions where the Stars inhabit, there is nothing that I look upon with more pleasure than the wide Sea and the Mountains of the Earth. There is something august and stately in the Air of these things that inspires the mind with great thoughts and passions: We do naturally upon such occasions think of God, and his greatness, and whatsoever hath but the shadow and appearance of INFINITE, as all things have that are too big for our comprehension, they fill and over-bear the mind with their Excess, and cast it into a pleasing kind of stupor and admiration.
Thomas Burnet, *The Sacred Theory of the Earth* (1684, first English edition)

10.1 Introduction

Would we really understand the Earth and its Life if we only understood the present? And if not, why not?

The fossil record is a very important – but so far highly underutilized – window that allows us to assess the impact of biological and beogeochemical processes over periods of time much longer than those normally considered by an ecologist. The fossil record suggests that there are likely to be processes – as indicated by patterns in the fossil record – that occur only on evolutionary or geological time scales. Moreover, time-averaging of fossil assemblages can actually enhance the expression of ecological signals because it damps (filters) out short-term noise. These patterns, and presumably the processes that produced them, are by no means strictly academic. They bear strongly upon man's impact on the health and biotic diversity of the planet. Paleontology and stratigraphy bear upon such environmental phenomena as population dynamics, speciation and extinction, the organization and resilience of biological communities to disturbance, and the occurrence of alternative community states in response to environmental disturbance, both natural and anthropogenic (Kidwell and Bosence, 1991; Kidwell and Flessa, 1995).

For many decades, applied paleontology has concentrated on finding energy resources (Chapter 7). It appears that in the future, though, applied paleontology will more frequently have to address "environmental" problems if the profession is to survive. Ironically, despite the extensive use of micropaleontology in petroleum exploration, foraminifera were used much earlier to date strata in a water well in 1877 near Vienna, Austria. Further studies followed in the U.S.A. (see Jones, 1956, pp. 1–6, for references), but it was primarily J. A. Udden of Augustana College (Illinois), who, in 1911, began to stress the importance of microfossils in correlating water wells. Udden would later forsake academe to become head of the newly organized Bureau of Economic Geology of Texas, where he shifted application of microfossils from water to petroleum.

We seem to have come full circle since Udden's time (Martin, 1991, 1995b, 1998a, 1999). Sloan (1995), for example, in an issue of the *Journal of Foraminiferal Research* devoted entirely to using foraminifera to solve environmental problems, describes the use of foraminifera (and other microfossils) to correlate discontinuous water-bearing units of alluvial origin in the San Francisco Bay area that serve as aquifers for municipal and agricultural water supplies. Otherwise, previous applications of paleontology to environmental problems are relatively scarce. Bandy et al. (1964a,b; 1965a,b) were among a handful of workers in the late 1950s and early 1960s who examined the response of foraminiferal populations to large inputs of sewage in shallow marine waters. Bandy et al.'s studies lay dormant, however, probably because the environmental movement had not yet fully developed and because of the heavy emphasis of applied micropaleontology on biostratigraphy in petroleum exploration. Foraminifera are ideally suited to environmental studies because their short generation times allow them to respond quickly to environmental change (Alve, 1995). Also, foraminifera appear to have fairly specific cellular defense mechanisms against toxic substances (Bresler and Yanko, 1995; Yanko et al., 1998), which, with further research, might prove useful in assessing pollution sources. Other recent practical applications of paleontology to environmental science and engineering include the use of foraminiferal biostratigraphic zonations in aligning the Channel Tunnel that now connects England and France as it was being excavated from both sides of the English Channel (Hart, 1996); the correlation of alluvial deposits at seismic hazard sites in the San Francisco Bay area (Sloan, 1995); and deciphering the earthquake history on active margins of the Pacific Northwest (Nelson and Jennings, 1993; Mathewes and Clague, 1994; Hyndman, 1995).

On longer time scales, a prime target of paleontological research could be the state of the environment prior to anthropogenic influences (deforestation, sewage, etc.), which may be subtle but far-reaching (e.g., Mackenzie and Morse, 1990; Powell, 1992; Alve, 1995; Barmawidjaja *et al.*, 1995; Kidwell and Flessa, 1995; Runnels, 1995; Jackson, 1997; Russell, 1997; Vitousek *et al.*, 1997, and related papers in same issue). How would such information affect legislation designed to preserve (or destroy) wetlands, for example? Such studies could also provide data on the potential influence of anthropogenic disturbance on actualistic models that are applied to ancient settings. Baseline studies might also prove invaluable to ecological and evolutionary theory because they would consider processes – including the effects of *natural* disturbance – that might occur over longer time scales. Haynes (1985), for example, concluded that increased subadult mortality of elephants in pre-anthropogenic death assemblages might be indicative of natural drought and not anthropogenic activities.

But in order to address these sorts of questions, and place paleontology in a more practical – and favorable – light to those who control the purse strings, taphonomists will have to increasingly bring to bear their energy and imagination on environmental problems. A number of instances of the application of taphonomy to environmental problems have been described in previous chapters. This chapter presents a brief sampling of other studies – more or less in order of decreasing scale – not only to demonstrate further the value of the fossil record in environmental studies, but also that the fossil record cannot be taken at face value. One ignores the quality of the fossil record and the pitfalls of preservation at one's peril.

10.2 Stratigraphic completeness: rates and patterns of evolution

One of the most contentious debates about the fossil record revolves around the tempo and mode of evolution (e.g., Eldredge, 1995). In the fossil record, rates of evolution are based on inferred rates of morphological change (Dingus and Sadler, 1982). Eldredge and Gould's (1972) original data for punctuated equilibrium came from shallow-water Devonian trilobites and Bermudan land snails (see also Gould and Eldredge, 1993, for update), but some of the best evidence for speciation and rapid evolution comes from deep-sea cores, because they presumably (and sometimes optimistically) represent long intervals of undisturbed sedimentation (Schiffelbein, 1984). Even in deep sea cores that exhibit relatively continuous and high sedimentation rates, however, there is severe attenuation of signals with periods shorter than a few thousand years (Schiffelbein, 1984; Chapter 4). Moreover, based on graphic correlation (Chapter 7), MacLeod

(1991) demonstrated that periods of apparently rapid evolution within Neogene (Miocene to Recent) planktonic lineages of foraminifera occurred during major changes in paleoceanographic regime and sediment accumulation rate. As accumulation rates slowed, apparent rates of morphological change increased because morphological change was concentrated within slow (condensed) intervals of sedimentation (cf. Figures 1.10, 1.11). In the case of macrofossils, "pseudoevolutionary" artifacts may be recognizable from changes in abrasion and fragmentation of shells, and encrustation by epizoans and bioerosion, all of which should – *intuitively* – increase when shells lie exposed at the SWI for long periods of time (Kidwell, 1986a). As discussed in Chapter 5, however, this is not necessarily the case; taphonomic grades of both macrofossils and microfossils may vary substantially with age.

Recent interpetations of the fossil record have suggested that global cooling (and drying) caused a major biotic turnover of mammals, including hominids, between about 2.5 and 2.7 Ma (late Pliocene) in Africa (Vrba, 1988; see also Vrba, 1980, 1985, 1993; Kerr, 1994a, 1996, 1997). Behrensmeyer *et al.* (1996, 1997) analyzed mammalian fossil assemblages from three separate sites in the Lake Turkana Basin (East and West Turkana, Omo) of northern Kenya that have been used to calibrate faunal change in other parts of Africa. They found that more than 60% of the species continued through the interval of climate change (2.8–2.5 Ma), which indicates no "major" biotic turnover; East Turkana had the highest turnover (54% of first and last appearances occur at approximately 2.5 Ma), but they attributed this to "a marked change in fossil productivity" (cf. Figure 1.8). Behrensmeyer *et al.* (1996) conclude that "even one of the best-documented vertebrate records must be subjected to rigorous quality control in order to reveal the extent of 'real' species turnover."

Similarly, based on evaluation of the stratigraphic record, Dingus and Sadler (1982) concluded that the stratigraphic record of 19 species of distantly related late Pliocene–Pleistocene molluscs of the Turkana Basin (Kubi Algi, Kubi Fora, and Guomde Formations) *may* indeed represent punctuated evolution associated with lake regression and geographic isolation of surviving populations (Williamson, 1981). However, the regressions also indicate widespread erosion at each of the levels of inferred rapid evolution (Williamson, 1981, his figure 4) which, coupled with subsequent transgression, could mimic punctuated speciation. Furthermore, the estimates of the durations of punctuated evolution (5–50 thousand years) appear to be based on extrapolation of long-term sediment accumulation rates to shorter intervals (cf. Chapters 5, 7). Sadler (1981) demonstrated that expected completeness of stratigraphic sections deteriorates as finer time scales are considered: the longer the section, the greater the likelihood that unconformities will be encountered and the slower will be the calculated sediment accumulation rates, whereas the shorter the interval, the greater the calculated sedimentation rate (Chapter 7). Although rapid evolution may occur (e.g., report by Johnson et al. [1996] of rapid evolution of cichlids in Lake Victoria), extrapolation of long-term sedimentation rates to short intervals (or vice versa) to determine the rates of geological or biological processes is fraught with pitfalls.

10.3 Extinction

10.3.1 The terrestrial record

Differential preservation of vertebrate assemblages across extinction horizons has also been much debated. In the case of the Late Cretaceous extinctions, an apparently complete sequence of terrestrial deposition is found in Montana and South Dakota in the Hell Creek Formation (Cretaceous), which consists of interfingering channel and floodplain deposits that reflect repeated erosion and potential reworking. Because of the apparent completeness of the sections, the occurrence of dinosaur remains (including teeth) in river channels scoured into the Hell Creek Formation and filled with Paleocene sediment and fossils has been used as evidence for the survival of dinosaurs into the Paleocene (e.g., Rigby et al., 1987). But based on palynological analyses, Lofgren et al. (1990) concluded that dinosaur teeth found in Paleocene silts deposited in channels had been reworked from the underlying Hell Creek Formation. These teeth may have been buffered against dissolution by moderately calcareous sediment (Retallack et al., 1987). Also, experimental studies indicate that dinosaur teeth can undergo considerable transport (equivalent of 360–480 km) but exhibit little or no wear (Argast et al., 1987; see also Eaton et al., 1989; Chapters 2, 3). Lofgren et al. (1990) concluded that "because of the potential for reworking, dinosaur remains derived from Paleocene fluvial deposits should not be assigned a Paleocene age unless they (1) are found in floodplain deposits [although these are subject to reworking; see Chapter 2], (2) are articulated, (3) are in channels that do not incise Cretaceous strata, or (4) are demonstrably reworked from Paleocene deposits" (see also Fastovsky, 1987; Eaton et al., 1989).

After intensive sampling across the boundary, Sheehan et al. (1991, p. 835) concluded that "family-level patterns of ecological diversity of dinosaurs ... are in agreement with an abrupt extinction event such as one caused by an asteroid impact"; as predicted by the impact hypothesis, fossil abundances were random below the K–T boundary and followed by abrupt truncation (cf. Signor–Lipps effect, Chapter 7). Although Sheehan *et al.* (1991) took great pains to omit any obviously transported or reworked remains, most of their samples came from channel, point bar, and floodplain paleosols, and their study is predicated on their premise that preservation had little effect across the boundary because "fluvial environments and diagenesis were relatively stable through the thickness of the Hell Creek formation" (Sheehan *et al.*, 1991, p. 836).

Sheehan *et al.* (1991) cite the study of modern mammalian assemblages of Amboseli Basin (Behrensmeyer and Boaz, 1980) in support of a second premise: namely, that in *fluvial* environments, preservation of big bones is favored over small ones (Law of Numbers; but see Chapters 1, 2). But Behrensmeyer and Boaz (1980) found that the number of species in living populations and in bone assemblages converged for animals of 50–100 kg or more body weight; below this weight, the number of species was underrepresented by fossil assemblages. Bone size and preservation is probably not so much the critical factor in the Hell Creek as the choice of a modern analog for comparison. Ironically, Amboseli Basin "serves as a model for *'pre- or non-transported'* conditions in fossil or recent settings" *because there are no major rivers into the basin* (Behrensmeyer and Boaz, 1980, p. 73; my italics). Although Amboseli Basin may serve as a model for other arid basins (Olduvai Gorge, Peninj and Chesowanja Basins; Behrensmeyer and Boaz, 1980), it is not the proper modern analog for study of preservation in the Hell Creek Formation. Rather, the choice lies between channel-lag and channel-fill assemblages, in which significant reworking and time-averaging of vertebrate remains may occur (Fastovsky, 1987; Eaton *et al.*, 1989; see also Chapter 2). Moreover, Amboseli Basin has alkaline soils conducive to bone preservation, whereas the bones in floodplain paleosols of the Hell Creek Formation were subject to leaching, as suggested by the absence of small mammal bones and teeth below the highest occurrences of dinosaur bones (Retallack *et al.*, 1987; see also Chapter 3), and the occurrence of the highest dinosaur bone in the Hell Creek Formation 60 cm below an iridium-rich layer at the apparent K–T boundary (Sheehan *et al.*, 1991).

Hulbert and Archibald (1995) concluded that the data used by Sheehan *et al.* (1991) cannot be used to argue for or against catastrophic or non-catastrophic extinction in the Hell Creek sections. Also, Williams (1994) argued that the distributions below the K–T boundary could merely have resulted from constant mortality, and that the impact hypothesis predicts *increased* hardpart input at the K–T boundary barring taphonomic loss. On the other hand, the proposed test(s) of the impact scenario may not be valid: increased catastrophic input might be so short that only a limited hardpart input could result from the

10.3 Extinction

mass-kill of a *biological* population, especially when coupled with time-averaging of the hardpart impulse (Fastovsky, 1990; Hunter, 1994).

10.3.2 The marine record

Deciphering extinction patterns in marine sections has proven no less complex. Extinction patterns among marine taxa are probably confounded by depositional environment, sampling effects, the different habitats and tolerances of taxa to environmental change, regional versus global climate change, physical and biological mixing, and the differential destruction of hardparts (see, for example, discussion of MacLeod's analysis of completeness of K–T boundary sections via graphic correlation in Chapter 7). Based on confidence intervals (CI; Chapter 7), Marshall and Ward (1996) found that just below the K–T boundary of sections in France and Spain (western Tethys), ammonites exhibited both background and abrupt extinction patterns (depending upon species), whereas inoceramids displayed a gradual extinction pattern. The abrupt change in ammonites may, however, have been associated with a sea-level regression (cf. MacLeod and Keller, 1991; see also section 7.4 on sequence stratigraphy and section 7.7 on graphic correlation).

Marshall (1995) also attempted to distinguish between sudden and gradual ammonite extinctions recorded in sections on Seymour Island, Antarctica. Although a literal reading of the Antarctic fossil record suggests gradual extinction of ammonites through 10–50 m of section below the K–T boundary, 50% CI show that the extinction pattern is consistent with a sudden mass extinction, and that an iridium anomaly falls within the confidence interval. In this approach, for a single taxon there is a 50% chance that the stratigraphic horizon corresponding to the time of extinction of the taxon lies between the uppermost occurrence and the end of the 50% CI, and that there is a 50% chance that the extinction horizon lies above the end of the 50% CI. "Hence, *if an ensemble of taxa became extinct simultaneously*, then on average half of the end points of the 50% confidence intervals on the stratigraphic ranges should lie below the extinction horizon, and half ... above the extinction horizon" (Marshall, 1995, p. 732; my italics). Computer simulations produced other extinction scenarios, however, including gradual ones over *less than 20 m of section*, that are consistent with the fossil record (Marshall, 1995). Thus, the observation that ammonites disappeared over 50 m is probably at least partially an artifact of abundance and preservation (ammonites consist of either aragonite or high-Mg calcite; Chapter 3) and their effect on sampling. On the other hand, there are assumptions of the CI method (continuity of section, etc.) that are often violated in stratigraphic sections

(MacLeod, 1995; see also Chapter 7). Consequently, Marshall (1995) advocated "saturation collecting" near the K–T boundary, without which he said it would be impossible to distinguish between sudden and gradual extinction in the Seymour Island ammonite record.

Microfossils are no less subject to the vagaries of preservation and sampling, even though bulk sediment samples of thousands or more specimens can be collected on a centimeter by centimeter basis. Based on graphic correlation, the El Kef (Tunisia) sections are the most complete and expanded sections known across the K–T boundary, probably because of their position on the outer shelf–upper slope, where sedimentation was more or less continuous (Figure 7.16; MacLeod and Keller, 1991; Macleod, 1995). Nevertheless, Keller et al. (1995) reported a drastic decline of $CaCO_3$ content in sediments across the K–T boundary in El Kef sections that should, intuitively, be associated with dissolution of small carbonate particles such as coccolithophorids. Thierstein (1980, 1981), however, noted that only a few species of coccolithophorids are affected by dissolution in El Kef and other sections and that small species tended to dissolve as larger ones were overgrown by secondary $CaCO_3$ (see Chapter 3). Thus, Thierstein (1981) concluded that a number of Cretaceous taxa *survived* into the Danian (Paleocene); other workers have also concluded that coccolithophorids survived into the Paleocene (for at least 10 ka) based on *bulk* stable isotope analyses that differed above and below the K–T boundary (see Pospichal, 1994, for review; cf. Chapter 5 on ^{14}C dates and time-averaging). By contrast to Thierstein (1981), Pospichal (1994) concluded that reworking of Cretaceous species into the Danian had occurred based on correlation of species abundances with short-term sea-level fluctuations in the earliest Paleocene and increased abundance of the dissolution-resistant *Micula decussata* above the K–T boundary. Pospichal (1994) and Gartner (1996) both concluded that coccolithophorids survived up to the K–T boundary (including El Kef) with no hint of previous biotic decline (although Gartner's sections consisted of mainly pelagic and hemipelagic sediments; cf. depositional model of MacLeod and Keller, 1991; MacLeod, 1995; Figure 7.16).

By contrast to the *apparent* extinction pattern of coccolithophorids, Keller et al. (1995) found a gradual decline of foraminifera beginning below the K–T boundary. Changes in benthic foraminiferal assemblages at El Kef were not seen in other sections and may represent a regional signal (Keller et al., 1995), whereas changes in planktonic assemblages reflect depth-stratification of species (see Chapter 2 for modern forms), which makes certain forms more susceptible to sea-level change and other environmental factors (Pospichal, 1994): all deep-dwelling planktic foraminifera (strongly stenotopic) and approximately one-half

10.3 Extinction

of all surface dwellers (somewhat more eurytopic) became extinct (Keller, 1996). Expansion of the oxygen minimum zone in conjunction with sea-level regression appears to have caused earlier turnover of planktonic (and benthic) foraminifera (Keller et al., 1995).

Strontium isotope ratios of foraminiferal tests also indicate extensive reworking (MacLeod and Huber, 1996; see Chapter 5). Reworked species of planktonic foraminifera were suspected from stable isotope analyses in El Kef sections and the preservational scales of Douglas (1971) and Malmgren (1987); MacLeod (1996) found that putative planktonic survivors consist largely of dissolution-susceptible species, but that the tests themselves do not record changes in taphonomic conditions (see also Chapters 3, 5). This is exactly the opposite of the deductive (*intuitive*) prediction based on dissolution-susceptibility of planktonic foraminifera (Chapter 3). Based on confidence intervals, MacLeod (1996, p. 108) concluded that there "seems to be no reason to suspect that a dissolution-based Signor–Lipps effect has artifically biased [the] fossil records."

The dispute over the rate, timing, and cause of the end-Permian extinction – the largest of all Phanerozoic mass extinctions – has been no less contentious (see Erwin, 1993, pp. 73–84, for review; cf. Chapter 9). Erwin (1993) concluded that the Permo-Triassic boundary is probably best placed based on graphic correlation of conodonts rather than ammonites (cf. Sweet, 1992, and Tozer, 1988; see Yin and Tong, 1998, for recent review). Correlation problems across the Permo-Triassic boundary appear to be the result of incomplete sections (although often assumed to be complete), biogeographic distributions, and Lazarus taxa. Potential problems of differential dissolution of conodonts (phosphatic) versus ammonoids (aragonitic) have not been considered, but reworking of conodonts does not appear to be a factor (Chapter 2). Correlation of high overall rates of extinction with background extinction (presumably related to an intensification of background extinction rates) previously suggested gradual but pulsed extinctions (Stanley and Yang, 1994). Gradual extinction patterns based on morphological trends in fusulinacean foraminifera do not appear to be a result of the Signor–Lipps effect: Rampino and Adler (1998) concluded that the end-Permian extinctions were abrupt based on confidence intervals. Most recently, based on U/Pb zircon dates, Bowring et al. (1998; see also Kerr, 1998) concluded that the end-Permian extinctions recorded in south China sections were also "catastrophic" (less than 1 myr in duration) and suggested several possible scenarios based on their geochronology; although the sections spanning the Permo-Triassic boundary appear complete, the strong negative $\delta^{13}C$ shift reported by them does not appear to be preceded by the much more gradual and prolonged shift to negative values (before the final negative spike) reported by Holser *et al.* (1991) and Magaritz *et al.* (1992) in the Austrian Alps and inferred by them to indicate much more gradual climate change preceding the final pulse of extinction (cf. Chapter 10).

10.4 Stasis and community unity?

The fossil record indicates significant change in ancient biotic communities through time (Chapter 9). Recently, some workers have suggested that a kind of "ecological locking" occurs, in which biological communities are resistant to environmental change below some threshold level (e.g., DiMichele, 1994; Morris and Ivany, 1994; Brett and Baird, 1995; Lieberman *et al.*, 1995). McKinney (1996) and McKinney *et al.* (1996) concluded that so-called stasis or locking may partly result from the enumeration of the most abundant (and presumably the most geographically widespread and eurytopic) species in the fossil record. Entrenched (eurytopic) species tend to recur together, are more likely to be sampled in the fossil record and to persist during times of environmental change, and may only exhibit significant turnover during major environmental crises; by contrast, rare species undergo much greater rates of turnover because they tend to be more stenotopic and suffer from greater population fragmentation (edge effects; Maurer and Nott, 1996). Although rare species typically far outnumber abundant ones (Koch, 1987; CoBabe and Allmon, 1994), rare species are more likely to be missing from stratigraphic sections because of destruction or lack of adequate sampling (McKinney, 1996; McKinney *et al.*, 1996). The general effect of decreased probability of preservation and sampling is to reduce the observability of less common taxa, and thus increase the perception of biotic turnover in the fossil record (McKinney *et al.*, 1996).

McKinney and Allmon (1995), McKinney *et al.* (1996), and McKinney (1996) simulated the distribution of fossils (in this case, benthic foraminifera) in hypothetical sections using a "particle model." This model specifies how widespread (and therefore abundant) species tend to recur in stratigraphic sections and to covary as suites or assemblages in the fossil record for long periods of time as rarer species come and go. Rare species migrate into the area of deposition but are only preserved infrequently. Somewhat more common species may migrate into the area only infrequently and are also unlikely to be preserved. Thus, through both rarity and taphonomic loss, rare species will be less abundant than very common ones in a stratigraphic section (McKinney *et al.*, 1996; McKinney, 1996). Although 80% or more of species in a region may be preserved (in this case the number of living molluscs in the Californian biogeographic province

10.4 Stasis and community unity?

found in Pleistocene sediments of the same region; Valentine, 1989), it is unlikely that all 80% will be found in every sample from a stratigraphic section in this province (McKinney *et al.*, 1996), and the high number of preservable rare species found in a sample must result from time-averaging of populations that exist and die over thousands of years (Chapter 5).

Pandolfi (1996) has found that the species composition of coral reef communities remained relatively constant during repeated sea-level changes of the latest Pleistocene (about 125–130 Ka). Although reef ecosystems are quite dynamic over short temporal and spatial scales, Pandolfi's findings suggest that reefs may be fundamentally different from other ecosystems or that "different ecological patterns and processes occur at different temporal scales" (Pandolfi, 1996, p. 152; see also Greenstein *et al.*, 1997). Pandolfi's results are, however, based on presence–absence data, and as demonstrated by the "particle model," if a species is present in a fossil assemblage, it is more likely to be common and therefore sampled, whereas a rare species is more likely to be absent from a fossil assemblage and left unrecorded.

Thus, the perception of community stasis may depend on the level of resolution (sample acuity, etc.) employed (Chapters 5, 7). Although community stasis probably does occur in the fossil record, it is measurable only for the most common species (McKinney *et al.*, 1996). Also, like sediment accumulation rates (Sadler, 1981), the degree of stasis increases with the duration of the interval examined (McKinney *et al.*, 1996). Long-term communities, which correspond to coarse levels of temporal resolution referred to as biofacies (i.e., community-type) by McKinney *et al.* (1996), and even coarser units called Ecological Evolutionary Units (EEUs) by Boucot (1983), will only turn over when significant disturbance occurs (e.g., extinction). These communities are the ones that have typically been studied by paleontologists. The biofacies and EEUs are, then, sufficiently coarse to damp any noise produced by the turnover of rare species (McKinney *et al.*, 1996).

The argument about stasis in the fossil record is related to a long-standing debate about whether biological communities consist of highly interdependent species or of a haphazard ensemble of species that just happen to occupy the same environment at the same time (Jackson, 1994, and references therein). Some workers have suggested that the latter view is simply a result of observations of local colonization and extinction made over short ecological time scales, and that longer time scales of observation (such as those available from the fossil record) demonstrate otherwise (Bretsky and Klofak, 1986). Buzas and Culver (1994) tested these hypotheses using the stratigraphic ranges and abundances (down to 1% or less) of benthic species of foraminifera in Cenozoic deposits of the Atlantic Coastal Plain formed during different highstands of sea-level (see also Buzas and Culver, 1998). According to them, whenever sea-level transgresses onto the continental shelf, new habitats are opened to invading species. Presumably these habitats are more or less equivalent for each transgression. If species composition of communities is relatively stable, then basically the same species groupings should occur during each highstand.

The same dynamic relationship is hinted at by the high species diversity of foraminifera in tropical carbonate environments. Up to about 150 species have been described from coral reef and related habitats of the Caribbean, many of which occur in the sediment of the same locale (e.g., Martin and Wright, 1988; Martin and Liddell, 1988). Some are dominant and persist (e.g., *Archais* and *Amphistegina*; Chapter 5), but others are quite rare. Interestingly, Pandolfi (1996) found that species composition of coral communities – the dead hardparts of which would be expected to behave differently from foraminiferal tests (Chapters 2, 3) – varied significantly through time because many species either emigrated (went extinct locally) or immigrated from elsewhere. Some emigrants returned during later transgressions, but many did not. It would appear that either niche space among foraminiferal species is highly subdivided (dependent, say, on critical levels of different micronutrients) and that foraminiferal "communities" behave differently from corals or that *perhaps* foraminiferal diversity has been artificially increased by time-averaging (cf. Peterson, 1977).

10.5 Disturbance and alternative community states

Despite apparent stasis in much of the fossil record, ecosystems are constantly recovering from disturbance (Reice, 1994). Based on the model of May (1977), Knowlton (1992) concluded that marine ecosystems are *metastable* because they exhibit different community states that persist under a *single* physical environmental regime (see also Knowlton *et al.*, 1990; Done, 1992). Alternative states (or *attractors* in the lingo of chaos theory; see Martin, 1998a) are usually the result of some disturbance to a system. Such behavior has very important implications for humankind's ability to predict the response of reefs and other ecosystems to natural versus anthropogenic disturbance for conservation and management efforts.

The more frequently a particular community state occurs (the larger its basin of attraction), the more often it should occur in the fossil record. Based on the record of Pleistocene reefs in the Caribbean, Jackson (1992; see also Aronson and Precht, 1997) found that reefs were dominated by the coral genera *Acropora*

10.5 Disturbance and alternative community states

and *Montastrea*, both of which are dominant in relatively undisturbed ("normal") reefs today. These community states, then, are typically stable and the fossil record seems to indicate that a massive disturbance is needed to shift ecosystems to a new community type (but see for "all-or-none" preservation of *Acropora* and other fragile coral genera in Chapter 2).

Nevertheless, several relatively minor disturbances to the reefs at Discovery Bay, Jamaica, beginning with Hurricane Allen in 1980 and followed by a massive die-off of sea urchins which feed on algae, have significantly altered the reef community there (Hughes *et al.*, 1985; Lessios *et al.*, 1984). Since macroalgae grow faster than corals (and their spat or settled larvae), loss of the urchins has allowed algae to overwhelm the reefs. As of this writing (1998), the Discovery Bay reefs are now dominated by algae, and it is uncertain if and when the reefs will return to their previous (and presumably normal) coral-dominated state. Apparently the disturbances were of sufficient magnitude to shift the community to an alternative community state, at least for the forseeable future (see also Greenstein, 1989; Greenstein *et al.*, 1995; Chapter 2).

Similarly, reefs on the north coast of Cuba were smothered with carbonate sands as part of a beach nourishment project. Over the succeeding 20 years, however, the sands have been eroded away and hard substrates, which are preferred by corals, are becoming increasingly available. The reefs are gradually restoring themselves, presumably through settlement of larvae from healthy populations located farther offshore (Manuel Iturralde-Vinent, personal communication; personal observations of the author in 1995). Unfortunately, these and other reefs in the vicinity may soon be threatened by hotel development and the accompanying sewage and sedimentation. Rising nutrient levels in many localities may also be threatening the health of coral reefs or have already largely destroyed them (e.g., Tomascik and Sander, 1985, 1987a,b). Increased nutrient levels stimulate plankton production and the invasion of rapidly growing suspension-feeding organisms that outcompete corals and shift communities toward ones dominated by macroalgae and sponges (Smith *et al.*, 1981; Hallock and Schlager, 1986; Hallock, 1987, 1988; Hallock *et al.*, 1993).

The effects of nutrients on reefs has been hotly disputed by some workers and may reflect the temporal scales of measurement of different studies. Although Szmant and Forrester (1996) recently concluded that nutrient levels are *not* elevated in waters off the Florida Keys, their samples were taken during a *1–2 year* period. In contrast, Cockey *et al.* (1996) found that from the years 1961 to 1992 there has been a shift in foraminiferal assemblages off Key Largo (Florida) from large, long-lived algal symbiont-bearing taxa to smaller, fast-growing heterotrophic taxa that may be the result of increased nutrient loading in the Keys. Also, working in Biscayne Bay (Florida), Hudson et al. (1994) found that growth rates calculated from colonies of *Montastrea annularis* (a common reef-building coral) as old as *200 years* began to decline after about 1950 and that the decline may be related to increasing sewage effluent. Bell (1991, 1992) and Swart et al. (1996) came to similar conclusions for the Great Barrier Reef (Australia) and Florida Bay, respectively (see also Chapter 2).

Other studies suggest permanent changes to whole regions following human settlement. It appears that the Chesapeake Bay has shifted to a new stable state as a result of human activities. Cooper (1995) used diatoms and pollen to reconstruct a 2000-year history of Chesapeake Bay. She found that sedimentation, eutrophication, and anoxia (lack of oxygen related to massive diatom blooms caused by eutrophication and then die-off) all increased dramatically following settlement of the region. Increased sedimentation and eutrophication were related to land use (deforestation), sewage input, and freshwater runoff.

10.6 Population dynamics and extinction in the fossil record

Species that are geographically widespread and therefore likely to be eurytopic are more likely to survive extinction than species that are stenotopic. Stenotopic species are more likely to be geographically restricted, and their populations may be smaller and undergo greater fluctuations in abundance than those of eurytopic species. Eurytopy is also more likely to promote a wider variety of trophic (feeding) strategies and other sorts of biotic interactions than stenotopy (McKinney et al., 1996).

McKinney and Frederick (1992) tested these hypotheses using the abundances of 25 species of benthic foraminifera collected at 1-m intervals from a single section of the Eo-Oligocene Ocala Limestone in Florida. They analyzed the population abundance fluctuations through time (upsection) using a fractal-based technique called rescaled-range analysis (Sugihara and May, 1990; see Chapter 7 for introduction to fractals). Despite the short-term incompleteness of the rock record (Chapter 7), the method works because it is fractal-based and therefore independent of scale; this approach may therefore prove useful in minimizing time-averaging and other types of taphonomic alteration.

Let $n(t)$ equal the abundance of a species during a particular time interval t, and let $n'(t)$ represent the normalized deviations of abundance

$$n'(t) = \bar{n}(t) - n(t) \tag{10.1}$$

for each t between 0 and T (the entire time interval). The rescaled range is then

$$R(T) = \max n'(t) - \min n'(t) \tag{10.2}$$

where $0 \leq t \leq T$. Variations in $R(T)$ through time are then analyzed with a fractal model

$$R(T) = cT^H \tag{10.3}$$

where c is a proportionality constant $[R(T) \propto T^H]$ and $H = 1/F$, and F is the fractal dimension ($F = 1$ for a straight line and $F = 2$ for a plane; Chapter 7). If $F = 2$, then $H = 0.5$, which means that the plane is "filled" by the curve and represents a random walk (Sugihara and May, 1990); in this case, the time series (the collection of n's over the entire time interval T) of rescaled abundance variations R also represents a random walk (Brownian motion) no matter what the scale of Δt (duration of observation). Thus, abundance fluctuations show no persistent trend in a particular direction (increase or decrease).

As F decreases from 2 to 1, H increases above 0.5, and R becomes less and less plane-filling (tending toward a line). In this case, there is a persistent trend of increasing variation in abundance at every scale of Δt. There will be higher highs and lower lows because as Δt increases, variation also increases. If, on the other hand, H decreases to <0.5, variation decreases (anti-persistent).

McKinney and Frederick (1992) calculated H for the abundance time series of each species of foraminifer. H was calculated from the slope of the line in the plot of $\log R(T)$ versus $\log T$ (because the logarithm of $R(T) = cT^H$ equals $\log R(T) = H \log T + \log c$, where H is the slope of the line $y = mx + b$, $x = \log T$ and $b = \log c$; Chapter 7). They then compared the H values of each species versus the stratigraphic range of the species in the fossil record to see if species with high (low) H values were in fact extant for short (long) intervals of geological time. They found a statistically significant negative correlation ($r = -0.60$; $P < 0.05$) between H of a species and its stratigraphic range; i.e., as H decreased, stratigraphic range increased. In other words, populations with greater abundance fluctuations go extinct sooner than those that maintain relatively stable populations. Whether or not this approach can be used to avoid time-averaging in other settings remains to be seen.

10.7 Holocene sea-level change

One area to which (micro)paleontology has contributed to understanding natural versus anthropogenic disturbance is the study of sea-level change. There are several reasons for the tremendous utility of microfossils, especially foraminifera, in high-resolution sea-level studies: (1) their tests are quite abundant in marsh facies, often numbering more than 1000 per 10 cm^3 (Scott and Medioli, 1980a); (2) they "provide the most sensitive parameter of marsh zonation because [they] live on the marsh surface or in the topmost ... sediment [and] are buried in an environment with limited or no bioturbation" (Nydick et al., 1995, pp. 139–140); and (3) marsh foraminiferal zones are said to be basically identical worldwide and occur within narrowly defined bands according to elevation (desiccation) and salinity (e.g., Scott and Medioli, 1980a, 1986). These zones may be less than 10 cm in vertical range and have been used to resolve high-frequency (decadal to millenial), low amplitude (<10 m) changes of Holocene sea-level to ±5–10 cm (Scott and Medioli, 1980a, 1986; Williams, 1989, 1994; Thomas and Varekamp, 1991; Varekamp et al., 1992; Gehrels, 1994; Nydick et al., 1995). In the Holocene, this resolution surpasses that of (a) marsh plants (e.g., *Spartina alterniflora*), which exhibit coarser zonations and which may decay beyond recognition (Goldstein, 1988; Gehrels, 1994); (b) the marine oxygen isotope curve; and (c) coral reef terraces (e.g., Edwards et al., 1993; Blanchon and Shaw, 1995).

Studies of marsh foraminifera hold important implications for: (1) substantiating postulated mechanisms of Holocene sea-level change, such as growth and decay of ice sheets or changes in ocean surface topography, over durations much longer than those afforded by synoptic observations; (2) correlating presumed Holocene sea-level fluctuations with climatic phenomena such as the Little Climate Optimum (AD 1000–1300) and Little Ice Age (AD 1400–1700) that are correlated with the rise and fall of ancient civilizations (Varekamp et al., 1992; van de Plaasche, 1991; van de Plassche et al., 1995); and (3) documenting the natural evolution of wetlands in response to sea-level rise (Fletcher et al., 1990, 1992, 1993; Fletcher, 1992). Some workers have suggested that the Holocene transgression has been basically smooth, and that Holocene coastal systems have evolved in a continuous manner, whereas other workers, often based on foraminiferal assemblages, have suggested five or six transgressions over approximately the past 7000 years (Williams, 1989, 1994; Thomas and Varekamp, 1991; Varekamp et al., 1992; van de Plaasche, 1991; van de Plassche et al., 1995) and that coasts have therefore evolved episodically. Which scenario is correct and what are the implications for rates of wetland loss – and management – in response to sea-level rise (Zedler, 1988; Stanley and Warne, 1993; Wheeler et al., 1995; see also van der Vink et al., 1998)?

Simply mapping marsh assemblages alone is unlikely detect the full range of climatic changes (and processes) because much of the dynamics of formation of marsh foraminiferal assemblages has been overlooked (Jonasson and Patterson, 1992;

Goldstein and Harben, 1993; Hoge, 1995; Goldstein *et al.*, 1995). Indeed, differential preservation of foraminiferal assemblages can, by itself, mimic the magnitude and frequency of sea-level change. In low marshes, fragile agglutinated (e.g., *Ammotium salsum*, *Miliammina fusca*) and calcareous (e.g., *Ammonia beccarii*) foraminifera are often lost via oxidation of organic test cements and dissolution, respectively, so that the fossil assemblage may be dominated by more preservable, typical high-marsh agglutinated species, such as *Jadammina macrescens* and *Trochammina inflata*, which normally comprise much smaller proportions of the original living population at the low-marsh SWI (Scott and Medioli, 1980a, 1986; Jonasson and Patterson, 1992; Goldstein and Harben, 1993; Patterson *et al.*, 1994; Goldstein *et al.*, 1995). Ironically, abundance changes of *Jadammina macrescens* and other robust species have been used to construct a number of sea-level curves which may, in some cases, mimic sea-level regressions when none have occurred. Low numbers of foraminifera in highest high marsh and non-marine facies have also been widely interpreted to indicate regression, although early diagenetic alteration of assemblages occurs in the high-marsh (e.g., Goldstein *et al.*, 1995). Such pronounced, but erroneous, signals of sea-level fall can falsely accentuate rates of sea-level rise and suggest erroneous causal mechanisms, as well (see also Chapter 5 for effects of infaunal taxa). Other microfossil taxa may occur too sporadically to be useful or may be subject to early diagenesis in marine sediments (Hoge, 1994; cf. John, 1995; Charman *et al.*, 1998), although testate amoebae and ostracodes are resistant to alteration in freshwater (lake) sediments of relatively low pH, including polluted environments (e.g., Patterson and Kumar, in press).

10.8 Paleophysiology

A relatively untapped field of research in paleontology is that of paleophysiology (Koch, 1998). One debate associated with this field, is the question of the body temperature of dinosaurs. Were they ectothermic (cold-blooded) or endothermic (warm-blooded)? Although recent evidence suggests that the the group lived at neither extreme but varied according to taxon (Chinsamy and Dodson, 1995), others have suggested that based on oxygen isotope ratios of bones, dinosaurs were warm-blooded.

Although the bones used in stable isotope analyses would appear to be unaltered, Kolodny *et al.* (1996) point out that the preservation of detailed bone microstructure – such as Haversian canals – is not an infallible criterion of the quality of preservation, as some of the best preserved vertebrate remains have been permineralized by silica, and apparently pristine shells of invertebrates may actually be chemically altered (Chapter 3). Kolodny *et al.* (1996) concluded that most apatitic vertebrate fossils are pseudomorphs replacing original skeletal structure because living bone is comprised by about one-third by weight organic matter; based on mass balance considerations, at least half of an apatitic fossil must, then, be new material added after death. Denys *et al.* (1996) came to similar conclusions based on electron microscopic analyses of small bones of Pleistocene (Olduvai) mammals in Tanzania (see also Grupe *et al.*, 1989).

Kolodny *et al.* (1996) found that although the correlation between oxygen isotope ratios of the phosphate in living fish bone ($\delta^{18}O_p$) and that in apatite is poor, it is much better in fossil fish. Moreover, overlap of $\delta^{18}O_p$ values in fish and mammals that lived in contact with the same waters suggests a strong environmental overprint on $\delta^{18}O_p$ during life, whereas the similarity between $\delta^{18}O_p$ in bones of fossil fish, dinosaurs and other reptiles that coexisted at different latitudes suggests that $\delta^{18}O_p$ is more indicative of the geochemistry of burial. Such changes may occur quite rapidly and therefore are not necessarily related to a taphonomic clock. Also, living bone contains almost no rare earth elements (REE) or uranium, whereas high concentrations of both may occur in fossils, which may reflect diagenetic alteration and reworking (Chapter 5; Trueman and Benton, 1997; see also exchange between Trueman and Palmer [1997] and Samoilov and Benjamini [1997]). Iacumin *et al.* (1996), for example, found substantial alteration of mammal bones exposed to meteoric waters in Arago cave (eastern Pyrénées, southeast France) that ranged in age from approximately 650 to 450 Ka, whereas Stuart-Williams *et al.* (1996) found changes in uranium concentrations in substantially younger human bones of Teotihuacan and Oaxaca (Mexico) that ranged in age from only about 300 BC to AD 750.

Although teeth intuitively seem more chemically resistant (Chapter 3), they also appear to reflect a combination of external environmental and internal physiological factors. Reinhard *et al.* (1996) found that, like bone, cave bear teeth from four sites in Spain reflected the oxygen isotope composition of ingested water, which is in turn correlated with mean annual temperature; thus, teeth may be used as paleothermometers in some cases. On the other hand, Fricke and O'Neill (1996) concluded that $\delta^{18}O_p$ values of teeth from modern sheep and fossil adult *Bison* reflect the chemistry of well water, surface water, and the mother's milk, seasonal variation in which was detected in samples taken down the length of teeth; they suggested that $\delta^{18}O_p$ of late-forming teeth (e.g., premolars) might be used as a proxy for $\delta^{18}O$ of local meteoric waters in studies of climate change.

Note added in proof:
Barrick *et al.* (1999, Oxygen isotopes from turtle bone: applications for terrestrial paleoclimates? *Palaios* 14: 186–191) indicate that oxygen isotopes of turtle bone may be used to reconstruct isotope ratios of local meteoric water.

11 Taphonomy as a historical science

I am still learning. Michelangelo

11.1 Major themes

Would we really understand the Earth and its Life if we only understood the present? Geology is the study of the history of the Earth and its Life. Taphonomy is no different. Taphonomy, and the more inclusive disciplines of paleontology and geology, are historical sciences (Chapter 1). The ultimate value of geology is that there are patterns and processes that can *only* be documented by the geological record (Martin, 1998a).

One of the major themes of this book is that biogenic particles and assemblages exhibit an incredible diversity of taphonomic pathways and histories (Chapters 2, 3). However, historical sciences find themselves in a nebulous realm: there are often too few parts to average their behavior and too many to account for separately with their own differential equation (Allen and Starr, 1982; see also Martin, 1998a). In such situations, it is very difficult to formulate laws, which presumably express unfailing *predictive* relationships between objects, processes, etc., except under the most restrictive of conditions (as is usually the case in the reductionist approach), much less rules or principles.

Other sciences have been forced, by their nature, to deal with similar conundrums. Small-number systems treat objects (e.g., planets) with the precision of differential equations (as Newton did). Nevertheless, the French mathematician Henri Poincaré demonstrated that although the behavior of two balls may be accurately predicted with mathematical equations, the motion of three or more (such as the Sun, Earth, and Moon) cannot (Three Ball Problem); hence, the motions of the planets are only an approximation – perhaps the best we have in all of the sciences – even though they are based on Newton's laws (Martin, 1998a). And then there are fractals, which represent highly unpredictable but nevertheless *deterministic* behavior generated by a handful of simple differential equations.

By contrast, large-number systems deal with so many objects (e.g., atoms, molecules), that the laws that are used to predict what will happen are *statistical* in nature and describe an *empirical* relationship (e.g., gas laws), not the ultimate *verae causae* of Herschel (1831; see also Ruse, 1979; Chapter 1). A canister of gas is an ergodic system, and its components will "explore" every possible configuration so quickly that its return time will be quite short. Even if the exact position of molecules depends on their previous interactions with other molecules, the overall macroscopic behavior of the system can be described quite accurately.

Geological and biological systems have much longer return times because they explore the possible configurations much more slowly. The concept of a steady state was a tenet near and dear to the heart of Charles Lyell (Chapter 1), and in most cases its simplifying assumptions seem to work quite well (Chapter 4). But, given enough time, natural systems *evolve*: on geological time scales of sufficient duration, the concepts of equilibrium and steady state are figments of reductionism (Chapters 8, 9; Martin, 1998*a*).

Not surprisingly, then, history and historical sciences have often been criticized as consisting of the "unique, unrepeatable, [and] unobservable" (Gould, 1986): if it can't be repeated, it isn't science. History can presumably demonstrate only tantalizing correlations (a relationship between two or more phenomena or sets of data), but it cannot demonstrate cause and effect (cf. above). Power laws are a prime example: these laws, like the gas laws, are statistical in nature and mostly ignore the effects of history, and, at least for earthquakes and mass extinctions, they cannot predict when the next catastrophe will happen (Bak, 1996; Martin, 1998*a*; see also Bambach, 1998).

Nevertheless, a large-number approach was adopted by Seilacher *et al.* (1985) in their classification of Lagerstätten (Chapter 6), and by Flessa and Kowalewski (1994) in their compilation of ^{14}C-dated shells (Chapter 5). And like the gas laws, such nomothetic (law-like) classifications of fossil assemblages provide important clues about depositional setting, rates of burial and bioturbation, durations of time-averaging, bottom-water oxygenation, and can be used to *predict* further occurrences of fossil assemblages and their traits.

Frodeman (1995, p. 965) argues that in historical sciences, "the specific causal circumstances surrounding the individual entity (what led up to it, and what its consequences were) are the main concern ... In geology, the goal is ... to chronicle the particular events that occurred at a given location ... *This means that hypotheses are not testable in the way they are in the experimental sciences*

... Although the geologist may be able to duplicate the laboratory conditions of another's experiment ... *the relationship of these experiments to the particularities of Earth's history ... remains uncertain*" [my italics].

Geology involves the methodology of *hermeneutics*, which originated in the nineteenth century as a means of reconciling contradictions in the Bible, and which was later applied to historical and legal documents to determine the original intent of the authors (Frodeman, 1995). In so doing, the interpreter must assign different values to different statements regarding their clarity or intent. In the same way, a paleontologist gives different weights to data, such as the the preservational state of hardparts (Chapters 2, 3) or the temporal acuity and continuity of strata (Chapters 4, 7), based on the clues of past events and processes in a way analogous to how a physician makes a diagnosis or a detective builds a circumstantial case against a defendant (Frodeman, 1995, p. 963). From a taphonomic standpoint, then, we are not so much interested in the fact that an assemblage formed, but in the *processes* of formation. In this sense, a taphonomist produces a narrative, but not in a pejorative sense; rather, narrative means explanation by integration into a larger history or *context* that gives meaning to the data; in other words, a *hierarchy* of parts that interacted at different spatiotemporal *scales* (e.g., Miller, 1988).

11.2 Some more rules

Ten rules of taphonomy were presented in Chapter 1 that often seem like so much common sense. For those who like to read the last chapter first, this book has enumerated a number of other guidelines, some of which are not *intuitively* obvious:

(11) Small hardparts do not necessarily outnumber large ones.
(12) Big hardparts do not necessarily last longer than small ones.
(13) The surface condition or taphonomic grade of a shell is not necessarily an accurate indicator of age, and hardparts that follow different pathways may come to resemble each other (*equifinality*), at least in shallow marine settings and perhaps in deeper ones as well. Not all biogenic remains are alike (Sorby Principle) and *the taphonomic clock may be reset repeatedly before final burial* (corollary 1 to Rule 13). This appears to be a general phenomenon, so that in actualistic studies involving reworking of late Quaternary sediments, many shells will exceed the range of ^{14}C dating and will have to be dated using amino acid racemization techniques (which can

be used back to approximately 1.5 myr; e.g., Powell and Davies, 1990; Wehmiller *et al.*, 1995; Martin *et al.*, 1996; Kowalewski *et al.*, 1998). Moreover (corollary 2), substantial variation in shell destruction may have occurred on different spatial (e.g., water depth) and temporal scales because of changes in shell mineralogy, architecture, environmental conditions, and the nature of taphonomic agents.

(14) Good preservation of fossil assemblages depends not just on rapid burial but also optimal sediment accumulation rates (Lagerstätten excepted; cf. Rule 1). If sedimentation is too slow, fossils of different stratigraphic zones are mixed together and information is lost, but if sediment accumulation is too fast, fossil assemblages are "diluted" and less likely to be preserved and discovered. Rule 14 raises issues about models of formation of fossil assemblages and their time-averaging. Studies of fossil assemblage formation are usually based on at least one very important simplifying assumption: that shell production and hardpart input to a fossil assemblage are constant, an assumption that has certainly proved useful when rates of time-averaging are on the order of hundreds of years or more. Moreover, an age distribution appears to be established for each assemblage, depending on the taphonomic setting: recent studies have found that age distributions of shells are skewed toward older ages, but assemblages appear to consist mainly of more recent remains (cf. principle of linear superposition in Chapter 5). If this is in fact the case, it has important consequences for coupling studies of the rates of taphonomic processes with rates of biogeochemical cycling.

(15) Fossil assemblages and their taphonomic traits formed along environmental gradients. The taphofacies concept is certainly one of the most powerful – perhaps *the* most powerful – predictive tools in taphonomy because environmental gradients of abrasion, dissolution, bioerosion, and mineralization are predictable. The taphofacies concept can be used to formulate hypotheses that can be tested via field observations, experiment, or a combination of both. But (corollary to Rule 15), similar taphofacies through time should be compared when evaluating cylic or secular trends, whereas comparison of isochronous – but different depositional settings – provides information on spatial variation of biotas (Behrensmeyer and Hook, 1992).

(16) Some fossil assemblages resemble some assemblages more than others, but they will likely never be *exactly* alike because there are so many pathways of fossil assemblage formation that each assemblage ("singularity") must

11.2 Some more rules

be examined on a case by case basis before generalizations are made. Nevertheless (corollary to Rule 16), fossil assemblages or their individual remains, when arranged in nomothetic classifications, may give important insights into taphonomic setting (cf. Section 11.1).

(17) Information gain about taphonomic settings (e.g., taphofacies; nomothetic-like classifications) and long-term community dynamics may result from the actions of taphonomic agents. Whenever possible, information gain should be stressed in future studies (Behrensmeyer and Kidwell, 1985) because it will lead to greater synergism between paleontologists, ecologists, and workers in other fields such as biogeochemistry.

(18) The fossil and stratigraphic record cannot be taken at face value, and stratigraphic sections and depositional settings must be evaluated with regard to the questions being asked. As has been documented many times, what seems obvious based on common sense, intuition, or cursory visual examination may very well be incorrect. It is no less true regarding the completeness and temporal acuity of stratigraphic sections.

(19) Taphonomic and stratigraphic agents are hierarchical (Allen and Starr, 1982; Martin, 1998a; see also Miller, 1988). Fossil assemblages form mainly by processes that occur at lower levels, although substantial variation may occur on spatial (regional down to highly localized patches) and temporal (ecological to geological) scales. Local processes would not appear to interact with those that occur at much higher levels – and vice versa – because the levels are too far apart in terms of time and space to interact strongly (Allen and Starr, 1982; Martin, 1998a). Nevertheless, if the temporal scale of observation is increased, the long-term cyclic and secular influences on the formation of fossil assemblages becomes apparent: tectonics, volcanism, atmospheric and oceanic pCO_2, Mg/Ca ratios, etc.

(20) In some – perhaps many – cases, the present may be the key to nothing at all (e.g., no analog conditions). The fossil record is replete with examples of biological and geochemical change, the rates of which cannot be measured over human time scales. Among other factors, fossil taxa and their diversity, including plankton; the chemical composition and external and internal morphology of hardparts; epifaunal and infaunal tiering (bioturbation); nutrient and food availability; rates of hardpart production; and predation intensity, not to mention abiotic factors such as the chemistry of seawater, have all changed through the Phanerozoic in cyclic and secular manners (Chapters 8, 9). Extrapolation from laboratory and field studies to the past, especially given the differences between

experimental methodologies and time scales of observation (Chapter 1; section 11.1), must be conducted in the context of ancient environments.

11.3 Final thoughts

Today, historical sciences such as geology – and especially paleontology – are under fire because of decreased employment opportunities and because of the historically anti-intellectual environment in the United States and elsewhere that presently hides behind such movements as creation science and balanced budget legislation. For most of this century the fortunes of paleontology have depended on the oil patch (Martin, 1991, 1995b, 1999; see also Hallock, 1995; Gastaldo et al., 1998). In the past decade, however, there has been a sea change, and paleontology has been largely swept aside: resource exploitation has been replaced by resource conservation and management. Even as this manuscript is being finalized, however, the fortunes of the petroleum industry are beginning to wax once again. Although enhanced employment in the petroleum industry will permit paleontology to regain some of its clout, if employment opportunities for paleontologists are to remain relatively constant instead of highly cyclical, paleontologists must accept the fact that this and related disciplines are more likely to thrive if their application to practical problems – whether those of resource exploitation or conservation – increases (Chapter 10; see also Hutchinson, 1997; Lane, 1997). What sorts of research might ameliorate the situation?

One of the main problems in bridging the gap between "environmental" (ecological) and paleontological scales is of course the phenomenon of time-averaging and the insidious problem of bioturbation. Bioturbation is one of the primary obstacles to interpreting the fossil record – not to mention anthropogenic and pre-anthropogenic environmental records – at *practicable levels of temporal resolution*. Unfortunately, bioturbation is probably the most studiously avoided of all aspects of taphonomy because of the mathematical complexities involved and the potential non-uniqueness of solutions. Better training in mathematical techniques – including statistics and numerical modeling – are required to overcome math-phobia.

Numerical simulations, although complex and tedious, would prove useful because both potential short-term and long-term results could be mimicked and then compared to the fossil record (e.g., Cutler and Flessa, 1990; Behrensmeyer and Chapman, 1993). The advantage of computer modeling is of course that one does not have to wait millions of years or more for the output. Such models are unlikely to provide *the* answer, but sensitivity analyses may give

11.3 Final thoughts

valuable insights about fossil assemblage formation and preservation over different spatiotemporal scales. The models could either be site-specific, regional, or perhaps even global in scale, depending on the data base and the hypotheses being posed. Site-specific numerical models will require accurate estimates of rates of bioturbation, sediment accumulation, burial and reworking, and shell input and loss for different taxa, whereas models of broader scope will have to include assessment of tectonics, sea-level change, productivity, etc.

Sampling programs and processing also come to mind in this regard, both of which vary from one laboratory to another and which may significantly affect the outcome of studies. In a much-publicized blind test on foraminiferal distributions below and above the K–T boundary at El Kef (Tunisia), for example, the results proved inconclusive (although they tended to lean against impact; Ginsburg, 1996, organizer; also, cf. Kerr, 1994b, and Keller, 1994; Officer and Page, 1996; see also discussion of similar controversies about the terrestrial K–T boundary in Chapter 10). The El Kef section has been demonstrated to be the most complete of all K–T sections based on graphic correlation. Nevertheless, as one participant mentioned, the experiment was probably doomed to fail because of the relatively coarse sampling regime that was used, which, along with different sample processing procedures, size fractions, and search times for rare species employed by different workers could mimic biological patterns of extinction and recovery (Kouwenhoven, 1997).

Most actualistic marine taphonomic studies have been confined to shallow water because of access and also probably the greater likelihood of funding. Some of these studies have dealt with heavily anthropogenically impacted communities like marshes and estuaries, in which the temporal resolution may be a few hundred years or less. Anthropogenic disturbance may also be affecting actualistic studies, a factor that has been almost entirely overlooked in taphonomic investigations. Powell (1992) suggested that an anthropogenically induced decline in shell input over the past century may account for carbonate production appearing to be insufficient to account for observed shelliness in modern habitats (cf. caveats in Chapter 5). He suggested that comparison of production and loss rates with carbonate content may be one way to assess possible anthropogenic impacts on Recent death assemblages (Powell, 1992; see also Reise and Schubert, 1987; Walter and Burton, 1990; Gattuso et al., 1998). Still, with the exception of Loubere and colleagues' work on the generation of foraminiferal assemblages, and the initial results of the Shelf-Slope Experimental Taphonomy Initiative (SSETI), taphonomy in shelf and slope settings has been mostly neglected; perhaps taphonomists could "piggyback" with researchers who are engaged in offshore studies that utilize research vessels, thereby decreasing expense and increasing feasibility in the eyes of funding agencies (S. M. Kidwell, 1994, personal communication, Geological Society of America meeting of the "Friends of Taphonomy"). Another possible offshore data base is the huge number of industrial wells; although samples from oil and gas wells have typically been dismissed as being too coarse (about 10-m composite ditch samples) and subject to downhole contamination ("caving" of younger sediments into older section as the well is drilled), if sedimentation rates are sufficiently high, sample mixing and contamination are minimized (Martin *et al.*, 1993; Martin and Fletcher, 1995). Indeed, micropaleontologists and petroleum geologists have refined their geochronology well beyond that anticipated by naysayers even 5 years ago. These sorts of studies could be coupled to depth-related diagenetic models (e.g., Middelburg *et al.*, 1997), durations of time-averaging, and sequence stratigraphic interpretations of sea-level and depositional environments (Flessa, 1993; Holland, 1995).

Taphonomists should also place more emphasis on the formation of fossil assemblages in relation to the rates of recycling of fossil elements, especially since the continental margins are important sites of carbon cycling (Walsh, 1991). Factors such as intertaxon differences in taphonomic behavior and their effect on differential preservation and time-averaging of assemblages are important not only to fossil preservation but also to porewater chemistry and the cycling of biogeochemically important elements, as well as the evolution of the cycles themselves. Quantitative estimates of such rates could also prove very revealing in terms of shell destruction and time-averaging in modern and ancient assemblages. Such studies would also complement investigations of the effect of adaptations and life habits on the evolution of hardparts, and the formation and preservation of taxa through time (e.g., Wetmore, 1987; Wood, 1993; Furbish and Arnold, 1997; Kontrovitz *et al.*, 1998).

The stratigraphic record is certainly imperfect. But it is not so imperfect that the stratigraphic ranges of fossil taxa have all been extended backward in time to a simultaneous origin, something that Charles Lyell wanted dearly to believe would eventually happen because of his religious convictions. At the other extreme (with some exceptions), paleontologists rarely get glimpses of "ecological" communities. But that does not mean that paleontologists must sit beneath the high table waiting for the crumbs to fall. Despite its imperfections, the fossil record has tremendous potential to provide accurate long-term records of paleobiological processes (e.g., Eldredge, 1995; Kidwell and Flessa, 1995), and taphonomists have begun to realize that their self-flagellation of the past was often unjustified. Although the geological record is imperfect, perhaps synergism between paleontologists and

11.3 Final thoughts

biologists would prove less problematical if workers were more cognizant of the limitations and benefits of taphonomic filters and temporal resolution.

Nevertheless, many non-taphonomists still ignore taphonomic processes. And although that often does not seem to make much difference in the results and interpretation of data, the potential pitfalls of such an approach are very real. This book will have served its purpose if it not only stimulates a greater appreciation of the nature of the stratigraphic record but also the application of taphonomy to academic and practical problems. Then taphonomy will truly come of age.

References

Abbott, S. T. and Carter, R. M. 1997. Macrofossil associations from mid-Pleistocene cyclothems, Castlecliff section, New Zealand: implications for sequence stratigraphy. *Palaios* 12:188–210.

Abel, O. 1912. *Grundzüge der Paläobiologie der Wirbeltiere*. Stuttgart: E. Schweitzerbartsche Verlagsbuchhandlung Nagele.

Abel. O. 1927. *Lebensbilder aus der Tierwelt der Vorzeit*. Jena: Verlag G. Fischer.

Abel, O. 1935. *Vorzeitliche Lebensspuren*. Jena: Verlag G. Fischer.

Abramovich, S., Almogi-Labin, A. and Benjamini, C. 1998. Decline of the Maastrichtian pelagic ecosystem based on planktic foraminifera assemblage change: implication for the terminal Cretaceous faunal crisis. *Geology* 26:63–66.

Acevedo, R. and Morelock, J. 1988. Effects of terrigenous sediment influx on coral reef zonation in southwestern Puerto Rico. *Proceedings 6th International Coral Reef Conference* 2:189–194.

Adelseck, C. G. 1977. Dissolution of deep-sea carbonate: preliminary calibration of preservational and morphologic aspects. *Deep-Sea Research* 24:1167–1187.

Adelseck, C. G. and Berger, W. H. 1977. On the dissolution of planktonic foraminifera and associated microfossils during settling and on the sea floor. In: W. V. Sliter, A. W. H. Bé and W. H. Berger (eds.) *Dissolution of Deep-Sea Carbonates*, pp. 70–81. Cambridge, MA: Cushman Foundation for Foraminiferal Research Special Publication No. 13.

Ahl, V. and Allen, T. F. H. 1996. *Hierarchy Theory: A Vision, Vocabulary, and Epistemology*. New York: Columbia University Press.

Aigner, T. 1982a. Calcareous tempestites: storm-dominated stratification in Upper Muschelkalk limestones (Middle Trias, SW-Germany). In: G. Einsele and A. Scilacher (eds.) *Cyclic and Event Stratification*, pp. 180–198. Berlin: Springer-Verlag.

Aigner, T. 1982b. Event-stratification in nummulite accumulations and in shell beds from the Eocene of Egypt. In: G. Einsele and A. Seilacher (eds.) *Cyclic and Event Stratification*, pp. 248–262. Berlin: Springer-Verlag.

Aigner, T. 1985. *Storm Depositional System: Dynamic Stratigraphy in Modern and Ancient Shallow-Marine Sequences*. Berlin: Springer-Verlag.

References

Albritton, C. C. 1986. *The Abyss of Time: Unraveling the Mystery of the Earth's Age.* Los Angeles: Jeremy P. Tarcher (Reprint).

Alexander, R. R. 1984. Comparative hydrodynamic stability of brachiopod shells on current-scoured arenaceous substrates. *Lethaia* 17:17–32.

Alexander, R. R. 1986. Life orientation and post-mortem reorientation of Chesterian brachiopod shells by paleocurrents. *Palaios* 1:303–311.

Alexander, R. R. 1990. Disarticulated shells of Late Ordovician brachiopods: inferences on strength of hinge and valve architecture. *Journal of Paleontology* 64:524–532.

Alexandersson, T. 1972. Micritization of carbonate particles: processes of precipitation and dissolution in modern shallow-marine sediments. *Bulletin of the Geological Institute, University of Uppsala* 3:201–236.

Algeo, T. J. 1993. Quantifying stratigraphic completeness: a probabilistic approach using paleomagnetic data. *Journal of Geology* 101:421–433.

Algeo, T. J. and Seslavinsky, K. B. 1995. The Paleozoic world: continental flooding, hypsometry, and sealevel. *American Journal of Science* 295:787–822.

Algeo, T., Berner, R. A., Maynard, J. B. and Scheckler, S. E. 1993. Late Devonian oceanic anoxic events and biotic crises: "Rooted" in the evolution of vascular plants? *GSA Today* 5(3):1.

Allen, J. R. L. 1984. Experiments on the settling, overturning and entrainment of bivalve shells and related models. *Sedimentology* 31:227–250.

Allen, T. F. H. and Hoekstra, T. W. 1992. *Toward a Unified Ecology.* New York: Columbia University Press.

Allen, T. F. H. and Starr, T. F. H. 1982. *Hierarchy: Perspectives for Ecological Complexity.* Chicago: University of Chicago Press.

Aller, J. Y. 1995. Molluscan death assemblages on the Amazon Shelf: implication for physical and biological controls on benthic populations. *Palaeogeography, Palaeoclimatology, Palaeoecology* 118:181–212.

Aller, R. C. 1982a. The effects of macrobenthos on chemical properties of marine sediment and overlying water. In: P. L. McCall and M. J. Tevesz (eds.) *Animal–Sediment Relations*, pp. 53–102. New York: Plenum Press.

Aller, R. C. 1982b. Carbonate dissolution in nearshore terrigenous muds: the role of physical and biological reworking. *Journal of Geology* 90:79–95.

Aller, R. C. and Cochran, J. K. 1976. ^{234}Th/^{238}Ur disequilibrium in near-shore sediment: particle reworking and diagenetic time scales. *Earth and Planetary Science Letters* 29:37–50.

Aller, R. C. and Dodge, R. E. 1974. Animal–sediment relations in a tropical lagoon, Discovery Bay, Jamaica. *Journal of Marine Research* 32:209–232.

Allison, P. A. 1986. Soft-bodied animals in the fossil record: the role of decay in fragmentation during transport. *Geology* 14:979–981.

Allison, P. A. 1988a. The role of anoxia in the decay and mineralization of proteinaceous macro-fossils. *Paleobiology* 14:139–154.

Allison, P. A. 1988b. Taphonomy of the Eocene London Clay biota. *Palaeontology* 31:1079–1110.

Allison, P. A. 1988c. Konservat-Lagerstätten: cause and classification. *Paleobiology* 14:331–344.

Allison, P. A. 1990. Variation in rates of decay and disarticulation of Echinodermata: implications for the application of actualistic data. *Palaios* 5:432–440.

Allison, P. A. and Brett, C. E. 1995. In situ benthos and paleo-oxygenation in the Middle Cambrian Burgess Shale, British Columbia, Canada. *Geology* 23:1079–1082.

Allison, P. A. and Briggs, D. E. G. 1991*a*. The taphonomy of soft-bodied animals. In: S. K. Donovan (ed.) *The Processes of Fossilization*, pp. 120–140. London: Belhaven Press.

Allison, P. A. and Briggs, D. E. G. 1991*b*. Taphonomy of nonmineralized tissues. In: P. A. Allison and D. E. G. Briggs (eds.) *Taphonomy: Releasing the Data Locked in the Fossil Record*, pp. 25–70. New York: Plenum Press.

Allison, P. A. and Briggs, D. E. G. 1993*a*. Burgess Shale biotas burrowed away? *Lethaia* 26:184–185.

Allison, P. A. and Briggs, D. E. G. 1993*b*. Exceptional fossil record: distribution of soft-tissue preservation through the Phanerozoic. *Geology* 21:527–530.

Allison, P. A. and Pye, K. 1994. Early diagenetic mineralization and fossil preservation in modern carbonate concretions. *Palaios* 9:561–575.

Allison, P. A., Smith, C. R., Kukert, H., Deming, J. and Bennett, B. A. 1991. Deep-water taphonomy of vertebrate carcasses: a whale skeleton in the bathyal Santa Catalina Basin. *Paleobiology* 17:78–89.

Allmon, W. D., Emslie, S. D., Jones, D. S. and Morgan, G. S. 1996. Late Neogene oceanographic change along Florida's west coast: evidence and mechanisms. *Journal of Geology* 104:143–162.

Almogi-Labin, A., Hemleben, C. and Meischner, D. 1998. Carbonate preservation and climatic changes in the central Red Sea during the last 380 kyr as recorded by pteropods. *Marine Micropaleontology* 33:87–107.

Alve, E. 1995. Benthic foraminiferal response to estuarine pollution: a review. *Journal of Foraminiferal Research* 25:190–203.

Amstutz, G. C. 1958. Coprolites: a review of the literature and a study of specimens from southern Washington. *Journal of Sedimentary Petrology* 28:498–508.

Anderson, D. M. 1994. Red tides. *Scientific American* 271(2):62–68.

Anderson, D. M., Lively, J. J., Reardon, E. M. and Price, C. A. 1985. Sinking characteristics of dinoflagellate cysts. *Limnology and Oceanography* 30:1000–1009.

Anderson, L. C. and McBride, R. A. 1996. Taphonomic and paleoenvironmental evidence of Holocene shell-bed genesis and history on the northeastern Gulf of Mexico shelf. *Palaios* 11:532–549.

Anderson, L. C., Sen Gupta, B. K. and Byrnes, M. R. 1997. Reduced seasonality of Holocene climate and pervasive mixing of Holocene marine section: northeastern Gulf of Mexico shelf. *Geology* 25:127–130.

Anderson, R. F., Bopp, R. F., Buesseler, K. O. and Biscaye, P. E. 1988. Mixing of particles and organic constituents in sediments from the continental shelf and slope off Cape Cod: SEEP-I results. *Continental Shelf Research* 8:925–946.

Andree, M. 1987. The impact of bioturbation on AMS ^{14}C dates on handpicked foraminifera: a statistical model. *Radiocarbon* 29:169–175.

Andree, M., Beer, J., Oeschger, H., Broecker, W., Mix, A., Ragano, N., O'Hara, P., Bonani, G., Hofman, H. J., Morenzoni, E., Nessi, M., Suter, M. and Wolfli, W. 1984.

^{14}C measurements on foraminifera of deep sea cores V28–238 and their preliminary interpretation. *Nuclear Instruments and Methods in Physics Research* B5:340–345.

Andrews, J. H. 1991. *Comparative Ecology of Microorganisms and Macroorganisms.* New York: Springer-Verlag.

Andrusevich, V. E., Engel, M. H., Zumberge, J. E. and Brothers, L. A. 1998. Secular, episodic changes in stable carbon isotope composition of crude oils. *Chemical Geology* 152:59–72.

Angell, R. W. 1967. The process of chamber formation in the foraminifer *Rosalina floridana* (Cushman). *Journal of Protozoology* 14:566–574.

Angell, R. W. 1979. Calcification during chamber development in *Rosalina floridana. Journal of Foraminiferal Research* 9:341–353.

Anonymous. 1997. Untitled pamphlet on the history and uses of amber. Copenhagen: Fehrn of Copenhagen (Ravhuset, Copenhagen Amber Museum).

Archer, D. and Maier-Reimer, E. 1994. Effect of deep-sea sedimentary calcite preservation on atmospheric CO_2 concentration. *Nature* 367:260–263.

Argast, S., Farlow, J. O., Gabet, R. M. and Brinkman, D. L. 1987. Transport-induced abrasion of fossil reptilian teeth: implications for the existence of Tertiary dinosaurs in the Hell Creek Formation, Montana. *Geology* 15:927–930.

Armentrout, J. D. 1996. High-resolution sequence biostratigraphy: Examples from the Gulf of Mexico Plio-Pleistocene. In: J. A. Howell and J. F. Aitken (eds.) *High Resolution Sequence Stratigraphy: Innovations and Applications*, pp. 65–86. London: The Geological Society Special Publication No. 104.

Arnaud, G., Arnaud, S., Ascenzi, A., Bonucci, E. and Graziani, G. 1978. On the problem of the preservation of human bone in sea-water. *Journal of Human Evolution* 7:409–420.

Aronson, R. B. 1992. Decline of the Burgess Shale fauna: ecologic or taphonomic restriction? *Lethaia* 25:225–229.

Aronson, R. B. 1993. Burgess Shale-type biotas were not just burrowed away: reply. *Lethaia* 26:185.

Aronson, R. B. and Precht, W. F. 1997. Stasis, biological disturbance, and community structure of a Holocene coral reef. *Paleobiology* 23:326–346.

Ascenzi, A. and Silvestrini, G. 1984. Bone-boring marine micro-organisms: an experimental design. *Journal of Human Evolution* 13:531–536.

Aslan, A. and Behrensmeyer, A. K. 1996. Taphonomy and time resolution of bone assemblages in a contemporary fluvial system: The East Fork River, Wyoming. *Palaios* 11:411–421.

Ausich, W. I. and Bottjer, D. J. 1991. History of tiering among suspension feeders in the benthic marine ecosystem. *Journal of Geological Education* 1991:313–318.

Baas-Becking, L. G. M., Kaplan, I. R. and Moore, D. 1960. Limits of the natural environment in terms of pH and oxidation–reduction potential. *Journal of Geology* 68:243–284.

Badgley, C. E. 1986a. Taphonomy of mammalian fossil remains from Siwalik rocks of Pakistan. *Paleobiology* 12:119–142.

Badgley, C. 1986b. Counting individuals in mammalian fossil assemblages from fluvial environments. *Palaios* 1:328–338.

Badgley, C. 1990. A statistical assessment of last appearances in the Eocene record of mammals. In: T. M. Bown and K. D. Rose (eds.) *Dawn of the Age of Mammals in the Northern Part of the Rocky Mountain Interior, North America*, pp. 153–167. Boulder: Geological Society of America Special Paper 243.

Bailey, C. M. and Peters, S. E. 1998. Glacially influenced sedimentation in the late Neoproterozoic Mechum River Formation, Blue Ridge province, Virginia. *Geology* 26:623–626.

Baird, G. C. and Brett, C. E. 1991. Submarine erosion on the anoxic seafloor: stratinomic, palaeoenvironmental, and temporal significance of reworked pyrite-bone deposits. In: R. V. Tyson and T. H. Pearson (eds.) *Modern and Ancient Continental Shelf Anoxia*, pp. 233–257. London: The Geological Society Special Publication No. 58.

Baird, G. C., Sroka, S. D., Shabica, C. W. and Beard, T. L. 1985. Mazon Creek-type fossil assemblages in the U.S. midcontinent Pennsylvanian: their recurrent character and paleoenvironmental significance. In: H. B. Whittington and S. Conway Morris (eds.) Extraordinary Fossil Biotas: Their Ecological and Evolutionary Significance. *Philosophical Transactions of the Royal Society of London B* 311:87–99.

Baird, G. C., Sroka, S. D., Shabica, C. W. and Kuecher, G. J. 1986. Taphonomy of Middle Pennsylvanian Mazon Creek area fossil localities, northeast Illinois: significance of exceptional fossil preservation in syngenetic concretions. *Palaios* 1:271–285.

Bak, P. 1996. *How Nature Works*. New York: Copernicus-Springer-Verlag.

Bambach, R. K. 1993. Seafood through time: changes in biomass, energetics, and productivity in the marine ecosystem. *Paleobiology* 19:372–397.

Bambach, R. K. 1998. Musings of an optimist – the study of past life should be on the brink of a great future. *Palaios* 13:511–513.

Bandel, K. 1974. Deep-water limestones from the Devonian-Carboniferous of the Carnic Alps, Austria. In: K. J. Hsü and H. C. Jenkyns (eds.) *Pelagic Sediments: on Land and Under the Sea*, pp. 93–115. Special Publication of the International Association of Sedimentology. Oxford: Blackwell.

Bandy, O. L. 1953. Ecology and paleoecology of some California foraminifera. Part 1. The frequency distribution of Recent foraminifera off California. *Journal of Paleontology* 27:161–182.

Bandy, O. L., Ingle, J. C. and Resig, J. M. 1964a. Foraminiferal trends, Laguna Beach outfall area, California. *Limnology and Oceanography* 9:112–123.

Bandy, O. L., Ingle, J. C. and Resig, J. M. 1964b. Foraminifera, Los Angeles County outfall area, California. *Limnology and Oceanography* 9:124–137.

Bandy, O. L., Ingle, J. C. and Resig, J. M. 1965a. Foraminiferal trends, Hyperion outfall, California. *Limnology and Oceanography* 10:314–332.

Bandy, O. L., Ingle, J. C. and Resig, J. M. 1965b. Modification of foraminiferal distribution by the Orange County outfall, California. *Ocean Science Engineering*, pp. 55–76.

Barbieri, R. 1996. Syndepositional taphonomic bias in foraminifera from fossil intertidal deposits, Colorado Delta (Baja California, Mexico). *Journal of Foraminiferal Research* 26:331–341.

Bard, E., Arnold, M., Duprat, J., Moyes, J. and Duplessy, J. C. 1987. Reconstruction of the last deglaciation: deconvolved records of

18O profiles, micropaleontological variations and accelerator mass spectrometric ^{14}C dating. *Climate Dynamics* 1:101–112.

Bareille, G., Labracherie, M., Mortlock, R. A., Maier-Reimer, E. and Froelich, P. N. 1998. A test of (Ge/Si)$_{opal}$ as a paleorecorder of (Ge/Si)$_{seawater}$. *Geology* 26:179–182.

Barghoorn, E. S. and Tyler, S. A. 1965. Microorganisms from the Gunflint Chert. *Science* 147:563–577.

Barker, R. 1997. *And the Waters Turned to Blood: The Ultimate Biological Threat*. New York: Simon and Schuster.

Barmawidjaja, D. M., van der Zwaan, G. J., Jorissen, F. J. and Puskaric, S. 1995. 150 years of eutrophication in the northern Adriatic Sea: evidence from a benthic foraminiferal record. *Marine Geology* 122:367–384.

Barrell, N. J. 1917. Rhythms and the measurements of geologic time. *Geological Society of America Bulletin* 28:745–904.

Barrera, E. 1994. Global environmental changes preceding the Cretaceous–Tertiary boundary: early–late Maastrichtian transition. *Geology* 22:877–880.

Barrera, E., Savin, S. M., Thomas, E. and Jones, C. E. 1997. Evidence for thermohaline-circulation reversals controlled by sea-level change in the latest Cretaceous. *Geology* 25:715–718.

Barron, E. J. 1989. Severe storms in Earth history. *Geological Society of America Bulletin* 101:601–612.

Barron, E. J. and Moore, G. T. 1994. *Climate Model Application in Paleoenvironmental Analysis*. Tulsa: SEPM (Society for Sedimentary Geology) Short Course No. 33.

Barron, J. A. 1993. Diatoms. In: J. H. Lipps (ed.) *Fossil prokaryotes and protists*, pp. 155–167. Boston: Blackwell Scientific Publications.

Barthel, K. W., Swinburne, N. H. M. and Conway Morris, S. 1994. *Solnhofen: a Study in Mesozoic Palaeontology*. Cambridge: Cambridge University Press.

Bartley, J. K. 1996. Actualistic taphonomy of cyanobacteria: implications for the Precambrian fossil record. *Palaios* 11:571–586.

Basan, P. B. and Frey, R. W. 1977. Actualpalaeontology and neoichnology of salt marshes near Sapelo Island, Georgia. In: T. P. Crimes and J. C. Harper (eds.) Trace Fossils 2, pp. 41–70. *Geological Journal* Special Issue 9.

Bates, N. R. and Brand, U. 1990. Secular variation of calcium carbonate mineralogy; an evaluation of ooid and micrite chemistries. *Geologische Rundschau* 79:27–46.

Bathurst, R. B. 1975. *Carbonate Sediments and Their Diagenesis*. Amsterdam: Elsevier.

Baumiller, T. K., Llewellyn, G., Messing, C. G. and Ausich, W. I. 1995. Taphonomy of isocrinid stalks: influence of decay and autotomy. *Palaios* 10:87–95.

Bé, A. W. H. 1977. An ecological, zoogeographic and taxonomic review of recent planktonic foraminifera. In: A. T. S. Ramsay (ed.) *Oceanic Micropaleontology*, pp. 1–100. London: Academic Press.

Bé, A. W. H. and Hutson, W. H. 1977. Ecology of planktonic foraminifera and biogeographic patterns of life and fossil assemblages in the Indian Ocean. *Micropaleontology* 23:369–414.

Bé, A. W. H., Morse, J. W. and Harrison, S. M. 1977. Progressive dissolution and ultrastructural breakdown of planktonic foraminifera. In: W. V. Sliter, A. W. H. Bé and W. H. Berger (eds.) *Dissolution of Deep-Sea Carbonates*, pp. 27–55. Cambridge,

MA: Cushman Foundation for Foraminiferal Research Special Publication No. 13.

Beck, R. A., Burbank, D. W., Sercombe, W. J., Olson, T. L. and Khan, A. M. 1995. Organic carbon exhumation and global warming during the early Himalayan collision. *Geology* 23:387–390.

Behrensmeyer, A. K. 1975. The taphonomy and paleoecology of Plio-Pleistocene vertebrate assemblages east of Lake Rudolf, Kenya. *Bulletin of the Museum of Comparative Zoology* 146:473–578.

Behrensmeyer, A. K. 1978. Taphonomic and ecologic information from bone weathering. *Paleobiology* 4:150–162.

Behrensmeyer, A. K. 1981. Vertebrate paleoecology in a Recent East African ecosystem. In: J. Gray, A. J. Boucot, and W. B. N. Berry (eds.) *Communities of the Past*, pp. 591–615. Stroudsburg, PA: Dowden, Hutchinson & Ross.

Behrensmeyer, A. K. 1982. Time resolution in fluvial vertebrate assemblages. *Paleobiology* 8:211–227.

Behrensmeyer, A. K. 1987. Miocene fluvial facies and vertebrate taphonomy in northern Pakistan. In: F. G. Ethridge, R. M. Flores and M. D. Harvey (eds.) *Recent Developments in Fluvial Sedimentology*, pp. 169–176. Tulsa: SEPM (Society for Sedimentary Geology) Short Course No. 39.

Behrensmeyer, A. K. 1988. Vertebrate preservation in fluvial channels. *Palaeogeography, Palaeoclimatology, Palaeoecology* 63:183–199.

Behrensmeyer, A. K. 1990. Bones. In: B. E. K. Briggs and P. R. Crowther (eds.) *Palaeobiology: A Synthesis*, pp. 232–235. Oxford: Blackwell.

Behrensmeyer, A. K. and Boaz, E. D. 1980. The recent bones of Amboseli National Park, Kenya, in relation to East African paleoecology. In: A. K. Behrensmeyer and A. P. Hill (eds.) *Fossils in the Making: Vertebrate Taphonomy and Paleoecology*, pp. 72–92. Chicago: University of Chicago Press.

Behrensmeyer, A. K. and Chapman, R. E. 1993. Models and simulations of time-averaging in terrestrial vertebrate accumulations. In: S. M. Kidwell and A. K. Behrensmeyer (eds.) *Taphonomic Approaches to Time Resolution in Fossil Assemblages*, pp. 125–149. Pittsburgh: Paleontological Society Short Courses in Paleontology No. 6.

Behrensmeyer, A. K. and Hook, R. W. 1992. Paleoenvironmental contexts and taphonomic modes. In: A. K. Behrensmeyer, J. D. Damuth, W. A. DiMichele, R. Potts, H.-D. Sues and S. L. Wing (eds.) *Terrestrial Ecosystems Through Time: Evolutionary Paleoecology of Terrestrial Plants and Animals*, pp. 15–136. Chicago: University of Chicago Press.

Behrensmeyer, A. K. and Kidwell, S. M. 1985. Taphonomy's contributions to paleobiology. *Paleobiology* 11:105–119.

Behrensmeyer, A. K., Todd, N. E., Potts, R. and McBrinn, G. 1996. Apparent versus real faunal turnover in the Late Pliocene vertebrate record of Africa. In: L. E. Babcock and W. I. Ausich (eds.) Sixth North American Paleontological Convention Abstracts of Papers, p. 28. Pittsburgh: Paleontological Society.

Behrensmeyer, A. K., Todd, N. E., Potts, R. and McBrinn, G. E. 1997. Late Pliocene faunal turnover in the Turkana Basin, Kenya and Ethiopia. *Science* 278:1589–1594.

Beier, J. A. and Hayes, J. M. 1989. Geochemical and isotopic evidence for paleoredox conditions during deposition of the Devonian-Mississippian New Albany

Shale, southern Indiana. *Geological Society of America Bulletin* 101:774–782.

Bell, M. A., Sadagursky, M. S. and Baumgartner, J. V. 1987. Utility of lacustrine deposits for the study of variation within fossil samples. *Palaios* 2:455–466.

Bell, M. A., Wells C. E. and Marshall, J. A. 1989. Mass-mortality layers of fossil stickleback fish: catastrophic kills of polymorphic schools. *Evolution* 43:607–619.

Bell, P. R. F. 1991. Status of eutrophication in the Great Barrier Reef Lagoon. *Marine Pollution Bulletin* 23:89–93.

Bell, P. R. F. 1992. Eutrophication and coral reefs: some examples in the Great Barrier Reef Lagoon. *Water Research* 26:553–568.

Beltrami, E. 1993. *Mathematical Models in the Social and Biological Sciences*. Boston: Jones and Bartlett.

Bender, M. L., Lorens, R. B. and Williams, D. F. 1975. Sodium, magnesium and strontium in the tests of planktonic foraminifera. *Micropaleontology* 21:448–459.

Bengston, S. and Conway Morris, S. 1992. Early radiation of biomineralizing phyla. In: J. H. Lipps and P. W. Signor (eds.) *Origin and Early Evolution of the Metazoa*, pp. 448–481. New York: Plenum Press.

Bengston, S. and Zhao, Y. 1997. Fossilized metazoan embryos from the earliest Cambrian. *Science* 277:1645–1648.

Benninger, L. K., Aller, R. C., Cochran, J. K. and Turekian, K. K. 1979. Effects of biological sediment mixing on the ^{210}Pb chronology and trace metal distribution in a Long Island Sound sediment core. *Earth and Planetary Science Letters* 43:241–259.

Benoit, G. J., Turekian, K. K. and Benninger, L. K. 1979. Radiocarbon dating of a core from Long Island Sound. *Estuarine and Coastal Marine Science* 9:171–180.

Benson, R. N. 1998. *Geology and Paleontology of the Lower Miocene Pollack Farm Fossil Site, Delaware*. Newark: Delaware Geological Survey Special Publication No. 21.

Benton, M. J. 1979. Increase in total global biomass over time. *Evolutionary Theory* 4:123–128.

Benton, M. J. 1995. Diversification and extinction in the history of life. *Science* 268:52–58.

Berg, H. C. 1993. *Random Walks in Biology*. Princeton: Princeton University Press.

Berger, W. H. 1967. Foraminiferal ooze: solution at depths. *Science* 156:383–385.

Berger, W. H. 1968. Planktonic foraminifera: selective solution and paleoclimate interpretation. *Deep-Sea Research* 15:31–43.

Berger, W. H. 1970. Planktonic foraminifera: selective solution and the lysocline. *Marine Geology* 8:111–138.

Berger, W. H. 1971. Sedimentation of planktonic foraminifera. *Marine Geology* 11:325–358.

Berger, W. H. 1977a. Dissolution of deep-sea carbonates: an introduction. In: W. V. Sliter, A. W. H. Bé and W. H. Berger (eds.) *Dissolution of Deep-Sea Carbonates*, pp. 7–10. Cambridge, MA: Cushman Foundation for Foraminiferal Research Special Publication No. 13.

Berger, W. H. 1977b. Deep-sea carbonates: dissolution profiles from foraminiferal preservation. In: W. V. Sliter, A. W. H. Bé and W. H. Berger (eds.) *Dissolution of Deep-Sea Carbonates*, pp. 82–86. Cambridge, MA: Cushman Foundation for Foraminiferal Research Special Publication No. 13.

Berger, W. H. 1978. Deep-sea carbonate: Pteropod distribution and the aragonite compensation depth. *Deep-Sea Research* 25:447–452.

Berger, W. H. 1989. Global maps of ocean productivity. In: W. H. Berger, V. S. Smetacek, and G. Wefer (eds.) *Productivity of the Ocean: Past and Present*, pp. 429–455. New York: Wiley.

Berger, W. H. and Heath, G. R. 1968. Vertical mixing in pelagic sediments. *Journal of Marine Research* 26:134–143.

Berger, W. H. and Mayer, L. A. 1978. Deep-sea carbonates: acoustic reflectors and lysocline fluctuations. *Geology* 6:11–15.

Berger, W. H. and Piper, D. J. W. 1972. Planktonic foraminifera: differential settling, dissolution, and redeposition. *Limnology and Oceanography* 17:275–287.

Berger, W. H. and Roth, P. H. 1975. Oceanic micropaleontology: progress and prospect. *Reviews of Geophysics and Space Physics* 13:561–585.

Berger, W. H. and Soutar, A. 1970. Preservation of plankton shells in an anaerobic basin off California. *Geological Society of America Bulletin* 81:275–282.

Berger, W. H., Johnson, R. F. and Killingley, J. S. 1977. "Unmixing" of the deep-sea record and the deglacial meltwater spike. *Nature* 269:661–663.

Berger, W. H., Ekdale, A.A. and Bryant, P. P. 1979. Selective preservation of burrows in deep-sea carbonates. *Marine Geology* 32:205–230.

Bergström, J. 1990. Hunsrück Slate. In: D. E. K. Briggs and P. R. Crowther (eds.) *Palaeobiology: a Synthesis*, pp. 277–279. Oxford: Blackwell.

Berner, R. A. 1969. Migration of iron and sulfur within anaerobic sediments during early diagenesis. *American Journal of Science* 267:19–42.

Berner, R. A. 1970. Sedimentary pyrite formation. *American Journal of Science* 268:1–23.

Berner, R. A. 1971. *Principles of Chemical Sedimentology*. New York: McGraw-Hill.

Berner, R. A. 1980. *Early Diagenesis: a Theoretical Approach*. Princeton: Princeton University Press.

Berner, R. A. 1981. A new geochemical classification of sedimentary environments. *Journal of Sedimentary Petrology* 51:359–365.

Berner, R. A. 1984. Sedimentary pyrite formation: an update. *Geochimica et Cosmochimica Acta* 48:605–615.

Berner, R. A. 1989. Biogeochemical cycles of carbon and sulfur and their effect on atmospheric oxygen over Phanerozoic time. *Palaeogeography, Palaeoclimatology, Palaeoecology* (Global and Planetary Change Section) 75:97–122.

Berner, R. A. 1990. Atmospheric carbon dioxide levels over Phanerozoic time. *Science* 249:1382–1386.

Berner, R. A. 1992. Weathering, plants, and the long-term carbon cycle. *Geochimica et Cosmochimica Acta* 56:3225–3231.

Berner, R. A. 1993. Paleozoic atmospheric CO_2: importance of solar radiation and plant evolution. *Science* 261: 68–70.

Berner, R. A. and Lasaga, A. 1989. Modeling the geochemical carbon cycle. *Scientific American* 60(3):74–81.

Berner, R. A. and Petsch, S. T. 1998. The sulfur cycle and atmospheric oxygen. *Science* 282:1426–1427.

Berner, R. A. and Raiswell, R. 1983. Burial of organic carbon and pyrite sulfur in sediments over Phanerozoic time: a new theory. *Geochimica et Cosmochimica Acta* 47:855–862.

Berner, R. A., Scott, R. A. and Thomlinson, C. 1970. Carbonate alkalinity in the pore waters of anoxic marine sediments. *Limnology and Oceanography* 15:544–549.

Berner, R. A., Baldwin, T. and Holdren, G. R. 1979. Authigenic iron sulfides as paleosalinity indicators. *Journal of Sedimentary Petrology* 49:1345–1350.

Bernhard, J. M. 1992. Benthic foraminiferal distribution and biomass related to pore-water oxygen content: central California continental slope and rise. *Deep-Sea Research* 39:585–605.

Berry, W. B. N. 1987. *Growth of a Prehistoric Time Scale: Based on Organic Evolution*. Palo Alto: Blackwell.

Berry, W. B. N. and Wilde, P. 1978. Progressive ventilation of the oceans: an explanation for the distribution of the lower Paleozoic black shales. *American Journal of Science* 278:257–275.

Berry, W. B. N., Wilde, P. and Quinby-Hunt, M. S. 1987. The oceanic non-sulfidic oxygen minimum zone: a habitat for graptolites? *Bulletin of the Geological Society of Denmark* 35:103–114.

Berthold, W.-U. 1976. Biomineralisation bei milioliden Foraminiferen und die Matrizen-Hypothese. *Naturwissenschaften* 63:196–197.

Bertness, M. D. and Miller, T. 1984. The distribution and dynamics of *Uca pugnax* (Smith) burrows in a New England salt marsh. *Journal of Experimental Marine Biology and Ecology* 83:211–237.

Bickart, K. J. 1984. A field experiment in avian taphonomy. *Journal of Vertebrate Paleontology* 4:525–535.

Birkeland, C. 1982. Terrestrial runoff as a cause of outbreaks of *Acanthaster planci* (Echinodermata: Asteroidea). *Marine Biology* 69:175–185.

Birkeland, C. 1987. Nutrient availability as a major determinant of differences among coastal hard-substratum communities in different regions of the tropics. In: C. Birkeland (ed.) *Comparison Between Atlantic and Pacific Tropical Marine Coastal Ecosystems: Community Structure, Ecological Processes, and Productivity*, pp. 45–97. Paris: UNESCO.

Birks, H. H., Whiteside, M. C., Stark, D. M. and Bright, R. C. 1976. Recent paleolimnology of three lakes in northwestern Minnesota. *Quaternary Research* 6:249–272.

Birks, H. J. B. 1981. Long-distance pollen in late Wisconsin sediments of Minnesota, U.S.A.: a quantitative analysis. *New Phytologist* 87:630–661.

Birks, H. J. B., Line, J. M., Juggins, S., Stevenson, A. C. and ter Braak, C. J. F. 1990. Diatoms and pH reconstruction. *Philosophical Transactions of the Royal Society of London* 327B:263–278.

Bishop, W. W. 1980. Paleogeomorphology and continental taphonomy. In: A. K. Behrensmeyer and A. P. Hill (eds.) *Fossils in the Making: Vertebrate Taphonomy and Paleoecology*, pp. 20–37. Chicago: University of Chicago Press.

Blackmon, P. D. and Todd, R. 1959. Mineralogy of some foraminifera as related to their classification and ecology. *Journal of Paleontology* 33:1–15.

Blake, D. F., Peacor, D. R. and Wilkinson, B. H. 1982. The sequence and mechanism of low-temperature dolomite formation: Calcian dolomites in a Pennsylvanian echinoderm. *Journal of Sedimentary Petrology* 52:59–70.

Blanchon, P. and Shaw, J. 1995. Reef drowning during the last deglaciation: evidence for catastrophic sea-level rise and ice-sheet collapse. *Geology* 23:4–8.

Blatt, H., Middleton, G. and Murray, R. 1972. *Origin of Sedimentary Rocks*. Englewood Cliffs, NJ: Prentice-Hall.

Blob, R. W. 1997. Relative hydrodynamic dispersal potentials of soft-shelled turtle elements: implications for interpreting skeletal sorting in assemblages of non-mammalian terrestrial vertebrates. *Palaios* 12:151–164.

Blob, R. W. and Fiorillo, A. R. 1996. The significance of vertebrate microfossil size and shape distributions for faunal abundance reconstructions: a Late Cretaceous example. *Paleobiology* 22:422–435.

Bluth, G. J. S. and Kump, L. R. 1991. Phanerozoic paleogeology. *American Journal of Science* 291:284–308.

Blyth Cain, J. D. 1968. Aspects of the depositional environment and palaeoecology of crinoidal limestones. *Scottish Journal of Geology* 4:191–208.

Boaz, N. T. 1982. Modern riverine taphonomy: its relevance to the interpretation of Plio-Pleistocene hominid paleoecology in the Omo Basin, Ethiopia. Ph.D. dissertation. University of California, Berkeley. Ann Arbor: University Microfilms International.

Bock, W. D. 1969. *Thalassia testudinum*, a habitat and means of dispersal for shallow water benthonic foraminifera. *Transactions, Gulf Coast Association of Geological Societies* 19:337–340.

Boersma, A. 1978. Foraminifera. In: B. U. Haq and A. Boersma (eds.) *Introduction to Marine Micropaleontology*, pp. 19–77. New York: Elsevier.

Boggs, S. 1987. *Principles of Sedimentology and Stratigraphy*. Columbus, OH: Merrill.

Bolles, E. B. 1997. *Galileo's Commandment: an Anthology of Great Science Writing*. New York: W. H. Freeman.

Bonner, J. T. 1988. *The Evolution of Complexity by Means of Natural Selection*. Princeton: Princeton University Press.

Bordeaux, Y. L. and Brett, C. E. 1990. Substrate specific association of epibionts on Middle Devonian brachiopods: implications for paleoecology. *Historical Biology* 4:203–220.

Boss, S. K. and Liddell, W. D. 1987. Back-reef and fore-reef analogs in the Pleistocene of north Jamaica: implications for facies recognition and sediment flux in fossil reefs. *Palaios* 2:219–228.

Boss, S. K. and Wilkinson, B. H. 1991. Planktogenic/eustatic control on cratonic oceanic carbonate accumulation. *Journal of Geology* 99:497–513.

Boston, W. B. and Mapes, R. H. 1991. Ectocochleate cephalopod taphonomy. In: S. K. Donovan (ed.) *The Processes of Fossilization*, pp. 220–240. London: Belhaven Press.

Bottjer, D. J. and Droser, M. L. 1994. The history of Phanerozoic bioturbation. In: S. K. Donovan (ed.) *The Paleobiology of Trace Fossils*. Chichester: Wiley.

Boucot, A. J. 1953. Life and death assemblages among fossils. *American Journal of Science* 251:25–40.

Boucot, A. J. 1983. Does evolution take place in an ecological vacuum? *Journal of Paleontology* 57:1–30.

Boucot, A. J., Brace, W. and DeMar, R. 1958. Distribution of brachiopod and pelecypod shells by currents. *Journal of Sedimentary Petrology* 28:321–332.

Boudreau, B. P. 1986a. Mathematics of tracer mixing in sediments. I. Spatially-dependent, diffusive mixing. *American Journal of Science* 286:161–198.

Boudreau, B. P. 1986b. Mathematics of tracer mixing in sediments. II. Nonlocal mixing and biological conveyor-belt phenomena. *American Journal of Science* 286:199–238.

Boudreau, B. P. 1997. *Diagenetic Models and Their Implementation: Modeling Transport and Reaction in Aquatic Sediments.* New York: Springer-Verlag.

Boudreau, B. P. and Imboden, D. M. 1987. Mathematics of tracer mixing in sediments. III. The theory of nonlocal mixing within sediments. *American Journal of Science* 287:693–719.

Boutilier, R. G., West, T. G., Pogson, G. H., Mesa, K. A., Wells, J. and Wells, M. J. 1996. Nautilus and the art of metabolic maintenance. *Nature* 382:534–536.

Bowler, P. J. 1976. *Fossils and Progress.* New York: Science History Publications.

Bowman, S. 1990. *Radiocarbon Dating.* Berkeley: University of California Press.

Bown, T. M. and Beard, K. C. 1990. Systematic lateral variation in the distribution of fossil mammals in alluvial paleosols, lower Eocene Willwood Formation, Wyoming. In: T. M. Bown and K. D. Rose (eds.) *Dawn of the Age of Mammals in the Northern Part of the Rocky Mountain Interior, North America*, pp. 135–151. Boulder: Geological Society of America Special Paper 243.

Bown, T. M. and Kraus, M. J. 1981a. Lower Eocene alluvial paleosols (Willwood Formation, northwest Wyoming, U.S.A.) and their significance for paleoecology, paleoclimatology, and basin analysis. *Palaeogeography, Palaeoclimatology, Palaeoecology* 34:1–30.

Bown, T. M. and Kraus, M. J. 1981b. Vertebrate fossil-bearing paleosol units (Willwood Formation, lower Eocene, northwest Wyoming, U.S.A.): implications for taphonomy, biostratigraphy, and assemblage analysis. *Palaeogeography, Palaeoclimatology, Palaeoecology* 34:31–56.

Bowring, S. A., Martin, M., Davidek, K., Erwin, D. H., Suter, S. J., Zu, Z. and Jin, Y. 1997. Geochronologic constraints on the end-Permian mass extinction. *Geological Society of America Abstracts with Programs* 29:A403.

Bowring, S. A., Erwin, D. H., Jin, Y. G., Martin, M. W., Davidek, K. and Wang, W. 1998. U/Pb Zircon geochronology and tempo of the end-Permian mass extinction. *Science* 280:1039–1045.

Boyle, E. A., Labeyrie, L. and Duplessy, J.-C. 1995. Calcitic foraminiferal data confirmed by cadmium in aragonitic *Hoeglundina*: application to the last glacial maximum in the northern Indian Ocean. *Paleoceanography* 10:881–900.

Bradley, W. H. 1946. Coprolites from the Bridger Formation of Wyoming: their composition and microorganisms. *American Journal of Science* 244:215–239.

Bradshaw, R. 1994. Quaternary terrestrial sediments and spatial scale: the limits to interpretation. In: A. Traverse (ed.) *Sedimentation of Organic Particles*, pp. 239–252. Cambridge: Cambridge University Press.

Brain, C. K. 1980. Some criteria for the recognition of bone-collecting agencies in African caves. In: A. K. Behrensmeyer and A. P. Hill (eds.) *Fossils in the Making: Vertebrate Taphonomy and Paleoecology*, pp. 108–130. Chicago: University of Chicago Press.

Bralower, H. J. and Thierstein, J. R. 1984. Low productivity and slow deep-water circulation in mid-Cretaceous oceans. *Geology* 12:614–618.

Brand, U. 1989*a*. Biogeochemistry of Late Paleozoic North American brachiopods and secular variation of seawater composition. *Biogeochemistry* 7:159–193.

Brand, U. 1989*b*. Aragonite–calcite transformation based on Pennsylvanian molluscs. *Geological Society of America Bulletin* 101:377–390.

Brand, U. 1990. Strontium isotope diagenesis of biogenic aragonite and low-Mg calcite. *Geochimica et Cosmochimica Acta* 55:505–513.

Brandt, D. S. 1986. Preservation of event beds through time. *Palaios* 1:92–96.

Brandt, D. S. 1989. Taphonomic grades as a classification for fossiliferous assemblages and implications for paleoecology. *Palaios* 4:303–309.

Brandt, D. S. and Elias, R. J. 1989. Temporal variations in tempestite thickness may be a geologic record of atmospheric CO_2. *Geology* 17:951–952.

Brasier, M. 1986. Why do lower plants and animals biomineralize? *Paleobiology* 12:241–250.

Brasier, M. 1992. Nutrient-enriched waters and the early skeletal fossil record. *Journal of the Geological Society of London* 149:621–629.

Brasier, M. 1995. Fossil indicators of nutrient levels. 2. Evolution and extinction in relation to oligotrophy. In: D. W. J. Bosence and P. A. Allison (eds.) *Marine Paleoenvironmental Analysis from Fossils*, pp. 133–150. Geological Society of London Special Publication No. 83.

Brass, G. W., Southam, J. R. and Peterson, W. H. 1982. Warm saline bottom water in the ancient ocean, *Nature* 296:620–623.

Bremer, M. L. and Lohmann, G. P. 1982. Evidence for primary control of the distribution of certain Atlantic Ocean benthonic foraminifera by degree of carbonate saturation. *Deep-Sea Research* 29:987–998.

Brenchley, P. J. and Newall, G. 1970. Flume experiments on the orientation and transport of models and shell valves. *Palaeogeography, Palaeoclimatology, Palaeoecology* 7:185–220.

Bresler, V. and Yanko, V. 1995. Chemical ecology: a new approach to the study of living benthic epiphytic foraminifera. *Journal of Foraminiferal Research* 25:267–279.

Bretsky, P. W. and Klofak, S. M. 1986. "Rules of assembly" for two Late Ordovician communities. *Palaios* 1:462–477.

Brett, C. E. 1990. Obrution deposits. In: D. E. K. Briggs and P. R. Crowther (eds.) *Palaeobiology: a Synthesis*, pp. 239–243. Oxford: Blackwell.

Brett, C. E. 1995. Sequence stratigraphy, biostratigraphy, and taphonomy in shallow marine environments. *Palaios* 10:597–616.

Brett, C. E. 1998. Sequence stratigraphy, paleoecology, and evolution: biotic clues and responses to sea-level fall. *Palaios* 13:241–262.

Brett, C. E. and Baird, G. C. 1986. Comparative taphonomy: a key to paleoenvironmental interpretation based on fossil preservation. *Palaios* 1:207–227.

Brett, C. E. and Baird, G. C. 1993. Taphonomic approaches to time resolution in stratigraphy: examples from Paleozoic marine mudrocks. In: S. M. Kidwell and A. K. Behrensmeyer (eds.) *Taphonomic Approaches to Time Resolution in Fossil Assemblages*, pp. 250–274. Pittsburgh: Paleontological Society Short Courses in Paleontology No. 6.

References

Brett, C. E. and Baird, G. C. 1995. Coordinated stasis and evolutionary ecology of Silurian to Middle Devonian faunas in the Appalachian Basin. In: D. H. Erwin and R. L. Anstey (eds.) *New Approaches to Speciation in the Fossil Record*, pp. 285–315. New York: Columbia University Press.

Brett, C. E. and Liddell, W. D. 1978. Preservation and palaeoecology of a Middle Ordovician hardground community. *Paleobiology* 4:329–348.

Brett, C. E. and Seilacher, A. 1991. Fossil Lagerstätten: a taphonomic consequence of event sedimentation. In: G. Einsele, W. Ricken and A. Seilacher (eds.) pp. 283–296. *Cycles and Events in Stratigraphy*. Berlin: Springer-Verlag.

Brett, C. E. and Speyer, S. E. 1990. Taphofacies. In: B. E. K. Briggs and P. R. Crowther (eds.) *Palaeobiology: a Synthesis*, pp. 258–263. Oxford: Blackwell.

Brett, C. E., Dick, V. B. and Baird, G. C. 1991. Comparative taphonomy and paleoecology of Middle Devonian dark grey and black shales from western New York. In: E. Landing and C. E. Brett (eds.) *Dynamic Stratigraphy and Depositional Environments of the Hamilton Group (Middle Devonian) in New York State*, Part 2, pp. 5–36. New York: New York State Museum Bulletin 469.

Bridgman, P. W. 1959. *The Way Things Are*. Cambridge, MA: Harvard University Press.

Briggs, D. E. G. 1990. Flattening. In: D. E. K. Briggs and P. R. Crowther (eds.) *Palaeobiology: a Synthesis*, pp. 244–247. Oxford: Blackwell.

Briggs, D. E. G. 1995. Experimental taphonomy. *Palaios* 10:539–550.

Briggs, D. E. G. and Gall, J.-C. 1990. The continuum in soft-bodied biotas from transitional environments: a quantitative comparison of Triassic and Carboniferous Konservat-Lagerstätten. *Paleobiology* 16:204–218.

Briggs, D. E. G. and Kear, A. J. 1993*a*. Decay and preservation of polychaetes: taphonomic thresholds in soft-bodied organisms. *Paleobiology* 19:107–135.

Briggs, D. E. G. and Kear, A. J. 1993*b*. Fossilization of soft-tissue in the laboratory. *Science* 259:1439–1442.

Briggs, D. E. G. and Kear, A. J. 1994*a*. Decay of the lancelet *Branchiostoma lanceolatum* (Cephalochordata): implications for the interpretation of soft-tissue preservation in conodonts and other primitive chordates. *Lethaia* 26: 275–287.

Briggs, D. E. G. and Kear, A. J. 1994*b*. Decay and mineralization of shrimps. *Palaios* 9:431–456.

Briggs, D. E. G. and Wilby, P. R. 1996. The role of the calcium carbonate/calcium phosphate switch in the mineralization of soft-bodied fossils. *Journal of the Geological Society of London* 153:665–668.

Briggs, D. E. G., Clarkson, E. N. K. and Aldridge, R. J. 1983. The conodont animal. *Lethaia* 16:1–14.

Briggs, D. E. G., Bottrell, S. H. and Raiswell, R. 1991. Pyritization of soft-bodied fossils: Beecher's Trilobite Bed, Upper Ordovician, New York State. *Geology* 19:1221–1224.

Briggs, D. E. G., Kear, A. J., Martill, D. M. and Wilby, P. R. 1993. Phosphatization of soft-tissue in experiments and fossils. *Journal of the Geological Society of London* 150:1035–1038.

Briggs, D. E. G., Erwin, D. H. and Collier, F. J. 1994. *The Fossils of the Burgess Shale*. Washington: Smithsonian Institution Press.

Briggs, D. E. G., Kear, A. J., Baas, M., De Leeuw, J. W. and Rigby, S. 1995. Decay and composition of the hemichordate *Rhabdopleura*: implications for the taphonomy of graptolites. *Lethaia* 28:15–23.

Briggs, D. E. G., Raiswell, R., Bottrell, S. H., Hatfield, D. and Bartels, C. 1996a. Controls on the pyritization of exceptionally preserved fossils: an analysis of the Lower Devonian Hunsrück Slate of Germany. *American Journal of Science* 296:633–663.

Briggs, D. E. G., Siveter, D. J. and Siveter, D. J. 1996b. Soft-bodied fossils from a Silurian volcaniclastic deposit. *Nature* 382:248–250.

Briggs, D. E. G., Wilby, P. R., Pérez-Moreno, B. P., Sanz, J. L. and Fregenal-Martínez, M. 1997. The mineralization of dinosaur soft tissue in the Lower Cretaceous of Las Hoyas, Spain. *Journal of the Geological Society of London* 154:587–588.

Briggs, D. E. G., Stankiewicz, B. A., Meischner, D., Bierstedt, A. and Evershed, R. P. 1998. Taphonomy of arthropod cuticles from Pliocene lake sediments, Willershausen, Germany. *Palaios* 13:386–394.

Broadhead, T. W. and Driese, S. G. 1994. Experimental and natural abrasion of conodonts in marine and eolian environments. *Palaios* 9:546–560.

Broadhead, T. W., Driese, S. G. and Harvey, J. L. 1990. Gravitational settling of conodont elements: implications for paleoecologic interpretations of conodont assemblages. *Geology* 18:850–853.

Broecker, W. S. 1982. Ocean chemistry during glacial time. *Geochimica et Cosmochimica Acta* 46:1689–1705.

Broecker, W. S., Mix, A., Andree, M. and Oeschger, H. 1984. Radiocarbon measurements on coexisting benthic and planktic foraminifera shells: potential for reconstructing ocean ventilation times over the past 20,000 years. *Nuclear Instruments and Methods in Physics Research* B5:331–339.

Broecker, W. S., Klas, M., Clark, E., Bonani, G., Ivy, S. and Wolfli, W. 1991. The influence of $CaCO_3$ dissolution on core top radiocarbon ages for deep-sea sediments. *Paleoceanography* 6:593–608.

Bromley, R. G., Hanken, N.-M. and Asgaard, U. 1990. Shallow marine bioerosion: preliminary results of an experimental study. *Bulletin of the Geological Society of Denmark* 38:85–99.

Brooks, D. R. and Wiley, E. O. 1988. *Evolution as Entropy*. Chicago: University of Chicago Press.

Brown, B., 1995 (ed.) Science and management. *Coral Reefs* 14:175–273.

Brown, G., Catt, J. A., Hollyer, S. E. and Ollier, C. D. 1969. Partial silicification of chalk fossils from the Chilterns. *Geological Magazine* 106:583–586.

Brown, S. J. and Elderfield, H. 1996. Variations in Mg/Ca and Sr/Ca ratios of planktonic foraminifera caused by postdepositional dissolution: evidence of shallow Mg-dependent dissolution. *Paleoceanography* 11:543–551.

Brown, T. A., Nelson, D. E., Mathewes, R. W. and Vogel, J. S. 1989. Radiocarbon dating of pollen by accelerator mass spectrometry. *Quaternary Research* 32:205–212.

Brunner, C. A. and Culver, S. J. 1992. Quaternary foraminifera from the walls of Wilmington, south Wilmington, and north Heyes canyons, U.S. east coast: implications for continental slope and rise evolution. *Palaios* 7:34–66.

Brunner, C. A. and Ledbetter, M. T. 1987. Sedimentological and micropaleontological detection of turbidite muds in hemipelagic sequences: an example from the Late Pleistocene levee of Monterey Fan, central California continental margin. *Marine Micropaleontology* 12:223–239.

Brunner, C. A. and Normark, W. R. 1985. Biostratigraphic implications for turbidite depositional processes on the Monterey deep-sea fan, central California. *Journal of Sedimentary Petrology* 55:495–505.

Brush, G. S. and Brush, L. M. 1972. Transport of pollen in a sediment-laden channel: a laboratory study. *American Journal of Science* 272:359–381.

Brush, G. S. and Brush, L. M. 1994. Transport and deposition of pollen in an estuary: signature of the landscape. In: A. Traverse (ed.) *Sedimentation of Organic Particles*, pp. 33–46. Cambridge: Cambridge University Press.

Brush, G. S. and Davis, F. W. 1984. Stratigraphic evidence of human disturbance in an estuary. *Quaternary Research* 22:91–108.

Brush, G. S. and DeFries, R. S. 1981. Spatial distributions of pollen in surface sediments of the Potomac Estuary. *Limnology and Oceanography* 26:295–309.

Bruton, D. L. 1991. Beach and laboratory experiments with the jellyfish *Aurelia* and remarks on some fossil "medusoid" traces. In: A. Simonetta and S. Conway Morris (eds.) *The Early Evolution of Metazoa and the Significance of Problematic Taxa*, pp. 125–129. Cambridge: Cambridge University Press.

Buatois, L. A., Mángano, G., Genise, J. F. and Taylor, T. N. 1998. The ichnologic invasion of the continental invertebrate invasion: evolutionary trends in environmental expansion, ecospace utilization, and behavioral complexity. *Palaios* 13:217–240.

Bullen, S. B. and Sibley, D. F. 1984. Dolomite selectivity and mimic replacement. *Geology* 12:655–658.

Burnham, R. J. 1989. Relationships between standing vegetation and leaf litter in a paratropical forest: implications for paleobotany. *Palaeogeography, Palaeoclimatology, Palaeoecology* 58:5–32.

Burnham, R. J. 1993*a*. Reconstructing richness in the plant fossil record. *Palaios* 8:376–384.

Burnham, R. J. 1993*b*. Time resolution in terrestrial macrofloras: guidelines from modern accumulations. In: S. M. Kidwell and A. K. Behrensmeyer (eds.) *Taphonomic Approaches to Time Resolution in Fossil Assemblages*, pp. 57–78. Pittsburgh: Paleontological Society Short Courses in Paleontology No. 6.

Burnham, R. J. and Spicer, R. A. 1986. Forest litter preserved by volcanic activity at El Chicón, Mexico: a potentially accurate record of the pre-eruption vegetation. *Palaios* 1:158–161.

Burnham, R. J., Wing, S. L. and Parker, G. C. 1992. The reflection of deciduous forest communities in leaf litter: implications for autochthonous litter assemblages from the fossil record. *Paleobiology* 18:30–49.

Butterfield, N. J. 1990. Organic preservation of non-mineralizing organisms and the taphonomy of the Burgess Shale. *Paleobiology* 16:272–286.

Butterfield, N. J. 1994. Burgess Shale-type fossils from a Lower Cambrian shallow shelf sequence in northwestern Canada. *Nature* 369:477–479.

Butterfield, N. J. 1995. Secular distribution of Burgess Shale-type preservation. *Lethaia* 28:1–13.

Butterfield, N. J. 1996. Fossil preservation in the Burgess Sale: reply. *Lethaia* 29:109–112.

Butterfield, N. J. 1997. Plankton ecology and the Proterozoic–Phanerozoic transition. *Paleobiology* 23:247–262.

Butterfield, N. J. and Nicholas, C. J. 1996. Burgess-Shale type preservation of both non-mineralizing and "shelly" Cambrian organisms from the Mackenzie Mountains, northwestern Canada. *Journal of Paleontology* 70:893–899.

Butterfield, N. J. and Rainbird, R. H. 1998. Diverse organic-walled fossils, including 'possible dinoflagellates', from the early Neoproterozoic of arctic Canada. *Geology* 26:963–966.

Buurman, P., van Breemen, N. and Henstra, S. 1973. Recent silicification of plant remains in acid sulphate soils. *Neues Jahrbuch für Mineralogie Monatshefte* 3:117–124.

Buzas, M. A. 1965. The distribution and abundance of foraminifera in Long Island Sound. *Smithsonian Miscellaneous Collections* 149(1):1–89.

Buzas, M. A. 1968. On the spatial distribution of foraminifera. *Contributions from the Cushman Foundation for Foraminiferal Research* 19:1–11.

Buzas, M. A. 1970. Spatial homogeneity: statistical analyses of unispecies and multispecies populations of foraminifera. *Ecology* 51:874–879.

Buzas, M. A. and Culver, S. J. 1994. Species pool and dynamics of marine paleocommunities. *Science* 264:1439–1441.

Buzas, M. A. and Culver, S. J. 1998. Assembly, disassembly, and balance in marine paleocommunities. *Palaios* 13:263–275.

Buzas, M. A., Culver, S. J. and Jorissen, F. J. 1993. A statistical evaluation of the microhabitats of living (stained) infaunal benthic foraminifera. *Marine Micropaleontology* 20:311–320.

Cadée, G. C. 1976. Sediment reworking by *Arenicola marina* on tidal flats in the Dutch Wadden Sea. *Netherlands Journal of Sea Research* 10:440–460.

Cadée, G. C. 1989. Size-selective transport of shells by birds and its palaeoecological implications. *Palaeontology* 32:429–437.

Cadée, G. C. 1991. The history of taphonomy. In: S. K. Donovan (ed.) *The Processes of Fossilization*, pp. 3–21. London: Belhaven Press.

Cadée, G. C. 1994. *Mya* shell manipulation by Turnstones (Aves) results in concave-up position and left/right sorting. *Palaios* 9:307–309.

Caldeira, H. 1995. Long-term control of atmospheric carbon dioxide: low-temperature seafloor alteration or terrestrial silicate-rock weathering? *American Journal of Science* 295:1077–1114.

Callender, W. R., Staff, G. M., Powell, E. N. and MacDonald, I. R. 1990. Gulf of Mexico hydrocarbon seep communities V. Biofacies and shell orientation of autochthonous shell beds below storm wave base. *Palaios* 5:2–14.

Callender, W. R., Powell, E. N., Staff, G. M. and Davies, D. J. 1992. Distinguishing autochthony, parautochthony and allochthony using taphofacies analysis: can cold seep assemblages be discriminated from assemblages of the nearshore and continental shelf? *Palaios* 7:409–421.

Callender, W. R., Powell, E. N. and Staff, G. M. 1994. Taphonomic rates of molluscan shells placed in autochthonous assemblages on the Louisiana continental slope. *Palaios* 9:60–73.

Calvert, S. E. 1974. Deposition and diagenesis of silica in marine sediments. In:

K. J. Hsü and H. C. Jenkyns (eds.) *Pelagic Sediments: On Land and Under the Sea*, pp. 273–299. Special Publication of the International Association of Sedimentology. Oxford: Blackwell.

Canfield, D. E. 1991. Sulfate reduction in deep-sea sediments. *American Journal of Science* 291:177–188.

Canfield, D. E. and Raiswell, R. 1991*a*. Pyrite formation and fossil preservation. In: P. A. Allison and D. E. G. Briggs (eds.) *Taphonomy: Releasing the Data Locked in the Fossil Record*, pp. 338–387. New York: Plenum Press.

Canfield, D. E. and Raiswell, R. 1991*b*. Carbonate precipitation and dissolution: its relevance to fossil preservation. In: P. A. Allison and D. E. G. Briggs (eds.) *Taphonomy: Releasing the Data Locked in the Fossil Record*, pp. 411–453. New York: Plenum Press.

Canfield, D. E. and Teske, A. 1996. Late Proterozoic rise in atmospheric oxygen concentration inferred from phylogenetic and sulfur isotope studies. *Nature* 382:127–132.

Caplan, M. L., Bustin, R. M. and Grimm, K. A. 1996. Demise of a Devonian-Carboniferous carbonate ramp by eutrophication. *Geology* 24:715–718.

Capone, D. G., Zehr, J. P., Paerl, H. W., Bergman, B. and Carpenter, E. J. 1997. *Trichodesmium*, a globally significant marine cyanobacterium. *Science* 276:1221–1229.

Carannante, G., Cherchi, A. and Simone, L. 1995. Chlorozoan versus foramol lithofacies in Upper Cretaceous rudist limestones. *Palaeogeography, Palaeoclimatology, Palaeoecology* 119:137–154.

Carpenter, K. 1982. Baby dinosaurs from the Late Cretaceous Lance and Hell Creek Formations and a description of a new species of theropod. *University of Wyoming Contributions to Geology* 20:123–134.

Carpenter, R., Peterson, M. L. and Bennett, J. T. 1982. ^{210}Pb-derived sediment accumulation and mixing rates for the Washington continental slope. *Marine Geology* 48:135–164.

Carpenter, S. J., Erickson, J. M., Lohmann, K. C. and Owen, M. R. 1988. Diagenesis of fossiliferous concretions from the Upper Cretaceous Fox Hills Formation, North Dakota. *Journal of Sedimentary Petrology* 58:706–723.

Carslaw, H. S. and Jaeger, J. C. 1959. *Conduction of Heat in Solids*. Oxford: Oxford University Press.

Carson, G. A. 1991. Silicification of fossils. In: P. A. Allison and D. E. G. Briggs (eds.) *Taphonomy: Releasing the Data Locked in the Fossil Record*, pp. 456–499. New York: Plenum Press.

Casey, R. E. 1993. Radiolaria. In: J. H. Lipps (ed.) *Fossil Prokaryotes and Protists*, pp. 249–284. Boston: Blackwell.

Cate, A. S. and Evans, I. 1994. Taphonomic significance of the biomechanical fragmentation of live molluscan shell material by a bottom-feeding fish (*Pogonias cromis*) in Texas coastal bays. *Palaios* 9:254–274.

Catto, N. R. 1985. Hydrodynamic distribution of palynomorphs in a fluvial succession, Yukon. *Canadian Journal of Earth Sciences* 22:1252–1256.

Causton, D. R. 1987. *A Biologist's Advanced Mathematics*. London: Allen and Unwin.

Cerling, T. E. 1989. Does the gas content of amber reveal the composition of palaeoatmospheres? *Nature* 339:695–696.

Chagué-Goff, C., Goodarzi, F. and Fyfe, W. S. 1996. Elemental distribution and pyrite occurrence in a freshwater peatland, Alberta. *Journal of Geology* 104:649–663.

Chamberlain, J. A. 1978. Mechanical properties of coral skeleton: compressive strength and its adaptive significance. *Paleobiology* 4:419–435.

Chamberlain, J. A. 1987. Locomotion of *Nautilus*. In: W. B. Saunders and N. H. Landman (eds.) *Nautilus: the Biology and Paleobiology of a Living Fossil*, pp. 489–526. New York: Plenum Press.

Chamberlain, J. A., Ward, P. D. and Weaver, J. S. 1981. Postmortem ascent of *Nautilus* shells: implications for cephalopod paleo-biogeography. *Paleobiology* 7:494–509.

Chaney, R. W. 1924. Quantitative studies of the Bridge Creek flora. *American Journal of Science* 8 (series 5):126–144.

Charman, D. J., Roe, H. M. and Gehrels, W. R. 1998. The use of testate amoebae in studies of sea-level change: a case study from the Taf Estuary, south Wales, UK. *The Holocene* 8:209–218.

Chave, K. E. 1954. Aspects of the biogeochemistry of magnesium. 1. Calcareous marine organisms. *Journal of Geology* 62:266–283.

Chave, K. E. 1964. Skeletal durability and preservation. In: J. Imbrie and N. D. Newell (eds.) *Approaches to Paleoecology*, pp. 377–387. New York: Wiley.

Cheetham, A. H. and Thomsen, E. 1981. Functional morphology of arborescent animals: strength and design of cheilostome bryozoan skeletons. *Paleobiology* 7:355–383.

Chinsamy, A. and Dodson, P. 1995. Inside a dinosaur bone. *American Scientist* 83:174–180.

Chou, L. and Wollast, R. 1984. Study of the weathering of albite at room temperature and pressure with a fluidizied bed reactor. *Geochimica et Cosmochimica Acta* 48:2205–2217.

Christensen, E. R. 1986. A model for radionuclides in sediments influenced by mixing and compaction. *Journal of Geophysical Research* 87:566–572

Christensen, E. R. and Bhunia, P. K. 1986. Modeling radiotracers in sediments: comparison with observations in Lakes Huron and Michigan. *Journal of Geophysical Research* 91:8559–8571.

Christensen, E. R. and Goetz, R. H. 1987. Historical fluxes of particle-bound pollutants from deconvolved sedimentary records. *Environmental Science and Technology* 21:1088–1096.

Christensen, E. R. and Klein, R. J. 1991. "Unmixing" of ^{137}Cs, Pb, Zn, and Cd records in Lake sediments. *Environmental Science and Technology* 35:1627–1637

Christensen, E. R. and Osuna, J. L. 1989. Atmospheric fluxes of lead, zinc, and cadmium from frequency domain deconvolution of sedimentary records. *Journal of Geophysical Research* 94:14 585–14 597.

Clark, G. R. and Lutz, R. A. 1980. Pyritization in the shells of living bivalves. *Geology* 8:268–271.

Clark, J., Beerbower, J. R. and Kietzke, K. K. 1967. Oligocene sedimentation, stratigraphy and paleoclimatology in the Big Badlands of South Dakota. *Fieldiana Geology* 5:1–158.

Claypool, G. E., Holser, W. T., Kaplan, I. R., Sakai, H. and Zak, I. 1980. The age curves of sulfur and oxygen isotopes in marine sulfate and their mutual interpretation. *Chemical Geology* 28:199–260.

Clifton, H. E. 1989. Sedimentologic approaches to paleobathymetry with applications to the Merced Formation of central California. *Palaios* 3:507–522.

Clifton, H. E. and Boggs, S. 1970. Concave-up pelecypod (*Psephidia*) shells in shallow marine sand, Elk River beds, southwestern Oregon. *Journal of Sedimentary Petrology* 40:888–897.

Cloetingh, S. 1988. Intraplate stresses: a tectonic cause for third-order cycles in apparent sea level? In: C. K. Wilgus, B. S. Hastings, C. G. St. C. Kendall, H. W. Posamentier, C. A. Ross and J. C. Van Wagoner (eds.) *Sea-Level Changes: An Integrated Approach*, pp. 19–29. Tulsa: Society of Economic Paleontologists and Mineralogists Special Publication No. 42.

CoBabe, E. A. and Allmon, W. D. 1994. Effects of sampling on paleoecologic and taphonomic analysis in high diversity fossil accumulations: an example from the Eocene Gosport Sand, Alabama. *Lethaia* 27:167–178.

Cochran, J. K. and Aller, R. C. 1979. Particle reworking in sediments from the New York Bight Apex: evidence fom ^{234}Th/^{238}Ur disequilibrium. *Estuarine and Coastal Marine Science* 9:739–747.

Cockey, E., Hallock, P. and Lidz, B. 1996. Decadal-scale changes in benthic foraminiferal assemblages off Key Largo, Florida. *Coral Reefs* 15:237–248.

Coffin, H. G. 1983. Erect floating stumps in Spirit Lake, Washington. *Geology* 11:298–299.

Coleman, M. L. 1985. Geochemistry of diagenetic non-silicate minerals: kinetic considerations. *Philosophical Transactions of the Royal Society of London* (A) 315:39–56.

Collinson, M. E. 1983. Accumulations of fruits and seeds in three small sedimentary environments in southern England and their palaeoecological implications. *Annals of Botany* 52:583–592.

Collinson, M. E. and Scott, A. C. 1987. Implications of vegetational change through the geological record on models for coal-forming environments. In: A. C. Scott (ed.) *Coal and Coal-Bearing Strata: Recent Advances*, pp. 67–85. Oxford: Geological Society Special Publication No. 32.

Compton, J. S., Hodell, D. A., Garrido, J. R. and Mallinson, D. J. 1993. Origin and age of phosphorite from the south-central Florida Platform. Relation of phosphogenesis to sea-level fluctuations and ^{13}C excursions. *Geochimica et Cosmochimica Acta* 57:131–146.

Conway Morris, S. 1985. Cambrian Lagerstätten: their distribution and significance. In: H. B. Whittington and S. Conway Morris (eds.) Extraordinary fossil biotas: their ecological and evolutionary significance. *Philosophical Transactions of the Royal Society of London* (B) 311:49–65.

Conway Morris, S. 1986. The community structure of the Middle Cambrian Phyllopod bed (Burgess Shale). *Palaeontology* 29:423–467.

Conway Morris, S. 1990. Burgess Shale. In: D. E. K. Briggs and P. R. Crowther (eds.) *Palaeobiology: A Synthesis*, pp. 270–274. Oxford: Blackwell.

Conway Morris, S. 1998. *The Crucible of Creation: The Burgess Shale and the Rise of Animals*. Oxford: Oxford University Press.

Cook, P. J. and Shergold, J. H. 1984. Phosphorus, phosphorites and skeletal evolution at the Precambrian–Cambrian boundary. *Nature* 308: 231–236.

Coombs, M. C. and Coombs, W. P. 1997. Analysis of the geology, fauna, and taphonomy of Morava Ranch quarry, Early

Miocene of northwest Nebraska. *Palaios* 12:165–187.

Cooper, S. L. 1995. Chesapeake Bay watershed historical land use: impact on water quality and diatom communities. *Ecological Applications* 5:703–723.

Cope, E. D. 1876. On some extinct reptiles and batrachians from the Judith River and Fox Hill beds of Montana. *Proceedings of the Academy of Natural Sciences of Philadelphia* 28:340–359.

Copper, P. 1974. Structure and development of early Paleozoic reefs. *Proceedings, 2nd International Coral Reef Symposium* 1:365–386.

Corliss, B. H. 1985. Microhabitats of benthic foraminifera within deep-sea sediments. *Nature* 314:435–438.

Corliss, B. H. and Chen, C. 1988. Morphotype patterns of Norwegian Sea deep-sea benthic foraminifera and ecological implications. *Geology* 16:716–719.

Corliss, B. H. and Emerson, S. 1990. Distribution of Rose Bengal stained deep-sea benthic foraminifera from the Nova Scotian continental margin and Gulf of Maine. *Deep-Sea Research* 37:381–400.

Corliss, B. H. and Fois, E. 1991. Morphotype analysis of deep-sea foraminifera from the northwest Gulf of Mexico. *Palaios* 6:589–605.

Corliss, B. H. and Honjo, S. 1981. Dissolution of deep-sea benthonic foraminifera. *Micropaleontology* 27:356–378.

Cottey, T. L. and Hallock, P. 1988. Test surface degradation in *Archaias angulatus*. *Journal of Foraminiferal Research* 18:187–202.

Courtillot, V. and Gaudemer, Y. 1996. Effects of mass extinctions on biodiversity. *Nature* 381:146–148.

Crank, J. 1975. *The Mathematics of Diffusion*. Oxford: Oxford University Press.

Crepet, W. L., Dilcher, D. L. and Potter, F. W. 1974. Eocene angiosperm flowers. *Science* 185:781–782.

Crimes, T. P., Insole, A. and Williams, B. P. J. 1995. A rigid-bodied Ediacaran biota from Upper Cambrian strata in Co. Wexford, Eire. *Geological Journal* 30:89–109.

Cross, A. T., Thompson, G. C. and Zaitzeff, J. B. 1966. Source and distribution of palynomorphs in bottom sediments, southern part of Gulf of California. *Marine Geology* 4:467–524.

Crowley, S. S., Dufek, D. A., Stanton, R. W. and Ryer, T. A. 1994. The effects of volcanic ash disturbances on a peat-forming environment: environmental disruption and taphonomic consequences. *Palaios* 9:158–174.

Crowley, T. J. and Baum, S. K. 1991. Towards reconciliation of Late Ordovician (~440 Ma) glaciation with very high CO_2 levels. *Journal of Geophysical Research* 96:22 597–22 610.

Culver, S. J. 1980. Differential two-way sediment transport in the Bristol Channel and Severn Estuary, United Kingdom. *Marine Geology* 34:39–43.

Culver, S. J. 1990. Benthic foraminifera of Puerto Rican mangrove-lagoon systems: potential for paleoenvironmental interpretations. *Palaios* 5:34–51.

Culver, S. J. 1991. Early Cambrian foraminifera from west Africa. *Science* 254:689–691.

Culver, S. J. 1994. Early Cambrian foraminifera from the southwestern Taoudeni Basin, West Africa. *Journal of Foraminiferal Research* 24:191–202.

Cummins, H., Powell, E. N., Stanton, R. J. and Staff, G. 1986. The rate of taphonomic loss in modern benthic habitats: how much of the potentially preservable community is preserved? *Palaeogeography, Palaeoclimatology, Palaeoecology* 52:291–320.

Curtis, C. D., Coleman, M. L. and Love, L. G. 1986. Pore water evolution during sediment burial from isotopic and mineral chemistry of calcite, dolomite and siderite concretions. *Geochimica et Cosmochimica Acta* 50:2321–2334.

Cutler, A. H. 1993. Mathematical models of temporal mixing in the fossil record. In: S. M. Kidwell and A. K. Behrensmeyer (eds.) *Taphonomic Approaches to Time Resolution in Fossil Assemblages*, pp. 169–187. Pittsburgh: Paleontological Society Short Courses in Paleontology No. 6.

Cutler, A. H. 1995. Taphonomic implications of shell surface textures. *Palaeogeography, Palaeoclimatology, Palaeoecology* 114:219–240.

Cutler, A. H. 1998. A note on the taphonomy of Lower Miocene fossil land mammals from the Miocene Calvert Formation at the Pollack Farm Site, Delaware. In: R. N. Benson (ed.) *Geology and Paleontology of the Lower Miocene Pollack Farm Fossil Site, Delaware*. Newark: Delaware Geological Survey Special Publication No. 21.

Cutler, A. H. and Flessa, K. W. 1990. Fossils out of sequence: computer simulations and strategies for dealing with stratigraphic disorder. *Palaios* 5:227–235.

Cutler, A. H. and Flessa, K. W. 1995. Bioerosion, dissolution and precipitation as taphonomic agents at high and low latitudes. *Senckenbergiana Maritima* 25(4/6):115–121.

Daley, G. M. 1993. Passive deterioration of shelly material: a study of the Recent eastern Pacific articulate brachiopod *Terebratalia transversa* Sowerby. *Palaios* 8:226–232.

Darwin, C. 1896. *The Formation of Vegetable Mold Through the Action of Worms, With Observations on Their Habits*. New York: D. Appleton and Company.

Davaud, E. and Septfontaine, M. 1995. Post-mortem onshore transportation of epiphytic foraminifera: recent example from the Tunisian coastline. *Journal of Sedimentary Research* A65:136–142.

Davies, D. J., Powell, E. N. and Stanton, R. J. 1989. Relative rates of shell dissolution and net sediment accumulation – a commentary: can shell beds form by the gradual accumulation of biogenic debris on the sea floor? *Lethaia* 22:207–212.

Davies-Vollum, K. S. and Wing, S. L. 1998. Sedimentological, taphonomic, and climatic aspects of Eocene swamp deposits (Willwood Formation, Bighorn Basin, Wyoming). *Palaios* 13:28–40.

Davis, J. C. 1986. *Statistics and Data Analysis in Geology*. New York: Wiley.

Davis, M. B. 1968. Pollen grains in lake sediment: redeposition caused by seasonal water circulation. *Science* 162:796–799.

Davis, M. B., Moeller, R. E. and Ford, J. 1984. Sediment focusing and pollen influx. In: E. Y. Haworth and J. W. G. Lund (eds.) *Lake Sediments and Environmental History*, pp. 261–293. Leicester: University of Leicester Press.

Davis, P. G. 1996. The taphonomy of *Archaeopteryx*. *Bulletin of the National Science Museum, Tokyo*, Series C (Geology & Paleontology) 22(3–4):91–106.

Davis, P. G. and Briggs, D. E. G. 1995. Fossilization of feathers. *Geology* 23:783–786.

Davis, P. G. and Briggs, D. E. G. 1998. The impact of decay and disarticulation on the preservation of fossil birds. *Palaios* 13:3–13.

Davis, R. B. 1974. Stratigraphic effects of tubificids in profundal lake sediments. *Limnology and Oceanography* 19:466–487.

Davis, R. B. and Webb, T. 1975. The contemporary distribution of pollen in eastern North America: a comparison with vegetation. *Quaternary Research* 5:395–434.

Demko, T. M., Dubiel, R. F. and Parrish, J. T. 1998. Plant taphonomy in incised valleys: implications for interpreting paleoclimate from fossil plants. *Geology* 26:1119–1122.

Denne, R. A. 1994. Operational applications of graphic correlation. In: H. R. Lane, G. Blake and N. R. MacLeod (eds.) *Graphic Correlation and the Composite Standard: the Methods and their Applications*. Houston: SEPM (Society for Sedimentary Geology) Research Conference. Abstracts with Programs.

Denne, R. A. and Sen Gupta, B. K. 1989. Effects of taphonomy and habitat on the record of benthic foraminifera in modern sediments. *Palaios* 4:414–423.

Denys, C., Williams, C. T., Dauphin, Y., Andrews, P. and Ferndandez-Jalvo, Y. 1996. Diagenetical changes in Pleistocene small mammal bones from Olduvai Bed I. *Palaeogeography, Palaeoclimatology, Palaeoecology* 126:121–134.

Desmond, A. 1982. *Archetypes and Ancestors: Palaeontology in Victorian London 1850–1875*. London: Blond and Briggs.

Desmond, A. and Moore, J. 1991. *Darwin: The Life of a Tormented Evolutionist*. New York: Time-Warner.

Deutsch, S. and Lipps, J. H. 1976. Test structure of the foraminifer *Carterina*. *Journal of Paleontology* 50:312–317.

Dhakar, S. P. and Burdige, D. J. 1996. A coupled, non-linear, steady state model for early diagenetic processes in pelagic sediments. *American Journal of Science* 296:296–330.

D'Hondt, S. D., Pilson, M. E. Q., Sigurdsson, H., Hanson, A. K. and Carey, S. 1994. Surface-water acidification and extinction at the Cretaceous–Tertiary boundary. *Geology* 22:983–986.

Dimbleby, G. W. 1957. Pollen analysis of soils. *New Phytologist* 56:12–28.

DiMichele, W. A. 1994. Ecological patterns in time and space. *Paleobiology* 20:89–92.

DiMichele, W. A. and Nelson, W. J. 1989. Small-scale spatial heterogeneity in Pennsylvanian-age vegetation from the roof shale of the Springfield Coal (Illinois Basin). *Palaios* 4:276–280.

Dingus, L. 1984. Effects of stratigraphic completeness on interpretations of extinction rates across the Cretaceous–Tertiary boundary. *Paleobiology* 10:420–438.

Dingus, L. and Sadler, P. M. 1982. The effects of stratigraphic completeness on estimates of evolutionary rates. *Systematic Zoology* 31:400–412.

Dodson, P. 1971. Sedimentology and taphonomy of the Oldman Formation (Campanian), Dinosaur Provincial Park, Alberta (Canada). *Palaeogeography, Palaeoclimatology, Palaeoecology* 10:21–74.

Dodson, P. 1973. The significance of small bones in paleoecological interpretation. *University of Wyoming Contributions to Geology* 12:15–19.

Dodson, P. 1975. Functional and ecological significance of relative growth in *Alligator*. *Journal of Zoology* 175:315–355.

Dodson, P. and Wexlar, D. 1979. Taphonomic investigations of owl pellets. *Paleobiology* 5:275–284.

Dodson, P., Behrensmeyer, A. K., Bakker, R. T. and McIntosh, J. S. 1980. Taphonomy and paleoecology of the dinosaur beds of the Jurassic Morrison Formation. *Paleobiology* 6:208–232.

Done, T. J. 1992. Phase shifts in coral reef communities and their ecological significance. *Hydrobiologia* 247:121–132.

Donovan, S. K. 1991. The taphonomy of echinoderms: calcareous multi-element skeletons in the marine environment. In: S. K. Donovan (ed.) *The Processes of Fossilization*, pp. 241–269. London: Belhaven Press.

Donovan, S. K. and Pickerell, R. K. 1995. Crinoid columns preserved in life position in the Silurian Arisaig Group of Nova Scotia, Canada. *Palaios* 10:362–370.

Donovan, S. K. and Veltkamp, C. J. 1994. Unusual preservation of late Quaternary millipedes from Jamaica. *Lethaia* 27:355–362.

Douglas, R. G. 1971. Cretaceous foraminifera from the northwest Pacific Ocean: Leg 6, Deep Sea Drilling Project. *Initial Reports of the Deep Sea Drilling Project* 6:1027–1053.

Douglas, R. G. and Savin, S. M. 1978. Oxygen isotopic evidence for the depth stratification of Tertiary and Cretaceous planktonic foraminifera. *Marine Micropaleontology* 3:175–196.

Douglas, R. G., Liestman, J., Walch, C., Blake, G. and Cotton, M. L. 1980. The transition from live to sediment assemblage in benthic foraminifera from the southern California borderland. In: M. E. Field, Bouma, A. H., I. P. Colburn, R. G. Douglas and Ingle, J. C. (eds.) *Quaternary Depositional Environments of the Pacific Coast*, pp. 257–280. Society of Economic Paleontologists and Mineralogists, Pacific Section 4.

Downing, K. F. and Park, L. E. 1998. Geochemistry and early diagenesis of mammal-bearing concretions from the Sucker Creek Formation (Miocene) of southeastern Oregon. *Palaios* 13:14–27.

Dozen, K. and Ishiga, H. 1997. Estimation of current velocity and direction from orientation of conical radiolarians in Lower Jurassic bedded cherts from southwest Japan: indications of eddy-driven sedimentation. *Marine Micropaleontology* 30:197–214.

Driscoll, E. G. 1967. Experimental field study of shell abrasion. *Journal of Sedimentary Petrology* 37:1117–1123.

Driscoll, E. G. 1970. Selective bivalve shell destruction in marine environments, a field study. *Journal of Sedimentary Petrology* 40:898–905.

Driscoll, E. G. and Weltin, T. P. 1973. Sedimentary parameters as factors in abrasive shell reduction. *Palaeogeography, Palaeoclimatology, Palaeoecology* 13:275–288.

Droser, M. L. and Bottjer, D. J. 1986. A semiquantitative field classification of ichnofabrics. *Journal of Sedimentary Petrology* 56:558–559.

Drummond, C. N. and Wilkinson, B. H. 1996. Stratal thickness frequencies and the prevalence of orderedness in stratigraphic sequences. *Journal of Geology* 104:1–18.

DuBois, L. G. and Prell, W. L. 1988. Effects of carbonate dissolution on the radiocarbon age structure of sediment mixed layers. *Deep-Sea Research* 35:1875–1885.

Dunbar, C. O. and Rodgers, J. 1957. *Principles of Stratigraphy*. New York: Wiley.

Dunwiddie, P. W. 1987. Macrofossil and pollen representation of coniferous trees in modern sediments from Washington. *Ecology* 68:1–11.

DuPont, L. M., Beug, H.-J., Stalling, H. and Tiedemann, R. 1989. First palynological results from site 658 at 21° N off northwest

Africa: pollen as climate indicators. *Proceedings of the Ocean Drilling Program, Scientific Results* 108:93–111.

Dyer, B. D. and Obar, R. A. 1994. *Tracing the History of Eukaryotic Cells: The Enigmatic Smile*. New York: Columbia University Press.

Dymond, J., Collier, R., McManus, J., Honjo, S. and Manganini, S. 1997. Can the aluminum and titanium contents of ocean sediments be used to determine the paleoproductivity of the oceans? *Paleoceanography* 12:586–593.

Eaton, J. G., Kirkland, J. I. and Doi, K. 1989. Evidence of reworked Cretaceous fossils and their bearing on the existence of Tertiary dinosaurs. *Palaios* 4:281–286.

Ebbestad, J. O. R. and Peel, J. S. 1997. Attempted predation and shell repair in Middle and Upper Ordovician gastropods from Sweden. *Journal of Paleontology* 71:1007–1019.

Eberth, D. A. 1990. Stratigraphy and sedimentology of vertebrate microfossil sites in the uppermost Judith River Formation (Campanian), Dinosaur Provincial Park, Alberta, Canada. *Palaeogeography, Palaeoclimatology, Palaeoecology* 78:1–36.

Edwards, L. E. 1989a. Quantitative biostratigraphy. In: N. L. Gilinsky and P. W. Signor (eds.) *Analytical Paleobiology*, pp. 39–58. Pittsburgh: Paleontological Society Short Courses in Paleontology No. 4.

Edwards, L. E. 1989b. Supplemented graphic correlation: a powerful tool for paleontologists and non-paleontologists. *Palaios* 4:127–143.

Edwards, R. L., Beck, J. W., Burr, G. S., Donahue, G. J., Chappell, J. M. A., Bloom, L., Druffel, E. R. M. and Taylor, F. W. 1993. A large drop in atmospheric $^{14}C/^{12}C$ and reduced melting in the Younger Dryas, documented with ^{230}Th ages of corals. *Science* 260:962–968.

Efremov, J. A. 1940. Taphonomy: new branch of paleontology. *Pan-American Geologist* 74:81–93.

Ekdale, A. A., Bromley, R. G. and Pemberton, S. G. 1984. *Ichnology: the Use of Trace Fossils in Sedimentology and Stratigraphy*. Tulsa: Society of Economic Paleontologists and Mineralogists Short Course Number 15.

Ekdale, A. A. and Stinnesbeck, W. 1998. Trace fossils in Cretaceous–Tertiary (KT) boundary beds in northeastern Mexico: implications for sedimentation during the KT boundary event. *Palaios* 13:593–602.

Elder, R. L. and Smith, G. R. 1988. Fish taphonomy and environmental inference in paleolimnology. *Palaeogeography, Palaeoclimatology, Palaeoecology* 62:577–592.

Eldredge, N. 1995. *Reinventing Darwin: the Great Debate at the High Table of Evolutionary Theory*. New York: Wiley.

Eldredge, N. and Gould, S. J. 1972. Punctuated equilibria: an alternative to phyletic gradualism. In: T. J. M. Schopf (ed.) *Models in Paleobiology*, pp. 82–115. San Francisco: W. H. Freeman.

Elick, J. M., Driese, S. G. and Mora, C. I. 1998. Very large plant and root traces from the Early to Middle Devonian: implications for early terrestrial ecosystems and atmospheric p(CO_2). *Geology* 26:143–146.

Ellison, S. P. 1987. Examples of Devonian and Mississippian lag concentrations from Texas. In: R. L. Austin (ed.) *Conodonts: Investigative Techniques and Applications*, pp. 77–93. Chichester: Ellis Horwood.

Elsik, W. C. 1971. Microbiological degradation of sporopollenin. In: J. Brooks, P. R. Grant, M. Muir, P. van Gijzel and

References

G. Shaw (eds.) *Sporopollenin*, pp. 480–511. New York: Academic Press.

Enos, P. 1991. Sedimentary parameters for computer modeling. In: E. K. Franseen, W. L. Watney, C. G. St. C. Kendall and W. Ross (eds.) *Sedimentary Modeling: Computer Simulations and Methods for Improved Parameter Definition*, pp. 63–99. Kansas Geological Survey Bulletin 233.

Enos, P. and Perkins, R. D. 1977. *Quaternary Sedimentation in South Florida*. Boulder: Geological Society of America Memoir 14.

Erdtman, G. 1969. *Handbook of Palynology*. New York: Hafner.

Erwin, D. H. 1993. *The Great Paleozoic Crisis: Life and Death in the Permian*. New York: Columbia University Press.

Erwin, D., Valentine, J. and Jablonski, D. 1997. The origin of animal body plans. *American Scientist* 85:126–137.

Evans, J. G. 1972. *Land Snails in Archaeology*. London: Seminar Press.

Fagerstrom, J. A. 1964. Fossil communities in paleoecology: Their recognition and significance. *Geological Society of America Bulletin* 75:1197–1216.

Fall, P. L. 1987. Pollen taphonomy in a canyon stream. *Quaternary Research* 28:393–406.

Fall, P. L. 1992. Pollen accumulation in a montane region of Colorado, USA: a comparison of moss polsters, atmospheric traps, and natural basins. *Review of Palaeobotany and Palynology* 72:169–197.

Farley, M. B. 1994. Modern pollen transport and sedimentation: an annotated bibliography. In: A. Traverse (ed.) *Sedimentation of Organic Particles*, pp. 503–524. Cambridge: Cambridge University Press.

Farrell, B. D. 1998. "Inordinate fondness" explained: why are there so many beetles? *Science* 281:555–559.

Farrell, J. W. and Prell, W. L. 1989. Climatic change and $CaCO_3$ preservation: An 800,000 year bathymetric reconstruction from the central equatorial Pacific Ocean. *Paleoceanography* 4:447–466.

Fastovsky, D. E. 1987. Paleoenvironments of vertebrate-bearing strata during the Cretaceous–Paleogene transition, eastern Montana and western North Dakota. *Palaios* 2:282–295.

Fastovsky, D. E. 1990. Rocks, resolution, and the record: a review of depositional constraints on fossil vertebrate assemblages at the terrestrial Cretaceous/Paleogene boundary, eastern Montana and western North Dakota. In: V. L. Sharpton and P. D. Ward (eds.) *Global Catastrophes in Earth History: an Interdisciplinary Conference on Impacts, Volcanism, and Mass Mortality*, pp. 541–548. Boulder: Geological Society of America Special Paper 247.

Fastovsky, D. E. and McSweeney, K. 1991. Paleocene paleosols of the petrified forests of Theodore Roosevelt National Park, North Dakota: a natural experiment in compound pedogenesis. *Palaios* 6:67–80.

Fedonkin, M. A. 1992. Vendian faunas and the early evolution of Metazoa. In: J. H. Lipps and P. W. Signor (eds.) *Origin and Early Evolution of the Metazoa*, pp. 87–129. New York: Plenum Press.

Feldman, H. R. 1989. Taphonomic processes in the Waldron Shale, Silurian, southern Indiana. *Palaios* 4:144–156.

Ferguson, D. K. 1985. The origin of leaf-assemblages: new light on an old problem. *Review of Palaeobotany and Palynology* 46:117–188.

Ferris, F. G., Fyfe, W. S. and Beveridge, T. J. 1988. Metallic ion binding by *Bacillus subtilis*: implications for the fossilization of microorganisms. *Geology* 16:149–152.

Filippelli, G. M. 1997. No evidence for preferential phosphorus release from anoxic sediments of the Sanich Inlet, British Columbia. *Geological Society of America Abstracts with Programs* 29:A340.

Filippelli, G. M. and Delaney, M. L. 1992. Similar phosphorus fluxes in ancient phosphorite deposits and a modern phosphogenic environment. *Geology* 20:709–712.

Filippelli, G. M. and Delaney, M. L. 1994. The oceanic phosphorus cycle and continental weathering during the Neogene. *Paleoceanography* 9:643–652.

Finney, S. C. and Berry, W. B. N. 1997. New perspectives on graptolite distributions and their use as indicators of platform margin dynamics. *Geology* 25:919–922.

Fischer, A. G. 1984. The two Phanerozoic supercycles. In: W. A. Berggren and J. A. van Couvering (eds.) *Catastrophes and Earth History*, pp. 129–150. Princeton: Princeton University Press.

Fischer, A. G. and Arthur, M. A. 1977. Secular variations in the pelagic realm. In: H. E. Cook and P. Enos (eds.) *Deep-Water Carbonate Environments*, pp. 19–50. Tulsa: Society of Economic Paleontologists and Mineralogists Special Publication No. 25.

Fishbein, E. and Patterson, R. T. 1993. Error-weighted maximum likelihood (EWML): a new statistically based method to cluster quantitative micropaleontological data. *Journal of Paleontology* 67:475–486.

Fisher, I. St. J. 1986. Pyrite replacement of mollusc shells from the Lower Oxford Clay (Jurassic) of England. *Sedimentology* 33:575–585.

Flajs, G. 1977. Die Ultrastrukturen des Kalkalgenskeletts. *Paläontographica* B 160:69–128.

Flessa, K. W. 1990. The "facts" of mass extinctions. In: V. L. Sharpton and P. D. Ward (eds.) *Global Catastrophes in Earth History: an Interdisciplinary Conference on Impacts, Volcanism, and Mass Mortality*. Boulder: Geological Society of America Special Paper 247.

Flessa, K. W. 1993. Time-averaging and temporal resolution in Recent marine shelly faunas. In: S. M. Kidwell and A. K. Behrensmeyer (eds.) *Taphonomic Approaches to Time Resolution in Fossil Assemblages*, pp. 9–33. Pittsburgh: Paleontological Society Short Courses in Paleontology No. 6.

Flessa, K. W. 1998. Well-traveled cockles: shell transport during the Holocene transgression of the southern North Sea. *Geology* 26:187–190.

Flessa, K. W. and Brown, T. J. 1983. Selective solution of macroinvertebrate calcareous hard parts: a laboratory study. *Lethaia* 16:193–205.

Flessa, K. W. and Kowalewski, M. 1994. Shell survival and time-averaging in nearshore and shelf environments: estimates from the radiocarbon literature. *Lethaia* 27:153–165.

Flessa, K. W., Kowalewski, M. and Walker, S. E. 1992. Post-collection taphonomy: shell destruction and the chevrolet. *Palaios* 7:553–554.

Flessa, K. W., Cutler, A. H. and Meldahl, K. H. 1993. Time and taphonomy: quantitative estimates of time-averaging and stratigraphic disorder in a shallow marine habitat. *Paleobiology* 19:266–286.

References

Fletcher, C. H. 1992. Sea-level trends and physical consequences: applications to the U.S. shore. *Earth Science Reviews* 33:73–109.

Fletcher, C. H., Knebel, H. J. and Kraft, J. C. 1990. Holocene evolution of an estuarine coast and tidal wetlands. *Geological Society of America Bulletin* 102:283–297.

Fletcher, C. H., Knebel, H. J. and Kraft, J. C. 1992. Holocene depocenter migration and sediment accumulation in Delaware Bay: a submerging marginal marine sedimentary basin. *Marine Geology* 103:165–183.

Fletcher, C. H., Van Pelt, J. E., Brush, G. S. and Sherman, J. 1993. Tidal wetland record of Holocene sea-level movements and climate history. *Palaeogeography, Palaeoclimatology, Palaeoecology* 102:177–213.

Fok-Pun, L. and Komar, P. D. 1983. Settling velocities of planktonic foraminifera: density variations and shape effects. *Journal of Foraminiferal Research* 13:60–68.

Folk, R. L. and Pittman, J. S. 1971. Length slow chalcedony: a new testament for vanished evaporites. *Journal of Sedimentary Petrology* 41:1045–1058.

Folk, R. L. and Robles, R. 1964. Carbonate sands of Isla Perez, Alacran reef complex, Yucatan. *Journal of Geology* 72:255–292.

Föllmi, K. B. 1996. The phosphorus cycle, phosphogenesis and marine phosphate-rich deposits. *Earth-Science Reviews* 40:55–124.

Föllmi, K. B., Weissert, H. and Lini, A. 1993. Nonlinearities in phosphogenesis and phosphorus–carbon coupling and their implications for global change. In: R. Wollast, F. T. Mackenzie and L. Chou (eds.) *Interactions of C, N, P and S Biogeochemical Cycles and Global Change*, pp. 447–474. Berlin: Springer-Verlag.

Foote, M. 1997a. Sampling, taxonomic description, and our evolving knowledge of morphological diversity. *Paleobiology* 23:181–206.

Foote, M. 1997b. Estimating taxonomic durations and preservation probability. *Paleobiology* 23:278–300.

Force, L. M. 1969. Calcium carbonate size distribution on the west Florida shelf and experimental studies on the microarchitectural control of skeletal breakdown. *Journal of Sedimentary Petrology* 39:902–934.

Fox, R. F. 1988. *Energy and the Evolution of Life*. New York: W. H. Freeman.

Frakes, L. A., Francis, J. E. and Syktus, J. I. 1992. *Climate Modes of the Phanerozoic*. Cambridge: Cambridge University Press.

Francis, S., Margulis, L. and Barghoorn, E. S. 1978. On the experimental silicification of microorganisms. II. On the time of appearance of eukaryotic organisms in the fossil record. *Precambrian Research* 6:65–100.

François, L. M. and Walker, J. C. G. 1992. Modelling the Phanerozoic carbon cycle and climate: constraints from the $^{87}Sr/^{87}Sr$ isotopic ratio of seawater. *American Journal of Science* 292:81–135.

Franzen, J. L. 1985. Exceptional preservation of Eocene vertebrates in the lake deposit of Grube Messel (West Germany). In: H. B. Whittington and S. Conway Morris (eds.) Extraordinary fossil biotas: their ecological and evolutionary significance. *Philosophical Transactions of the Royal Society of London* (B) 311:181–186.

Franzen, J. L. 1990. Grube Messel. In: D. E. K. Briggs and P. R. Crowther (eds.) *Palaeobiology: A Synthesis*, pp. 289–294. Oxford: Blackwell.

Freiwald, A. 1995. Bacteria-induced carbonate degradation: a taphonomic case study of *Cibicides lobatulus* from a high-boreal carbonate setting. *Palaios* 10:337–346.

Frey, R. W. and Basan, P. B. 1981. Taphonomy of relict Holocene salt marsh deposits, Cabretta Island, Georgia. *Senckenbergiana Maritima* 13(4/6):111–155.

Frey, R. W. and Howard, J. D. 1986. Taphonomic characteristics of offshore mollusk shells, Sapelo Island, Georgia. *Tulane Studies in Geology* 19:51–61.

Frey, R. W., Howard, J. D. and Pryor, W. A. 1978. Ophiomorpha: its morphologic, taxonomic, and environmental significance. *Palaeogeography, Palaeoclimatology, Palaeoecology* 23:199–229.

Fricke, H. C. and O'Neill, J. R. 1996. Inter- and intra-tooth variation in the oxygen isotope composition of mammalian tooth enamel phosphate: implications for palaeoclimatological and palaeobiological research. *Palaeogeography, Palaeoclimatology, Palaeoecology* 126:91–99.

Frodeman, R. 1995. Geological reasoning: geology as an interpretive and historical science. *Geological Society of America Bulletin* 107:960–968.

Furbish, D. J. and Arnold, A. J. 1997. Hydrodynamic strategies in the morphological evolution of spinose planktonic foraminifera. *Geological Society of America Bulletin* 109:1055–1072.

Fürsich, F. T. 1971. Hartgründe und Kondensation im Dogger von Calvados. *Neues Jahrbuch für Geologie und Paläontologie Abhandlung* 138:313–342.

Fürsich, F. T. 1978. The influence of faunal condensation and mixing on the preservation of fossil benthic communities. *Lethaia* 11:243–250.

Fürsich, F. T. 1982. Rhythmic bedding and shell bed formation in the Upper Jurassic of East Greenland. In: G. Einsele and A. Seilacher (eds.) *Cyclic and Event Stratification*, pp. 209–222. Berlin: Springer-Verlag.

Fürsich, F. T. and Aberhan, M. 1990. Significance of time-averaging for paleocommunity analysis. *Lethaia* 23:143–152.

Fürsich, F. T. and Flessa, K. W. 1987. Taphonomy of tidal flat molluscs in the northern Gulf of California: paleoenvironmental analysis despite the perils of preservation. *Palaios* 2:543–559

Fürsich, F. T. and Flessa, K. W. 1991. The origin and interpretation of Bahia la Choya (northern Gulf of Califronia) taphocoenoses: implications for paleoenvironmental analysis. *Zitteliana* 18:165–169.

Fürsich, F. T. and Kauffman, E. G. 1984. Palaeoecology of marginally marine sedimentary cycles in the Albian Bear River Formation of southwestern Wyoming (USA). *Palaeontology* 27:501–536.

Futterer, E. 1974. Significance of boring sponge *Cliona* for the origin of fine-grained material of carbonate sediments. *Journal of Sedimentary Petrology* 44:79–80.

Futterer, E. 1978a. Hydrodynamic behavior of biogenic particles. *Neues Jahrbuch für Geologie und Paläontologie Abhandlung* 157:37–42.

Futterer, E. 1978b. Studien über die Einregelung, Anlagerung und Einbettung biogener Hartteile im Strömungskanal. *Neues Jahrbuch für Geologie und Paläontologie Abhandlung* 156:87–131.

Futterer, E. 1978c. Untersuchungen über die Sink- und Transportgeschwindigkeit biogener Hartteile. *Neues Jahrbuch für Geologie und Paläontologie Abhandlung* 155:318–359.

Gabbott, S. E., Aldridge, R. J. and Theron, J. N. 1995. A giant conodont with preserved muscle tissue from the Upper Ordovician of South Africa. *Nature* 374: 800–803.

Gajewski, K., Winkler, M. G. and Swain, A. M. 1985. Vegetation and fire history from three lakes with varved sediments in northwestern Wisconsin (U.S.A.). *Review of Palaeobotany and Palynology* 44:277–292.

Gall, J.-C. 1990. Les voiles microbiens: leur contribution à la fossilisation des organismes au corps mou. *Lethaia* 23:21–28.

Gandolfo, M. A., Nixon, K. C., Crepet, W. L., Stevenson, D. W. and Friis, E. M. 1998. Oldest known fossil monocotyledons. *Nature* 394:532–533.

Gao, S. and Collins, M. 1995. Net sand transport direction in a tidal inlet, using foraminiferal tests as natural tracers. *Estuarine, Coastal and Shelf Science* 40:681–697.

Garrels, R. M., Lerman, A. and Mackenzie, F. T. 1976. Controls of atmospheric O_2 and CO_2: past, present, and future: *American Scientist* 64:306–315.

Garrels, R. M. and Mackenzie, F. T. 1971. *Evolution of Sedimentary Rocks*. New York: Norton.

Garrison, R. 1981. Diagenesis of ocean carbonate sediments: a review of the DSDP perspective. In: J. E. Warme, R. G. Douglas and E. L. Winterer (eds.) *The Deep-Sea Drilling Project: a Decade of Progress*, pp. 181–208. Tulsa: SEPM (Society for Sedimentary Geology) Special Publication No. 32.

Gartner, S. 1996. Calcareous nannofossils at the Cretaceous–Tertiary boundary. In: N. MacLeod and G. Keller (eds.) *Cretaceous Tertiary Mass Extinctions: Biotic and Environmental Changes*, pp. 27–47. New York: Norton.

Gastaldo, R. A. 1989. Preliminary observations on phytotaphonomic assemblages in a subtropical/temperate Holocene bayhead delta: Mobile Delta, Gulf Coastal Plain, Alabama. *Review of Palaeobotany and Palynology* 58:61–83.

Gastaldo, R. A. 1990. The paleobotanical character of log assemblages necessary to differentiate blow-downs resulting from cyclonic winds. *Palaios* 5:472–478.

Gastaldo, R. A. 1992*a*. Taphonomic considerations for plant evolutionary investigations. *Paleobotanist* 41:211–223.

Gastaldo, R. A. 1992*b*. Regenerative growth in fossil horsetails following burial by alluvium. *Historical Biology* 6:203–219.

Gastaldo, R. A. 1997. What is the fidelity of autochthonous fossil leaf litter assemblages for the evaluation of paleoclimate? *Geological Society of America Abstracts with Programs* 29:A430.

Gastaldo, R. A. and Huc, A.-Y. 1992. Sediment facies, depositional environments, and distribution of phytoclasts in the Recent Mahakam River Delta, Kalimantan, Indonesia. *Palaios* 7:574–590.

Gastaldo, R. A., Douglass, D. P. and McCarroll, S. M. 1987. Origin, characteristics, and provenance of plant macrodetritus in a Holocene crevasse splay, Mobile Delta, Alabama. *Palaios* 2:229–240.

Gastaldo, R. A., Bearce, S. C., Degges, C. W., Hunt, R. J., Peebles, M. W. and Violette, D. L. 1989. Biostratinomy of a Holocene oxbow lake: a backswamp to mid-channel transect. *Palaeogeography, Palaeoclimatology, Palaeoecology* 58:47–59.

Gastaldo, R. A., Ashley, G., Lane, H. R., MacLeod, N., O'Neill, B. J. and Cheng-Yuan, W. 1998. Paleontology in the 21st century: an international workshop 3–9 September 1997. *Palaios* 13:87–90.

Gattuso, J.-P., Frankignoulle, M., Bourge, I., Romaine, S. and Buddemeier, R. W. 1998. Effect of calcium carbonate saturation of seawater on coral calcification. *Global and Planetary Change* 18:37–46.

Gautier, D. L. 1982. Siderite concretions: indicators of early diagenesis in the Gammon Shale (Cretaceous). *Journal of Sedimentary Petrology* 52:859–871.

Gehrels, W. R. 1994. Determining relative sea-level change from salt-marsh foraminifera and plant zones on the coast of Maine, U.S.A. *Journal of Coastal Research* 10:990–1009.

Gersonde, R. and Wefer, G. 1987. Sedimentation of biogenic siliceous particles in Antarctic waters from the Atlantic sector. *Marine Micropaleontology* 11:311–332.

Gibbs, R. J., Mathews, M. D. and Link, D. A. 1971. The relationship between sphere size and settling velocity. *Journal of Sedimentary Petrology* 41:7–18.

Ginsburg, R. N. (organizer) 1996. The Cretaceous–Tertiary boundary: the El Kef blind test. *Marine Micropaleontology* 29:65–103.

Ginsburg, R. N., Lloyd, R. M., Stockman, K. W. and McCallum, J. 1963. Shallow water carbonate sediments. In: M. N. Hill (ed.) *The Sea*, pp. 554–582. New York: Wiley.

Glass, B. P. 1969. Reworking of deep-sea sediments as indicated by the vertical dispersion of the Australasian and Ivory Coast microtektite horizons. *Earth and Planetary Science Letters* 6:409–415.

Glass, B. P. and Hazel, J. E. 1990. Chronostratigraphy of Upper Eocene microspherules: comment & reply. *Palaios* 5:387–390.

Glob, P. V. 1969. *The Bog People*. Ithaca: Cornell University Press.

Glover, C. P. and Kidwell, S. M. 1993. Influence of organic matrix on the post-mortem destruction of mollusc shells. *Journal of Geology* 101:729–747.

Glynn, P.-W. (ed.) 1996. Coral reefs of the eastern Pacific. *Coral Reefs* 15:69–147.

Goldstein, S. T. 1988. Foraminifera of relict salt marsh deposits, St Catherines Island, Georgia: taphonomic implications. *Palaios* 3:327–334.

Goldstein, S. T. and Harben, E. B. 1993. Taphofacies implications of infaunal foraminiferal assemblages in a Georgia salt marsh, Sapelo Island. *Micropaleontology* 39:53–62.

Goldstein, S. T., Watkins, G. T. and Kuhn, R. M. 1995. Microhabitats of salt marsh foraminifera: St Catherines Island, Georgia, USA. *Marine Micropaleontology* 26:17–29.

Golenberg, E. M., Giannasi, D. E., Clegg, M. T., Smiley, C. J., Durbin, M., Henderson, D. and Zurawski, G. 1990. Chloroplast DNA sequence from a Miocene *Magnolia* species. *Nature* 344:656–658.

Golubic, S., Perkins, R. D. and Lukas, K. J. 1975. Boring microorganisms and microborings in carbonate substrates. In: R. W. Frey (ed.) *The Study of Trace Fossils*, pp. 229–259. Berlin: Springer-Verlag.

Goodfriend, G. A. 1987. Chronostratigraphic studies of sediments in the Negev Desert using amino acid epimerization analysis of land snails. *Quaternary Research* 28:374–392.

Goodwin, R. G. 1988. Pollen taphonomy in Holocene glaciolacustrine sediments, Glacier Bay, Alaska: a cautionary note. *Palaios* 3:606–611.

Gordon, C. C. and Buikstra, J. E. 1981. Soil pH, bone preservation, and sampling bias at mortuary sites. *American Antiquity* 46:566–571.

References

Gould, S. J. 1965. Is uniformitarianism necessary? *American Journal of Science* 263:223–228.

Gould, S. J. 1986. Evolution and the triumph of homology, or why history matters. *American Scientist* 74:60–69.

Gould, S. J. 1996. *Full House: The Spread of Excellence from Plato to Darwin*. New York: Harmony Books.

Gould, S. J. and Eldredge, N. 1993. Punctuated equilibrium comes of age. *Nature* 366:223.

Grabert, B. 1971. Zur Eigenung von Foraminiferen als Indikatoren für Sandwanderung. *Deutsche Hydrographische Zeitschrift* 24:1–14.

Graham, B. F. 1957. Labelling pollen of woody plants with radioactive isotopes. *Ecology* 38:156–158.

Graham, J. B., Dudley, R., Aguilar, N. M. and Gans, C. 1995. Implications of the late Palaeozoic oxygen pulse for physiology and evolution. *Nature* 375:117–120.

Graham, R. W. 1993. Processes of time-averaging in the terrestrial vertebrate record. In: S. M. Kidwell and A. K. Behrensmeyer (eds.) *Taphonomic Approaches to Time Resolution in Fossil Assemblages*, pp. 102–124. Pittsburgh: Paleontological Society Short Courses in Paleontology No. 6.

Grayson, D. K. 1984. *Quantitative Zooarcheology*. Orlando: Academic Press.

Green, M. A., Aller, R. C. and Aller, J. Y. 1992. Experimental evaluation of the influences of biogenic reworking on carbonate preservation in nearshore sediments. *Marine Geology* 107:175–181.

Green, M. A., Aller, R. C. and Aller, J. Y. 1993. Carbonate dissolution and temporal abundances of foraminifera in Long Island Sound sediments. *Limnology and Oceanography* 38:331–345.

Greene, M. T. 1982. *Geology in the Nineteenth Century: Changing Views of a Changing World*. Ithaca, New York: Cornell University Press.

Greenstein, B. J. 1989. Mass mortality of the West-Indian echinoid *Diadema antillarum* (Echinodermata: Echinoidea): a natural experiment in taphonomy. *Palaios* 4:487–492.

Greenstein, B. J. 1991. An integrated study of echinoid taphonomy: predictions for the fossil record of four echinoid families. *Palaios* 6:519–540.

Greenstein, B. J. 1992. Taphonomic bias and the evolutionary history of the family Cidaridae (Echinodermata: Echinoidea). *Paleobiology* 18:50–79.

Greenstein, B. J. and Curran, H. A. 1997. How much ecological information is preserved in fossil coral reefs and how reliable is it? Proceedings, 8th International Coral Reef Symposium, vol. 1, pp. 417–422.

Greenstein, B. J. and Moffat, H. A. 1996. Comparative taphonomy of modern and Pleistocene corals, San Salvador, Bahamas. *Palaios* 11:57–63.

Greenstein, B. J. and Pandolfi, J. M. 1997. Preservation of community structure in modern reef coral and death assemblages of the Florida Keys: implications for the Quaternary fossil record of coral reefs. *Bulletin of Marine Science* 61:431–452.

Greenstein, B. J., Pandolfi, J. M. and Moran, P. J. 1995. Taphonomy of crown-of-thorns starfish: implications for recognizing ancient population outbreaks. *Coral Reefs* 14:91–97.

Greenstein, B. J., Pandolfi, J. M. and Curran, H. A. 1997. The completeness of the Pleistocene fossil record: implications for stratigraphic adequacy. In: S. K. Donovan and C. R. C. Paul (eds.) *Adequacy of the Fossil Record*. Chichester: Wiley.

Greenwood, D. R. 1991. The taphonomy of plant macrofossils. In: S. K. Donovan (ed.) *The Processes of Fossilization*, pp. 141–169. London: Belhaven Press.

Gregory, M. R., Ballance, P. F., Gibson, G. W. and Ayling, A. M. 1979. On how some rays (Elasmobranchia) excavate feeding depressions by jetting water. *Journal of Sedimentary Petrology* 49:1125–1130.

Greiner, G. O. G. 1970. Distribution of major benthonic foraminiferal groups on the Gulf of Mexico continental shelf. *Micropaleontology* 16:83–101.

Grierson, J. D. 1976. *Leclercqia complexa* (Lycopsida, Middle Devonian): its anatomy, and the interpretation of pyrite petrifaction. *American Journal of Botany* 63:1184–1202.

Grimaldi, D. 1996. *Amber: Window to the Past*. New York: Abrams.

Grimaldi, D., Bonwich, E., Delannoy, M. and Doberstein, S. 1994. Electron microscopic studies of mummified tissues in amber fossils. American Museum Novitates, no. 3097.

Grobe, H. and Fütterer, D. 1981. Zur Fragmentierung benthischer Foraminiferen in der Kieler Bucht (Westliche Ostsee). *Meyniana* 33:85–96.

Grossman, E. L. 1992. Isotope studies of Paleozoic paleoceanography: opportunities and pitfalls. *Palaios* 7:241–243.

Grotzinger, J. P. and Kasting, J. F. 1993. New constraints on Precambrian ocean composition. *Journal of Geology* 101:235–243.

Grotzinger, J. P. and Knoll, A. H. 1995. Anomalous carbonate precipitates: is the Precambrian the key to the Permian? *Palaios* 10:578–596.

Grupe, G., Piepenbrink, H. and Schoeninger, M. J. 1989. Note on microbial influence on stable carbon and nitrogen isotopes in bone. *Applied Geochemistry* 4:299.

Guinasso, N. L. and Schink, D. R. 1975. Quantitative estimates of biological mixing rates in abyssal sediments. *Journal of Geophysical Research* 80:3032–3043.

Gutiérrez-Marco, J. C. and Lenz, A. C. 1998. Graptolite synrhabdosomes: biological or taphonomic entities? *Paleobiology* 24:37–48.

Hageman, S. A. and Kaesler, R. L. 1998. Wall structure and growth of fusulinacean foraminifera. *Journal of Paleontology* 72:181–190.

Hall, S. A. 1981. Deteriorated pollen grains and the interpretation of Quaternary pollen diagrams. *Review of Palaeobotany and Palynology* 32:193–206.

Hall, S. A. 1989. Pollen analysis and paleoecology of alluvium. *Quaternary Research* 31:435–438.

Hallam, A. 1992. *Phanerozoic Sea-Level Changes*. New York: Columbia University Press.

Hallock, P. 1982. Evolution and extinction in larger foraminifera: *Proceedings of the Third North American Paleontological Convention* 1:221–225.

Hallock, P. 1987. Fluctuations in the trophic resource continuum: a factor in global diversity cycles? *Paleoceanography* 2:457–471.

Hallock, P. 1988. The role of nutrient availability in bioerosion: consequences to carbonate buildups. *Palaeogeography, Palaeoclimatology, Palaeoecology* 63:275–291.

Hallock, P. 1995. Promoting foraminiferal research. *Journal of Foraminiferal Research* 25:186–187.

References

Hallock, P. and Schlager, W. 1986. Nutrient excess and the demise of coral reefs and carbonate platforms. *Palaios* 1:389–398.

Hallock, P., Premoli Silva, I. and Boersma, A. 1991. Similarities between planktonic and larger foraminiferal evolutionary trends through Paleogene paleoceanographic changes. *Palaeogeography, Palaeoclimatology, Palaeoecology* 83:49–64.

Hallock, P., Müller-Karger, F. E. and Halas, J. C. 1993. Coral reef decline. *National Geographic Exploration and Research* 9:358–378.

Hanley, J. H. and R. M. Flores. 1987. Taphonomy and paleoecology of nonmarine mollusca: indicators of alluvial plain lacustrine sedimentation, upper part of the Tongue River Member, Fort Union Formation (Paleocene), northern Powder River basin, Wyoming and Montana. *Palaios* 2:479–476.

Hanson, C. B. 1980. Fluvial taphonomic processes: models and experiments. In: A. K. Behrensmeyer and A. P. Hill (eds.) *Fossils in the Making: Vertebrate Taphonomy and Paleoecology*, pp. 156–181. Chicago: University of Chicago Press.

Hanson, D. B. and Buikstra, J. E. 1987. Histomorphological alteration in buried human bone from the Lower Illinois Valley: implications for paleodietary research. *Journal of Archaeological Science* 14:549–563.

Haq, B. U., Hardenbol, J. and Vail, P. R. 1988. Mesozoic and Cenozoic chronostratigraphy and cycles of sea-level change. In: C. K. Wilgus, B. S. Hastings, C. G. St. C. Kendall, H. W. Posamentier, C. A. Ross and J. C. Van Wagoner (eds.) *Sea-Level Changes: an Integrated Approach*, pp. 71–108. Tulsa: Society of Economic Paleontologists and Mineralogists Special Publication No. 42.

Hardie, L. A. 1996. Secular variation in seawater chemistry: an explanation for the coupled secular variation in the mineralogies of marine limestones and potash evaporites over the past 600 m.y. *Geology* 24:279–283.

Harding, G. C. H. 1973. Decomposition of marine copepods. *Limnology and Oceanography* 18:670–673.

Hare, P. E. 1974. Amino acid dating of bone: the influence of water. *Carnegie Institution of Washington Yearbook* 73:576–581.

Hare, P. E. 1980. Organic geochemistry of bone and its relation to the survival of bone in the natural environment. In: A. K. Behrensmeyer and A. P. Hill (eds.) *Fossils in the Making: Vertebrate Taphonomy and Paleoecology*, pp. 131–152. Chicago: University of Chicago Press.

Harper, H. E. and Knoll, A. H. 1975. Silica, diatoms, and Cenozoic radiolarian evolution. *Geology* 3:175–177.

Harper, E. M., Forsythe, G. T. W. and Palmer, T. 1998. Taphonomy and the Mesozoic marine revolution: preservation state masks the importance of boring predators. *Palaios* 13:352–360.

Hart, M. B. 1996. The geology and micropaleontology of the Channel Tunnel. In: J. E. Repetski (ed.) *Sixth North American Paleontological Convention*, Abstracts and Program, p. 162. Pittsburgh: Paleontological Society Special Publication No. 8.

Harwood, D. M. and Gersonde, R. 1990. Lower Cretaceous diatoms from ODP Leg 113 Site 693 (Weddell Sea). Part 2: Resting spores, chrysophycean cysts, an endoskeletal dinoflagellate, and notes on the origin of diatoms. In: P. F. Barker *et al.* (eds.) *Proceedings of the Ocean Drilling Program, Scientific Results*, pp. 403–425. Washington, DC: U.S. Government Printing Office.

Harwood, D. M. and Webb, P.-N. 1998. Glacial transport of diatoms in the Antarctic Sirius Group: Pliocene refrigerator. *GSA Today* 8(4):1.

Haszeldine, R. S. 1984. Muddy deltas in freshwater lakes, and tectonism in the Upper Carboniferous Coalfield in NE England. *Sedimentology* 31:811–822.

Havinga, A. J. 1967. Palynology and pollen preservation. *Review of Palaeobotany and Palynology* 2:81–98.

Havinga, A. J. 1971. An experimental investigation into the decay of pollen and spores in various soil types. In: J. Brooks, P. R. Grant, M. Muir, P. van Gijzel, and G. Shaw (eds.) *Sporopollenin*, pp. 446–479. New York: Academic Press.

Hay, W. W. 1995. Paleoceanography of marine organic-carbon-rich sediments. In: A. Y. Huc (ed.) *Paleogeography, Paleoclimate and Source Rocks*, pp. 21–59. Tulsa: American Association of Petroleum Geologists Memoir 40.

Hay, W. W., Sloan, J. L. and Wold, C. N. 1988. Mass/age distribution and composition of sediments on the ocean floor and the global rate of sediment subduction. *Journal of Geophysical Research* 93 (B12):14 933–14 940.

Hay, W. W. and Wold, C. N. 1990. Relation of selected mineral deposits to the mass/age distribution of Phanerozoic sediments. *Geologische Rundschau* 79:495–512.

Haynes, G. 1985. Age profiles in elephant and mammoth bone assemblages. *Quaternary Research* 24:333–345.

Hayward, J. L., Folsom, S. D., Elmendorf, D. L., Tambrini, A. A. and Cowles, D. L. 1997. Experiments on the taphonomy of amniote eggs in marine environments. *Palaios* 12:482–488.

Heath, G. R. 1974. Dissolved silica and deep-sea sediments. In: W. W. Hay (ed.) *Studies in Paleo-oceanography*, pp. 77–93. Tulsa: SEPM (Society for Sedimentary Geology) Special Publication No. 20.

Hecht, A. D., Eslinger, E. V. and Garmon, L. B. 1977. Experimental studies on the dissolution of planktonic foraminifera. In: W. V. Sliter, A. W. H. Bé and W. H. Berger (eds.) *Dissolution of Deep-Sea Carbonates*, pp. 56–69. Cambridge, MA: Cushman Foundation for Foraminiferal Research Special Publication No. 13.

Hecht, F. 1933. Der Verbleib der organischen Substanz der Tiere bei meerischer Einbettung. *Senckenbergiana* 15(3-4):165–249.

Hedges, J. I., Cowie, G. L., Ertel, J. W., Barbour, R. J. and Hatcher, P. G. 1985. Degradation of carbohydrates and lignins in buried woods. *Geochimica et Cosmochimica Acta* 49:701–711.

Heikoop, J. M., Tsujita, C. J., Heikoop, C. E., Risk, M. J. and Dickin, A. P. 1997. Effects of volcanic ashfall recorded in ancient marine benthic communities: comparison of a nearshore and an offshore environment. *Lethaia* 29:125–139.

Henderson, R. A. 1984. Diagenetic growth of euhedral megaquartz in the skeleton of a stromatoporoid. *Journal of Sedimentary Petrology* 54:1138–1146.

Henderson, S. W. and Frey, R. W. 1986. Taphonomic redistribution of mollusk shells in a tidal inlet channel. *Palaios* 1:3–16.

Henrich, R. and Wefer, G. 1986. Dissolution of biogenic carbonates: effects of skeletal structure. *Marine Geology* 71:341–362.

Henwood, A. 1992. Exceptional preservation of dipteran flight muscle and the taphonomy of insects in amber. *Palaios* 7:203–212.

References

Henwood, A. 1993. Recent plant resins and the taphonomy of organisms in amber: a review. *Modern Geology* 19:35–59.

Herbert, T. D. and Sarmiento, J. L. 1991. Ocean nutrient distribution and oxygenation: limits on the formation of warm saline bottom water over the past 91 m.y. *Geology* 19:702–705.

Herschel, J. F. W. 1831. *A Preliminary Discourse on the Study of Natural Philosophy*. London: Longman, Rees, Orme, Brown and Green. (Reprint, University of Chicago Press, 1987).

Hesselbo, S. P. 1987. The biostratinomy of *Dikelocephalus* sclerites: implications for the use of trilobite attitude data. *Palaios* 2:605–608.

Heusser, C. J. and Florer, L. E. 1973. Correlation of marine and continental Quaternary pollen records from the northeast Pacific and western Washington. *Quaternary Research* 3:661–670.

Heusser, L. 1978. Spores and pollen in the marine realm. In: B. U. Haq and A. Boersma (eds.) *Introduction to Marine Micropaleontology*, pp. 327–339. New York: Elsevier.

Heusser, L. E. 1983. Pollen distribution in the bottom sediments of the western North Atlantic Ocean. *Marine Micropaleontology* 8:77–88.

Heusser, L. E. 1988. Pollen distribution in marine sediments on the continental margin off northern California. *Marine Geology* 80:131–147.

Hewitt, R. A. 1988. Nautiloid shell taphonomy: interpretations based on water pressure. *Palaeogeography, Palaeoclimatology, Palaeoecology* 63:15–25.

Highsmith, R. C. 1980. Geographic patterns of coral bioerosion: a productivity hypothesis. *Journal of Experimental Marine Biology and Ecology* 46:177–196.

Hill, A. P. 1980. Early postmortem damage to the remains of some contemporary East African mammals. In: A. K. Behrensmeyer and A. P. Hill (eds.) *Fossils in the Making: Vertebrate Taphonomy and Paleoecology*, pp. 131–152. Chicago: University of Chicago Press.

Hippensteel, S. P. and Martin, R. E. 1999. Foraminifera as an indicator of overwash deposits, barrier island sediment supply, and barrier island evolution: Folly Beach, South Carolina. *Palaeogeography, Palaeoclimatology, Palaeoecology*. In press.

Ho, C. and Coleman, J. M. 1969. Consolidation and cementation of recent sediments in the Atchafalya Basin. *Geological Society of America Bulletin* 80:183–192.

Hodder, I. 1982. *The Present Past: an Introduction to Anthropology for Archaeologists*. New York: Pica Press.

Hof, C. and Briggs, D. E. G. 1997. Decay and mineralization of mantis shrimps (Stomatopoda: Crustacea): a key to their fossil record. *Palaios* 12:420–438.

Hoffman, P. F., Kaufman, A. J., Halverson, G. P. and Schrag, D. P. 1998. A Neoproterozoic snowball Earth. *Science* 281:1342–1346.

Hoffman, R. 1988. The contribution of raptorial birds to patterning in small mammal assemblages. *Paleobiology* 14:81–90.

Hoge, B. E. 1994. Wetland ecology and paleoecology: relationships between biogeochemistry and preservable taxa. *Current Topics in Wetland Biogeochemistry* 1:48–67.

Hoge, B. E. 1995. Wetland microfossil taphonomy: a model for the interpretation of fine-scale sea-level fluctuations. *Geological*

Society of America Abstracts with Programs 27:A28.

Holdaway, H. K. and Clayton, C. J. 1982. Preservation of shell microstructure in silicified brachiopods from the Upper Cretaceous Wilmington Sands of Devon. *Geological Magazine* 119:371–382.

Holland, S. M. 1988. Taphonomic effects of sea-floor exposure on an Ordovician brachiopod assemblage. *Palaios* 3:588–597.

Holland, S. M. 1995. The stratigraphic disribution of fossils. *Paleobiology* 21:92–109.

Holmden, C., Creaser, R. A., Muehlenbachs, K., Leslie, S. A. and Bergström, S. M. 1998. Isotopic evidence for geochemical decoupling between ancient epeiric seas and bordering oceans: implications for secular curves. *Geology* 26:567–570.

Holmes, P. L. 1994. The sorting of spores and pollen by water: experimental and field evidence. In: A. Traverse (ed.) *Sedimentation of Organic Particles*, pp. 9–32. Cambridge: Cambridge University Press.

Holser, W. T. and Magaritz, M. 1987. Events near the Permian–Triassic boundary. *Modern Geology* 11:155–280.

Holser, W. T. and Magaritz, M. 1992. Cretaceous/Tertiary and Permian/Triassic boundary events compared. *Geochimica et Cosmochimica Acta* 56:3297–3309.

Holser, W. T., Schidlowski, M., Mackenzie, F. T. and Maynard, J. B. 1988. Geochemical cycles of carbon and sulfur. In: C. B. Gregor, R. M. Garrels, F. T. Mackenzie and J. B. Maynard (eds.) *Chemical Cycles in the Evolution of the Earth*, pp. 105–173. New York: Wiley.

Holser, W. T., Schönlaub, H.-P., Boeckelmann, K. and Magaritz, M. 1991. The Permian–Triassic of the Gartnerkofel-1 Core (Carnic Alps, Austria): synthesis and conclusions. *Abhandlungen der Geologischen Bundesanstalt* 45:213–232.

Honjo, S. 1977. Dissolution of suspended coccoliths in the deep-sea water column and sedimentation of coccolith ooze. In: W. V. Sliter, A. W. H. Bé and W. H. Berger (eds.) *Dissolution of Deep-Sea Carbonates*, pp. 114–128. Cambridge, MA: Cushman Foundation for Foraminiferal Research Special Publication No. 13.

Hood, K. 1986. GraphCor: Interactive Graphic Correlation for Microcomputers. Houston, Texas.

Hooghiemstra, H. 1988. Palynological records from northwest African marine sediments: a general outline of the interpretation of the pollen signal. *Philosophical Transactions of the Royal Society of London* (B) 318:431–449.

Hooykaas, R. 1963. *The Principle of Uniformity in Geology, Biology and Theology*. Leiden: Brill.

Hopkins, J. S. 1950. Differential flotation and deposition of coniferous and deciduous tree pollen. *Ecology* 31:633–641.

Horner, J. R. 1982. Evidence of colonial nesting and site fidelity among ornithischian dinosaurs. *Nature* 297:675–676.

Horner, J. R. and Weishampel, D. B. 1988. A comparative embryological study of two ornithischian dinosaurs. *Nature* 358:59–61.

Horowitz, A. S. and Pachut, J. F. 1994. Lyellian bryozoan percentages and the fossil record of the recent bryozoan fauna. *Palaios* 9:500–505.

Hotinski, R. M., Kump, L. R., Bice, K. L., Najar, R. G. and Arthur, M. A. 1998. Quantitative assessment of ocean stagnation hypotheses for end-Permian extinction. *Geological Society of America Abstracts with Programs* 30:A310.

References

Hubbert, M. K. 1967. Critique of the principle of uniformity. In: C. C. Albritton (ed.) *Uniformity and Simplicity*, pp. 3–33. Boulder: Geological Society of America Special Paper No. 89.

Hudson, J. D. 1982. Pyrite in ammonite-bearing shales from the Jurassic of England and Germany. *Sedimentology* 29:639–667.

Hudson, J. H., Hanson, K. J., Halley, R. B. and Kindinger, J. L. 1994. Environmental implications of growth rate changes in Montastrea annularis: Biscayne Bay Park, Florida. *Bulletin of Marine Science* 54:647–669.

Hughes, T. P., Keller, B. D., Jackson, J. B. C. and Boyle, M. J. 1985. Mass mortality of the echinoid *Diadema antillarum* Phillipi in Jamaica. *Bulletin of Marine Science* 36:377–384.

Hulbert, S. H. and Archibald, J. D. 1995. No statistical support for sudden (or gradual) extinction of dinosaurs. *Geology* 23:881–884.

Hungerbühler, A. 1998. Taphonomy of the prosauropod dinosaur *Sellosaurus*, and its implications for carnivore faunas and feeding habits in the Late Triassic. *Palaeogeography, Palaeoclimatology, Palaeoecology*, 143:1–29.

Hunter, J. 1994. Lack of a high body count at the K–T boundary. *Journal of Paleontology* 68:1158.

Hutchinson, P. J. 1997. Environmental taphonomy. *Palaios* 12:403–404.

Hutson, W. H. 1980. Bioturbation of deep-sea sediments: oxygen isotopes and stratigraphic uncertainty. *Geology* 8:127–130.

Hyndman, R. D. 1995. Giant earthquakes of the Pacific northwest. *Scientific American* 273 (December):68–75.

Iacumin, P., Cominotto, D. and Longinelli, A. 1996. A stable isotope study of mammal skeletal remains of mid-Pleistocene age, Arago cave, eastern Pyrénées, France. Evidence of taphonomic and diagenetic effects. *Palaeogeography, Palaeoclimatology, Palaeoecology* 126:151–160.

Imbrie, J., Hays, J. D., Martinson, D. G., McIntyre, A. I., Mix, A., Morley, J. J., Pisias, N. G., Prell, W. L. and Shackleton, N. J. 1984. The orbital theory of Pleistocene climate: support from a revised chronology of the marine $\delta^{18}O$ record. In: A. Berger et al. (eds.) *Milankovitch and Climate*, part I, pp. 269–305. Dordrecht: Reidel.

Ingall, E. D., Bustin, R. M. and Van Cappellen, P. 1993. Influence of water column anoxia on the burial and preservation of carbon and phosphorus in marine shales. *Geochimica et Cosmochimica Acta* 57:303–316.

Isozaki, Y. 1997. Permo-Triassic boundary superanoxia and stratified superocean: records from lost deep sea. *Science* 276:235–238.

Iturralde-Vinent, M. A. and MacPhee, R. D. 1996. Age and paleogeographical origin of Dominican amber. *Science* 273:1850–1852.

Izuka, S. K. and Kaesler, R. L. 1986. Biostratinomy of ostracode assemblages from a small reef flat in Maunalua Bay, Oahu, Hawaii. *Journal of Paleontology* 60:347–360.

Jacka, A. D. 1974. Replacement of fossils by length-slow chalcedony and associated dolomitzation. *Journal of Sedimentary Petrology* 44:421–427.

Jackson, J. B. C. 1992. Pleistocene perspectives on coral reef community structure. *American Zoologist* 32:719–731.

Jackson, J. B. C. 1994. Community unity? *Science* 264:1412–1413.

Jackson, J. B. C. 1997. Reefs since Columbus. *Coral Reefs* 16 (Suppl):S23–S32.

Jackson, T. A. and Moore, C. B. 1976. Secular variations in kerogen structure and carbon, nitrogen and phosphorus concentrations in pre-Phanerozoic and Phanerozoic sedimentary rocks. *Chemical Geology* 18:107–136.

Jacobs, D. K. 1996. Chambered cephalopod shells, buoyancy, structure and decoupling: history and red herrings. *Palaios* 11:610–614.

Jacobs, D. K. and Landman, N. H. 1994. *Nautilus*: model or muddle? *Lethaia* 27:95–96.

James, N. P. 1983. Reef. In: P. Scholle, D. G. Bebout and C. H. Moore (eds.) *Carbonate Depositional Environments*, pp. 346–440. Tulsa: American Association of Petroleum Geologists.

James, N. P. 1996. Reading oceanographic change from neritic carbonates: the cool-water carbonate problem. In: M. Mutti, T. Simo, H. Weissert and P. Baker (eds.) *Carbonates and Global Change: an Interdisciplinary Approach*. SEPM/IAS Research Conference, Wildhaus, Switzerland. Abstract Book, pp. 74–75.

James, N. P. and Bone, Y. 1989. Petrogenesis of Cenozoic, temperate water calcarenites, south Australia: a model for meteoric/shallow burial diagenesis of shallow water calcite sediments. *Journal of Sedimentary Petrology* 59:191–203.

James, N. P., Wray, J. L. and Ginsburg, R. N. 1988. Calcification of encrusting aragonitic algae (Peyssonneliaceae): implications for the origin of late Paleozoic reefs and cements. *Journal of Sedimentary Petrology* 58:291–303.

Janssen, C. R. 1966. Recent pollen spectra from the deciduous and coniferous–deciduous forests of northeastern Minnesota: a study in pollen dispersal. *Ecology* 37:804–825.

Joachimski, M. M. and Buggisch, W. 1993. Anoxic events in the late Frasnian: causes of the Frasnian-Famennian faunal crisis? *Geology* 21:675–678.

John, S. 1995. Microfaunal assemblages: their use in interpreting depositional environments within a transgressive valley-fill sequence. *Geological Society of America Abstracts with Programs* 27:A28.

Johnson, K. R. 1993. Time resolution and the study of Late Cretaceous and early Tertiary megafloras. In: S. M. Kidwell and A. K. Behrensmeyer (eds.) *Taphonomic Approaches to Time Resolution in Fossil Assemblages*, pp. 210–227. Pittsburgh: Paleontological Society Short Courses in Paleontology No. 6.

Johnson, R. G. 1957. Experiments on the burial of shells. *Journal of Geology* 65:527–535.

Johnson, R. G. 1960. Models and methods for analysis of the mode of formation of fossil assemblages. *Geological Society of America Bulletin* 71:1075–1086.

Johnson, R. G. 1962. Mode of formation of marine fossil assemblages of the Pleistocene Millerton Formation of California. *Bulletin of the Geological Society of America* 73:113–130.

Johnson, T. C., Scholz, C. A., Talbot, M. R., Kelts, K., Ricketts, R. D., Ngobi, G., Beuning, K., Ssemmanda, I. and McGill, J. W. 1996. Late Pleistocene desiccation of Lake Victoria and rapid evolution of cichlid fishes. *Science* 273:1091–1093.

Johnston, C. A. 1995. Effects of animals on landscape pattern. In: L. Hansson, L. Fahrig and G. Merriam (eds.) *Mosaic Landscapes and Ecological Processes*, pp. 57–80. London: Chapman and Hall.

Jonasson, K. E. and Patterson, R. T. 1992. Preservation potential of salt marsh foraminifera from the Fraser River delta, British Columbia. *Micropaleontology* 38:289–301.

References

Jones, D. 1998. Defossilization. *Nature* 394:727.

Jones, D. J. 1956. *Introduction to Microfossils.* New York: Hafner (reprinted 1969).

Jones, G. A. and Ruddiman, W. F. 1982. Assessing the global meltwater spike. *Quaternary Research* 17:148–172.

Jorissen, F. J., De Stigter, H. C. and Widmark, J. G. V. 1995. A conceptual model explaining benthic foraminiferal microhabitats. *Marine Micropaleontology* 26:3–15.

Julson, A. P. and Rack, F. R. 1992. The relationship between sediment fabric and planktonic microfossil taphonomy: how do plankton skeletons become pelagic ooze? *Palaios* 7:167–177.

Jumars, P. A. and Wheatcroft, R. A. 1989. Responses of benthos to changing food quality and quantity, with a focus on deposit feeding and bioturbation. In: W. H. Berger, V. S. Smetacek and G. Wefer (eds.) *Productivity of the Ocean: Present and Past,* pp. 235–253. New York: Wiley.

Karowe, A. L. and Jefferson, T. H. 1987. Burial of trees by eruptions of Mt St Helens, Washington: implications for the interpretation of fossil forests. *Geological Magazine* 124:191–204.

Katz, B. J. and Man, E. H. 1980. The effects and implications of ultrasonic cleaning on the amino acid geochemistry of foraminifera. In: P. E. Hare, T. C. Hoering and K. King (eds.) *Biogeochemistry of Amino Acids,* pp. 215–222. New York: Wiley.

Kauffman, E. G. 1981. Ecologic reappraisal of the German Posidonienschiefer. In: J. Gray, A. J. Boucot and W. B. N. Berry (eds.) *Communities of the Past,* pp. 311–381. Stroudsburg, PA: Dowden, Hutchinson & Ross.

Kauffman, S. A. 1993. *The Origins of Order.* Oxford: Oxford University Press.

Kauffman, S. A. 1995. *At Home in the Universe.* Oxford: Oxford University Press.

Kaufman, A. J., Jacobsen, S. B. and Knoll, A. H. 1993. The Vendian record of Sr and C isotopic variations in seawater: implications for tectonics and paleoclimate. *Earth and Planetary Science Letters* 120:409–430.

Kaushik, N. K. and Hynes, H. B. N. 1971. The fate of the dead leaves that fall into streams. *Archivs für Hydrobiologie* 68:465–515.

Kaye, C. A. and Barghoorn, E. S. 1964. Late Quaternary sea-level change and crustal rise at Boston, Massachusetts, with notes on the autocompaction of peat. *Geological Society of America Bulletin* 75:63–80.

Kaźmierczak, J., Coleman, M. L., Gruszczyński, M. and Kempe, S. 1996. Cyanobacterial key to the genesis of micritic and peloidal limestones in ancient seas. *Acta Palaeontologica Polonica* 41:319–338.

Kaźmierczak, J., Ittekot, V. and Degens, E. T. 1985. Biocalcification through time: environmental challenge and cellular response. *Palaontologisches Zeitschrift* 59:15–33.

Keafer, B. A., Buesseler, K. O. and Anderson, D. M. 1992. Burial of living dinoflagellate cysts in estuarine and nearshore sediments. *Marine Micropaleontology* 20:147–161.

Keigwin, L. D. 1979. Late Cenozoic stable isotope stratigraphy and paleoceanography of Deep Sea Drilling Project sites from the east equatorial and central north Pacific Ocean. *Earth and Planetary Science Letters* 45:361–382.

Keir, R. S. and Hurd, D. C. 1983. The effect of encapsulated fine grain sediment and test morphology on the resistance of planktonic foraminifera to dissolution. *Marine Micropaleontology* 8:193–214.

Keith, M. L. 1982. Violent volcanism, stagnant oceans and some inferences regarding petroleum, strata-bound ores and mass extinctions. *Geochimica et Cosmochimica Acta* 46:2621–2637.

Keller, C. K. and Wood, B. D. 1993. Possibility of chemical weathering before the advent of vascular land plants. *Nature* 364:223–225.

Keller, G. 1994. K–T boundary issues. *Science* 264:641.

Keller, G. 1996. The Cretaceous–Tertiary mass extinction in planktonic foraminifera: biotic constraints for catastrophe theories. In: N. MacLeod and G. Keller (eds.) *Cretaceous Tertiary Mass Extinctions: Biotic and Environmental Changes*, pp. 49–84. New York: Norton.

Keller, G. and Barron, J. A. 1983. Paleoceanographic implications of Miocene deep-sea hiatuses. *Geological Society of America Bulletin* 94:590–613.

Keller, G., Li, L. and MacLeod, N. 1995. The Cretaceous/Tertiary boundary stratotype section at El Kef, Tunisia: how catastrophic were the mass extinctions? *Palaeogeography, Palaeoclimatology, Palaeoecology* 119:221–254.

Kellert, S. H. 1993. *In the Wake of Chaos*. Chicago: University of Chicago Press.

Kelley, P. H. and Hansen, T. A. 1996a. Recovery of the naticid gastropod predator–prey system from the Cretaceous–Tertiary and Eocene–Oligocene extinctions. In: M. B. Hart (ed.) *Biotic Recovery from Mass Extinction Events*, pp. 373–386. London: Geological Society Special Publication No. 102.

Kelley, P. H. and Hansen, T. A. 1996b. Naticid gastropod prey selectivity through time and the hypothesis of escalation. *Palaios* 11:437–445.

Kemp, R. A. and Unwin, D. M. 1997. The skeletal taphonomy of *Archaeopteryx*: a quantitative approach. *Lethaia* 30:229–238.

Kempe, S. and Degens, E. T. 1985. An early soda ocean? *Chemical Geology* 53:95–108.

Kempe, S. and Kaźmierczak, J. 1994. The role of alkalinity in the evolution of ocean chemistry, organization of living systems, and biocalcification processes. *Bulletin de l'Institut Océanographique, Monaco*, no. spécial 13:61–117.

Kennett, J. P. 1982. *Marine Geology*. Englewood Cliffs, NJ: Prentice-Hall.

Kenrick, P. and Edwards, D. 1988. The anatomy of Lower Devonian *Gosslingia breconensis* Heard based upon pyritized axes, with some comments on the permineralization process. *Botanical Journal of the Linnaean Society* 97:95–123.

Kerr, R. A. 1994a. An ice age nudge for human evolution in Africa? *Science* 263:173–174.

Kerr, R. A. 1994b. Testing an ancient impact's punch. *Science* 263:1371–1372.

Kerr, R. A. 1996. New mammal data challenge evolutionary pulse theory. *Science* 273:431–432.

Kerr, R. A. 1997. Climate–evolution link weakens. *Science* 276:1968.

Kerr, R. A. 1998a. Biggest extinction looks catastrophic. *Science* 280:1007.

Kerr, R. A. 1998b. Earliest animals old once more? *Science* 282:1020.

Ketcher, K. and Allmon, W. D. 1993. Environment and mode of deposition of a Pliocene coral bed: coral thickets and storms in the fossil record. *Palaios* 8:3–17.

Kidwell, S. M. 1986a. Models for fossil concentrations: paleobiologic implications. *Paleobiology* 12:6–24.

Kidwell, S. M. 1986b. Taphonomic feedback in Miocene assemblages: testing the role of dead hardparts in benthic communities. *Palaios* 1:239–255.

Kidwell, S. M. 1988. Taphonomic comparison of passive and active continental margins: Neogene shell beds of the Atlantic coastal plain and northern Gulf of California. *Palaeogeography, Palaeoclimatology, Palaeoecology* 63:201–223.

Kidwell, S. M. 1989. Stratigraphic condensation of marine transgressive records: origins of major shell deposits in the Miocene of Maryland. *Journal of Geology* 97:1–24.

Kidwell, S. M. 1990. Phanerozoic evolution of macroinvertebrate shell accumulations: preliminary data from the Jurassic of Britain. In: W. Miller (ed.) *Paleocommunity Temporal Dynamics: the Long-Term Development of Multispecies Assemblies*, pp. 309–327. Pittsburgh: Paleontological Society Special Publication No. 5.

Kidwell, S. M. 1991. The stratigraphy of shell concentrations. In: P. A. Allison and D. E. G. Briggs (eds.) *Taphonomy: Releasing the Data Locked in the Fossil Record*, pp. 212–290. New York: Plenum Press.

Kidwell, S. M. 1993a. Time-averaging and temporal resolution in Recent marine shelly faunas. In: S. M. Kidwell and A. K. Behrensmeyer (eds.) *Taphonomic Approaches to Time Resolution in Fossil Assemblages*, pp. 9–33. Pittsburgh: Paleontological Society Short Courses in Paleontology No. 6.

Kidwell, S. M. 1993b. Taphonomic expressions of sedimentary hiatuses: field observations on bioclastic concentrations and sequence anatomy in low, moderate and high subsidence settings. *Geologische Rundschau* 82:189–202.

Kidwell, S. M. and Baumiller, T. 1989. Experimental disintegration of regular echinoids: roles of temperature, oxygen, and decay thresholds. *Paleobiology* 16:247–271.

Kidwell, S. M. and Behrensmeyer, A. K. 1993a. Taphonomic approaches to time resolution in fossil assemblages. In: S. M. Kidwell and A. K. Behrensmeyer (eds.) *Taphonomic Approaches to Time Resolution in Fossil Assemblages*, pp. 1–8. Pittsburgh: Paleontological Society Short Courses in Paleontology No. 6.

Kidwell, S. M. and Behrensmeyer, A. K. 1993b. Summary: estimates of time-averaging. In: S. M. Kidwell and A. K. Behrensmeyer (eds.) *Taphonomic Approaches to Time Resolution in Fossil Assemblages*, pp. 301–302. Pittsburgh: Paleontological Society Short Courses in Paleontology No. 6.

Kidwell, S. M. and Bosence, D. W. J. 1991. Taphonomy and time-averaging of marine shelly faunas. In: P. A. Allison and D. E. G. Briggs (eds.) *Taphonomy: Releasing the Data Locked in the Fossil Record*, pp. 116–209. New York: Plenum Press.

Kidwell, S. M. and Brenchley, P. J. 1994. Patterns in bioclastic accumulation through the Phanerozoic: changes in input or in destruction? *Geology* 22:1139–1143.

Kidwell, S. M. and Flessa, K. W. 1995. The quality of the fossil record: populations, species, and communities. *Annual Review of Ecology and Systematics* 26:269–299.

Kidwell, S. M. and Jablonski, D. J. 1983. Taphonomic feedback: Ecological consequences of shell accumulation. In: M. J. S. Tevesz and P. L. McCall (eds.) *Biotic Interactions in Recent and Fossil Benthic Communities*, pp. 195–248. New York: Plenum Press.

Kidwell, S. M., Fürsich, F. T. and Aigner, T. 1986. Conceptual framework for the analysis and classification of fossil concentrations. *Palaios* 1:228–238.

Kilham, P. and Kilham, S. S. 1980. The evolutionary ecology of phytoplankton. In: I. Morris (ed.) *The Physiological Ecology of Phytoplankton*, pp. 571–597. Berkeley: University of California Press.

King, D. G., Soller, M. and Kashi, Y. Evolutionary tuning knobs. *Endeavour* 21:36–40.

Kitts, D. B. 1977. *The Structure of Geology*. Dallas: Southern Methodist University Press.

Klein, R. G. and Cruz-Uribe, K. 1984. *The Analysis of Animal Bones from Archeological Sites*. Chicago: University of Chicago Press.

Kluessendorf, J. 1994. Predictability of Silurian *Fossil-Konservat-Lagerstätten* in North America. *Lethaia* 27:337–344.

Knauth, L. P. 1979. A model for the origin of chert in limestone. *Geology* 7:274–277.

Knoll, A. H. 1985. Exceptional preservation of photosynthetic organisms in silicified carbonates and silicified peats. In: H. B. Whittington and S. Conway Morris (eds.) Extraordinary fossil biotas: their ecological and evolutionary significance. *Philosophical Transactions of the Royal Society of London* (B) 311:111–122.

Knoll, A. H. 1989. Evolution and extinction in the marine realm: some constraints imposed by phytoplankton. *Philosophical Transactions of the Royal Society of London* (B) 325:279–290.

Knoll, A. H. and Swett, K. 1987. Micropaleontology across the Precambrian–Cambrian boundary in Spitsbergen. *Journal of Paleontology* 61:898–926.

Knoll, A. H., Bambach, R. K., Canfield, D. E. and Grotzinger, J. P. 1996. Comparative Earth history and Late Permian mass extinction. *Science* 273:452–457.

Knoll, M. A. and James, W. C. 1987. Effect of the advent and diversification of vascular land plants on mineral weathering through geologic time. *Geology* 15:1099–1102.

Knowlton, N. 1992. Thresholds and multiple stable states in coral reef community dynamics. *American Zoologist* 32:674–682.

Knowlton, N., Lang, J. C. and Keller, B. D. 1990. Case study of natural population collapse: post-hurricane predation on Jamaican staghorn corals. *Smithsonian Contributions to Marine Sciences* 31:1–125.

Kobluk, D. R. and Kahle, C. F. 1977. Bibliography of the endolithic (boring) algae and fungi and related geological processes. *Bulletin of Canadian Petroleum Geology* 25:208–223.

Koch, C. F. 1987. Prediction of sample size effects on the measured temporal and geographic distribution patterns of species. *Paleobiology* 13:100–107.

Koch, P. L. 1998. Isotope paleoecology and land vertebrates: individuals, species, and ecosystems. *Palaios* 13:309–310.

Kolodny, Y., Luz, B., Sander, M. and Clements, W. A. 1996. Dinosaur bones: fossils or pseudomorphs? The pitfalls of physiology reconstruction from apatitic fossils. *Palaeogeography, Palaeoclimatology, Palaeoecology* 126:161–171.

Kontrovitz, M. 1975. A study of the differential transportation of ostracodes. *Journal of Paleontology* 49:937–941.

Kontrovitz, M. 1987. Comment and reply on "Use of ostracodes to recognize downslope contamination in paleobathymetry and a preliminary reappraisal of the paleodepth of the Prasás Marls (Pliocene), Crete, Greece." *Geology* 15:377–378.

Kontrovitz, M., Snyder, S. W. and Brown, R. J. 1978. A flume study of the movement of foraminifera tests. *Palaeogeography, Palaeoclimatology, Palaeoecology* 23:141–150.

Kontrovitz, M., Kilmartin, K. C. and Snyder, S. W. 1979. Threshold velocities of tests of planktic foraminifera. *Journal of Foraminiferal Research* 9:228–232.

Kontrovitz, M., Pani, E. A. and Bray, H. 1998. Experimental crushing of some podocopid ostracode valves: an aspect of taphonomy. *Palaios* 13:500–507.

Korth, W. W. 1979. Taphonomy of microvertebrate fossil assemblages. *Annals of the Carnegie Museum* 48:235–285.

Kotler, E., Martin, R. E. and Liddell, W. D. 1992. Experimental analysis of abrasion and dissolution resistance of modern reef-dwelling foraminifera: implications for the preservation of biogenic carbonate. *Palaios* 7:244–276.

Kouwenhoven, T. J. 1997. The El Kef blind test: how blind was it? Comment. *Marine Micropaleontology* 32:397–398.

Kowalewski, M. 1996a. Time-averaging, overcompleteness, and the geological record. *Journal of Geology* 104:317–326.

Kowalewski, M. 1996b. Taphonomy of a living fossil: the linguide brachiopod *Glottidia palmeri* Dall from Baja California, Mexico. *Palaios* 11: 244–265.

Kowalewski, M. 1997. The reciprocal taphonomic model. *Lethaia* 30:86–88.

Kowalewski, M. and Demko, T. M. 1997. Trace fossils and population paleoecology: comparative analysis of size–frequency distributions derived from burrows. *Lethaia* 29:113–124.

Kowalewski, M., Dulai, A. and Fürsich, F. T. 1998. A fossil record full of holes: the Phanerozoic history of drilling predation. *Geology* 26:1091–1094.

Kowalewski, M. and Flessa, K. W. 1996. Improving with age: the fossil record of lingulide brachiopods and the nature of taphonomic megabiases. *Geology* 24:977–980.

Kowalewski, M., Goodfriend, G. and Flessa, K. W. 1998. High-resolution estimates of temporal mixing within shell beds: the evils and virtues of time-averaging. *Paleobiology* 24:287–304.

Kozlova, I. M. 1986. Transportation of radiolarian shells by currents (calculations based on the example of the Kuroshio). *Marine Micropaleontology* 11:197–201.

Kramm, U. and Wedepohl, K. H. 1991. The isotopic composition of strontium and sulfur in seawater of Late Permian (Zechstein) age. *Chemical Geology* 90:253–262.

Krantz, D. E. 1990. Mollusk-isotope records of Plio-Pleistocene marine paleoclimate, U.S. Middle Atlantic Coastal Plain. *Palaios* 5:317–335.

Krassilov, V. A. 1996. A general scheme of ecosystem evolution. *Paleontological Journal (Moscow)* 30:626–633.

Krishnaswami, S., Beninger, L. K., Aller, R. C. and Vondamm, K. L. 1980. Atmospherically derived radionuclides as tracers of sediment mixing and accumulation in near shore marine and lake sediments: evidence from ^{7}Be, ^{210}Pb, and ^{239,240}Pu. *Earth and Planetary Science Letters* 47:307–318.

Kuenen, Ph. H. 1950. *Marine Geology*. New York: Wiley.

Kump, L. R. 1991. Interpeting carbon-isotope excursions: Strangelove oceans. *Geology* 19:299–302.

Labandeira, C. C., Phillips, T. L. and Norton, R. A. 1997. Oribatid mites and the decomposition of plant tissues in Paleozoic coal-swamp forests. *Palaios* 12:319–353.

Lal, D. and Peters, B. 1967. Cosmic-ray produced radioactivity on the earth. *Handbuch der Physik* 46:551–612.

LaMontagne, R. W., Murray, R. W., Wei, K.-Y., Leinen, M. and Wang, C.-H. 1996. Decoupling of carbonate preservation, carbonate concentration, and biogenic accumulation: a 400-kyr record from the central equatorial Pacific Ocean. *Paleoceanography* 11:553–562.

Lane, H. R. 1997. Paleontology in the 21st century or which way ought paleontology proceed from here? *Palaios* 12:95–97.

Langenheim, J. and Beck, C. W. 1965. Infrared spectra as a means of determining botanical sources of amber. *Science* 149:52–55.

Larson, D. W. and Rhoads, D. C. 1983. The evolution of infaunal communities and sedimentary fabrics. In: M. J. S. Tevesz and P. L. McCall (eds.) *Biotic Interactions in Recent and Fossil Benthic Communities*, pp. 627–648. New York: Plenum Press.

Lawrence, D. R. 1968. Taphonomy and information losses in fossil communities. *Geological Society of America Bulletin* 79:1315–1330.

Lask, P. B. 1993. The hydrodynamic behavior of sclerites from the trilobite *Flexicalymene meeki*. *Palaios* 8:219–225.

Le, J. and Thunell, R. C. 1996. Modelling planktic foraminiferal assemblage changes and application to sea surface temperature estimation in the western equatorial Pacific. *Marine Micropaleontology* 28:211–229.

LeCalvez, J. 1936. Sur le genre *Iridia*. *Archives de Zoologie Expérimentale et Générale* 78:115–131.

LeClair, E. E. 1993. Effects of anatomy and environment on the relative preservability of asteroids: a biomechanical comparison. *Palaios* 8:233–243.

Leidy, J. 1856. Notices of the remains of extinct reptiles and fishes discovered by Dr. F. V. Hayden in the badlands of the Judith River, Nebraska Territory. *Proceedings of the Academy of Natural Sciences of Philadelphia* 8:72–73.

Leo, R. F. and Barghoorn, E. S. 1976. Silicification of wood. *Botanical Museum Leaflets, Harvard University* 25:1–47.

Lessios, H. A., Robertson, D. R. and Cubit, J. D. 1984. Spread of *Diadema* mass mortality through the Caribbean. *Science* 226:335–337.

Levinton, J. S. 1996. Trophic group and the end-Cretaceous extinction: did deposit feeders have it made in the shade? *Paleobiology* 22:104–112.

Lewis, R. D. 1986. Relative rates of skeletal disarticulation in modern ophiuroids and Paleozoic crinoids. *Geological Society of America Abstracts with Programs* 18:672.

Lewis, R. D. 1987. Post-mortem decomposition of ophiuroids from the Mississippi Sound. *Geological Society of America Abstracts with Programs* 19:94–95.

Li, X. and Droser, M. L. 1997. Nature and distribution of Cambrian shell concentrations: evidence from the Basin and Range province of the western United States (California, Nevada, and Utah). *Palaios* 12:111–126.

Licari, G. R. 1978. Biogeology of the late Pre-Phanerozoic Beck Spring Dolomite of eastern California. *Journal of Paleontology* 52:767–792.

Liddell, W. D. 1975. Recent crinoid biostratinomy. *Geological Society of America Abstracts with Programs* 7:1169.

Liddell, W. D. and Martin, R. E. 1989. Taphofacies in modern carbonate environments: implications for the formation of foraminiferal assemblages. *International Geological Congress, Abstracts* 2:299.

Liddell, W. D., Wright, S. H. and Brett, C. E. 1997. Sequence stratigraphy and paleoecology of the Middle Cambrian Spence Shale in northern Utah and southern Idaho. *Brigham Young University Geology Studies* 42(I):59–78.

Lieberman, B. S., Brett, C. E. and Eldredge, N. 1995. A study of stasis and change in two species lineages from the Middle Devonian of New York state. *Paleobiology* 21:15–27.

Lin, S. and Morse, J. W. 1991. Sulfate reduction and iron sulfide mineral formation in Gulf of Mexico anoxic sediments. *American Journal of Science* 291:55–89.

Lindberg, D. R. and Kellogg, M. G. 1982. Bathymetric anomalies in the Neogene fossil record: the role of diving marine birds. *Paleobiology* 8:402–407.

Linke, P. and Lutze, G. F. 1993. Microhabitat preferences of benthic foraminifera: a static or a dynamic adaptation to optimize food acquisition? *Marine Micropaleontology* 20:215–234.

Lipps, J. H. 1970. Plankton evolution. *Evolution* 24:1–22.

Lipps, J. H. 1973. Test structure in Foraminifera. *Annual Review of Microbiology* 27:471–488.

Lipps, J. H. 1985. Earliest Foraminifera and Radiolaria from North America: evolutionary and geological implications. *Geological Society of America Abstracts with Programs* 17:644–645.

Lipps, J. H. 1987. Diversification of the Late Proterozoic–Cambrian marine biota. *International Geological Congress Abstracts* 2:306.

Lipps, J. H. 1988. Predation on foraminifera by coral reef fish: taphonomic and evolutionary implications. *Palaios* 3:315–326.

Lipps, J. H. and McCartney, K. 1993. Chrysophytes. In: J. H. Lipps (ed.) *Fossil Prokaryotes and Protists*, pp. 141–154. Boston: Blackwell Scientific Publications.

Lipps, J. H. and A. Yu. Rozanov. 1996. The late Precambrian–Cambrian agglutinated fossil *Platysolenites*. *Paleontological Journal* 30:679–687.

Lirman, D. and Fong, P. 1997. Patterns of damage to the branching coral *Acropora palmata* following Hurricane Andrew: damage and survivorship of hurricane-generated asexual recruits. *Journal of Coastal Research* 13:67–72.

Liu, Y. and Gastaldo, R. A. 1992. Characteristics and provenance of log-transported gravels in a Carboniferous channel deposit. *Journal of Sedimentary Petrology* 62:1072–1083.

Lockley, M. G. and Conrad, K. 1989. The paleoenvironmental context, preservation and paleoecological significance of dinosaur tracksites in the western USA. In: D. D. Gillette and M. G. Lockley (eds.) *Dinosaur Tracks and Traces*, pp. 121–134. Cambridge: Cambridge University Press.

Loeblich, A. R. and Tappan, H. 1989. Implications of wall composition and structure in agglutinated foraminifers. *Journal of Paleontology* 63:769–777.

Lofgren, D. L., Hotton, C. L. and Runkel, A. C. 1990. Reworking of Cretaceous dinosaurs into Paleocene channel deposits, upper Hell Creek Formation, Montana. *Geology* 18:874–877.

Logan, G. A., Collins, M. J. and Eglinton, G. 1991. Preservation of organic biomolecules. In: P. A. Allison and D. E. G. Briggs (eds.) *Taphonomy: Releasing the Data Locked in the Fossil Record*, pp. 1–24. New York: Plenum Press.

Logan, G. A., Hayes, J. M., Hieshima, G. B. and Summons, R. E. 1995. Terminal Proterozoic reorganization of biogeochemical cycles. *Nature* 376:53–56.

Lord, A. R. and Hamilton, G. B. 1982. Paleozoic calcareous nannofossils. In: A. R. Lord (ed.) *A Stratigraphical Index of Calcareous Nanofossils*, p. 16. New York: Halsted Press.

Lord, C. J. and Church, T. M. 1983. The geochemistry of salt marshes: sedimentary ion diffusion, sulfate reduction, and pyritization. *Geochimica et Cosmochimica Acta* 47:1381–1391.

Loubere, P. 1982. The western Mediterranean during the last glacial: attacking a no-analog problem. *Marine Micropaleontology* 7:311–325.

Loubere, P. 1989. Bioturbation and sedimentation rate control of benthic microfossil taxon abundances in surface sediments: a theoretical approach to the analysis of species microhabitats. *Marine Micropaleontology* 14:317–325.

Loubere, P. 1991. Deep-sea benthic foraminiferal assemblage response to a surface ocean productivity gradient: a test. *Paleoceanography* 6:193–204.

Loubere, P. 1997. Benthic foraminiferal assemblage formation, organic carbon flux, and oxygen concentrations on the outer continental shelf and slope. *Journal of Foraminiferal Research* 27:93–100.

Loubere, P. and Gary, A. 1990. Taphonomic process and species microhabitats in the living to fossil assemblage transition of deeper water benthic foraminifera. *Palaios* 5:375–381.

Loubere, P., Gary, A. and Lagoe, M. 1993a. Generation of the benthic foraminiferal assemblage: theory and preliminary data. *Marine Micropaleontology* 20:165–181.

Loubere, P., Gary, A. and Lagoe, M. 1993b. Sea-bed biogeochemistry and benthic foraminiferal zonation on the slope of the northwest Gulf of Mexico. *Palaios* 8:439–449.

Loubere, P., Meyers, P. and Gary, A. 1995. Benthic foraminiferal microhabitat selection, carbon isotope values, and association with larger animals: a test with *Uvigerina peregrina*. *Journal of Foraminiferal Research* 25:83–95.

Lowenstam, H. A. and Margulis, L. 1980. Evolutionary prerequisites for early Phanerozoic calcareous skeletons. *Biosystems* 12:27–41.

Lowenstam, H. A. and Weiner, S. 1983. Mineralization by organisms and the evolution of biomineralization. In: M. Westbrock and E. W. de Jong (eds.) *Biomineralization and Biological Metal Accumulation*, pp. 191–203. Dordrecht: Reidel.

Lucas, J. and Prévôt, L. E. 1991. Phosphates and fossil preservation. In: P. A. Allison and D. E. G. Briggs (eds.) *Taphonomy: Releasing the Data Locked in the Fossil Record*, pp. 389–409. New York: Plenum Press.

Ludvigsen, R. 1989. The Burgess Shale: not in the shadow of the Cathedral Escarpment. *Geoscience Canada* 16:51–59.

Ludwig, W. and Probst, J.-L. 1998. River sediment discharge to the oceans: present-day controls and global budgets. *American Journal of Science* 298:265–295.

Lundquist, J. J., Culver, S. J. and Stanley, D. J. 1997. Foraminiferal and lithologic indicators of depositional processes in Wilmington and South Heyes submarine canyons, U.S. Atlantic continental slope. *Journal of Foraminiferal Research* 27:209–231.

Luther, G. W., Ferdelman, T. G., Kostka, J. E., Tsamakis, E. J. and Church, T. M. 1991. Temporal and spatial variability of reduced sulfur species (FeS_2, $S_2O_3^{2-}$) and porewater parameters in salt marsh sediments. *Biogeochemistry* 14:57–88.

Lyle, M. 1997. Could early Cenozoic thermohaline circulation have warmed the poles? *Paleoceanography* 12:161–167.

Lyman, R. L. 1994a. *Vertebrate Taphonomy*. Cambridge: Cambridge University Press.

Lyman, R. L. 1994b. Relative abundances of skeletal specimens and taphonomic analysis of vertebrate remains. *Palaios* 9:288–298.

Mackenzie, F. T. and Agegian, C. 1989. Biomineralization and tentative links to plate tectonics. In: R. E. Crick (ed.) *Origin, Evolution, and Modern Aspects of Biomineralization in Plants and Animals*, pp. 11–27. New York: Plenum Press.

Mackenzie, F. T. and Morse, J. W. 1990. *Geochemistry of Sedimentary Carbonates*. Amsterdam: Elsevier.

Mackenzie, F. T. and Morse, J. W. 1992. Sedimentary carbonates through Phanerozoic time. *Geochemica et Cosmochimica Acta* 56:3281–3295.

Mackenzie, F. T. and Pigott, J. D. 1981. Tectonic controls of Phanerozoic sedimentary rock cycling. *Journal of the Geological Society of London* 138:183–196.

Mackenzie, F. T., Bischoff, W. D., Bishop, F. C., Loyens, M., Schoonmaker, J. and Wollast, R. 1983. Magnesian calcites: low temperature co-occurrence, solubility and solid solution behavior. In: R. J. Reeder (ed.) *Reviews in Mineralogy* 11:97–144. Washington, D.C.: Mineralogical Society of America.

Mackenzie, F. T., Kulm, L. D., Cooley, R. L. and Barnhart, J. T. 1965. *Homotrema rubrum* (Lamarck), a sediment transport indicator. *Journal of Sedimentary Petrology* 35:265–272.

MacLeod, K. G. and Huber, B. T. 1996. Strontium isotopic evidence for extensive reworking in sediments spanning the Cretaceous–Tertiary boundary at ODP site 738. *Geology* 24:463–466.

MacLeod, N. 1991. Punctuated anagenesis and the importance of stratigraphy to paleobiology. *Paleobiology* 17:167–188.

MacLeod, N. 1995. Graphic correlation of new Cretaceous/Tertiary (K/T) boundary sections/cores from Denmark, Alabama, Mexico, and the southern Indian Ocean: Implications for a global sediment accumulation model. In: K. O. Mann, H. R. Lane and J. A. Stein (eds.) *Graphic Correlation and the Composite Standard Approach*, pp. 215–233. Tulsa: SEPM (Society for Sedimentary Geology) Special Publication No. 53.

MacLeod, N. 1996. Nature of the Cretaceous–Tertiary planktonic foraminiferal record: stratigraphic confidence intervals, Signor–Lipps effect, and patterns of survivorship. In: N. MacLeod and G. Keller (eds.) *Cretaceous Tertiary Mass Extinctions: Biotic and Environmental Changes*, pp. 85–138. New York: Norton.

MacLeod, N. and Keller, G. 1991. How complete are Cretaceous/Tertiary boundary sections? A chronostratigraphic estimate based on graphic correlation. *Geological Society of America Bulletin* 103:1439–1457.

MacLeod, N. and Sadler, P. 1995. Estimating the line of correlation. In: K. O. Mann, H. R. Lane, and J. A. Stein (eds.) *Graphic Correlation and the Composite Standard Approach*, pp. 51–64. Tulsa: SEPM (Society for Sedimentary Geology) Special Publication No. 53.

Magaritz, M., Holser, H. T. and Kirschvink, J. L. 1986. Carbon-isotope events across the Precambrian–Cambrian boundary on the Siberian platform. *Nature* 320:258–259.

Magaritz, M., Kristhnamurthy, R. V. and Holser, W. T. 1992. Parallel trends in organic and inorganic carbon isotopes across the Permian/Triassic boundary. *American Journal of Science* 292:727–739.

Magwood, J. P. A. and Ekdale, A. A. 1994. Computer-aided analysis of visually complex ichnofacies in deep-sea sediments. *Palaios* 9:102–115.

Maiklem, W. R. 1968. Some hydraulic properties of bioclastic carbonate grains. *Sedimentology* 10:101–109.

Malinky, J. M. and Heckel, P. H. 1998. Paleoecology and taphonomy of faunal assemblages in gray "core" (offshore) shales in midcontinent Pennsylvanian cyclothems. *Palaios* 13:311–334.

Maliva, R. G. and Siever, R. 1988. Mechanism and controls of silicification of fossils in limestones. *Journal of Geology* 96:387–398.

Maliva, R. G., Knoll, A. H. and Siever, R. 1989. Secular change in chert distribution: a reflection of evolving biological participation in the silica cycle. *Palaios* 4:519–532.

Małkowski, K., Gruszczyński, M., Hoffman, A. and Halas, S. 1989. Oceanic stable isotope composition and a scenario for the Permo-Triassic crisis. *Historical Biology* 2:289–309.

Malmgren, B. A. 1987. Differential dissolution of Upper Cretaceous planktonic foraminifera from a temperate region of the South Atlantic Ocean. *Marine Micropaleontology* 11:251–271.

Mandelbrot, B. 1983. *The Fractal Geometry of Nature*. New York: W. H. Freeman.

Manjunatha, B. R. and Shankar, R. 1996. Signature of non-steady-state diagenesis in continental shelf sediments. *Estuarine, Coastal and Shelf Science* 42:361–369.

Mann, K. O., Lane, H. R. and Stein, J. A. (eds.) 1995. *Graphic Correlation and the Composite Standard Approach*. Tulsa: SEPM (Society for Sedimentary Geology) Special Publication No. 53.

Mapes, R. H. and Mapes, G. 1997. Biotic destruction of terrestrial plant debris in the Late Paleozoic marine environment. *Lethaia* 29:157–169.

Maples, C. G. and Archer, A. W. 1989. Paleoecological and sedimentological significance of bioturbated crinoid calyces. *Palaios* 4:379–383.

Margalef, R. 1968. *Perspectives in Ecological Theory*. Chicago: University of Chicago Press.

Margalef, R. 1971. The pelagic ecosystem of the Caribbean Sea. *Symposium on Investigations and Resources of the Caribbean Sea and Adjacent Regions*, pp. 483–498. Paris: UNESCO.

Margulis, L. E. 1991. Symbiosis and symbionticism. In: L. Margulis and R. Fester (eds.) *Symbiosis as a Source of Evolutionary Innovation*, pp. 1–14. Cambridge, MA: MIT Press.

Margulis, L. E. 1992. Symbiosis theory: cells as microbial communities. In: L. Margulis and L. Olendzenski (eds.) *Environmental Evolution: Effects of the Origin and Evolution of Life on Planet Earth*, pp. 149–173. Cambridge, MA: MIT Press.

References

Margulis, L. E., Barghoorn, E. S., Ashendorf, D., Banerjee, S., Chase, D., Francis, S., Giovannoni, S. and Stolz, S. 1980. The microbial community in the layered sediments at Laguna Figueroa, Baja, California, Mexico. Does it have Precambrian analogues? *Precambrian Research* 11:92–123.

Marshall, C. R. 1990. Confidence intervals on stratigraphic ranges. *Paleobiology* 16:1–10.

Marshall, C. R. 1995. Distinguishing between sudden and gradual extinctions in the fossil record: predicting the position of the Cretaceous–Tertiary iridium anomaly using the ammonite fossil record on Seymour Island, Antarctica. *Geology* 23:731–734.

Marshall, C. R. 1997. Confidence intervals on stratigraphic ranges with nonrandom distributions of fossil horizons. *Paleobiology* 23:165–173.

Marshall, C. R. and Ward, P. D. 1996. Sudden and gradual molluscan extinctions in the latest Cretaceous of western European Tethys. *Science* 274:1360–1363.

Martill, D. M. 1985. The preservation of marine vertebrates in the Lower Oxford Clay (Jurassic) of central England. In: H. B. Whittington and S. Conway Morris (eds.) Extraordinary fossil biotas: their ecological and evolutionary significance. *Philosophical Transactions of the Royal Society of London* (B) 311:155–165.

Martill, D. M. 1987a. A taphonomic and diagenetic case study of a partially articulated ichthyosaur. *Palaeontology* 30:543–555.

Martill, D. M. 1987b. Prokaryote mats replacing soft-tissues in Mesozoic marine reptiles. *Modern Geology* 11:265–269.

Martill, D. M. 1988. Preservation of fish in the Cretaceous Santana Formation. *Palaeontology* 31:1–18.

Martill, D. M. 1990. Macromolecular resolution of fossilized muscle tissue from an elopomorph fish. *Nature* 346:171–172.

Martill, D. M. 1991. Bones as stones: The contribution of vertebrate remains to the lithologic record. In: S. K. Donovan (ed.) *The Processes of Fossilization*, pp. 270–292. London: Belhaven Press.

Martin, P. S. and Gray, J. 1962. Pollen analysis and the Cenozoic. *Science* 137:103–111.

Martin, R. E. 1986. Habitat and distribution of the foraminifer *Archaias angulatus* (Fichtel and Moll) (Miliolina, Soritidae). *Journal of Foraminiferal Research* 16:201–206.

Martin, R. E. 1988. Benthic foraminiferal zonation in deep-water carbonate platform margin environments, northern Little Bahama Bank. *Journal of Paleontology* 62:1–8.

Martin, R. E. 1991. Beyond biostratigraphy: micropaleontology in transition? *Palaios* 6:437–438.

Martin, R. E. 1993. Time and taphonomy: actualistic evidence for time-averaging of benthic foraminiferal assemblages. In: S. M. Kidwell and A. K. Behrensmeyer (eds.) *Taphonomic Approaches to Time Resolution in Fossil Assemblages*, pp. 34–56. Pittsburgh: Paleontological Society Short Course No. 6.

Martin, R. E. 1995a. Cyclic and secular variation in microfossil biomineralization: clues to the biogeochemical evolution of the oceans. *Global and Planetary Change* 11:1–23.

Martin, R. E. 1995b. The once and future profession of micropaleontology. *Journal of Foraminiferal Research* 25:372–373.

Martin, R. E. 1996a. Secular increase in nutrient levels through the Phanerozoic: implications for productivity, biomass, and diversity of the marine biosphere. *Palaios* 11:209–219.

Martin, R. E. 1996b. Biomineralization and endosymbiosis in foraminifera in response to ocean chemistry. *Paleontological Journal (Moscow)* 30:662–668.

Martin, R. E. 1997. Secular rise in marine C:P burial ratios during the last ~0.9 by: biomass and diversity of the marine biosphere. *Geological Society of America Abstracts with Programs* 29:A66–67.

Martin, R. E. 1998a. *One Long Experiment: Scale and Process in Earth History*. New York: Columbia University Press.

Martin, R. E. 1998b. Catastrophic fluctuations in nutrient levels as an agent of mass extinction: upward scaling of ecological processes? In: M. L. McKinney and J. A. Drake (eds.) *Biodiversity Dynamics: Origination and Extinction of Populations, Species, Communities, and Higher Taxa*. New York: Columbia University Press.

Martin, R. E. 1999. Introduction. In: R. E. Martin (ed.) *Environmental Micropaleontology*. New York: Plenum Press (in press).

Martin, R. E. and Fletcher, R. R. 1995. Graphic correlation of Plio-Pleistocene sequence boundaries, Gulf of Mexico: oxygen isotopes, ice volume, and sea level. In: K. O. Mann, H. R. Lane and J. A. Stein (eds.) *Graphic Correlation and the Composite Standard Approach*, pp. 235–248. Tulsa: SEPM (Society for Sedimentary Geology) Special Publication No. 53.

Martin, R. E. and Liddell, W. D. 1988. Foraminiferal biofacies on a north coast fringing reef (1–75 m), Discovery Bay, Jamaica. *Palaios* 3:298–314.

Martin, R. E. and Liddell, W. D. 1989. Relation of counting methods to taphonomic gradients and biofacies zonation of foraminiferal sediment assemblages. *Marine Micropaleontology* 15:67–89.

Martin, R. E. and Liddell, W. D. 1991. Taphonomy of foraminifera in modern carbonate environments: implications for the formation of foraminiferal assemblages. In: S. K. Donovan (ed.) *The Processes of Fossilization*, pp. 170–193. London: Belhaven Press.

Martin, R. E. and Wright, R. C. 1988. Information loss in the transition from life to death assemblages of foraminifera in back reef environments, Key Largo, Florida. *Journal of Paleontology* 62:399–410.

Martin, R. E., Johnson, G. W., Neff, E. D. and Krantz, D. E. 1990. Quaternary planktonic foraminiferal assemblage zones of the northeast Gulf of Mexico, Colombia Basin (Caribbean Sea), and tropical Atlantic Ocean: graphic correlation of microfossil and oxygen isotope datums. *Paleoceanography* 5:531–555.

Martin, R. E., Neff, E. D., Johnson, G. W. and Krantz, D. E. 1993. Biostratigraphic expression of Pleistocene sequence boundaries, Gulf of Mexico. *Palaios* 8:155–171.

Martin, R. E., Harris, M. S. and Liddell, W. D. 1995. Taphonomy and time-averaging of foraminiferal assemblages in Holocene tidal flat sediments, Bahia la Choya, Sonora, Mexico (northern Gulf of California). *Marine Micropaleontology* 26:187–206.

Martin, R. E., Wehmiller, J. F., Harris, M. S. and Liddell, W. D. 1996. Comparative taphonomy of foraminifera and bivalves in Holocene shallow-water carbonate and siliciclastic regimes: taphonomic grades and temporal resolution. *Paleobiology* 22:80–90.

Mathewes, R. W. and Clague, J. J. 1994. Detection of large prehistoric earthquakes in the Pacific Northwest by microfossil analysis. *Science* 264:688–691.

References

Matisoff, G. 1982. Mathematical models of bioturbation. In: P. L. McCall and M. J. S. Tevesz (eds.) *Animal–Sediment Relations*, pp. 289–330. New York: Plenum Press.

Maurer, B. A. and Nott, M. P. 1996. Geographic range fragmentation and the evolution of biological diversity. In: M. L. McKinney and J. A. Drake (eds.) *Biodiversity Dynamics: Origination and Extinction of Populations, Species, Communities, and Higher Taxa*. New York: Columbia University Press.

May, J. A. and Perkins, R. D. 1979. Endolithic infestation of carbonate substrates below the sediment–water interface. *Journal of Sedimentary Petrology* 49:357–378.

May, J. A., Macintyre, I. G. and Perkins, R. D. 1982. Distribution of microborers within planted substrates along a barrier reef transect, Carrie Bow Bay, Belize. In: K. Rützler and I. G. Macintyre (eds.) *The Atlantic Barrier Reef Ecosystem at Carrie Bow Cay, Belize*, pp. 93–107. Washington, D.C.: Smithsonian Institution Press.

May, R. M. 1977. Thresholds and breakpoints in ecosystems with a multiplicity of stable states. *Nature* 269:471–477.

May, R. M. and Oster, G. F. 1976. Bifurcations and dynamic complexity in simple ecological models. *American Naturalist* 110:573–599.

McCartan, L., Owens, J. P., Blackwelder, B. W., Szabo, B. J., Belknap, D. F., Kriausakul, N., Mitterer, R. M. and Wehmiller, J. F. 1982. Comparison of amino acid racemization geochronometry with lithostratigraphy, biostratigraphy, uranium-series coral dating, and magnetostratigraphy in the Atlantic coastal plain of the southeastern United States. *Quaternary Research* 18:337–359.

McCave, I. N. 1988. Biological pumping upwards of the coarse fraction of deep sea sediments. *Journal of Sedimentary Petrology* 58:148–158.

McCorkle, D. C., Keigwin, L. D., Corliss, B. H. and Emerson, S. R. 1990. The influence of microhabitats on the carbon isotopic composition of deep-sea benthic foraminifera. *Paleoceanography* 5:161–185.

McCorkle, D. C., Martin, P. A., Lea, D. W. and Klinkhammer, G. P. 1995. Evidence of a dissolution effect on benthic foraminiferal shell chemistry: $\delta^{13}C$, Cd/Ca, Ba/Ca, and Sr/Ca results from the Ontong Java Plateau. *Paleoceanography* 10:699–714.

McGoff, H. J. 1991. The hydrodynamics of conodont elements. *Lethaia* 24:235–247.

McGowan, J. A. 1971. Oceanic biogeography of the Pacific. In: Funnell, B. M. and Riedel, W. R. (eds.) *The Micropalaeontology of Oceans*, pp. 3–74. Cambridge: Cambridge University Press.

McGowan, J. A. and Walker, P. W. 1993. Pelagic diversity patterns. In: Ricklefs, R. E. and Schluter, D. (eds.) *Species Diversity in Ecological Communities: Historical and Geographical Perspectives*, pp. 203–214. Chicago: University of Chicago Press.

McKinney, M. L. 1989. Periodic mass extinctions: product of biosphere growth dynamics? *Historical Biology* 2:273–287.

McKinney, M. L. 1991. Completeness of the fossil record: an overview. In: S. K. Donovan (ed.) *The Processes of Fossilization*, pp. 66–83. London: Belhaven Press.

McKinney, M. L. 1996. The biology of fossil abundance. *Revista Española de Paleontologia* 11:125–133.

McKinney, M. L. and Allmon, W. D. 1995. Metapopulations and disturbance: from patch dynamics to biodiversity dynamics.

In: D. H. Erwin and R. L. Anstey (eds.) *New Approaches to Speciation in the Fossil Record*. New York: Columbia University Press.

McKinney, M. L. and Frederick, D. 1992. Extinction and population dynamics: new methods and evidence from Paleogene foraminifera. *Geology* 20:343–346.

McKinney, M. L., Lockwood, J. L. and Frederick, D. R. 1996. Rare species and scale-dependence in ecosystem stasis. In: L. C. Ivany and K. M. Schopf (eds.) New perspectives on faunal stability in the fossil record. *Palaeogeography, Palaeoclimatology, Palaeoecology* 127:191–207.

McMenamin, M. A. S. 1998. *The Garden of Ediacara: Discovering the First Complex Life*. New York: Columbia University Press.

McMenamin, M. A. S. and McMenamin, D. L. S. 1994. *Hypersea: Life on Land*. New York: Columbia University Press.

McNeill, D. F., Budd, A. F. and Borne, P. F. 1997. Earlier (late Pliocene) first appearance of the Caribbean reef-building coral *Acropora palmata*: stratigraphic and evolutionary implications. *Geology* 25:891–894.

McNichol, A. P., Lee, C. and Druffel, E. R. M. 1988. Carbon cycling in coastal sediments, 1. A quantitative estimate of the remineralization of organic carbon in the sediments of Buzzards Bay, MA. *Geochimica et Cosmochimica Acta* 52:1531–1543.

Meldahl, K. H. 1987. Sedimentologic and taphonomic implications of biogenic stratification. *Palaios* 2:350–358.

Meldahl, K. 1990. Sampling, species abundance, and the stratigraphic signature of mass extinction: a test using Holocene tidal flat molluscs. *Geology* 18:890–893.

Meldahl, K. 1993. Geographic gradients in the formation of shell concentrations: Plio-Pleistocene marine deposits, Gulf of California. *Palaeogeography, Palaeoclimatology, Palaeoecology* 101:1–25.

Meldahl, K. and Cutler, A. H. 1992. Neotectonics and taphonomy: Pleistocene molluscan shell accumulations in the northern Gulf of California. *Palaios* 7:187–197.

Meldahl, K. and Flessa, K. W. 1990. Taphonomic pathways and comparative biofacies and taphofacies in a Recent intertidal/shallow shelf environment. *Lethaia* 23:43–60.

Meldahl, K., Flessa, K. W. and Cutler, A. H. 1997. Time-averaging and postmortem skeletal survival in benthic fossil assemblages: quantitative comparisons among Holocene environments. *Paleobiology* 23:207–229.

Meldahl, K. H., Scott, D. and Carney, K. 1995. Autochthonous leaf assemblages as records of deciduous forest communities: an actualistic study. *Lethaia* 28:383–394.

Mellett, J. S. 1974. Scatalogical origin of microvertebrate fossil accumulations. *Science* 185:349–350.

Menard, H. W. and Boucot, A. J. 1951. Experiments on the movement of shells by water. *American Journal of Science* 249:131–151.

Metzger, R. A. 1989. Upper Devonian (Frasnian-Famennian) conodont biostratigraphy in the subsurface of north-central Iowa and southeastern Nebraska. *Journal of Paleontology* 63:503–524.

Metzler, C. V., Wenkam, C. R. and Berger, W. H. 1982. Dissolution of foraminifera in the eastern equatorial Pacific: an in situ experiment. *Journal of Foraminiferal Research* 12:362–368.

Meyer, C. 1991. Burial experiments with marine turtle carcasses and their paleoecological significance. *Palaios* 6:89–96.

References

Meyer, D. L. 1971. Post-mortem disarticulation of Recent crinoids and ophiuroids under natural conditions. *Geological Society of America Abstracts with Programs* 3:645.

Meyer, D. L. and Meyer, K. B. 1986. Biostratinomy of Recent crinoids (Echinodermata) at Lizard Island, Great Barrier Reef, Australia. *Palaios* 1:294–302.

Meyer, D. L., Ausich, W. I. and Terry, R. E. 1989. Comparative taphonomy of echinoderms in carbonate facies: Fort Payne Formation (Lower Mississippian) of Kentucky and Tennessee. *Palaios* 4:533–552.

Miall, A. D. 1990. *Principles of Sedimentary Basin Analysis*. New York: Springer-Verlag.

Middelburg, J. J., Soetaert, K. and Herman, P. M. J. 1997. Empirical relationships for use in global diagenetic models. *Deep-Sea Research* 44:327–344.

Middleton, G. V. 1967. The orientation of concave–convex particles deposited from experimental turbidity currents. *Journal of Sedimentary Petrology* 37:229–232.

Mii, H.-S., Grossman, E. L. and Yancey, T. E. 1997. Stable carbon and oxygen isotope shifts in Permian seas of West Spitsbergen: global change or diagenetic artifact? *Geology* 25:227–230.

Miller, A. I. 1988. Spatial resolution in subfossil molluscan remains: implications for paleobiological analyses. *Paleobiology* 14:91–103.

Miller, A. I. 1998. Biotic transitions in global marine diversity. *Science* 281:1157–1160.

Miller, F. X. 1977. The graphic correlation method in biostratigraphy. In: E. G. Kauffman and J. E. Hazel (eds.) *Concepts and Methods of Biostratigraphy*, pp. 165–186. Stroudsburg: Dowden, Hutchinson, and Ross.

Miller, M. and Smail, S. E. 1997. A semiquantitative field method for evaluating bioturbation on bedding planes. *Palaios* 12:391–396.

Mitchell, L. and Curry, G. B. 1997. Diagenesis and survival of intracrystalline amino acids in fossil and Recent mollusc shells. *Palaeontology* 40:855–874.

Moberly, R. 1968. Loss of Hawaiian littoral sand. *Journal of Sedimentary Petrology* 38:17–34.

Moldowan, J. M. and Talyzina, N. M. 1998. Biogeochemical evidence for dinoflagellate ancestors in the early Cambrian. *Science* 281:1168–1170.

Moldowan, J. M., Dahl, J., Jacobson. S. R., Huizinga, B. J., Fago, F. J., Shetty, R., Watt, D. S. and Peters, K. E. 1996. Chemostratigraphic reconstruction of biofacies: molecular evidence linking cyst-forming dinoflagellates with pre-Triassic ancestors. *Geology* 24:159–162.

Moore, D. G. and Scruton, P. C. 1957. Minor internal structures of some recent unconsolidated sediments. *Bulletin of the American Association of Petroleum Geologists* 41:2723–2751.

Moore, G. T., Hayashida, D. N. and Ross, C. A. 1993. Late Early Silurian (Wenlockian) general circulation model-generated upwelling, graptolitic black shales, and organic-rich source rocks: an accident of plate tectonics? *Geology* 21:17–20.

Moore, J. A. 1993. *Science as a Way of Knowing: the Foundations of Modern Biology*. Cambridge, MA: Harvard University Press.

Mora, C. I., Driese, S. G. and Colarusso, L. A. 1996. Middle to late Paleozoic atmospheric CO_2 levels from soil carbonate and organic matter. *Science* 271:1105–1107.

Morris, P. J. and Ivany, L. C. 1994. Response to environmental change in light of ecological locking. *Eos* (Suppl) 75:80–81.

Morse, J. W., Wang, Q. and Tsio, M. Y. 1997. Influences of temperature and Mg:Ca ratio on $CaCO_3$ precipitates from seawater. *Geology* 25:85–87.

Muhs, D. R. 1992. The last interglacial/glacial transition in North America: evidence from Uranium-series dating of coastal deposits. In: P. Clark and P. Lea (eds.) *The Last Interglaciation/Glaciation Transition in North America*, pp. 31–51. Boulder: Geological Society of America Special Paper 270.

Muller, J. 1959. Palynology of Recent Orinoco delta and shelf sediments. *Micropaleontology* 5:1–32.

Müller, K. J. 1985. Exceptional preservation in calcareous nodules. In: H. B. Whittington and S. Conway Morris (eds.) Extraordinary fossil biotas: their ecological and evolutionary significance. *Philosophical Transactions of the Royal Society of London* (B) 311:67–73.

Murata, K. J. 1940. Volcanic ash as a source of silica for the silicification of wood. *American Journal of Science* 238:586–596.

Murray, J. W. 1965. Significance of benthic foraminiferids in plankton samples. *Journal of Paleontology* 39:156–157.

Murray, J. W. 1989. Syndepositional dissolution of calcareous foraminifera in modern shallow-water sediments. *Marine Micropaleontology* 15:117–121.

Murray, J. W. 1991. *Ecology and Palaeoecology of Benthic Foraminifera*. New York. Wiley.

Murray, J. W. and Wright, C. A. 1970. Surface textures of calcareous foraminiferids. *Palaeontology* 13:184–187.

Murray, J. W., Sturrock, S. and Weston, J. 1982. Suspended load transport of foraminiferal tests in a tide- and wave-swept sea. *Journal of Foraminiferal Research* 12:51–65.

Murray-Wallace, C. V. and Belperio, A. P. 1995. Identification of remanié fossils using amino acid racemisation. *Alcheringa* 18:219–227.

Murray-Wallace, C. V., Ferland, M. A., Roy, P. S. and Sollar, A. 1996. Unravelling patterns of reworking in lowstand shelf deposits using amino acid racemisation and radiocarbon dating. *Quaternary Science Reviews (Quaternary Geochronology)* 15:685–697.

Myers, A. C. 1977. Sediment processing in a marine subtidal sandy bottom community. *Journal of Marine Research* 35:609–647.

Mylroie, J. E. 1993. Return of the coral reef hypothesis: basin to shelf partitioning of $CaCO_3$ and its effect on atmospheric CO_2: comment. *Geology* 21:475.

Nagle, J. S. 1967. Wave and current orientation of shells. *Journal of Sedimentary Petrology* 37:1124–1138.

Narbonne, G. M. 1998. The Ediacara biota: a terminal Neoproterozoic experiment in the evolution of life. *GSA Today* 2:1–6.

Nelson, A. R. and Jennings, A. E. 1993. Intertidal foraminifera and earthquake hazard assessment in the Cascadia subduction zone. *Geological Society of America Abstracts with Programs* 25:A138.

Neumann, A. C. 1966. Observations on coastal erosion in Bermuda and measurements of the boring rate of the sponge *Cliona lampa*. *Limnology and Oceanography* 11:92–108.

Neumann, A. C. and Land, L. S. 1975. Lime mud deposition and calcareous algae in the

Bight of Abaco, Bahamas: a budget. *Journal of Sedimentary Petrology* 45:763–786.

Neumann, A. C., Gebelein, C. D. and Scoffin, T. P. 1970. The composition, structure and erodability of subtidal mats, Abaco, Bahamas. *Journal of Sedimentary Petrology* 40:274–297.

Newton, C. and Laporte. L. 1989. *Ancient Environments*. Englewood Cliffs, NJ: Prentice-Hall.

Nicol, D. 1966. Cope's Rule and Precambrian and Cambrian invertebrates. *Journal of Paleontology* 40:1397–1399.

Nicolis, G. and Prigogine, I. 1989. *Exploring Complexity: an Introduction*. San Francisco, W. H. Freeman.

Nittrouer, C. A. and Sternberg, R. W. 1981. The formation of sedimentary strata in an allochthonous shelf environment: the Washington continental shelf. *Marine Geology* 42:201–232.

Norris, R. D. 1986. Taphonomic gradients in shelf fossil assemblages: Pliocene Purisima Formation, California. *Palaios* 1:252–266.

Norris, R. D. 1989. Cnidarian taphonomy and affinities of the Ediacara biota. *Lethaia* 22:381–393.

Nozaki, Y., Cochran, J. K. and Turekian, K. K. 1977. Radiocarbon and ^{210}Pb distribution in submersible taken deep sea cores from project Famous. *Earth and Planetary Science Letters* 34:167–173.

Nydick, K. R., Bidwell, A. B., Thomas, E. and Varekamp, J. C. 1995. A sea-level rise curve from Guilford, Connecticut, USA. *Marine Geology* 124:137–159.

Nykvist, N. 1962. Experiments on leaf litter of *Alnus glutinosa*, *Fagus silvatica*, and *Quercus ruber* V. Leaching and decomposition of litter. *Oikos* 13:232–248.

Oehler, J. H. and Schopf, J. W. 1971. Artificial microfossils: experimental studies of permineralization of blue-green algae in silica. *Science* 174:1229–1231.

Officer, C. B. 1982. Mixing, sedimentation rates and age dating for sediment cores. *Marine Geology* 46:261–278.

Officer, C. B. and Lynch, D. R. 1982. Interpretation procedures for the determination of sediment parameters from time-dependent flux inputs. *Earth and Planetary Science Letters* 61:55–62.

Officer, C. B. and Lynch, D. R. 1983. Determination of mixing parameters from tracer distributions in deep-sea sediment cores. *Marine Geology* 52:59–74.

Officer, C. B. and Page, J. 1996. *The Great Dinosaur Extinction Controversy*. Reading, MA: Addison-Wesley.

Okubo, A. and Levin, S. A. 1989. A theoretical framework for data analysis of wind dispersal of seeds and pollen. *Ecology* 70:329–338.

Oliver, J. S. and Graham, R. W. 1994. A catastrophic kill of ice-trapped coots: time-averaged versus scavenger-specific disarticulation patterns. *Paleobiology* 20:229–244.

Olsen, P. E. 1986. A forty-million-year lake record of early Mesozoic orbital climatic forcing. *Science* 234:842–848.

Olson, E. C. 1952. The evolution of a Permian vertebrate chronofauna. *Evolution* 6:181–196.

Olson, E. C. 1962. Late Permian terrestrial vertebrates, U.S.A. and U.S.S.R. *American Philosophical Society, Transactions* 52:3–224.

Olson, E. C. 1980. Taphonomy: its history and role in community evolution. In: A. K. Behrensmeyer and A. P. Hill (eds.) *Fossils in the Making: Vertebrate Taphonomy and*

Paleoecology, pp. 5–19. Chicago: University of Chicago Press.

Olszewski, T. 1997. Accounting for bias due to time-averaging in fossil data: when is enough enough? *Geological Society of America Abstracts with Programs* 29:A265.

Opdyke, B. N. and Walker, J. C. G. 1992. Return of the coral reef hypothesis: basin to shelf partitioning of $CaCO_3$ and its effect on atmospheric CO_2. *Geology* 20:733–736.

Opdyke, B. N. and Wilkinson, B. H. 1988. Surface area control of shallow cratonic to deep marine carbonate accumulation. *Paleoceanography* 3:685–783.

Opdyke, B. N. and Wilkinson, B. H. 1993. Carbonate mineral saturation state and cratonic limestone accumulation. *American Journal of Science* 293:217–234.

Orr, P. J., Briggs, D. E. G. and Kearns, S. L. 1998. Cambrian Burgess Shale animals replicated in clay minerals. *Science* 281:1173–1175.

Ozarko, D. L., Patterson, R. T. and Williams, H. F. L. 1997. Marsh foraminifera from Nanaimo, British Columbia (Canada): implications of infaunal habitat and taphonomic biasing. *Journal of Foraminiferal Research* 27:51–68.

Paabo, S., Higuchi, R. G. and Wilson, A. C. 1989. Ancient DNA and the polymerase chain reaction. *Journal of Biological Chemistry* 264:9709–9712.

Pandolfi, J. M. 1996. Limited membership in Pleistocene reef coral assemblages from the Huon Peninsula, Papua New Guinea: constancy during global change. *Paleobiology* 22:152–176.

Pandolfi, J. M. and Greenstein, B. J. 1997a. Taphonomic alteration of reef corals: effects of reef environment and coral growth form. I. The Great Barrier Reef. *Palaios* 12:27–42.

Pandolfi, J. M. and Greenstein, B. J. 1997b. Preservation of community structure in death assemblages of deep water Caribbean reef corals. *Limnology and Oceanography* 42:1505–1516.

Pandolfi, J. M. and Minchin, P. R. 1995. A comparison of taxonomic composition and diversity between reef coral life and death assemblages in Madang Lagoon, Papua New Guinea. *Palaeogeography, Palaeoclimatology, Palaeoecology* 119:321–341.

Parfit, M. 1995. The floods that carved the west. *Smithsonian* 26(1):48–58.

Parker, F. L. and Berger, W. H. 1971. Faunal and solution patterns of planktonic foraminifera in surface sediments of the South Pacific. *Deep-Sea Research* 18:73–107.

Parker, R. B. and Toots, H. 1980. Trace elements in bones as paleobiological indicators. In: A. K. Behrensmeyer and A. P. Hill (eds.) *Fossils in the Making: Vertebrate Taphonomy and Paleoecology*, pp. 197–207. Chicago: University of Chicago Press.

Parrish, J. T. 1982. Upwelling and petroleum source beds, with reference to Paleozoic. *American Association of Petroleum Geologists Bulletin* 66:750–774.

Parrish, J. T. 1987. Palaeo-upwelling and the distribution of organic-rich rocks. In: J. Brooks and A. J. Fleet (eds.) *Marine Petroleum Source Rocks*, pp. 199–205. London: Geological Society of London.

Parrish, J. T. 1993. Climate of the supercontinent Pangea. *Journal of Geology* 101:215–233.

Passier, H. F., Bosch, H.-J., Nijenhuis, I. A., Lourens, L. J., Böttcher, M. E., Leenders, A., Sinninghe Damsté, J. S., de Lange, G. J. and de Leeuw, J. W. 1999. Sulphidic Mediterranean surface waters during Pliocene sapropel formation. *Nature* 397:146–149.

Pattee, H. H. 1978. The complementarity principle in biological and social structures. *Journal of Social and Biological Structures* 1:191–200.

Patterson, R. T. 1990. Intertidal benthic foraminiferal biofacies on the Fraser River delta, British Columbia: modern distributions and paleoecological importance. *Micropaleontology* 36:229–244.

Patterson, R. T. and Fishbein, E. 1989. Re-examination of the statistical methods used to determine the number of point counts needed for micropaleontological quantitative research. *Journal of Paleontology* 63:245–248.

Patterson, R. T. and Kumar, A. 1999. Use of Arcellacea (thecamoebians) to gauge levels of contamination and remediation in industrially polluted lakes. In: R. E. Martin (ed.) *Environmental Micropaleontology*. New York: Plenum Press.

Patterson, R. T., Ozarko, D. L., Guilbault, J.-P. and Clague, J. J. 1994. Distribution and preservation potential of marsh foraminiferal biofacies from the lower mainland and Vancouver Island, British Columbia. *Geological Society of America Abstracts with Programs* 26:530.

Patzkowsky, M. E. 1995. Gradient analysis of Middle Ordovician brachiopod biofacies: biostratigraphic, biogeographic, and macroevolutionary implications. *Palaios* 10:154–179.

Paul, C. R. C. 1992. How complete does the fossil record have to be? *Revista Española de Paleontologia* 7:127–133.

Peak, D. and Frame, M. 1994. *Chaos Under Control: the Art and Science of Complexity*. New York: W. H. Freeman.

Peebles, M. W. and Lewis, R. D. 1988. Differential infestation of shallow-water benthic foraminifera by microboring organisms: possible biases in preservation potential. *Palaios* 3:345–351.

Peebles, M. W. and Lewis, R. D. 1991. Surface textures of benthic foraminifera from San Salvador, Bahamas. *Journal of Foraminiferal Research* 21:285–292.

Pemberton, S. G., Risk, M. J. and Buckley, D. E. 1976. Supershrimp: deep bioturbation in the strait of Canso, Nova Scotia. *Science* 192:790–791.

Peng, T.-H. and Broecker, W. S. 1984. The impacts of bioturbation on the age difference between benthic and planktonic foraminifera in deep sea sediments. *Nuclear Instruments and Methods in Physics Research* B5:346–352.

Peng, T.-H., Broecker, W. S. and Berger, W. H. 1979. Rates of benthic mixing in deep-sea sediment as determined by radio-active tracers. *Quaternary Research* 11:141–149.

Peng, T.-H., Broecker, W. S., Kipphut, G. and Shackleton, N. 1977. Benthic mixing in deep sea cores as determined by ^{14}C dating and its implications regarding climate stratigraphy and the fate of fossil fuel CO_2. In: N. R. Andersen and A. Malahoff (eds.) *The Fate of Fossil Fuel CO_2 in the Oceans*, pp. 355–373. New York: Plenum Press.

Pennington, W., Hayworth, E. Y., Bonny, A. P. and Lishman, J. P. 1972. Lake sediments in northern Scotland. *Philosophical Transactions of the Royal Society of London* 264:191–294.

Pennisi, E. 1998. How the genome readies itself for evolution. *Science* 281:1131–1134.

Perkins, R. D. and Halsey, S. D. 1971. Geological significance of microboring fungi and algae in Carolina shelf sediments. *Journal of Sedimentary Petrology* 41:843–853.

Perry, C. T. 1996. Distribution and abundance of macroborers in an Upper

Miocene reef system, Mallorca, Spain: implications for reef development and framework destruction. *Palaios* 11:40–56.

Peterson, C. H. 1976. Relative abundance of living and dead molluscs in two California lagoons. *Lethaia* 9:137–148.

Peterson, C. H. 1977. The paleoecological significance of undetected short-term temporal variability. *Journal of Paleontology* 51:976–981.

Peterson, C. H. and Black, R. 1988. Density-dependent mortality caused by physical stress interacting with biotic history. *American Naturalist* 131:257–270.

Petsch, S. T. and Berner, R. A. 1998. Coupling the geochemical cycles of C, P, Fe, and S: the effect on atmospheric O_2 and the isotopic records of carbon and sulfur. *American Journal of Science* 298:246–262.

Phillips, F. J. 1986. A review of graphic correlation. *Computer Oriented Geological Society Computer Contributions* 2:73–91.

Piepenbrink, H. 1989. Examples of chemical changes during fossilization. *Applied Geochemistry* 4:273–280.

Pigott, J. D. and Land, L. S. 1986. Interstitial water chemistry of Jamaican reef sediment: sulfate reduction and submarine cementation. *Marine Chemistry* 19:355–378.

Pike, J. and Kemp, A. E. S. 1997. Early Holocene decadal-scale ocean variability recorded in Gulf of California laminated sediments. *Paleoceanography* 12:227–238.

Pimm, S. L. 1991. *The Balance of Nature?: Ecological Issues in the Conservation of Species and Communities*. Chicago: University of Chicago Press.

Pitrat, C. W. 1970. Phytoplankton and the late Paleozoic wave of extinction. *Palaeogeography, Palaeoclimatology, Palaeoecology* 8:49–55.

Pizzuto, J. E. and Schwendt, A. E. 1997. Mathematical modeling of autocompaction of a Holocene transgressive valley-fill deposit, Wolfe Glade, Delaware. *Geology* 25:57–60.

Plotnick R. E. 1986*a*. Taphonomy of a modern shrimp: implications for the arthropod fossil record. *Palaios* 1:286–293.

Plotnick, R. E. 1986*b*. A fractal model for the distribution of stratigraphic hiatuses. *Journal of Geology* 94:885–890.

Plotnick, R. E. and M. L. McKinney. 1993. Ecosystem organization and ecosystem dynamics. *Palaios* 8:202–212.

Plotnick, R. E., Baumiller, T. and Wetmore, K. L. 1988. Fossilization potential of the mud crab, *Panopeus* (Brachyura: Xanthidae) and temporal variability in crustacean taphonomy. *Palaeogeography, Palaeoclimatology, Palaeoecology* 63:27–43.

Podlaha, O. G., Mutterlose, J. and Veizer, J. 1998. Preservation of $\delta^{18}O$ and $\delta^{13}C$ in belemnite rostra from the Jurassic/early Creatceous successions. *American Journal of Science* 298:324–347.

Popp, B. N., Anderson, T. F. and Sandberg, P. A. 1986*a*. Textural, elemental and isotopic variations among constituents in Middle Devonian limestones. *Journal of Sedimentary Petrology* 56:715–727.

Popp, B. N., Anderson, T. T. and Sandberg, P. A. 1986*b*. Brachiopods as indicators of original isotopic compositions in some Paleozoic limestones. *Geological Society of America Bulletin* 97:1262–1269.

Popper, K. R. 1959. *The Logic of Scientific Discovery*. London: Hutchinson.

Pospichal, J. J. 1994. Calcareous nannofossils at the K–T boundary, El Kef: no evidence for stepwise, gradual, or sequential extinctions. *Geology* 22:99–102.

Potter, L. D. and Rowley, J. 1960. Pollen rain and vegetation, San Augustin plains, New Mexico. *Botanical Gazette* 122:1–25.

Powell, E. N. 1992. A model for assemblage death formation: can sediment shelliness be explained? *Journal of Marine Research* 50:229–265.

Powell, E. N. and Davies, D. J. 1990. When is an "old" shell really old? *Journal of Geology* 98:823–844.

Powell, E. N., Cummins, H., Stanton, R. J. and Staff, G. 1984. Estimation of the size of molluscan larval settlement using the death assemblage. *Estuarine and Coastal Shelf Science* 18:367–384.

Powell, E. N., Logan, A., Stanton, R. J., Davies, D. J. and Hare, P. E. 1989. Estimating time-since-death from the free amino acid content of the mollusc shell: a measure of time averaging in modern death assemblages? Description of the technique. *Palaios* 4:16–31.

Prentice, I. C. 1985. Pollen representation, source area, and basin size: toward a unified theory of pollen analysis. *Quaternary Research* 23:76–86.

Price, T. D., Blitz, J., Burton, J. and Ezzo, J. A. 1992. Diagenesis in prehistoric bone: problems and solutions. *Journal of Archaeological Science* 19:513–529.

Purdy, E. D. 1968. Carbonate diagenesis: an environmental survey. *Geologica Romana* 7:183–228.

Pye, K. 1984. SEM analysis of siderite cements in intertidal marsh sediments, Norfolk, England. *Marine Geology* 56:1–12.

Pye, K., Dickson, J. A. D., Schiavon, N., Coleman, M. L. and Cox, M. 1990. Formation of siderite-Mg-calcite-iron sulphide concretions in intertidal marsh and sandflat sediments, north Norfolk, England. *Sedimentology* 37:325–343.

Racki, G. 1998. Frasnian–Famennian biotic crisis: undervalued tectonic control? *Palaeogeography, Palaeoclimatology, Palaeoecology* 141:177–198.

Railsback, L. B. 1993. Original mineralogy of Carboniferous worm tubes: evidence for changing marine chemistry and biomineralization. *Geology* 21:703–706.

Railsback, L. B. and Anderson, T. F. 1987. Control of Triassic seawater chemistry and temperature on the evolution of post-Paleozoic aragonite-secreting faunas. *Geology* 15:1002–1005.

Railsback, L. B., Ackerly, S. C., Anderson, T. F. and Cisne, J. L. 1990. Palaeontological and isotope evidence for warm saline deep waters in Ordovician seas. *Nature* 343:156–159.

Raiswell, R. 1971. The growth of Cambrian and Liassic concretions. *Sedimentology* 17:147–171.

Raiswell, R. 1987. Non-steady state microbiological diagenesis and the origin of concretions and nodular limestones. In: J. D. Marshall (ed.) *Diagenesis of Sedimentary Sequences*, pp. 41–54. London: Geological Society of London Special Publication 36.

Raiswell, R. and Berner, R. A. 1986. Pyrite and organic matter in Phanerozoic normal marine shales. *Geochimica et Cosmochimica Acta* 50:1967–1976.

Raiswell, R. and Canfield, D. E. 1998. Sources of iron for pyrite formation in marine sediments. *American Journal of Science* 298:219–245.

Rampino, M. R. and Adler, A. C. 1998. Evidence for abrupt latest Permian mass extinction of foraminifera: results of tests for the Signor–Lipps effect. *Geology* 26:415–418.

Ramsey, K. W., Leathers, D. J., Wells, D. V. and Talley, J. H. 1998. *Summary Report: The*

Coastal Storms of January 27–29 and February 4–6, 1998, Delaware and Maryland. Newark: Delaware Geological Survey Open File Report No. 40.

Rathburn, A. E. and Miao, Q. 1995. The taphonomy of deep-sea benthic foraminifera: comparison of living and dead assemblages from box and gravity cores taken in the Sulu Sea. *Marine Micropaleontology* 25:127–149.

Rau, G. H. 1976. Dispersal of terrestrial plant litter into a subalpine lake. *Oikos* 27:153–160.

Rau, G., Arthur, M. A. and Dean, W. E. 1987. $^{15}N/^{14}N$ variations in Cretaceous Atlantic sedimentary sequences: implications for past changes in marine nitrogen biogeochemistry. *Earth and Planetary Science Letters* 82:269–279.

Raup, D. M. 1972. Taxonomic diversity during the Phanerozoic. *Science* 177:1065–1071.

Raup, D. M. 1976a. Species diversity in the Phanerozoic: a tabulation. *Paleobiology* 2:279–288.

Raup, D. M. 1976b. Species diversity in the Phanerozoic: an interpretation. *Paleobiology* 2:289–297.

Raup, D. M. 1977. Species diversity in the Phanerozoic: systematists follow the fossils. *Paleobiology* 3:328–329.

Raup, D. M. 1989. The case for extraterrestrial causes of extinction. *Philosophical Transactions of the Royal Society of London* (B) 325:421–435.

Raymo, M. E. 1991. Geochemical evidence supporting T. C. Chamberlin's theory of glaciation. *Geology* 19:344–347.

Raymo, M. E. 1994. The Himalayas, organic carbon burial, and climate in the Miocene. *Paleoceanography* 9:399–404.

Raymo, M. E., Ruddiman, W. F. and Froelich, P. N. 1988. Influence of late Cenozoic mountain building on ocean geochemical cycles. *Geology* 16:649–653.

Raymond, A. 1987. Interpreting ancient swamp communities: can we see the forest in the peat? *Review of Palaeobotany and Palynology* 52:217–231.

Reaves, C. M. 1984. The migration of iron and sulfur during the early diagenesis of marine sediments. Ph.D. thesis. New Haven: Yale University.

Reaves, C. M. 1986. Organic matter metabolizability and calcium carbonate dissolution in nearshore marine muds. *Journal of Sedimentary Petrology* 56:486–494.

Redfield, A. C. 1958. The biological control of chemical factors in the environment. *American Journal of Science* 48:206–226.

Redfield, A. C. 1972. Development of a New England salt marsh. *Ecological Monographs* 42:201–237.

Reice, S. R. 1994. Nonequilibrium determinants of biological community structure. *American Scientist* 82:424–435.

Reinhard, E., de Torres, T. and O'Neil, J. O. 1996. $^{18}O/^{16}O$ ratios of cave bear tooth enamel: a record of climate variability during the Pleistocene. *Palaeogeography, Palaeoclimatology, Palaeoecology* 126:45–59.

Reise, K. and Schubert, A. 1987. Macrobenthic turnover in the subtidal Wadden Sea: the Norderaue revisited after 60 years. *Helgolander Meeresuntersuchungen* 41:69–82.

Reiss, Z. and Schneidermann, N. 1969. Ultramicrostructure of *Hoeglundina*. *Micropaleontology* 15:135–144.

Renard, M. 1986. Pelagic carbonate chemostratigraphy (Sr, Mg, ^{18}O, ^{13}C). *Marine Micropaleontology* 10:117–164.

References

Renaut, R. W., Jones, B. and Rosen, M. R. 1996. Primary silica oncoids from Orakeikorako Hot Springs, North Island, New Zealand. *Palaios* 11:446–458.

Renne, P. R., Zichao, Z., Richards, M. A., Black, M. T. and Basu, A. R. 1995. Synchrony and causal relations between Permian–Triassic boundary crises and Siberian flood volcanism. *Science* 269:1413–1416.

Retallack, G. 1984. Completeness of the rock and fossil records: some estimates using fossil soils. *Paleobiology* 10:59–78.

Retallack, G. J. 1988. Down-to-earth approaches to vertebrate paleontology. *Palaios* 3:335–344.

Retallack, G. 1994. Were the Ediacaran fossils lichens? *Paleobiology* 20:523–544.

Retallack, G., Leahy, G. D. and Spoon, M. D. 1987. Evidence from paleosols for ecosystem change across the Cretaceous/Tertiary boundary in eastern Montana. *Geology* 15:1090–1093.

Rex, G. 1985. A laboratory flume investigation of the formation of fossil stem fills. *Sedimentology* 32:245–255.

Rex, G. M. and Chaloner, W. G. 1983. The experimental formation of plant compression fossils. *Palaeontology* 26:231–252.

Reyment, R. A. 1958. Some factors in the distribution of fossil cephalopods. *Stockholm Contributions to Geology* 1:97–184.

Reyment, R. A. 1980. Floating orientation of cephalopod shell models. *Palaeontology* 23:931–936.

Rhoads, D. C. and Stanley, D. J. 1965. Biogenic graded bedding. *Journal of Sedimentary Petrology* 35:956–963.

Rhodes, M. C. and Thayer, C. W. 1991. Mass extinctions: ecological selectivity and primary production. *Geology* 19:877–880.

Rhodes, M. C. and Thompson, R. J. 1993. Comparative physiology of suspension-feeding in living brachiopods and bivalves: evolutionary implications. *Paleobiology* 19:322–334.

Richter, D. K. and Fuchtbauer, H. 1978. Ferroan calcite replacement indicates former magnesian calcite skeletons. *Sedimentology* 25:843–860.

Richter, R. 1928. Aktuopalaontologie und Palaobiologie, eine Abgrenzung. *Senckenbergiana* 10:285–292.

Riding, R. 1993. Phanerozoic patterns of marine $CaCO_3$ precipitation. *Naturwissenschaften* 80:513–515.

Rigby Jr, J. K., Newman, K. R., Smit, J., van der Kaars, S. and Rigby, J. K. 1987. Dinosaurs from the Paleocene part of the Hell Creek Formation, McCone County, Montana. *Palaios* 2:296–302.

Riggs, S. R. 1984. Paleoceanographic model of Neogene phosphorite deposition, U.S. Atlantic continental margin. *Science* 223:123–131.

Risk, M. J., Sammarco, P. W. and Edinger, E. N. 1995. Bioerosion in *Acropora* across the continental shelf of the Great Barrier Reef. *Coral Reefs* 14:79–86.

Risk, M. J., Sayer, B. G., Tevesz, M. J. S. and Karr, C. D. 1997. Comparison of the organic matrix of fossil and recent bivalve shells. *Lethaia* 29:197–202.

Risk, M. J., Venter, R. D., Pemberton, S. G. and Buckley, D. E. 1978. Computer simulation and sedimentological implications of burrowing by *Axius serratus*. *Canadian Journal of Earth Sciences* 15:1370–1374.

Rivkin, R. B. and Anderson, M. R. 1997. Inorganic nutrient limitation of oceanic bacterioplankton. *Limnology and Oceanography* 42:730–740.

Robbins, J. A. 1978. Geochemical and geophysical applications of radioactive lead. In: J. O. Nriagu (ed.) *The Biogeochemistry of Lead in the Environment*, pp. 285–393. Amsterdam: Elsevier.

Robbins, J. A. 1982. Stratigraphic and dynamic effects of sediment reworking by Great Lakes zoobenthos. *Hydrobiologia* 92:611–622.

Robbins, J. A. 1986. A model for particle-selective transport of tracers in sediments with conveyor belt deposit feeders. *Journal of Geophysical Research* 91:8542–8558.

Robinson, J. M. 1990. Lignin, land plants and fungi: biological evolution affecting Phanerozoic oxygen balance. *Geology* 15:607–610.

Rogers, R. R. 1993. Systematic patterns of time-averaging in the terrestrial vertebrate record: a Cretaceous case study. In: S. M. Kidwell and A. K. Behrensmeyer (eds.) *Taphonomic Approaches to Time Resolution in Fossil Assemblages*, pp. 228–249. Pittsburgh: Paleontological Society Short Course No. 6.

Ronov, A. B. 1968. Probable changes in the composition of sea water during the course of geological time. *Sedimentology* 10:25–43.

Ronov, A. B. 1976. Global carbon geochemistry, volcanism, carbonate accumulation, and life. *Geochemistry International* 13:172–195.

Ronov, A. B. 1994. Phanerozoic transgressions and regressions on the continents: a quantitative approach based on areas flooded by the sea and areas of marine and continental deposition. *American Journal of Science* 294:777–801.

Rosenzweig, M. L. and Abramsky, Z. 1993. How are diversity and productivity related? In: R. E. Ricklefs and D. Schluter (eds.) *Species Diversity in Ecological Communities: Historical and Geographical Perspectives*, pp. 52–65. Chicago: University of Chicago Press.

Roth, P. H. and Berger, W. H. 1977. Distribution and dissolution of coccoliths in the south and central Pacific. In: W. V. Sliter, A. W. H. Bé and W. H. Berger (eds.) *Dissolution of Deep-Sea Carbonates*, pp. 87–113. Cambridge, MA: Cushman Foundation for Foraminiferal Research Special Publication No. 13.

Round, F. E. and Crawford, R. M. 1981. The lines of evolution of the Bacillariophyta. I. Origin. *Proceedings of the Royal Society of London* (B) 211:237–260.

Round, F. E. and Crawford, R. M. 1984. The lines of evolution of the Bacillariophyta. II. The centric series. *Proceedings of the Royal Society of London* (B) 221:169–188.

Rubey, W. W. 1933. Settling velocities of gravel, sand, and silt particles. *American Journal of Science* 25:325–338.

Rubey, W. W. 1951. Geologic history of sea water. An attempt to state the problem. *Geological Society of America Bulletin* 62:1111–1148.

Ruddiman, W. F. and Glover, L. K. 1972. Vertical mixing of ice-rafted volcanic ash in North Atlantic sediments. *Geological Society of America, Bulletin* 83:2817–2836.

Ruddiman, W. F., Jones, G., Peng, T.-H., Glover, L., Glass, B. and Liebertz, P. 1980. Tests for size and shape dependency in deep-sea mixing. *Sedimentary Geology* 25:257–276.

Runnegar, B. 1982. The Cambrian explosion: animals or fossils? *Journal of the Geological Society of Australia* 29:395–411.

Runnegar, B. 1996. Proterozoic progress to megascopic complexity. In: J. E. Repetski (ed.) *Sixth North American Paleontological*

Convention, Abstracts and Program, p. 334. Pittsburgh: Paleontological Society Special Publication No. 8.

Runnels, C. N. 1995. Environmental degradation in ancient Greece. *Scientific American* 272 (March):96–99.

Ruse, M. 1979. *The Darwinian Revolution: Science Red in Tooth and Claw*. Chicago: University of Chicago Press.

Russell, E. W. B. 1997. *People and the Land Through Time: Linking Ecology and History*. New Haven: Yale University Press.

Rutherford, S. L. and Lindquist, S. 1998. Hsp90 as a capacitor for morphological evolution. *Nature* 396:336–342.

Ruttenberg, K. C. and Berner, R. A. 1993. Authigenic apatite formation and burial in sediments from non-upwelling, continental margin environments. *Geochimica et Cosmochimica Acta* 57:991–1007.

Rützler, K. 1975. The role of burrowing sponges in bioerosion. *Oecologia* 19:203–216.

Sadler, P. M. 1981. Sediment accumulation rates and the completeness of stratigraphic sections. *Journal of Geology* 89:569–584.

Sadler, P. M. 1993. Models of time-averaging as a maturation process: How soon do sedimentary sections escape reworking? In: S. M. Kidwell and A. K. Behrensmeyer (eds.) *Taphonomic Approaches to Time Resolution in Fossil Assemblages*, pp. 188–209. Pittsburgh: Paleontological Society Short Course No. 6.

Sadler, P. M. and Strauss, D. J. 1990. Estimation of completeness of stratigraphical sections using empirical data and theoretical models. *Journal of the Geological Society of London* 147:471–485.

Saffert, H. and Thomas, E. 1998. Living foraminifera and total populations in salt marsh peat cores: Kelsey Marsh (Clinton, CT) and the Great Marshes (Barnstable, MA). *Marine Micropaleontology* 33:175–202.

Salazar-Jiménez, A., Frey, R. W. and Howard, J. D. 1982. Concavity orientations of bivalve shells in estuarine and nearshore shelf sediments, Georgia. *Journal of Sedimentary Petrology* 52:565–586.

Salmon, W. C. 1967. *The Foundations of Scientific Inference*. Pittsburgh: University of Pittsburgh Press.

Salthe, S. N. 1985. *Evolving Hierarchical Systems: Their Structure and Representation*. New York: Columbia University Press.

Samoilov, V. S. and Benjamini, C. 1997. On possible causes of trace element enrichment in dinosaur remains: reply. 1997. *Palaios* 12:497–498.

Sandberg, P. A. 1975. Bryozoan diagenesis: bearing on the nature of the original skeleton of rugose corals. *Journal of Paleontology* 49:587–604.

Sandberg, P. A. 1983. An oscillating trend in Phanerozoic non-skeletal carbonate mineralogy. *Nature* 305:19–22.

Saunders, W. B. and Ward, P. D. 1994. *Nautilus* is not a model for the function and behavior of ammonoids. *Lethaia* 27:47–48.

Savarese, M. 1994. Taphonomic and paleoecologic implications of flow-induced forces on concavo-convex articulate brachiopods: an experimental approach. *Lethaia* 27:301–312.

Savin, S. M. and Douglas, R. G. 1973. Stable isotope and magnesium geochemistry of Recent planktonic foraminifera from the South Pacific. *Geological Society of America Bulletin* 84:2327–2342.

Savrda, C. E. 1991. Ichnology in sequence stratigraphic studies: an example from the Lower Paleocene of Alabama. *Palaios* 6:39–54.

Savrda, C. E. 1995. Ichnologic applications in paleoceanographic, paleoclimatic, and sea-level studies. *Palaios* 10:565–577.

Schäfer, W. 1972. *Ecology and Palaeoecology of Marine Environments*. Chicago: University of Chicago Press.

Scheihing, M. H. and Pfefferkorn, H. W. 1984. The taphonomy of land plants in the Orinoco Delta: a model for the incorporation of plant parts in clastic sediments of Late Carboniferous age of Euramerica. *Review of Palaeobotany and Palynology* 41:205–240.

Schidlowski, M. 1991. Quantitative evolution of global biomass through time: biological and geochemical constraints. In: S. H. Schneider and P. J. Boston (eds.) *Scientists on Gaia*, pp. 211–222. Cambridge, MA: MIT Press.

Schieber, J. 1998. Possible indicators of microbial mat deposits in shales and sandstones: examples from the mid-Proterozoic Belt Supergroup, Montana, U.S.A. *Sedimentary Geology* 120:105–124.

Schiffelbein, P. 1984. Effect of benthic mixing on the information content of deep-sea stratigraphic signals. *Nature* 311:651–653.

Schiffelbein, P. 1985. Extracting the benthic mixing impulse response function: a constrained deconvolution technique. *Marine Geology* 64:313–336.

Schiffelbein, P. 1986. The interpretation of stable isotopes in deep-sea sediments: an error analysis case study. *Marine Geology* 70:313–320.

Schindel, D. E. 1980. Microstratigraphic sampling and the limits of paleontologic resolution. *Paleobiology* 6:408–426.

Schindel, D. E. 1982. Resolution analysis: a new approach to the gaps in the fossil record. *Paleobiology* 8:340–353.

Schlager, W. 1974. Preservation of cephalopod skeletons and carbonate dissolution on ancient Tethyan sea floors. In: K. J. Hsü and H. C. Jenkyns (eds.) *Pelagic Sediments: On Land and Under the Sea*, pp. 49–70. Special Publication of the International Association of Sedimentology. Oxford: Blackwell.

Schlanger, S. O. and Douglas, R. G. 1974. The pelagic ooze–chalk–limestone transition and its implications for marine stratigraphy. In: K. J. Hsü and H. C. Jenkyns (eds.) *Pelagic Sediments: On Land and Under the Sea*, pp. 117–148. Special Publication of the International Association of Sedimentology. Oxford: Blackwell.

Schlanger, S. O., Douglas, R. G., Lancelot, Y., Moore, T. C. and Roth, P. H. 1973. Fossil preservation and diagenesis of pelagic carbonates from the Magellan Rise, central north Pacific Ocean. *Initial Reports of the Deep Sea Drilling Project* 17:407–426.

Schlüter, T. 1990. Baltic amber. In: D. E. K. Briggs and P. R. Crowther (eds.) *Palaeobiology: a Synthesis*, pp. 294–297. Oxford: Blackwell.

Schmitt, J. G. and Boyd, D. W. 1981. Patterns of silicification in Permian pelecypods and brachiopods from Wyoming. *Journal of Sedimentary Petrology* 51:1297–1308.

Schopf, J. W. 1975. Modes of fossil preservation. *Review of Palaeobotany and Palynology* 20:27–53.

Schopf, J. W. 1996. Precambrian: the age of microscopic life. In: J. E. Repetski (ed.) *Sixth North American Paleontological Convention, Abstracts and Program*, p. 345. Pittsburgh: Paleontological Society Special Publication No. 8.

Schrader, H.-J. 1971. Fecal pellets: role in sedimentation of pelagic diatoms. *Science* 174:55–57.

Schröder, C. J. 1986. Deep-water arenaceous foraminifera in the northwest Atlantic Ocean. *Canadian Technical Report of Hydrography and Ocean Science* 71:1–191.

Schröder, C. J., Scott, D. B. and Medioli, F. S. 1987. Can smaller benthic foraminifera be ignored in paleoenvironmental analysis? *Journal of Foraminiferal Research* 17:101–105.

Schroeder, J. O., Murray, R. W., Leinen, M., Pflaum, R. C. and Janecek, T. R. 1997. Barium in equatorial Pacific carbonate sediment: terrigenous, oxide, and biogenic associations. *Paleoceanography* 12:125–146.

Schubert, J. K., Kidder, D. L. and Erwin, D. H. 1997. Silica-replaced fossils through the Phanerozoic. *Geology* 25:1031–1034.

Schwarcz, H. P., Hedges, R. E. M. and Ivanovich, M. (eds.) 1989. First International Workshop on Bone. *Applied Geochemistry* 4:211–343.

Schwarzacher, W. 1963. Orientation of crinoids by current action. *Journal of Sedimentary Petrology* 33:580–586.

Scoffin, T. P. 1970. The trapping and binding of subtidal carbonate sediments by marine vegetation in Bimini Lagoon, Bahamas. *Journal of Sedimentary Petrology* 40:248–273.

Scoffin, T. P. 1992. Taphonomy of coral reefs: a review. *Coral Reefs* 11:57–77.

Scott, A. C. and Rex, G. 1985. The formation and significance of Carboniferous coal balls. In: H. B. Whittington and S. Conway Morris (eds.) Extraordinary fossil biotas: their ecological and evolutionary significance. *Philosophical Transactions of the Royal Society of London* (B) 311:123–137.

Scott, D. B. and Medioli, F. S. 1980*a*. *Quantitative Studies on Marsh Foraminiferal Distributions in Nova Scotia: Implications for Sea Level Studies*. Cambridge, MA: Cushman Foundation for Foraminiferal Research Special Publication 17.

Scott, D. B. and Medioli, F. S. 1980*b*. Living vs. total foraminiferal populations: their relative usefulness in paleoecology. *Journal of Paleontology* 54:814–831.

Scott, D. B. and Medioli, F. S. 1986. Foraminifera as sea-level indicators. In: O. van de Plassche (ed.) *Sea Level Research: a Manual for the Collection of Data*, pp. 435–455. IGCP Projects 61 and 200. Norwich, England: Geo Books.

Seastedt, T. R. and Crossley, D. A. 1984. The influence of arthropods on ecosystems. *Bioscience* 34:157–161.

Seastedt, T. R. and Tate, C. M. 1981. Decomposition rates and nutrient contents of arthropod remains in forest litter. *Ecology* 62:13–19.

Seilacher, A. 1970. Begriff und Bedeutung der Fossil-Lagerstätten. *Neues Jahrbuch für Geologie und Paläontologie, Monatshefte* 1970:34–39.

Seilacher, A. 1973. Biostratinomy: the sedimentology of biologically standardized particles. In: R. N. Ginsburg (ed.) *Evolving Concepts in Sedimentology*, pp. 159–177. Baltimore: Johns Hopkins University Press.

Seilacher, A. 1976. Sonderforschungsbereich 53, "Palökologie: Arbeitsbericht 1970–75. *Zentralblatt für Geologie und Paläontologie* II:206–210.

Seilacher, A. 1982. General remarks about event deposits. In: G. Einsele and A. Seilacher (eds.) *Cyclic and Event Stratification*, pp.161–174. Berlin: Springer-Verlag.

Seilacher, A. 1985. The Jeram model: event condensation in a modern intertidal environment. In: U. Bayer and A. Seilacher (eds.) *Sedimentary and Evolutionary Cycles*, pp. 33–34. Berlin: Springer-Verlag.

Seilacher, A. 1989. Vendozoa: organismic construction in the Proterozoic biosphere. *Lethaia* 22:229–239.

Seilacher, A. 1990. Taphonomy of Fossil-Lagerstätten: overview. In: D. E. K. Briggs and P. R. Crowther (eds.) *Palaeobiology: a Synthesis*, pp. 266–270. Oxford: Blackwell.

Seilacher, A., Andalib, F., Dietl, G. and Gocht, H. 1976. Preservational history of compressed Jurassic ammonites from southern Germany. *Neues Jahrbuch für Geologie und Paläontologie Abhandlung* 152:307–356.

Seilacher, A., Reif, W.-E. and Westphal, F. 1985. Sedimentological, ecological and temporal patterns of fossil Lagerstätten. In: H. B. Whittington and S. Conway Morris (eds.) Extraordinary fossil biotas: their ecological and evolutionary significance. *Philosophical Transactions of the Royal Society of London* (B) 311:5–23.

Sen Gupta, B. K., Shin, I.-C. and Wendler, S. T. 1987. Relevance of specimen size in distribution studies of deep-sea benthic foraminifera. *Palaios* 2:332–338.

Sepkoski, J. J. 1992. Phylogenetic and ecologic patterns in the Phanerozoic history of marine diversity. In: N. Eldredge (ed.) *Systematics, Ecology, and the Biodiversity Crisis*, pp. 77–100. New York: Columbia University Press.

Sepkoski, J. J., Bambach, R. K., Raup, D. M. and Valentine, J. W. 1981. Phanerozoic marine diversity and the fossil record. *Nature* 293:435–437.

Sepkoski, J. J., Bambach, R. K. and Droser, M. L. 1991. Secular changes in Phanerozoic event bedding and the biological overprint. In: G. Einsele, W. Ricken, and A. Seilacher (eds.) *Cycles and Events in Stratigraphy*, pp. 298–312. Berlin: Springer-Verlag.

Shabica, C. W. and Hay, A. A. (eds.) 1997. *Richardson's Guide to the Fossil Fauna of Mazon Creek*. Chicago: Northeastern Illinois University.

Shackleton, N. J. 1986. Paleogene stable isotope events. *Palaeogeography, Palaeoclimatology, and Palaeoecology* 57:91–102.

Sharma, P., Gardner, L. R., Moore, W. S. and Bollinger, M. S. 1987. Sedimentation and bioturbation in a salt marsh as revealed by ^{210}Pb, ^{137}Cs, and ^{7}Be studies. *Limnology and Oceanography* 32:313–326.

Shaw, A. B. 1964. *Time in Stratigraphy*. New York: McGraw-Hill.

Sheehan, P. M. 1977. Species diversity in the Phanerozoic: a reflection of labor by systematists? *Paleobiology* 3:325–328.

Sheehan, P. and Hansen, T. A. 1986. Detritus feeding as a buffer to extinction at the end of the Cretaceous. *Geology* 14:868–870.

Sheehan, P. M., Fastovsky, D. E., Hoffman, R. G., Berghaus, C. B. and Gabriel, D. L. 1991. Sudden extinction of the dinosaurs: Latest Cretaceous, Upper Great Plains, U.S.A. *Science* 254:835–839.

Sheldon, R. P. 1980. Episodicity of phosphate deposition and deep ocean circulation – an hypothesis. In: Y. K. Bentor (ed.) *Marine Phosphorites – Geochemistry, Occurrence, Genesis*, pp. 239–247. Tulsa: Society of Economic Paleontologists and Mineralogists Special Publication No. 29.

Sherrod, B. L. 1995. Taphonomy of salt marsh diatom assemblages: Late Holocene examples from Georgia and the Pacific northwest. *Geological Society of America Abstracts with Programs* 27:A29.

Shipman, P. 1975. Implications of drought for vertebrate fossil assemblages. *Nature* 257:667–668.

Shipman, P. 1981. *Life History of a Fossil: an Introduction to Taphonomy and Paleoecology*. Cambridge, MA: Harvard University Press.

Shipman, P. and Rose, J. 1983. Early hominid hunting, butchering, and carcass-processing behaviors: approaches to the fossil record. *Journal of Anthropological Archaeology* 2:57–98.

Shipman, P. and Rose, J. 1988. Bone tools: an experimental approach. In: S. L. Olsen (ed.) *Scanning Electron Microscopy in Archaeology*, pp. 303–335. British Archaeological Reports International Series 452.

Shipman, P. and Walker, A. 1980. Bone-collecting by harvesting ants. *Paleobiology* 6:496–502.

Shotwell, J. A. 1955. An approach to the paleoecology of mammals. *Ecology* 36:327–337.

Showstack, R. 1998. Whale falls may provide important stepping stone habitat between deep sea vents and seeps. *Eos* 79(4):45.

Shroba, C. S. 1993. Taphonomic features of benthic foraminifera in a temperate setting: experimental and field observations on the role of abrasion, solution and microboring in the destruction of foraminiferal tests. *Palaios* 8:250–266.

Sigleo, A. C. 1978. Organic geochemistry of silicified wood, Petrified Forest National Park, Arizona. *Geochimica et Cosmochimica Acta* 42:1397–1405.

Sigleo, A. C. 1979. Geochemistry of silicified wood and associated sediments, Petrified Forest National Park, Arizona. *Chemical Geology* 26:151–163.

Signor, P. W. 1982. Species richness in the Phanerozoic: compensating for sampling bias. *Geology* 10:625–628.

Signor, P. 1985. Real and apparent trends in species richness through time. In: J. W. Valentine (ed.) *Phanerozoic Diversity Patterns: Profiles in Macroevolution*, pp. 129–150. Princeton: Princeton University Press.

Signor, P. W. 1992. Taxonomic diversity and faunal turnover in the Early Cambrian: did the most severe mass extinction of the Phanerozoic occur in the Botomian Stage? In: S. Lidgard and P. R. Crane (eds.) *Fifth North American Paleontological Convention, Abstracts and Program*, p. 272. Pittsburgh: Paleontological Society Special Publication No. 6.

Signor, P. W. and Brett, C. W. 1984. The mid-Paleozoic precursor to the Mesozoic marine revolution. *Paleobiology* 10:229–245.

Signor, P. W. and Lipps, J. H. 1982. Sampling bias, gradual extinction patterns and catastrophes in the fossil record. In: L. T. Silver and P. H. Schulz (eds.) *Geological Implications of Impacts of Large Asteroids and Comets on the Earth*, pp. 291–296. Boulder: Geological Society of America Special Paper 190.

Signor, P. W. and Vermeij, G. J. 1994. The plankton and the benthos: origins and early history of an evolving relationship. *Paleobiology* 20:297–319.

Silverman, R. H. and Williams, B. R. G. 1999. Stress responses: translational control perks up. *Nature* 397:208–209.

Simkiss, K. 1989. Biomineralisation in the context of geological time. *Transactions of the Royal Society of Edinburgh* 80:193–199.

Simon, A., Poulicek, M., Velimirov, B. and Mackenzie, F. T. 1994. Comparison of anaerobic and aerobic biodegradation of mineralized skeletal structures in marine and estuarine conditions. *Biogeochemistry* 25:167–195.

Simon, W. 1986. *Mathematical Techniques for Biology and Medicine*. New York: Dover.

Simone, L. and Carannante, G. 1988. The fate of foramol ("temperate-type") carbonate platforms. *Sedimentary Geology* 60:347–354.

Simonson, B. M. 1987. Early silica cementation and subsequent diagenesis in arenites from four early Proterozoic iron formations in North America. *Journal of Sedimentary Petrology* 57:494–511.

Skyrms, B. 1966. *Choice and Chance: an Introduction to Inductive Logic*. Belmont, CA: Dickinson Publishing.

Sloan, D. 1995. Use of foraminiferal biostratigraphy in mitigating pollution and seismic problems, San Francisco, California. *Journal of Foraminiferal Research* 25:260–266.

Sloan, L. C., Bluth, G. J. S. and Filippelli, G. M. 1997. A comparison of spatially resolved and global mean reconstructions of continental denudation under ice-free and present conditions. *Paleoceanography* 12:147–160.

Sloss, L. L. 1963. Sequences in the cratonic interior of North America. *Geological Society of America Bulletin* 74:93–114.

Smayda, T. J. 1971. Normal and accelerated sinking of phytoplankton in the sea. *Marine Geology* 11:105–122.

Smil, V. 1997. Global population and the nitrogen cycle. *Scientific American* 277(1):76–81.

Smith, A. M. 1992. Bioerosion of bivalve shells in Hauraki Gulf, North Island, New Zealand. In: C. N. Battershill *et al*. (eds.) *Proceedings of the Second International Temperate Reef Symposium (Auckland, New Zealand)*, pp. 175–181. Wellington: National Institute of Water and Atmospheric Research.

Smith, A. M. 1995. Palaeoenvironmental interpretation using bryozoans: a review. In: D. W. J. Bosence and P. A. Allison (eds.) *Marine Paleoenvironmental Analysis from Fossils*, pp. 231–243. London: Geological Society of London Special Publication No. 83.

Smith, A. M. and Nelson, C. S. 1994. Selectivity in sea-floor processes: taphonomy of bryozoans. In: P. J. Hayward, J. S. Ryland and P. D. Taylor (eds.) *Biology and Palaeobiology of Bryozoans*, pp. 177–180. Fredensborg: Olsen & Olsen.

Smith, A. M. and Nelson, C. S. 1996. Differential abrasion of bryozoan skeletons: taphonomic implications for paleoenvironmental interpretation. In: D. P. Gordon, A. M. Smith and J. A. Grant-Mackie (eds.) *Bryozoans in Space and Time*, pp. 305–313. Wellington: National Institute of Water and Atmospheric Research Ltd.

Smith, A. M., Nelson, C. S. and Danaher, P. J. 1992. Dissolution behavior of bryozoan sediments: taphonomic implications for nontropical shelf carbonates. *Palaeogeography, Palaeoclimatology, Palaeoecology* 93:213–226.

Smith, A. M., Nelson, C. S. and Spencer, H. G. 1998. Skeletal carbonate mineralogy of New Zealand bryozoans. *Marine Geology* 151:27–46.

Smith, C. R. 1985. Food for the deep sea: utilization, dispersal, and flux of nekton falls at the Santa Catalina basin floor. *Deep-Sea Research* 32:417–442.

Smith, C. R., Pope, R. H., DeMaster, D. J. and Magaard, L. 1993. Age-dependent mixing of deep-sea sediments. *Geochimica et Cosmochimica Acta* 57:1473–1478.

Smith, G. R., Stearley, R. F. and Badgley, C. E. 1988. Taphonomic bias in fish diversity from Cenozoic floodplain environments. *Palaeogeography, Palaeoclimatology, Palaeoecology* 63:263–273.

Smith, J. D. 1977. Modelling of sediment transport on continental shelves. In: E. D. Goldberg, I. N. McCave, J. J. Obrien and J. H. Steele (eds.) *The Sea*, vol. 6, pp. 539–577. New York: Wiley.

Smith, J. N. and Schafer, C. T. 1984. Bioturbation processes in continental slope and rise sediments delineated by Pb-210, microfossil and textural indicators. *Journal of Marine Research* 42:1117–1145.

Smith, P. E., Evensen, N. M., York, D. and Odin, G. S. 1998. Single-grain ^{40}Ar-^{39}Ar ages of glauconies: implications for the geologic time scale and global sea level variations. *Science* 279:1517–1519.

Smith, R. K. 1987. Fossilization potential in modern shallow-water benthic foraminiferal assemblages. *Journal of Foraminferal Research* 17:117–122.

Smith, R. M. H. 1993. Vertebrate taphonomy of Late Permian floodplain deposits in the southwestern Karoo Basin of South Africa. *Palaios* 8:45–67.

Smith, S. V. 1971. Budget of calcium carbonate, southern California continental borderland. *Journal of Sedimentary Petrology* 41:798–808.

Smith, S. V., Kimmerer, W. J., Laws, E. A., Brock, R. E. and Walsh, T. W. 1981. Kaneohe Bay sewage diversion experiment: perspectives on ecosystem responses to nutritional perturbation. *Pacific Science* 35:279–395.

Snyder, S. W., Hale, W. R. and Kontrovitz, M. 1990*a*. Distributional patterns of modern benthic foraminifera on the Washington continental shelf. *Micropaleontology* 36:245–258.

Snyder, S. W., Hale, W. R. and Kontrovitz, M. 1990*b*. Assessment of postmortem transportation of modern benthic foraminifera of the Washington continental shelf. *Micropaleontology* 36:259–282.

Solow, A. R. and Smith, W. 1997. On fossil preservation and the stratigraphic ranges of taxa. *Paleobiology* 23:271–277.

Sorby, H. C. 1879. The structure and origin of limestones. *Geological Society of London, Proceedings* 35:56–95.

Sorhannus, U. 1997. The origination time of diatoms: an analysis based on ribosomal RNA data. *Micropaleontology* 43:215–218.

Speyer, S. E. 1991. Trilobite taphonomy: a basis for comparative studies of arthropod preservation, functional anatomy and behavior. In: S. K. Donovan (ed.) *The Processes of Fossilization*, pp. 194–219. London: Belhaven Press.

Speyer, S. E. and Brett, C. E. 1986. Trilobite taphonomy and Middle Devonian taphofacies. *Palaios* 1: 312–327.

Speyer, S.E and Brett, C. E. 1988. Taphofacies models for epeiric sea environments: Middle Paleozoic examples. *Palaeogeography, Palaeoclimatology, Palaeoecology* 63:225–262.

Spicer, R. A. 1981. The sorting and deposition of allochthonous plant material in a modern environment at Silwood Lake, Silwood Park, Berkshire, England. *United States Geological Survey Professional Paper* 1143:1–77.

Spicer, R. A. 1989. The formation and interpretation of plant fossil assemblages. In: J. A. Callow (ed.) *Advances in Botanical Research*, vol. 16, pp. 95–191. London: Academic Press.

Spicer, R. A. 1991. Plant taphonomic processes. In: P. A. Allison and D. E. G. Briggs (eds.) *Taphonomy: Releasing the Data Locked in the Fossil Record*, pp. 72–113. New York: Plenum Press.

Spicer, R. A. and Wolfe, J. A. 1987. Plant taphonomy of late Holocene deposits in Trinity (Clair Engle) Lake, northern California. *Paleobiology* 13:227–245.

Springer, D. A. and Flessa, K. W. 1996. Faunal gradients in surface and subsurface shelly accumulations from a recent clastic tidal flat, Bahia la Choya, northern Gulf of California, Mexico. *Palaeogeography, Palaeoclimatology, Palaeoecology* 126:261–279.

Staff, G., Powell, E. N., Stanton, R. J. and Cummins, H. 1985. Biomass: is it a useful tool in paleocommunity reconstruction? *Lethaia* 18:209–232.

Staff, G. M., Stanton, R. J., Powell, E. N. and Cummins, E. H. 1986. Time-averaging, taphonomy, and their impact on paleocommunity reconstruction: death assemblages in Texas bays. *Geological Society of America Bulletin* 97:428–443.

Stankiewicz, B. A., Briggs, D. E. G., Evershed, R. P. and Duncan, I. J. 1997*a*. Chemical preservation of insect cuticle from the Pleistocene asphalt deposits of California, USA. *Geochimica et Cosmochimica Acta* 11:2247–2252.

Stankiewicz, B. A., Briggs, D. E. G., Evershed, R. P., Flannery, M. B. and Wuttke, M. 1997*b*. Preservation of chitin in 25-million-year-old fossils. *Science* 276:1541–1543.

Stankiewicz, B. A., Poinar, H. N., Briggs, D. E. G., Evershed, R. P. and Poinar, G. O., Jr. 1998. Chemical preservation of plants and insects in natural resins. *Proceedings of the Royal Society of London, Series B* 265:641–647.

Stanley, D. J. and Bernasconi, M. P. 1998. Relict and palimpsest depositional patterns on the Nile shelf recorded by molluscan faunas. *Palaios* 13:79–86.

Stanley, D. J. and Warne, A. G. 1993. Nile Delta: Recent geological evolution and man impact. *Science* 260:628–634.

Stanley, S. M. 1990. Delayed recovery and the spacing of major extinctions. *Paleobiology* 16:401–414.

Stanley, S. M. and Yang, X. 1994. A double mass extinction at the end of the Paleozoic Era. *Science* 266:1340–1344.

Stapleton, R. P. 1973. Ultrastructure of tests of some Recent benthic hyaline foraminifera. *Palaeontographica* 142(A):16–49.

Steel, R. J. 1974. Cornstone (fossil caliche): its origin, stratigraphic, and sedimentologic importance in the New Red Sandstone, western Scotland. *Journal of Geology* 82:351–369.

Stein, C. L. 1982. Silica recrystallization in petrified wood. *Journal of Sedimentary Petrology* 52:1277–1282.

Strauss, D. and Sadler, P. M. 1989. Classical confidence intervals and Bayesian probability estimates for ends of local taxon ranges. *Mathematical Geology* 21:411–427.

Stringham, G. E., Simons, D. B. and Guy, H. P. 1969. The behavior of large particles falling in quiescent fluids. *United States Geological Survey Professional Paper* 562-C:1–36.

Stroeven, A. P., Burckle, L. H., Kleman, J. and Prentice, M. L. 1998. Atmospheric transport of diatoms in the Antarctic Sirius Group: Pliocene deep freeze. *GSA Today* (4):1.

Stroeven, A. P., Prentice, M. L. and Kleman, J. 1996. On marine microfossil transport and pathways in Antarctica during the late Neogene: evidence from the Sirius Group at Mount Fleming. *Geology* 24:727–730.

Stuart-Williams, H. Le Q., Schwarcz, H. P., White, C. D. and Spence, M. W. 1996. The isotopic composition and diagenesis of human bone from Teotihuacan and Oaxaca, Mexico. *Palaeogeography, Palaeoclimatology, Palaeoecology* 126:1–14.

Stumpf, R. P. 1983. The process of sedimentation on the surface of a salt marsh. *Estuarine, Coastal and Shelf Science* 17:495–508.

Suess, E. 1981. Phosphate regeneration from sediments of the Peru continental margin by dissolution of fish debris. *Geochimica et Cosmochimica Acta* 45:577–588.

Sugihara, G. and May, R. M. 1990. Applications of fractals in ecology. *Trends in Ecology and Evolution* 5:79–86.

Sugita, S. 1993. A model of pollen source area for an entire lake surface. *Quaternary Research* 39:239–244.

Summerson, C. H. (ed.) 1978. *Sorby on Geology: A Collection of Papers from 1853 to 1906 by Henry Clifton Sorby*. Miami: Comparative Sedimentology Laboratory, Rosenstiel School of Marine and Atmospheric Sciences, University of Miami.

Sverdrup, H. U., Johnson, M. W. and Fleming, R. H. 1942. *The Oceans: Their Physics, Chemistry, and General Biology*. Englewood Cliffs, NJ: Prentice-Hall.

Swart, P. K., Healy, G. F., Dodge, R. E., Kramer, P., Hudson, J. H., Halley, R. B. and Robblee, M. B. 1996. The stable oxygen and carbon isotopic record from a coral growing in Florida Bay: a 160-year record of climatic and anthropogenic influence. *Palaeogeography, Palaeoclimatology, Palaeoecology* 123:219–237.

Sweet, W. C. 1992. A conodont based high-resolution biostratigraphy for the Permo-Triassic boundary interval. In: W. C. Sweet, Y. Zunyi, J. M. Dickins and Y. Hongfu (eds.) *Permo-Triassic Events in the Eastern Tethys*, pp. 120–133. Cambridge: Cambridge University Press.

Swinchatt, J. P. 1965. Significance of constituent composition, texture and skeletal breakdown in some Recent carbonate sediments. *Journal of Sedimentary Petrology* 35:71–90.

Szmant, A. M. and Forrester, A. 1996. Water column and sediment nitrogen and phosphorus distribution patterns in the Florida Keys, U.S.A. *Coral Reefs* 15:21–41.

Tanabe, K., Inazumi, A., Tamahama, K. and Katsuta, T. 1984. Taphonomy of half and compressed ammonites from the Lower Jurassic black shales of the Toyora area, west Japan. *Palaeogeography, Palaeoclimatology, Palaeoecology* 47:329–346.

Tappan, H. 1980. *Paleobiology of Plant Protists*. San Francisco: W. H. Freeman.

Tappan, H. 1982. Extinction or survival: selectivity and causes of Phanerozoic crises. In: L. T. Silver and P. H. Schultz (eds.) *Geological Implications of Impacts of Large Asteroids and Comets on the Earth*, pp. 265–276. Boulder: Geological Society of America.

Tappan, H. 1986. Phytoplankton: below the salt at the global table. *Journal of Paleontology* 60:545–554.

Tappan, H. and Loeblich, A. R. 1988. Foraminiferal evolution, diversification, and extinction. *Journal of Paleontology* 62:695–714.

Tardy, Y., N'Kounkou, and Probst, J.-L. 1989. The global water cycle and continental erosion during Phanerozoic time (570 my). *American Journal of Science* 289:455–483.

Tauber, H. 1967. Investigation of the mode of pollen transfer in forested areas. *Review of Palaeobotany and Palynology* 3:227–286.

Taylor, P. D. and Allison, P. A. 1998. Bryozoan carbonates through time and space. *Geology* 26:459–462.

Taylor, W. L. and Brett, C. E. 1996. Taphonomy and paleoecology of echinoderm Lagerstätten from the Silurian (Wenlockian) Rochester Shale. *Palaios* 11:118–140.

Tegelaar, E. W., DeLeeuw, J. W., Derenne, S. and Largeau, C. 1989. A reappraisal of kerogen formation. *Geochimica et Cosmochimica Acta* 53:3103–3106.

Tegelaar, E. W., Kerp, H., Visscher, H., Schenck, P. A. and DeLeeuw, J. W. 1991. Bias of the paleobotanical record as a consequence of variations in the chemical composition of higher vascular plant cuticles. *Paleobiology* 17:133–144.

Teichert, C. and Serventy, D. L. 1947. Deposits of shells transported by birds. *American Journal of Science* 245:322–328.

Thayer, C. W. 1977. Recruitment, growth, and mortality of a living articulate brachiopod, with implications for the interpretation of survivorship curves. *Paleobiology* 3:98–109.

Thayer, C. W. 1983. Sediment-mediated biological disturbance and the evolution of marine benthos. In: M. J. S. Tevesz and P. L. McCall (eds.) *Biotic Interactions in Recent and Fossil Benthic Communities*, pp. 479–625. New York: Plenum Press.

Thayer, C. W. 1992. Escalating energy budgets and oligotrophic refugia: winners and drop-outs in the Red Queen's race. In: S. Lidgard and P. R. Crane (eds.) *Fifth North American Paleontological Convention, Abstracts and Program*, p. 290. Pittsburgh: Paleontological Society Special Publication No. 6.

Thickpenny, A. and Leggett, J. K. 1987. Stratigraphic distribution and palaeo-oceanographic significance of European early Paleozoic organic-rich sediments. In: J. Brooks and A. J. Fleet (eds.) *Marine Petroleum Source Rocks*, pp. 231–247. London: Geological Society of London.

Thierstein, H. R. 1980. Selective dissolution of late Cretaceous and earliest Tertiary calcareous nanofossils: experimental evidence. *Cretaceous Research* 2:165–176.

Thierstein, H. R. 1981. Late Cretaceous nanoplankton and the change at the Cretaceous–Tertiary boundary. In: J. E. Warme, R. G. Douglas and E. L. Winterer (eds.) *The Deep-Sea Drilling Project: A Decade of Progress*, pp. 355–394. Tulsa: SEPM (Society for Sedimentary Geology) Special Publication No. 32.

Thomas, E. and Varekamp, J. C. 1991. Paleo-environmental analyses of marsh sequences (Clinton, Connecticut): evidence for punctuated rise in sealevel during the latest Holocene. *Journal of Coastal Research* (Special Issue) 11:125–158.

Thompson, E. I. and Schmitz, B. 1997. Barium and the late Paleocene $\delta^{13}C$ maximum: evidence of increased marine surface productivity. *Paleoceanography* 12:239–254.

Thompson, P. R. and Whelan, J. K. 1980. Fecal pellets at Deep Sea Drilling Project Site 436. In: R. von Huene et al. (eds.) *Initial Reports of the Deep-Sea Drilling Project*, pp. 921–935. Washington, D.C.: U.S. Government Printing Office.

Thorne, J. A., Grace, E., Swift, D. J. P. and Niedoroda, A. 1991. Sedimentation on continental margins. III. The depositional fabric: an analytical approach to stratification and facies identification. In: D. J. P. Swift, G. F. Oertel, R. W. Tillman and J. A. Thorne (eds.) *Shelf Sand and Sandstone Bodies: Geometry, Facies and Sequence Stratigraphy*, pp. 59–87. International Association of Sedimentologists Special Publication 14. Oxford: Blackwell.

References

Thunell, R. C. 1976. Optimum indices of calcium carbonate dissolution in deep-sea sediments. *Geology* 4:525–528.

Tissot, B. 1979. Effect on prolific petroleum source rocks and major coal deposits caused by sea-level changes. *Nature* 277:462–465.

Tomascik, T. and Sander, F. 1985. Effects of eutrophication on reef-building corals. I. Growth rate of the reef-building coral *Montastrea annularis*. *Marine Biology* 87:143–155.

Tomascik, T. and Sander, F. 1987*a*. Effects of eutrophication on reef-building corals. II. Structure of scleractinian coral communities on fringing reefs, Barbados, West Indies. *Marine Biology* 94:53–75.

Tomascik, T. and Sander, F. 1987*b*. Effects of eutrophication on reef-building corals. III. Reproduction of the reef-building coral *Porites porites*. *Marine Biology* 94:77–94.

Toots, H. 1965. Sequence of disarticulation in mammalian skeletons. *University of Wyoming Contributions to Geology* 4:37–39.

Towe, K. M. 1996. Fossil preservation in the Burgess Shale. *Lethaia* 29:107–108.

Towe, K. M. and Hemleben, C. 1976. Diagenesis of magnesian calcite: evidence from miliolacean foraminifera. *Geology* 4:337–339.

Tozer, E. T. 1988. Towards a definition of the Permian–Triassic boundary. *Episodes* 11:251–255.

Trauth, M. H., Sarnthein, M. and Arnold, A. 1997. Bioturbational mixing depth and carbon flux at the seafloor. *Paleoceanography* 12:517–526.

Traverse, A. 1990. Studies of pollen and spores in rivers and other bodies of water, in terms of source-vegetation and sedimentation, with special reference to Trinity River and Bay, Texas. *Review of Palaeobotany and Palynology* 64:297–303.

Traverse, A. 1994. Sedimentation of land-derived palynomorphs in the Trinity–Galveston Bay area, Texas. In: A. Traverse (ed.) *Sedimentation of Organic Particles*, pp. 69–102. Cambridge: Cambridge University Press.

Tréguer, P., Nelson, D. M., Van Bennekom, A. J., DeMaster, D. J., Leynaert, A. and Quéguiner, B. 1995. The silica balance in the world ocean: a reestimate. *Science* 268:375–379.

Trewin, N. H. 1986. Palaeoecology and sedimentology of the Achanarras fish bed of the Middle Old Red Sandstone, Scotland. *Transactions of the Royal Society of Edinburgh, Earth Sciences* 77:21–46.

Trewin, N. H. 1989. The Rhynie hot-spring deposit. *Earth Science Conservation* 26:10–12.

Trueman, C. N. and Benton, M. J. 1997. A geochemical method to trace the taphonomic history of reworked bones in sedimentary settings. *Geology* 25:263–266.

Trueman, C. and Palmer, M. R. 1997 Diagenetic origin of REE in vertebrate apatite: a reconsideration of Samoilov and Benjamini, 1996. *Palaios* 12:495–497.

Tucker, M. E. 1974. Sedimentology of Palaeozoic pelagic limestones: the Devonian Griotte (southern France) and Cephalopodenkalk (Germany). In: K. J. Hsü and H. C. Jenkyns (eds.) *Pelagic Sediments: On Land and Under the Sea*, pp. 71–92. Special Publication of the International Association of Sedimentology. Oxford: Blackwell.

Tucker, M. E. 1991. The diagenesis of fossils. In: S. K. Donovan (ed.) *The Processes of Fossilization*, pp. 84–104. London: Belhaven Press.

Tucker, M. E. 1992. The Precambrian–Cambrian boundary: seawater chemistry, ocean circulation and nutrient suppy in metazoan evolution, extinction and biomineralization. *Journal of the Geological Society of London* 149:655–668.

Tudhope, A. W. and Risk, M. J. 1985. Rate of dissolution of carbonate sediments by microboring organisms, Davies Reef, Australia. *Journal of Sedimentary Petrology* 55:440–447.

Ulanowicz, R. E. 1980. An hypothesis on the development of natural communities. *Journal of Theoretical Biology* 85:223–245.

Ulanowicz, R. E. 1997. *Ecology, The Ascendent Perspective*. New York: Columbia University Press.

Underwood, C. J. 1993. The position of graptolites within Lower Palaeozoic planktic ecosystems. *Lethaia* 26:189–202.

Vail, P. R., Mitchum, R. M., Todd, R. G., Widmier, J. M., Thompson, S., Sangree, J. B., Bubb, J. N. and Hatelid, W. G. 1977. Seismic stratigraphy and global changes of sea level. In: C. E. Payton (ed.) *Seismic Stratigraphy: Applications to Hydrocarbon Exploration*, pp. 49–212. Tulsa: American Association of Petroleum Geologists Memoir 26.

Valentine, J. W. 1973. *Evolutionary Paleoecology of the Marine Biosphere*. Englewood Cliffs, NJ: Prentice-Hall.

Valentine, J. W. 1989. How good was the fossil record? Clues from the Californian Pleistocene. *Paleobiology* 15:83–94.

Valentine, J. W. 1995. Why no new phyla after the Cambrian? Genome and ecospace hypotheses revisited. *Palaios* 10:190–194.

Van Andel, Tj. H. 1975. Mesozoic/Cenozoic calcite compensation depth and the global distribution of calcareous sediments. *Earth and Planetary Science Letters* 26:187–195.

Van Andel, Tj. H. 1994. *New Views on an Old Planet: a History of Global Change*. Cambridge: Cambridge University Press.

Van Bergen, P. F., Collinson, M. E., Briggs, D. E. G., De Leeuw, J. W., Scott, A. C., Evershed, R. P. and Finch, P. 1995. Resistant biomacromolecules in the fossil record. *Acta Botanica Neerlandica* 44:319–342.

Van Cappellen, P. and Berner, R. A. 1988. A mathematical model for the early diagenesis of phosphorus and fluorine in marine sediments: apatite precipitation. *American Journal of Science* 288:289–333.

Van Cappellen, P. and Ingall, E. D. 1994. Benthic phosphorus regeneration, net primary production, and ocean anoxia: a model of the coupled marine biogeochemical cycles of carbon and phosphorus. *Paleoceanography* 9:677–692.

Van de Plaasche, O. 1991. Late Holocene sea-level fluctuations on the shore of Connecticut inferred from transgressive and regressive overlap boundaries in salt-marsh deposits. *Journal of Coastal Research* (Special Issue) 11:159–179.

Van de Plaasche, O., Chrzastowski, M. J., Orford, J. D., Hinton, J. A. and Long, A. J. (eds.) 1995. Coastal evolution in the Quaternary: IGCP Project 274. *Marine Geology* 124:1–339.

Van de Poel, H. M. and Schlager, W. 1994. Variations in Mesozoic–Cenozoic skeletal carbonate mineralogy. *Geologie en Mijnbouw* 73:31–51.

van der Vink, G., Allen, R. M., Chapin, J., Crooks, M., Fraley, W., Krantz, J., Lavigne, A. M., LeCuyer, A., MacColl, E. K., Morgan, W. J., Ries, B., Robinson, E., Rodriquez, K., Smith, M. and Sponberg, K. 1998. Why the United States is becoming more vulnerable to natural disasters. *Eos* 79:533.

References

Van Harten, D. 1986. Use of ostracodes to recognize downslope contamination in paleobathymetry and a preliminary reappraisal of the paleodepth of the Prasás Marls (Pliocene), Crete, Greece. *Geology* 14:856–859.

Van Harten, D. 1987. Comment and reply on "Use of ostracodes to recognize downslope contamination in paleobathymetry and a preliminary reappraisal of the paleodepth of the Prasás Marls (Pliocene), Crete, Greece." *Geology* 15:377–378.

Van Straaten, L. M. J. U. 1952. Biogenic textures and formation of shell beds in the Dutch Wadden Sea. *Proceedings of the Koninklijke Nederlandse Akademie Van Wetenschappen* (B) 55:500–516.

Van Valen, L. 1976. Energy and evolution. *Evolutionary Theory* 1:179–229.

Varekamp, J. C., Thomas, E. and van de Plassche, O. 1992. Relative sea-level rise and climate change over the last 1500 years. *Terra Nova* 4:293–304.

Vasconcelos, C., McKenzie, J. A., Bernasconi, S., Grujic, D. and Tien, A. J. 1995. Microbial mediation as a possible mechanism for natural dolomite formation at low temperatures. *Nature* 377:220–222.

Veizer, J. 1988. The evolving exogenic cycle. In: C. B. Gregor, R. M. Garrels, F. T. Mackenzie and J. B. Maynard (eds.) *Chemical Cycles in the Evolution of the Earth*, pp. 175–220. New York: Wiley.

Veizer, J., Fritz, P. and Jones, B. 1986. Geochemistry of brachiopods: oxygen and carbon isotopic records of Paleozoic oceans. *Geochimica et Cosmochimica Acta* 50:1679–1696.

Velbel, M. A. and Brandt, D. S. 1989. Differential preservation of brachiopod valves: taphonomic bias in *Platystrophia ponderosa*. *Palaios* 4:193–195.

Vénec-Peyré, M.-T. 1996. Bioeroding foraminifera: a review. *Marine Micropaleontology* 28:19–30.

Vermeij, G. J. 1987. *Evolution and Escalation: an Ecological History of Life*. Princeton: Princeton University Press.

Vermeij, G. J. 1989. The origin of skeletons. *Palaios* 6:585–589.

Vermeij, G. J. 1990, Asteroids and articulates: is there a causal link? *Lethaia* 23:431–432.

Vermeij, G. J. 1994. The evolutionary interaction among species: selection, escalation, and coevolution. *Annual Review of Ecology and Systematics* 25:219–236.

Vermeij, G. J. 1995. Economics, volcanoes, and Phanerozoic revolutions. *Paleobiology* 21:125–152.

Vidal, G. and Knoll, A. H. 1982. Radiations and extinctions of plankton in the late Proterozoic and early Cambrian. *Nature* 296:57–60.

Vidal, G. and Moczydłowska-Vidal, M. 1997. Biodiversity, speciation, and extinction trends of Proterozoic and Cambrian phytoplankton. *Paleobiology* 23:230–246.

Viohl, G. 1990. Solnhofen lithographic limestones. In: D. E. K. Briggs and P. R. Crowther (eds.) *Palaeobiology: a Synthesis*, pp. 285–289. Oxford: Blackwell.

Vitousek, P. M., Mooney, H. A., Lubchenco, J. and Melillo, J. M. 1997. Human domination of Earth's ecosystems. *Science* 277:494–499.

Volk, T. 1989a. Sensitivity of climate and atmospheric CO_2 to deep-ocean and shallow-ocean carbonate burial. *Nature* 337:637–640.

Volk, T. 1989b. Rise of angiosperms as a factor in long-term climatic cooling. *Geology* 17:107–110.

Voorhies, M. R. 1969. Taphonomy and population dynamics of an early Pliocene vertebrate fauna, Knox County, Nebraska. *University of Wyoming Contributions to Geology Special Paper* 1:1–69.

Vrba, E. S. 1980. Evolution, species and fossils: how does life evolve? *South African Journal of Science* 76:61–84.

Vrba, E. S. 1985. Environment and evolution: alternative causes of the temporal distribution of evolutionary events. *South African Journal of Science* 81:229–236.

Vrba, E. S. 1988. Late Pliocene climatic events and hominid evolution. In: F. E. Grine (ed.) *The Evolutionary History of Robust Australopithecines*, pp. 405–426. New York: Aldine.

Vrba, E. 1993. Turnover-pulses, the Red Queen, and related topics. *American Journal of Science* 293A:418–452.

Waage, K. M. 1964. Origin of repeated concretion layers in the Fox Hills Formation. *Kansas Geological Survey Bulletin* 169:541–563.

Wade, M. 1968. Preservation of soft-bodied animals in Precambrian sandstones at Ediacara, South Australia. *Lethaia* 1:238–267.

Walker, K. R. and Bambach, R. K. 1971. The significance of fossil assemblages from fine-grained sediments. *Geological Society of America Abstracts with Programs* 3:A783–784.

Walker, K. R. and Diehl, W. W. 1985. The role of marine cementation in the preservation of Lower Paleozoic assemblages. In: H. B. Whittington and S. Conway Morris (eds.) Extraordinary fossil biotas: their ecological and evolutionary significance. *Philosophical Transactions of the Royal Society of London* (B) 311:143–153.

Walker, S. E. 1988. Taphonomic significance of hermit crabs (Anomura: Paguridea): Epifaunal hermit crab – infaunal gastropod example. *Palaeogeography, Palaeoclimatology, Palaeoecology* 63:45–71.

Walker, S. E. 1989. Hermit crabs as taphonomic agents. *Palaios* 4:439–452.

Walker, S. E. 1992. Criteria for recognizing marine hermit crabs in the fossil record using gastropod shells. *Journal of Paleontology* 66:535–558.

Walker, S. E. and Carlton, J. T. 1995. Taphonomic losses become taphonomic gains: an experimental approach using the rocky shore gastropod, *Tegula funebralis*. *Palaeogeography, Palaeoclimatology, Palaeoecology* 114:197–217.

Wall, D., Dale, B., Lohmann, G. P. and Smith, W. K. 1977. The environmental and climatic distribution of dinoflagellate cysts in the North and South Atlantic Oceans and adjacent seas. *Marine Micropaleontology* 2:121–200.

Wallin, I. E. 1927. *Symbionticism and the Origin of Species*. Baltimore: Williams and Wilkins.

Walossek, D. 1993. The Upper Cambrian *Rehbachiella* and the phylogeny of Branchiopoda and Crustacea. *Fossils and Strata* 32:1–202.

Walsh, J. J. 1991. Importance of continental margins in the marine biogeochemical cycling of carbon and nitrogen. *Nature* 350:53–55.

Walter, L. M. 1985. Relative reactivity of skeletal carbonates during dissolution: implications for diagenesis. In: N. Schneiderman and P. M. Harris (eds.) *Carbonate Cements*, pp. 3–16. Tulsa: SEPM (Society for Sedimentary Geology) Special Publication No. 36.

Walter, L. M. and Burton, E. A. 1990. Dissolution of recent platform carbonate sediments in marine pore fluids. *American Journal of Science* 290:601–643.

Walter, L. M. and Morse, J. W. 1984. Reactive surface area of skeletal carbonates during dissolution: effect of grain size. *Journal of Sedimentary Petrology* 54:1081–1090.

Walther, J. 1904. Die Fauna der Solnhofener Plattenkalk, bionomisch betrachtet. *Jenaische Denkschriften* 11:133–214.

Walther, J. 1910. Die Sedimente der Taubenbank im Golfe von Neapel. *Abhandlungen der koeniglich Preussischen Akademie der Wissenschaften, Philosophisch-Historische Classe* 1910:1–49.

Wasmund, E. 1926. Biocoenose und Thanatocoenose. *Archiv fur Hydrobiologie* 17:1–116.

Watson, J. P. 1967. A termite mound in an Iron Age burial ground in Rhodesia. *Journal of Ecology* 55:663–669.

Weaver, C. E. 1967. Potassium, illite, and the ocean. *Geochimica et Cosmochimica Acta* 31:2181–2196.

Webb, T. 1993. Constructing the past from late-Quaternary pollen data: temporal resolution and a zoom lens space–time perspective. In: S. M. Kidwell and A. K. Behrensmeyer (eds.) *Taphonomic Approaches to Time Resolution in Fossil Assemblages*, pp. 79–101. Pittsburgh: Paleontological Society Short Courses in Paleontology No. 6.

Wehmiller, J. F. and Belknap, D. F. 1982. Amino acid estimates, Quaternary Atlantic coastal plain: comparison with U-series dates, biostratigraphy, and paleomagnetic control. *Quaternary Research* 18:311–336.

Wehmiller, J. F., York, L. L. and Bart, M. L. 1995. Amino acid racemization geochronology of reworked Quaternary mollusks on U.S. Atlantic coast beaches: implications for chronostratigraphy, taphonomy, and coastal sediment transport. *Marine Geology* 124:303–337.

Weigelt, J. 1928. Die Pflanzenreste des mitteldeutschen Kupferschiefers und ihre Einschaltung ins Sediment. *Fortschritte der Geologie und Palaontologie* 6:395–592.

Weigelt, J. 1989. *Recent Vertebrate Carcasses and Their Paleobiological Implications*. Chicago: University of Chicago Press.

Werner, D. (ed.) 1977. *The Biology of Diatoms*. Berkeley: University of California Botanical Monograph 13.

West, O. L. O. 1995. A hypothesis for the origin of fibrillar bodies in planktic foraminifera by bacterial endosymbiosis. *Marine Micropaleontology* 26:131–135.

Westbroek, P. 1997. The origins of animal skeletons. *Geological Society of America Abstracts with Programs* 29:A54.

Western, D. 1980. Linking the ecology of past and present mammal communities. In: A. K. Behrensmeyer and A. P. Hill (eds.) *Fossils in the Making: Vertebrate Taphonomy and Paleoecology*, pp. 41–54. Chicago: University of Chicago Press.

Westrop, S. R. 1986. Taphonomic versus ecologic controls on taxonomic relative abundance patterns in tempestites. *Lethaia* 19:123–132.

Wetmore, K. L. 1987. Correlations between test strength, morphology and habitat in some benthic foraminifera from the coast of Washington. *Journal of Foraminiferal Research* 17:1–13.

Wetzel, A. and Uchman, A. 1998. Deep-sea benthic food content recorded by ichnofacies: a conceptual model based on observations from Paleogene flysch, Carpathians, Poland. *Palaios* 13:533–546.

Whatley, R. C., Trier, K. and Dingwall, P. M. 1982. Some preliminary observations on certain mechanical and biophysical properties of the ostracod carapace. In: Bate, R. H., Robinson, E. and Sheppard, L. M. (eds.) *Fossil and Recent Ostracodes*, pp. 76–104. Chichester: Ellis Horwood.

Wheatcroft, R. A. 1990. Preservation potential of sedimentary event layers. *Geology* 18:843–845.

Wheatcroft, R. A. 1992. Experimental tests for particle size-dependent bioturbation in the deep ocean. *Limnology and Oceanography* 37:90–104.

Wheatcroft, R. A. and Jumars, P. A. 1987. Statistical re-analysis for size dependency in deep-sea mixing. *Marine Geology* 77:157–163.

Wheatcroft, R. A., Jumars, P. A., Smith, C. R. and Noewell, A. R. M. 1990. A mechanistic view of the particulate biodiffusion coefficient: step lengths, rest periods and transport directions. *Journal of Marine Research* 48:177–207.

Wheeler, B. D., Shaw, S. C., Fojt, W. J. and Robertson, R. A. 1995. *Restoration of Temperate Wetlands*. New York: Wiley.

White, P. D., Fastovsky, D. E. and Sheehan, P. M. 1998. Taphonomy and suggested structure of the dinosaurian assemblage of the Hell Creek Formation (Maastrichtian), Eastern Montana and western North Dakota. *Palaios* 13:41–51.

Whitmer, A. M., Ramenofsky, A. F., Thomas, J., Thibodeaux, L. J., Field, S. D. and Miller, B. J. 1989. Stability or instability: the role of diffusion in trace element studies. In: M. B. Schiffer (ed.) *Archaeological Method and Theory*, vol. 1, pp. 205–273. Tucson: University of Arizona Press.

Whittington, H. B. and Conway Morris, S. (eds.) 1985. Extraordinary fossil biotas: their ecological and evolutionary significance. *Philosophical Transactions of the Royal Society of London* (B) 311:1–192.

Wicken, J. S. 1980. A thermodynamic theory of evolution. *Journal of Theoretical Biology* 87:9–23.

Wignall, P. B. and Hallam, A. 1996. Facies change and the end-Permian mass extinction in S.E. Sichuan, China. *Palaios* 11:587–596.

Wignall, P. B. and Twitchett, R. J. 1996. Oceanic anoxia and the end Permian mass extinction. *Science* 272:1155–1158.

Wilby, P. R., Briggs, D. E. G., Bernier, P. and Gaillard, C. 1996a. Role of microbial mats in the fossilization of soft tissues. *Geology* 24:787–790.

Wilby, P. R., Briggs, D. E. G. and Riou, B. 1996b. Mineralization of soft-bodied invertebrates in a Jurassic metalliferous deposit. *Geology* 24:847–850.

Wild, R. 1990. Holzmaden. In: D. E. K. Briggs and P. R. Crowther (eds.) *Palaeobiology: a Synthesis*, pp. 282–285. Oxford: Blackwell.

Wilde, P. and Berry, W. B. N. 1982. Progressive ventilation of the oceans: potential for return to anoxic conditions in the post-Paleozoic. In: S. O. Schlanger and M. B. Cita (eds.) *Nature and Origin of Cretaceous Carbon-Rich Facies*, pp. 209–224. London: Academic Press.

Wilde, P. and Berry, W. B. N. 1984. Destabilization of the oceanic density structure and its significance to marine "extinction" events. *Palaeogeography, Palaeoclimatology, Palaeoecology* 48:143–162.

Wilgus, C. K., Hastings, B. S., Kendall, C. G. St. C., Posamentier, H. W., Ross, C. A. and Van Wagoner, J. C. (eds.) 1988. *Sea-Level Changes: An Integrated Approach*. Tulsa: Society of Economic Paleontologists and Mineralogists Special Publication No. 42.

References

Wilkin, R. T. and Barnes, H. L. 1997. Pyrite formation in an anoxic estuarine basin. *American Journal of Science* 297:620–650.

Wilkinson, B. H. 1979. Biomineralization, paleoceanography, and the evolution of calcareous marine organisms. *Geology* 7:524–527.

Wilkinson, B. H. and Algeo, T. J. 1989. Sedimentary carbonate record of calcium–magnesium cycling. *American Journal of Science* 289:1158–1194.

Wilkinson, B. H. and Given, R. K. 1986. Secular variation in abiotic marine carbonates. Constraints on Phanerozoic atmospheric carbon dioxide contents and oceanic Mg/Ca ratios. *Journal of Geology* 94:321–333.

Wilkinson, B. H., Owen, R. M. and Carroll, A. R. 1985. Submarine hydrothermal weathering, global eustasy, and carbonate polymorphism in Phanerozoic marine oolites. *Journal of Sedimentary Petrology* 55:171–183.

Wilkinson, B. H. and Walker, J. C. G. 1989. Phanerozoic cycling of sedimentary carbonate. *American Journal of Science* 289:525–548.

Wilkinson, C. R. 1987. Sponge biomass as an indication of reef productivity in two oceans. In: C. Birkeland (ed.) *Comparison Between Atlantic and Pacific Tropical Marine Coastal Ecosystems: Community Structure, Ecological Processes, and Productivity*, pp. 99–103. Paris: UNESCO.

Williams, H. F. L. 1989. Foraminiferal zonations on the Fraser River delta and their application to paleoenvironmental interpretations. *Palaeogeography, Palaeoclimatology, Palaeoecology* 73:39–50.

Williams, H. F. L. 1994. Intertidal benthic foraminiferal biofacies on the central Gulf Coast of Texas: modern distribution and application to sea level reconstruction. *Micropaleontology* 40:169–183.

Williams, L. A., Parks, G. A. and Crerar, D. A. 1985. Silica diagenesis. I. Solubility controls. *Journal of Sedimentary Petrology* 55:301–311.

Williams, M. E. 1994. Catastrophic versus noncatastrophic extinction of the dinosaurs: testing, falsifiability, and the burden of proof. *Journal of Paleontology* 68:183–190.

Williamson, P. G. 1981. Palaeontological documentation of speciation in Cenozoic molluscs from Turkana Basin. *Nature* 293:437–443.

Wilson, M. V. H. 1977. Paleoecology of Eocene lacustrine varves at Horsefly, British Columbia. *Canadian Journal of Earth Sciences* 14:953–962.

Wilson, M. V. H. 1980. Eocene lake environments: depth and distance-from-shore variation in fish, insect, and plant assemblages. *Palaeogeography, Palaeoclimatology, Palaeoecology* 32:21–44.

Wilson, M. V. H. 1987. Predation as source of fish fossils in Eocene lake sediments. *Palaios* 2:497–504.

Wilson, M. V. H. 1988a. Reconstruction of ancient lake environments using both autochthonous and allochthonous fossils. *Palaeogeography, Palaeoclimatology, Palaeoecology* 62:609–623.

Wilson, M. V. H. 1988b. Taphonomic processes: information loss and information gain. *Geoscience Canada* 15:131–148.

Wilson, M. V. H. and Barton, D. G. 1996. Seven centuries of taphonomic variation in Eocene freshwater fishes preserved in varves: paleoenvironments and temporal averaging. *Paleobiology* 22:535–542.

Wing, S. L. 1984. Relation of paleovegetation to geometry and cyclicity of some fluvial carbonaceous deposits. *Journal of Sedimentary Petrology* 54:52–66.

Wnuk, C. and Pfefferkorn, H. W. 1984. The life-habits and paleoecology of Middle Pennsylvanian medullosan pteridosperms based on an in situ assemblage from the Bernice Basin (Sullivan County, Pennsylvania, U.S.A.). *Review of Palaeobotany and Palynology* 41:329–351.

Wnuk, C. and Pfefferkorn, H. W. 1987. A Pennsylvanian-age terrestrial storm deposit: using plant fossils to characterize the history and process of sediment accumulation. *Journal of Sedimentary Petrology* 57:212–221.

Wold, C. N. and Hay, W. W. 1990. Estimating ancient sediment fluxes. *American Journal of Science* 290:1069–1089.

Wold, C. N. and Hay, W. W. 1993. Reconstructing the age and lithology of eroded sediment. *Geoinformatics* 4:137–144.

Wolfe, J. A. 1971. Tertiary climatic fluctuations amd methods of analysis of Tertiary floras. *Palaeogeography, Palaeoclimatology, Palaeoecology* 9:27–57.

Wood, J. M., Thomas, R. G. and Visser, J. 1988. Fluvial processes and vertebrate taphonomy: the Upper Cretaceous Judith River Formation, south-central Dinosaur Provincial Park, Alberta, Canada. *Palaeogeography, Palaeoclimatology, Palaeoecology* 66:127–143.

Wood, R. 1993. Nutrients, predation, and the history of reef-building. *Palaios* 8:526–543.

Woodland, B. G. and Stenstrom, R. C. 1979. The occurrence and origin of siderite concretions in the Francis Creek Shale (Pennsylvanian) of northeastern Illinois. In: M. H. Nitecki (ed.) *Mazon Creek Fossils*, pp. 69–104. New York: Academic Press.

Woodward, J. and Goodstein, D. 1996. Conduct, misconduct and the structure of science. *American Scientist* 84:479–490.

Worsley, T. R., Nance, R. D. and Moody, J. B. 1986. Tectonic cycles and the history of the Earth's biogeochemical and paleoceanographic record. *Paleoceanography* 1:233–263.

Wright, D. H., Currie, D. J. and Maurer, B. A. 1993. Energy supply and patterns of species richness on local and regional scales. In: R. E. Ricklefs and D. Schluter (eds.) *Species Diversity in Ecological Communities*, pp. 66–74. Chicago: University of Chicago Press.

Wright, J., Schrader, H. and Holser, W. T. 1987. Paleoredox variations in ancient oceans recorded by rare earth elements in fossil apatite. *Geochimica et Cosmochimica Acta* 51:631–644.

Wulff, J. I. 1990. Biostratinomic utility of *Archimedes* in environmental interpretation. *Palaios* 5:160–166.

Wyatt, A. R. 1984. Relationship between continental area and elevation. *Nature* 311:370–372.

Yanko, V., Ahmad, M. and Kaminski, M. 1998. Morphological deformities of benthic foraminiferal tests in response to pollution by heavy metals: implications for pollution monitoring. *Journal of Foraminiferal Research* 28:177–200.

Yin, H. and Tong, J. 1998. Multidisciplinary high-resolution correlation of the Permian–Triassic boundary. *Palaeogeography, Palaeoclimatology, Palaeoecology* 143:199–211.

Zangerl, R. and Richardson, E. S. 1963. The paleoecological history of two Pennsylvanian black shales. *Fieldiana* (Geology Memoirs, Chicago Museum of Natural History) 4.

References

Zedler, J. B. 1988. Restoring diversity in salt marshes? Can we do it? In: E. O. Wilson and F. M. Peter (eds.) *Biodiversity*, pp. 317–325. Washington, D.C.: National Academy Press.

Zeuthen, E. 1947. Body size and metabolic rate in the animal kingdom with special regard to the marine micro-fauna. *Comptes Rendus des Travaux du Laboratoire Carlsberg*, Séries Chemie 26:17–161.

Zhang, L., Liddell, W. D. and Martin, R. E. 1993. Hydraulic properties of foraminifera from shallow-water siliciclastic environments: a possible transport indicator in the stratigraphic record. *Geological Society of America Abstracts with Programs* 25:A428.

Zonneveld, K. A. F., Versteegh, G. J. M. and de Lange, G. J. 1997. Preservation of organic-walled dinoflagellate cysts in different oxygen regimes: a 10 000-year natural experiment. *Marine Micropaleontology* 29:393–405.

Index

Abel, O. 1
abrasion 27ff., 133
    fragmentation and 41, 372
    of propagules 231
abscission of leaves; *see* plants
acid rain 315
acritarchs; *see* plankton
actualism 6–10; *see also* aktuo-paläontologie; artifacts, taphonomic; methodology in historical sciences
advection; *see* bioturbation
aerobic decay 113
    see also bacterial oxidation; bioturbation; dissolution and precipitation of CaCO$_3$
aktuo-paläontologie 7
    Lagerstätten and 258
    modern analogs and 10, 101, 275, 338, 371
    research and 6ff., 51, 133, 389
    see also anthropogenic disturbance; artifacts, taphonomic; methodology in historical sciences
albedo 314
algae
    calcareous 38, 40, 45, 46, 47, 48, 117, 129, 130, 131
    filamentous green 48; *see also* bacteria; cyanobacteria
aliasing 197; *see also* artifacts, taphonomic; sampling

alkalinity 113, 116, 122, 148, 150, 213, 239, 319, 323, 339, 340; *see also* anoxia; dissolution and precipitation of CaCO$_3$
"all-or-none" preservation; *see* Cnidaria
alternative community states; *see* applications of taphonomy
Alumina 338
amino acids
    alloisoleucine/isoleucine values 207, 213
    Chione (Pelecypoda) and 198, 207, 208, 209, 213
    free 203
    Mercenaria (Pelecypoda) 209
    racemization 132, 203, 209, 389
Ammonoids; *see* Cephalopods
ammonia 113; *see also* dissolution and precipitation of CaCO$_3$, sequence of
amoebae, testate 157, 385
amphibians 92ff., 160
    frog and toad skin 92
anaerobic decay 113; *see also* dissolution and precipitation of CaCO$_3$
analogy, reasoning from; *see* methodology in historical sciences
annelids 55ff., 146
    Nephrops 58
    Nereis 58; *see also* polychaetes; worms and worm tubes

anoxia 54, 57, 58, 64, 66, 115, 314, 333, 348, 350, 350, 382
  anoxic (euxinic) basins 115, 215, 341, 347, 359, 364
  bacterial sulfate reduction 58, 113, 114, 115, 116, 135, 144, 148, 149, 150, 151, 213, 215, 216, 239, 240, 247, 249, 250, 251, 262, 252, 258 267, 314, 318, 323, 333, 339, 341, 343, 341
  cerium curve and 314, 343
  gyttja conditions 253
  hydrocarbon source rocks 314, 360
  Jet Rock 340
  Kellwasser horizons 351
  Mesozoic versus Paleozoic 359
  organic matter, metabolizability 137, 339, 341
  productivity and 217
  stagnation and 235, 237, 239, 240, 247, 249, 250, 251, 253, 255, 262
  superanoxia 358
  see also black shales; nutrients; oceanic anoxia events (OAES); oceanic circulation; sulfur isotopes
Antarctic bottom water (AABW); see oceanic circulation
antecedent topography; see assemblages, fossil
anthropogenic disturbance 25, 114, 161, 192, 233, 266, 369, 370, 371, 380, 382, 383, 385, 392, 393
  changes in land use 102
  deforestation 256, 370, 370, 382
  erosion 102, 366
  eutrophication 50, 71, 101, 327, 364, 366, 370, 382
  shallow-water dissolution and 114
  see also applications of taphonomy; artifacts, taphonomic apatite 145, 152, 240
  chlorapatite 146
  dahllite (carbonate hydroxyapatite) 146
  francolite (carbonate fluorapatite or carfap) 145, 146, 240

hydroxyapatite 146, 152, 153, 155
vivianite 240, 261
applications of taphonomy 1, 4, 369ff., 392
  alternative community states, recognition of 25, 369, 378, 380
  biological surveys 4, 14
  biotic turnover of mammals 372
  conservation and management 50, 71, 370, 380, 384
  co-ordinated stasis 191, 378, 379
  earthquake history 370, 388
  ecological and evolutionary studies 3, 4, 25, 26, 161, 190, 191, 235, 296, 337, 369, 370, 371, 373, 378, 379, 382, 391, 392, 394
  employment opportunities and 370, 392
  fidelity of assemblages 4, 191, 192
  insurance companies and storm frequency 182
  mass extinction 375, 376
  paleophysiology 385, 386
  population dynamics and extinction 379, 382–383
  rates and patterns of evolution 371–373
  sea-level change during holocene 383
  wetlands 371, 384
  see also information; taphonomy; time-averaging
aragonite 113, 116, 117, 156
  undersaturation 113
  supersaturation 113
  see also dissolution and precipitation of $CaCO_3$
archaeocyathids 17; see also sponges
architecture, effects on dissolution 118ff.
arthropods 55ff., 146, 251, 252
  Anomalocaris 237
  Balanus 131
  branchiopods 252
  Callianassa 57, 64, 168, 213
  copepods 57, 252
  crabs (unspecified) 59
  crickets 59
  crustaceans (unspecified) 255–257, 259, 262

Index

ecdysis 59
fiddler crabs 64
flies 266
harvester ants 162
hermit crabs 20, 53, 64
horseshoe crab 255
insects (unspecified) 244, 255, 256, 259, 262, 263, 265, 266,
lobsters 59
millipedes 59
mites 264
molt assemblages 62, 204, 249
pycnogonids 252
ostracodes 157, 252, 253, 259
Palaemon 58, 238
Pandalus 57
Panopeus 58
shrimp (unspecified) 57, 58, 59, 64, 152
spiders 264
springtails 264
stomatopods 152
termites 155
trilobites 59, 62, 108, 116, 146, 238, 247, 249, 251, 352, 361, 371
   enrolled in response to anoxia 249
   molt ensembles 204, 249
Upogebia 57, 64, 168
weevils 262
articulation, degree of 57, 205, 238, 239
artifacts, taphonomic 25
  aliasing 197
  analytical 164, 183, 184
  analytical time-averaging 108, 190, 229
  bias 3, 72, 236
  bias of deep-sea sections toward instantaneous change 304
  cell walls and nuclei 143
  completeness of fossil representation (Shotwell method) 75
  compressional 243
  cores, exposure to air 129
  cyanobacteria and 142, 143
  cyclicity induced by sampling regime 197
  dietary preferences of vertebrates 155
  dissolution and anthropogenic disturbance 114
  diversity 296
  flattening 243
  industrial well cuttings and mixing 108, 116, 394
  juveniles, loss of 191
  meltwater spikes 183, 184
  monographic effects 336
  morphometric trends 24
  negative isotope shifts 128
  Nyquist frequency 197
  offsets, stratigraphic 183
  outcrop area 4, 333
  overshoots 184
  pitfalls of laboratory protocols 41, 62, 116
  pooling (resampling) 193, 198
  printing through 102
  pseudoevolutionary 4, 372
  sampling 197, 198, 233, 271, 296, 393
  sea-surface temperatures (SSTs) and dissolution 125
  sea-level change and compaction 185
  sieving 92, 93, 108
  Signor–Lipps effect 288, 296ff., 373, 377
  species-specific morphologic traits 125
  stacking oxygen isotope records 197
  ultrasonication and amino acids 130
  upward scaling from experiments 392
  see also anthropogenic disturbance; counting procedures; megabiases; time-averaging
assemblages, fossil
  abundance and 191
  active continental margins and 4, 206
  allochthonous 13, 15, 19, 27, 69, 83, 89, 97, 99, 101, 204, 230, 292
  amalgamated 270
  antecedent topography and 17, 213
  autochthonous 13, 15, 19, 27, 41, 42, 68, 90, 94, 100, 101, 103, 192, 204, 231, 296
  background versus catastrophic processes and 85, 220

assemblages, fossil (*contd*)
  bedding surfaces and 190
  biocoenosis 7, 13, 24
  biofabric 16
  biomass and 191
  biostratigraphically condensed 201, 206
  bird-generated 19
  bone beds 21, 199, 202, 221, 223, 286, 290
  census 14, 15, 199, 200, 204, 235
  complex 18
  composite (within-habitat time-averaged) 200, 205, 211, 291
  condensed 286, 289, 290
  diagenetic 20
  environmentally condensed (telescoped) 201, 206
  event 163, 183, 195, 204, 223, 238, 269, 291, 296
  extrinsic biogenic 19
  fidelity of 4, 191, 192
  geometry of 17
  hermit crab generated 20
  heterozoan 322
  hiatal 206, 211, 290–292, 296
  high-energy (within-habitat time-averaged) 15
  high subsidence settings and 290, 296
  inductive models of formation 14ff., 25, 62
  internal structure 18
  intrinsic biogenic 19
  Johnson's models of formation 15ff., 20
  Kidwell's models of formation 22ff.
  lag 206, 211, 289, 292
  linear superposition, principle of, and 186, 207, 209
  low energy (within-habitat time-averaged) 15, 200, 201, 205
  low subsidence settings and 289
  mixed 20
  models and classifications of 14, ff., 390
  moderate subsidence settings and 291, 296
  molt 59, 62, 204, 249
  monotypic 16
  parautochthonous 13, 15, 19, 27, 37, 41, 90, 133, 192, 214, 296
  passive continental margins and 4, 206, 289
  photozoan 322
  polytypic 16
  productivity and 372
  sedimentological 20
  simple 18
  simple event 291
  skewed ages 188
  taphocoenoses 7
  taphofacies 24ff.
  taxonomic composition and diversity 192
  thanatocoenoses 7, 13, 41, 55, 56, 121
  within-habitat time-averaged 15, 200, 201, 205
attractors 380; *see also* fractals
authigenic mineralization 239, 240, 243, 246
autocompaction; *see* compaction
autotomy 66; *see also* echinoderms

Bacon, Francis 9
bacteria 256
  autolithification and 263
  chitinoclastic 57
  films 243, 253
  flowing mat structures 263
  lithified 254
  see also algae; cyanobacteria; Lagerstätten
banded iron formations (BIFs or jaspilites) 142, 338
  see also oxygenation
basidiomycetes 141
Behrensmeyer, Anna K.
  stages of weathering developed by 74, 82, 83, 87, 89, 160
Berger–Heath (box) model of bioturbation, 182, 183, 184, 194, 195, 196, 216, 218, 182ff., 194
  climatic fluctuations and 196
  stratigraphic distribution of tracer and 183
  see also biostratigraphy; bioturbation

Index

bias, taphonomic; *see* artifacts, taphonomic
bioclastic packing; *see* assemblages, biofabric
biocoenosis; *see* assemblages, fossil
bioerosion 22, 24, 41, 48–50, 55, 91, 133, 146, 322, 372
   endolithic boring 132
   fungal 48, 96
   microbial 96
biofabric; *see* assemblages, fossil
biofacies 24, 201, 379
biogeochemical cycles 4, 6, 101, 110, 143, 330, 348, 359, 369, 390, 391, 394, 394
   carbon 394
   carbon and silica cycles, evolution of 366ff.
   carbon and sulfur cycles 338, 341–345
   silica 143, 338, 368
bioherms 286, 314; biostromes; *see also* Cnidaria and associated biota; sequence stratigraphy
biostratigraphy
   biogeography and 307
   biozones 201, 301, 307
   confidence intervals (CIs) and 299ff., 304, 375, 377
      limitations of 299
   construction of the Channel Tunnel and 370
   correlation 268, 301, 370
   correlation of water wells and 370
   ecostratigraphic zonation 302, 305
   first and last appearance datums 180, 194, 195, 199, 238, 301ff., 326
   graphic corrrelation and 301ff., 371, 375–377
   leaking 198
   lowest and highest occurrences of fossil datums 238, 287, 301, 326
   piping 198
   zonation 301, 307, 370
   see also assemblages, fossil; bioturbation; graphic correlation
biostratinomic classifications 16ff.; *see also* assemblages, fossil

biostratinomy 2
biostromes 391
   algal and crinoidal 235, 316
   see also bioherms; Cnidaria and associated biota; echinoderms; sequence stratigraphy
biotic turnover
   mammals of Lake Turkana basin 372–373
   see also assemblages, fossil; mass extinction; mass mortality
bioturbation 13, 17, 37, 54, 58, 62, 114, 115, 132, 137, 144, 194, 241, 246, 251, 258, 268, 269, 299, 303, 332, 333, 339, 340, 347, 352, 353, 355, 366, 361, 384, 388, 392, 393
   advection 163, 165, 179, 181
   analytical artifacts 164, 244, 245
   analytical models 184
   apparent burial and 179
   Berger–Heath (box) model 182, 183, 184, 194, 195, 216, 218
   biodiffusion coefficient (D) 163, 164, 167, 169, 170, 171, 173, 179, 180
   biodiffusion coefficient (D), estimation of 177, 178, 179
   biogenic graded bedding and 19, 20, 181, 213
   bioirrigation 114, 340
   Boltzmann's constant and 165
   Brownian movement 165, 273
   bulldozing of suspension feeders 356
   Burgess Shale-type faunas 107, 236, 243, 247, 332, 333
   burial by 213
   by roots 100
   by skates and rays 19, 20
   caveats of diffusion models 179ff.
   composite-layer, mixing + sedimentation model 178
   concentration distribution, Gaussian 173
   confounding of true sedimentation rate by 177
   conveyor belt deposit feeders (CDFs) 19, 20, 133, 179, 181, 213

bioturbation (*contd*)
  core x-rays and 175
  deconvolution 183ff.
  deep infaunal 247
  dimensionless parameter (G) 173, 181, 182
  dinoturbation 162
  disarticulation of carcasses and 57
  drift 165
  earthworms 162
  error function 181
  escape burrows 200, 204, 238, 249
  Fick's equations, derivation of 169ff.
  first and last appearance datums (FADs and LADs) 13, 180, 301
  Guinasso–Schink (diffusion) model 162, 163, 172, 175, 182, 183
    general solution 172
    Newton–Raphson method of solution 175
    optimization 175
  harvester ants and 162
  historical layer 182
  horizontal mixing 179
  ichnofacies 163, 215
  impulse inputs 173, 177, 181, 183, 184
  inhibition by anoxia 247
  lacustrine environments 181, 184
  leaking 198
  local mixing 168
  low-pass filters and 161, 196
  lysozyme 165, 179
  marshes 182, 184, 384
  mesozoic 359
  microtektites 163, 180, 183, 199, 302
  mima mounds 168
  mixed layer thickness, estimation of 175ff.
  mucus linings of burrows 137
  non-local mixing 168, 179
  Péclet number 181
  phosphorus scavenging and 351
  piping 177, 198
  pocket gophers 168
  pollutants 183
  prairie dog "towns" 162
  random walk model 165ff., 534
  rodents 162
  roots, as impediment to 161, 162
  scavenging and 58
  sedimentary parameters, estimation of, and 175ff.
  sedimentation rate, estimation of, and 175ff.
  semi-quantitative estimation of 161
  shrimp and 179
  Signor–Lipps effect 288, 296ff., 373, 377
  size-selective feeding 17, 180, 199
  step length 165
  surface mixed layer 49, 57, 58, 163, 196, 269
  taphonomically active zone (TAZ) 49, 57, 58
  terrestrial environments 162
  trampling 75, 82, 83, 153, 162, 223
  uprooting of trees and 95, 162
  versus physical reworking 198
  see also artifacts, taphonomic; biostratigraphy; filters; numerical modeling; reworking; sampling; sediment-water interface (SWI)
birds 92ff.
  Archaeopteryx 254, 255
  as taphonomic agents 19
  "avetheropod" 254
  Diatryma 160
  feathers 254, 263
  nests 20
Bishop, William 306
bivalves (pelecypods) 1, 15, 16, 17, 21ff., 38, 40, 45, 47, 51ff., 90, 117, 118, 131, 131, 145, 203, 207, 209–211, 238, 253, 257, 299, 322, 356, 361
  articulated 204
  Chione 132, 198, 207ff.
  Exogyra 145
  Gryphaea 91
  inoceramids 199, 375
  juveniles 255, 238

Index

lithophagid 48
Mercenaria 209
modern fauna, bivalve-rich 361, 362
oysters 3, 17, 20, 47, 55, 56
pectinids 209
venerid 55
black shales 314, 333, 339, 346, 347, 351, 358; *see also* anoxia; sulfur isotopes
bogs and mires 232, 234, 267; *see also* paleosols; soils
Boltzmann's constant; *see* Bioturbation
bone 3, 10, 25, 133, 146, 153, 202, 221
   beds 21, 235, 304, 286, 290
   cancellous (spongy) 155
   compact bone 155
   cracking and splintering by salt crystallization 75
   juvenile 74, 90, 107
   mineralogy, microstructure, and age 153
   oxygen isotope ratios and 385
   release of phosphate 240
   weathering stages developed by Behrensmeyer 74, 82, 83, 87, 89, 160
   see also assemblages, fossil; amphibians; birds; mammals; reptiles; applications of taphonomy; sequence stratigraphy
boundary conditions 12
   see also bioturbation; hierarchy theory; numerical modeling
brachiopods 51ff., 117, 118, 144–146, 211, 328, 330, 332, 352, 356, 361
   cheniers and 207
   Glottidia palmeri 54, 211
   lingulide 332
   paleozoic brachiopod-rich fauna 352, 361, 362
   pedicle:brachial valve ratios 54–55
   Terebratalia transversa 54
Brett, Carl
   taphofacies models and 24ff.
   time-averaging criteria for ancient settings 203ff.

Brownian motion 273, 383; *see also* bioturbation; fractals
bryozoans 45, 47, 48, 117, 144, 204
bryozoan-rich carbonates 361
burial 165, 171
   catastrophic; *see* obrution; storms; tempestites; turbidites
   velocity 181
   see also bioturbation; sedimentation (accumulation) rates
bypassing, of surface mixed layer (TAZ) 214; *see also* foraminifera, differential preservation of infaunal species

calcareous nanoplankton; *see* oozes; plankton
calcite and related minerals 113, 117, 152, 157
   ankerite 150
   ferroan 116, 258
   high-mg calcite (HMC) 116
   low-Mg 116, (LMC)
   siderite 135, 150, 151, 153, 157, 239, 257, 258, 261, 263
   supersaturation 147
   undersaturation 146
calcite compensation depth (CCD) 121, 123, 125, 127, 321, 332, 354
   calcite-pull model 321
   calcite-push model 321
   corrosiveness of bottom waters and 320
   fluctuations in 320, 321
   red clay and 121
   see also dissolution and precipitation of $CaCO_3$; oozes; Plankton
calcitization 117; *see also* replacement
calcium toxicity 323, 328
cannibalization of rocks; *see* megabiases
carbon and sulfur reservoirs
   redox conditions and 366
   ratios in sediments 341
   shifts between 341
   see also anoxia; biogeochemical cycles; carbon isotopes; nutrients; sulfur isotopes

carbon dioxide, atmospheric 110, 114, 310, 313, 314, 315, 320, 322, 326; *see also* cements; Cnidaria; weathering
carbon isotopes 210, 216, 343, 349, 350, 355, 357, 358, 359, 360, 365, 366
   end-Permian record 355, 377
   terrestrial carbon reservoirs and 351
   see also oceanic anoxic events (OAEs)
carbon-14, dating; *see* radiocarbon dates
carbonate equilibria 112, 116; *see also* dissolution and precipitation of $CaCO_3$
carbonate regimes, versus terrigenous 27ff.
case-hardening 55
catastrophism; *see* methodology in historical sciences; *see also* actualism; aktuo-paläontologie
cathodoluminescence; *see* microstructure
caves 221, 352, 386
cellular morphology, retention of 240, 254, 258, 264; *see also* artifacts, taphonomic
cements, abiotic carbonate 317ff., 318, 338, 357
   atmospheric pCO_2, and primary mineralogy 318
   coral reef hypothesis 320
   crystal habits 318, 328
   long-term regulation 310, 313
   Mg/Ca ratios and primary mineralogy 317
   negative feedback and 315
   ooids and oolitic ironstones and 314, 317
   short-term regulation 309, 320ff.
cephalopods 51
   ammonoids 117, 201, 253, 255, 340, 375, 377
   aptychi 51, 117, 255
   belemnites 253, 255
   coleoids 252, 253
   half-ammonites 117, 340
   ink 256
   limestones 202
   nautiloids 330, 352

   octopods 252, 330
   pearly *Nautilus* 51, 330
   postmortem drift and 330
   squid 240
cerium curve; *see* anoxia; sulfur isotopes
chalcedony; *see* silica
Chaney, R.W. 3
charcoalification ("fusainization"), forest fires and 267
chemocline; *see* nutrients; oceanic circulcation
cheniers; *see* brachiopods
cherts
   biogenic 364, 367
   Paleozoic 143
   see also oozes; plankton
chitin 56, 252, 262
chitinophosphate 146
clay 85, 133, 333
   adsorption and preservation 249
   balls 99
   cohesiveness 33
   illite 153, 250, 333, 334, 353
   kaolinite 153, 333, 353
   montmorillonite-rich 85
   smectite 160, 250, 333
Cnidaria and associated biota 45ff., 244
   alyconarian spicules 38
   Aurelia 246
   coral reef hypothesis 320
   corals (unspecified) 193, 203, 204, 257
   jellyfish (unspecified) 255
   Medusae (unspecified) 241, 256
   rugose corals 116
   scleractinia 38, 40, 47, 48, 117, 130, 131, 361, 379, 380, 384
      Acropora 45, 47, 48, 49, 131, 363, 380
      Agaricia 50
      "all-or-none" preservation 50
      Fungia 49, 131
      Montastrea 50, 381, 382
      nutrients and 381
      Pocillopora 363
      Porites 50, 363
      sedimentation and 381

Index

Septastrea 48
sewage 370, 381, 382
Siderastrea 50
coal 3, 93, 100, 107, 142, 258–259, 262
    cannel 267
    swamps 102, 256, 258, 314, 316, 347, 353, 356
    see also Lagerstätten; lignin; peat; plants
coccolithophorids; *see* oozes; plankton
collagen 146, 155
colloids; *see* sediment transport
compaction, sediment 125, 127, 139, 153, 183, 251, 258
    autocompaction 100, 184, 185
    see also artifacts, taphonomic
completeness of fossil representation (Shotwell method) 75
compression; *see* artifacts, taphonomic
concretions 147ff., 239, 243, 247, 253, 257, 286, 340
    cone-in-cone structures 147
    criteria for relative age of formation 147
    freeze–thaw method for cracking 256
    in marginal marine sediments 150
    in terrestrial environments 152
    siderite 85, 256, 257, 258, 263
    see also calcite; pyrite
cone-in-cone structures; *see* concretions
confidence intervals (CIs); *see* biostratigraphy;
    see also graphic correlation; mass extinction
conodonts; *see* plankton
conservation and management; *see* applications of taphonomy constraint; *see* hierarchy theory
contingency; *see* methodology in historical sciences
conveyor belt deposit feeders (CDFs); *see* bioturbation
coordinated stasis; *see* applications of taphonomy
coprolites 74, 91, 157, 160, 255; *see also* dietary preferences
coquinas; *see* sequence stratigraphy

coral reef hypothesis; *see* carbon dioxide; cements; Cnidaria and associated biota
counting procedures 104ff.
    "completeness" of fossil representation (Shotwell method) 75
    biomass and 109
    minimum number of elements (MNE) 106
    minimum number of individuals (MNI) 105
    number of identified specimens per taxon (NISP) 105
    sieve method 108
    small sample sizes 108, 109
    sudden death assemblages and 105
    see also artifacts, taphonomic
cratonic sequences; *see* sea-level change; sequence stratigraphy
ctenophores; *see* Lagerstätten
cutan 157
cuticle 56, 252, 262
cutin 157
Cuvier, Georges von 1, 5
cyanobacteria 48, 142, 143, 237, 240, 246, 255
    films 240
    mats 246, 255
    sealing 237
    see also algae; bacteria; Lagerstätten
cycles of sedimentation and climate 309ff.
cyclic trends in preservation; *see* megabiases

d'Orbigny, Alcide 1
da Vinci, Leonardo 1, 7
Darwin, Charles 6, 162, 268
deconvolution
    pollutants and 183
    see also bioturbation; numerical modeling
DEEP site (Long Island Sound) 114
deforestation; *see* anthropogenic disturbance
"defossilization" 133
degassing of the earth 334
deltas 3, 101, 258, 268, 271, 279, 284, 306
    lobe-switching and 279, 306

denitrification 113, 348; *see also* biogeochemical cycles
Descartes, René 10
dessication 246, 264
   cracks 85
deterministic behavior; *see* fractals; methodology in historical sciences
devitrification 142
dewatering 243, 258; *see also* compaction
diagenesis 110ff., 268
   bone and teeth 146, 386
   history 139, 147
   non-steady state 208
   potential 125; *see also* assemblages, fossil; artifacts, taphonomic; authigenic mineralization; cements; dissolution and precipitation of $CaCO_3$
diastems 276; *see also* unconformities
diatoms; *see* oozes; plankton
dietary preferences of vertebrates 74, 155; *see also* applications of taphonomy; artifacts, taphonomic; coprolites; diagenesis
diffusion models of bioturbation; *see* bioturbation
dimensionless reworking rate (DWR) 269ff. *see also* reworking; stratigraphic acuity, completeness; stratigraphic maturation
dinoflagellates; *see* plankton
dinosaurs
   body temperature of based on bone microstructure 385
   experiments on preservation of teeth 373
   preservation in Hell Creek secton 373–374
   skin 263
   see also reptiles
disarticulation 94, 205, 236, 239
   prevention 54, 57–58; *see also* articulation; necrolysis; time-averaging
dissolution and precipitation of $CaCO_3$ 110ff., 268, 339, 367, 385
   bioirrigation and 114, 340
   carbonate equilibria 112, 116
   concurrent silicification and 145
   deep atlantic 219
   deep pacific 219
   dissolution-resistance of benthic foraminifera versus planktonic 219
   homogeneous 218
   incongruent 116
   interface 218
   latitude and 116
   LeChatelier's Principle and 111
   local dissolution of shells 138
   planktonic/benthic ratios 215
   pressure solution 20, 145, 188
   sequential 218, 219
   sequence of 110ff.
   size and 118ff.
   see also calcite compensation depth (CCD); foraminifera; lysocline; microstructure; mineralogy
diversity; *see* artifacts, taphonomic; assemblages, fossil dolomite 117, 144, 150, 318
dolomitization 117, 338
   microdolomite 117
downhole contamination (caving); *see* artifacts, taphonomic; petroleum wells

earthquakes and seismic hazards; *see* applications of taphonomy
ecdysis 59
echinoderms 47, 65, 117, 144, 146, 206, 251, 255
   autotomy and 66
   Acanthaster (crown-of-thorns starfish) 71
   blastoids 108, 358
   Clypeaster 130, 131
   crinoids 65, 66, 108, 204, 235, 253, 255, 331
   crinoidal limestones 235, 316
   Diadema 71
   echinoids (unspecified) 45, 48, 65, 91, 117, 130, 131, 145, 238, 239, 253, 322, 330, 332, 381

Index

edrioasteroids 204
holothurians 256, 257
mutable collagenous tissues 65
ophiuroids 66, 204, 238, 251, 253
Pisaster 46
spines 38
sterom 65
Strongylocentrotus 46, 66
see also applications of taphonomy; artifacts, taphonomic; biostromes
ecological and evolutionary studies; see applications of taphonomy; artifacts, taphonomic; information
ecostratigraphic zonation; see biostratigraphy; graphic correlation
edge effects 378
Efremov, J.A. 1, 3, 4, 11
elephants; see applications of taphonomy; mammals
elvis taxa; see mass extinction
embryos 240, 253, 353; see also Lagerstätten
encrustation 48, 50, 51, 54, 148, 153, 238, 331, 372
endolithic boring; see bioerosion; encrustation; fouling
energy and evolution 348ff.
    Belousov–Zhabotinsky reaction 362
    bifurcations 363
    diversification of anti-fouling predators and grazers 361
    durophagous (shell-crushing) predation 351, 352, 353, 355
    entropy 362
    immanent properties 363
    lengthening of food chains 355
    marine carbon-to-phosphorus (MCP) burial ratios 351, 352, 354, 359, 360, 362, 364, 365
    Meso-Cenozoic fauna, mollusc-rich 361, 362
    Mesozoic marine revolution 497
    metabolic rates and 355, 361–363, 366
    modern fauna, bivalve-rich 361, 362
    oligotrophic refugia 352, 353

    Paleozoic fauna, brachiopod-rich 352, 361
    predation and 331, 352, 353, 362ff.
    second law of thermodynamics 362
    self-organization 362
    symbiosis 354, 360, 365
    thickening shell beds 361
    tiering 352, 355
    trace fossil diversity 336
    trends on land 364; see also nutrients; productivity
    trilobite-rich fauna and 352, 361, 362
epifaunal versus infaunal habit 54, 58
equifinality 10, 389; see also taphonomic grades
ergodic systems; see methodology in historical sciences
erosion 269, 275, 360, 366; see also diastems; stratigraphic maturation; unconformities
estuaries 232, 393
eukaryotes, advent of 364
eurytopic taxa, biotic turnover and 382; see also particle model of fossil populations
eutrophic conditions; see nutrients
euxinic basins; see anoxia
event horizons; see assemblages, fossil
experiments, taphonomic 37, 44, 51, 55, 57, 66, 85, 90, 95, 142, 152 157, 388, 390
    abrasion with quartz sand 41
    acid baths 47, 130
    airtight jar 57, 58
    aluminum glitter 37
    cage 92
    carcasses 57
    color-coded seeds 97, 102
    dental acrylic models 59
    dissolution 129
    dolomitization 117
    eggshell 90
    field 7, 10, 57, 71, 80, 81, 102
    fish decomposition 92
    flat paper shapes 96

experiments, taphonomic (contd)
  flume studies 27, 36, 42, 44, 51–53, 59, 76, 77, 80, 104
  glass bead tracers 37
  implantation of substrates 130
  jar experiments 57, 58
  laboratory 10, 116, 240
  Lagerstätten 236, 240, 244, 266
  pollen 3, 102
  settling experiments 36, 43
  shaker tables 41
  spray-painted (tracer) leaves 96, 102
  taphonomic settings and 30
  teeth 373
  tumblers 41, 45, 47, 55, 58, 66, 72, 83, 96, 97, 238, 239
  upward scaling from experiments 392
exposure effects 15, 23; *see also* assemblages, fossil; Johnson models of assemblage formation; Kidwell models of assemblage formation; residence time at sediment–water interface (SWI); weathering
external molds 117, 146
extinction; *see* mass extinction

facies 24ff., 199, 201, 379
feathers; *see* birds
fecal pellets 42, 74, 101, 123, 127, 368
  oxygenation of atmosphere and 323
  see also biogeochemical cycles; plankton
fermentation 91, 115; *see also* dissolution and precipitation of $CaCO_3$
Fick's equations, derivation of 169ff.
  steady state nature of 163, 170, 171
  see also bioturbation
fidelity, of fossil assemblages 4, 191ff.; *see also* applications of taphonomy; assemblages, fossil
filters 310
  attenuation of signals 371
  cutoff frequency 263
  enhancement of signals 369
  impulse response function (h) 187
  low-pass 161, 196
  passband region 196
  signal-to-noise ratio (s/n) 197
  stopband region 197
  taphonomic 25
  see also artifacts, taphonomic; bioturbation; time-averaging
finite-consolidation strain theory 185; *see also* compaction
first and last appearance datums (FADs and LADs); *see* Berger–Heath (box) model; biostratigraphy; bioturbation
fish 72, 92ff., 160, 240, 252, 256, 259, 262, 332
  agnatha 251
  cichlids 373
  coelacanth 352
  placoderms 251
  sharks 1, 253, 255
  skates and rays 19, 20
  teleosts 253
floating vegetation (flotant) 243
floods 80, 85, 89, 93, 157, 159; *see also* storms; obrution; tempestites; turbidites
flotation, of carcasses 59, 77, 85, 87, 91, 92, 97
flowing mat structures; *see* bacteria; Lagerstätten
fluid movement; *see* sediment transport
fluxes 169, 170, 171; *see also* bioturbation; Fick's equations
FOAM (Friends of Anoxic Muds) site (Long Island Sound) 114, 115
focusing, sediment 234
foraminifera 35, 38, 41, 43, 103, 117–119, 123, 125, 127–129, 132, 182, 184, 194, 199, 206, 210, 212–217, 253, 299, 301, 312, 321, 340, 348, 354, 360, 366, 369, 370, 372, 376–380, 382, 384, 385, 393
  agglutinated 127, 129, 214, 217, 323, 325, 326
  as sediment tracers 35, 182

Index

delay in calcification 327
correlation of microstructure with abiotic cements 326, 328
differential preservation of infaunal species 214, ff.
dissolution of benthics versus planktonics 119, 123, 127, 128, 219
dissolution of marsh species 385
dissolution resistance and mass extinction 377
flume studies 36, 104
foraminiferal lysocline 123
fusulinids 118, 144
Komokiacea 127
marsh 37, 129, 214, 385
microhabitat preferences 214, 216
nummulites 215
origins of test 327, 329
oxidation of organic cements 129, 385
patchy reproduction 37
planktonic 42, 119
planktonic/benthic ratios 215
production of $CaCO_3$, by 123
reef 36, 38, 39, 40, 41, 108, 123, 127, 128, 129, 130, 131, 212, 214, 380
sequence of dissolution of benthics 127ff.
settling velocities 33, 36, 38, 43, 45
terrigenous regimes and 41
versus calcification in metazoa 327, 329
fossil recovery potential function 300
fouling 50, 361; *see also* bioerosion; energy and evolution; megabiases
fractals 271ff., 382, 387
  attractors 380
  Cantor bar model of hiatus distribution 276ff.
  dimensions and 273
  hiatus distribution and 276
  Koch's snowflake 273, 275
  power laws 157, 201, 319, 388
  rescaled-range analysis 382
  self-similarity and 274, 275, 306
  see also methodology in historical sciences; biostratigraphy; Signor–Lipps effect

free energy yield 112
freeze–thaw cycles 162; *see also* reworking; Lagerstätten
freshwater ecosystems; *see* lacustrine environments; bioturbation; Lagerstätten

gambler's ruin; *see* reworking; stratigraphic maturation
gastropods 20, 38, 39, 45, 47, 51, 53, 55, 64, 90, 130, 131, 156, 158, 159, 162, 202, 253
  land 156, 159, 207, 297, 371
  limacid slugs 156
  turritellids 20, 53, 133
glaciation 42, 348, 350, 351, 355, 360, 361
  southern hemisphere 314, 315, 326, 354
  see also anoxia; icehouse mode; oceanic circulation; upwelling
gleying; *see* soils
gradients 330, 362, 390
  bathymetric 16, 66, 109
  cyclic and secular trends and 390, 391
  decay and 91
  latitudinal 66
  oligotrophic-to-eutrophic 349
  (paleo) environmental 14, 20, 24, 390
  taphofacies and 24ff., 41, 62, 83, 89, 90, 257, 390–391
  taphonomic 15, 134, 215, 256, 262, 288
  temperature 66
graphic correlation 199, 288, 301, 371ff., 375–377, 393
  composite section 299, 302, 304
  composite standard units (CSUs) 304
  drilling wells and 305
  line of correlation (LOC) 302
  range extension 304
  Shaw plots 302, 305
  standard reference section (SRS) 303
  see also biostratigraphy; bioturbation; stratigraphic acuity; temporal acuity
graptolites; *see* plankton

grasslands, expansion of 325, 361; *see also* oozes; plankton
graywackes 338
greenhouse conditions 313, 314, 318, 319, 321, 325, 326, 333, 334
Gressly, Armand 1
Guinasso-Schink bioturbation model; *see* bioturbation
gypsum 153, 160
gyttja conditions; *see* anoxia; Lagerstätten

halocline; *see* Lagerstätten; *see also* oceanic circulation hardgrounds 238
hematite and related iron oxides 114, 135, 152, 153
   ferrihydrite 135
   goethite 153
   lepidocrocite 135
   limonite 153, 157, 160
   oxyhydroxides 135
herding behavior; *see* mass mortality
hermeneutics; *see* methodology in historical sciences
Herschel, John 6, 388
hiatus 276ff., 300, 306; *see also* diastems; fractals; graphic correlation; sequence stratigraphy; unconformities
hierarchy theory
   boundary conditions 12
   constraint 12
   cyclic and secular change and 391
   governors 310
   holons 11, 12, 309, 310
   initiating conditions 12
   perception and 309, 310
   return times and 195, 310, 317
   scale 10, 168, 188, 233, 271, 309, 389, 390, 391, 392
   taphonomic processes and 306
   time-stratigraphic subdivisions and 307
   upward scaling 5, 391
   see also methodology in historical sciences
historical conditioning of fossils 55
Hjulström diagram; *see* sediment transport

hominids 372, 386; *see also* assemblages, fossil
Hooke, Robert 1
Hume's (David) paradox 8; *see also* methodology in historical sciences
Hutton, James 5, 9
hydraulic behavior; *see* sediment transport
hydrocarbon source rocks 314, 360; *see also* anoxia
hydrothermal vents and sulfide mineralization 141
hypersalinity 143, 235, 241, 254, 255
hypothetico-deductive method; *see* methodology in historical sciences

icehouse conditions 313, 316, 318, 319, 326, 357
ichnofabric 286
ichnofacies 163, 215
ichnofossils (Lebensspuren) 7, 241, 246
   Chondrites 253
   death march 204, 255
   escape burrows 200, 204, 238, 249
   nonmarine record 347
   Zoophycos 62
   see also bioturbation
improbability of preservation 235
infauna
   "by-passing" of surface mixed layer 214
   differential preservation 214ff.
information
   gain 14, 24, 391
   loss 3, 13–14, 24
internal molds (steinkerns) 117, 137, 201, 290
iridium; *see* mass extinction
iron oxides; *see* hematite
iron reduction 113; *see also* dissolution and precipitation of $CaCO_3$
iron sulfides; *see* pyrite

Jeram model 205; *see also* infauna
Jet Rock 340; *see also* anoxia
Johnson, Ralph Gordon 3

Index

models of assemblage formation 15ff.
k-strategist 94
karst 155, 262
    permineralization and 264; *see also* Lagerstätten
Kellwasser horizons; *see* anoxia; mass extinction
Kepler, Johann 6
Kidwell, Susan
    models of assemblage formation 22ff.

lacustrine environments 95, 97, 100, 101, 103, 184, 231, 234, 259, 261, 385
    analogs for ancient lakes 101, 259
    nutrients in 261
    pollutants and 183
    see also bioturbation; Lagerstätten
Lagerstätten 1, 14, 55, 59, 65, 68, 91, 200, 235, 236, 268, 332, 388, 390
    achanarras fish bed 263
    agnatha 251
    alum shale formation 252
    amber 90, 255, 264, 265, 332
        Baltic 265, 266
        "blau Erde" 266
        Dominican Republic 266
    Annularia (sphenopsid) 257
    Anomalocaris canadensis 237
    anoxia and 247
    apatite 240
    Araucariaceae 265, 332
    Archaeopteryx 254, 255
    arthropod 251
    asphyxiation 238
    Aurelia 246
    authigenic mineralization 239, 240, 243, 246
    "avetheropod" 254
    bark 229, 257
    bats 263
    Beecher's trilobite bed 247, 251
    biases, secular 247
    branchiopods 251, 252
    Burgess Shale 107, 236, 243, 247, 332, 333

    cell structure 240, 258, 264
    cellulose 240
    cephalopods 252
    cerin limestones 243
    chitin 252
    Clarkia Formation 262
    clays, preservation by 249
    clothing 266
    Cnidaria 244, 246
    coal balls 258
    coccolithophorids 253
    coleoid cephalopods 252, 253
    collapse calderas 255
    collapse of soft-bodied organisms 243
    Compsognathus 254
    conifers 253
    conodonts 236, 240
    Copaifera 265
    Copal 265
    crinoids 255
    criteria for evaluation 245
    crocodiles 253
    crustaceans 252, 262
    ctenophores 251
    cuticle 252, 262
    cyanobacterial mats and preservation 237, 240, 246, 255
    cycads 253
    dessication 246, 264
    diatom-like structures from late Proterozoic 368
    dinosaur skin 263
    Diptera 266
    DNA 262, 264
    Ediacaran biota 244
    embryos 240, 253, 353
    Emu Bay Shale 247, 250
    experimental simulation 237, 240, 244, 266
    feathers 254, 263
    fish 329, 252
    flattening (printing through) 243
    flies 266
    flowing mat structures 263
    Frankfort Shale 251

Lagerstätten (*contd*)
  freezing 264
  fruits 231, 237, 257, 263
  fungal spores and hyphae 264
  genesis of 237ff.
  Ginkgo 253
  glycocalyx 254
  Green River Formation 261
  Grès-à-Voltzia shales 241, 256
  Grube Messel 262
  gyttja conditions 253
  hair 266
  halocline 255
  holothurians 351, 354
  horses pregnant with preserved embryos 263
  horseshoe crab 255
  horsetails 253
  hot springs 263
  human remains 266
  Hunrsrückschiefer (Hunsrück Slate) 247, 251
  hymenoptera 266
  hypersalinity 241, 254, 255
  ichthyosaurs 253
  ink (cephalopod) 256
  insects 244, 255, 256, 259, 262, 263, 265, 266
  Karroo 261
  karst 262, 264
  konservat (conservation) 58, 147, 235, 244, 286
  konzentrat (concentration) 235, 236
  Kupferschiefer 3
  La Voulte-sur-Rhône 252
  LaBrea tarpits 264
  lacustrine 259, 261
  Magnolia 262
  Mazon Creek 147, 241, 243, 256, 258, 333
  Medusae 241, 256
  Messel Shale 241
  microbial mats 333
  microfossil (coccolithophorid) 321, 366, 367
  mites 264
  Montceau-les-Mines 256
  moors 267
  mummification 264
  Neuropteris 257
  nomothetic classifications 5, 12, 244
  octopods 252
  Old Red Sandstone 263
  ophiuroids 238, 251, 253
  orsten 252
  Oxford Clay 241, 340
  peat 257, 258, 264
  peat bogs 264, 266, 267
  permafrost 264
  phosphate and 240, 252
  Pinus succinifera 265
  pitfalls of 235
  placoderms 251
  Plattenkalk 239, 253
  plesiosaurs 253
  Pollack Farm site 83
  Posidonienschiefer (Holzmaden) 243, 250, 253, 339, 340
  prediction of occurrence 237, 244, 286
  Pseudolarix (pine) 265
  pterosaurs 252, 253, 254
  pycnogonids 252
  Rhynie Chert 263
  Santana Formation 252, 261
  Saurischia 253
  sequence stratigraphy and 286
  sharks 253, 55
  skin 259, 266
  slime molds 266
  Solnhofen Limestone 1, 241, 253, 339
  Spence Shale 247
  sphenopids 257
  spiders 264
  Sporangia 264
  spores 263, 257
  squid 240
  stickleback fish 236
  stomata 264
  tannic and fulvic acids 266

tar 117, 162, 264
teleosts 253
traps 107, 236, 264
Vendobionta (Vendozoa) 244
vertebrates 255
weevils 262
Wheeler Formation 247
Willershausen Lagerstätte 262
window on ecological and evolutionary dynamics 235
large number systems; *see* methodology in historical sciences
latitude, and dissolution 116; *see also* lysocline
law
   of numbers 9, 75, 226, 374
   of the lower jaw 72
   of the ribs 73
   of carbonate deposition 314, 321
   of constant proportions (Dittmar's) 111
Lawrence, D.R. 3, 56
laws; *see* methodology in historical sciences
Lazarus taxa; *see* mass extinction
LeChatelier's Principle 111
lignin 96, 141, 229, 230, 240, 347, 353, 365; *see also* coal; organic matter; plants
line of correlation (LOC); *see* graphic correlation
linear superposition, principle of 186, 207, 209, 390; *see also* time-averaging
Little Ice Age 384; *see also* sea-level change
loess 271
Lyell, Charles 5, 6, 268, 388
lysocline 121, 123, 125, 127, 219
   aragonite (pteropod) 123
   coccolith (low-Mg) 123
   deep-sea 320
   latitudinal 322
   shallowing near continental margins 123, 215

macroalgae 381; *see also* mass mortality
magnetic reversals 301, 302

estimating completeness of terrestrial sections with 278, 282, 283, 302
see also graphic correlation; stratigraphic completeness
mammals 152, 155, 160, 162, 198, 296
   assemblages 104, 371, 372, 374, 381, 386
   availability of 72, 85
   bats 263
   Bison 221, 386
   cow 226, 227
   disarticulation 72
   elephants 72, 235, 371
   hair 266
   horses, pregnant 263
   humans 266
   prairie dogs 162
   sheep 386
   skin 92
   small mammals, number of 75
   wildebeest 226, 227
manganese oxides 114
manganese reduction 113; *see also* dissolution and precipitation of $CaCO_3$
marine carbon-to-phosphorus (MCP) burial ratios; *see* energy and evolution; nutrients; productivity
marshes 37, 64, 100, 129, 135, 161, 177, 182, 184, 258, 339, 384, 393
   levee effects 177
   low rates of bioturbation in 384; *see also* foraminifera
mass extinction 10, 45, 288, 296, 308, 350, 355, 388
   apparent survivorship following 376, 377
   asteroid impacts as cause of 373
   bias of deep-sea sections and 304
   carbon isotope excursions, end-Permian 377
   Elvis taxa 296
   end-Cretaceous 296, 360, 376
   end-Permian 313, 355, 356, 358, 377
   eustatic sea-level model and 304
   iridium 356, 374–375
   Kellwasser horizons 351

mass extinction (*contd*)
  K–T boundary 304, 374–376, 393
  Lazarus taxa 296, 377
  nutrient collapse and 347, 348, 355,
  palynological data 373
  patterns of 296, 297
  periodicity 316
  rebound effect 317
  recovery from 317
  Siberian Traps and 356
  Signor–Lipps effect 288, 296ff., 373, 377
  stable isotope analyses and 362
  superanoxia 358
  superoligotrophic conditions 348
  Ur/Pb zircon dates 377
mass mortality 2, 20, 71, 72, 92, 200, 204, 223, 227, 235, 252, 259, 263, 306, 347
  african elephant and 235
  drought and 235
  herding behavior and 347
  macroalgae and 381
  sea urchins and 381
  stickleback fish 248
maximum likelihood estimation 300
megabiases 4, 10, 309ff., 330ff.
  among-taxon 331
  bioturbation 332ff.
  cannibalization of rocks and 238, 334, 338
  constant mass model 334ff.
  cyclic preservation and biomineralization 4, 309ff.
  diversification of anti-fouling predators and grazers 361
  diversification of epifaunal suspension feeders 351
  dolomitization 338
  early cementation and preservation 338
  global 332, 334
  linear accumulation model and 334ff.
  lithology 338
  mass/age distribution of rocks 334ff.
  Mg/Ca ratios 338
  outcrop area 4, 334ff.
  predation 331ff., 362ff.
  pull of the Recent 334, 336, 337
  pyritization 341ff.
  scavenging 331, 332
  secular trends in preservation 330ff.
  secular increase in biomass and diversity 4, 362
  static and dynamic distortions 331
  transfer of $CaCO_3$, to deep ocean 338
  within-taxon megabias 330
  see also artifacts, taphonomic
megamonsoons 316; *see also* storms; tempestites
methodological uniformitarianism; methodology in historical sciences
mesotrophic conditions; *see* nutrients
metazoa, rise of 322ff., 364
  bauplans 337; *see also* foraminifera
meteoric waters 386
methanogenesis 91, 115, 139, 148, 149, 258; *see also* dissolution and precipitation of $CaCO_3$
methodology in historical sciences 5
  actualism 6–10
  analogy and, 6–7, 17
  catastrophism 5, 7
  causal ordering and 5
  contingency 12
  deduction 7ff.
  deterministic behavior 388
  empirical relationships 6, 388
  ergodic systems 388
  falsifiability of hypotheses 8, 9
  formal analogies 10
  fractals 387
  hermeneutics 389
  Hume's paradox 8
  hypothetico-deductive method 7, 8
  idiographic approach 12
  induction 7, 9
  inductive taphonomic models and 23ff.
  large number systems 388
  laws 11, 12, 387
  laws, rules and hierarchy and 11, 387

Index

methodological uniformitarianism 5–7, 204
no analog conditions and assemblages 190, 204, 391, 392
nomothetic classifications 5, 12, 244, 388, 391
non-steady-state conditions 208
Occam's Razor (principle of parsimony) 6, 25; see also quotation on p. vi
power laws 157, 201, 388
principles 11, 387
rationalists 9, 10
reductionism 12, 387, 388
relational analogies 10
return time 195, 270, 310, 317, 388
rules 11, 389
scale 10, 188, 233, 390, 391, 392
small-number systems 387
steady state view of the earth and 5, 163, 170, 218, 219, 388
substantive uniformitarianism 6, 204
synthetic a priori statement 8
three ball problem 387
unique objects 5
verae causae 6, 388
see also actualism; aktuo-paläontologie; hierarchy theory
Mg/Ca ratios 116, 318, 338; see also dissolution and precipitation of $CaCO_3$; dolomite; megabiases; weathering
micritized envelopes 132, 253
microarchitecture, skeletal 47; see also Sorby principle
microstructure 110, 137, 145, 146
  bone 385
  cathodoluminescence 117
  effects on dissolution 118ff.
  silica replacement and 386
  see also Sorby principle
microtektites 163, 180, 183, 199, 302; see also bioturbation; graphic correlation
microvertebrate accumulations 157, 223
Milankovich frequencies 262, 313, 320; see also varves

mineralization
  biocontrolled 329
  bioinduced 329
mineralogy, shell 46, 110, 144, 151
  effects on dissolution 116ff.
models and classifications of fossil assemblages 14ff.
molecular evolution 247, 266
mollusca
  chitons 49
  Meso-Cenozoic fauna, mollusc-rich 361, 362
  miscellaneous 114, 146, 251, 259, 372
  thylacocephalans 252
  see also bivalves; cephalopods; gastropods
Monte Carlo simulations 283, 296
mummification; see Lagerstätten

necrolysis 2, 27ff., 238
  temperature and 54
nest sites 30, 90, 222
Newton, Isaac 6, 387
nitrate reduction 113; see also dissolution and precipitation of $CaCO_3$
no analog conditions; see methodology in historical sciences
nomothetic classifications; see Lagerstätten; methodology in historical sciences
North Atlantic deep water (NADW); see oceanic circulation
numerical modeling 100, 184, 210, 392
  autocompaction 184
  averaged inputs 184
  continuous inputs 184
  eutrophication 366
  finite-consolidation strain theory 185
  of increased rates of anthropogenic erosion 366
  sensitivity analyses 392
  see also bioturbation; compaction
nutrients 50, 71, 91, 316, 381ff.
  availability 332, 348, 349
  Ba, Cd, and Sr as indicators 128, 351
  charophytes and 356

nutrients (contd)
  chemocline 347
  collapse 358
  cyanobacteria (*Trichodesmium*) and 348
  cycling on shelves 359
  denitrification 113, 348
  embryonic development and 353
  eutrophic levels 101, 261, 351, 360
  "Euzoic" era 362
  hypothesis 348, 364
  inventory 348
  life history strategies and 50, 353, 363
  marine carbon-to-phosphorus (MCP) burial ratios 351, 352, 354, 359, 360, 362, 364, 365
  mesotrophic levels 354, 360, 361, 367
  "Mesozoic" era 362
  oligotrophic levels 50, 101, 348, 354
  oligotrophic oceanic gyres 348, 349
  oligotrophic refugia 352, 353
  oligozoic "era" 362
  phosphate 347, 348
  phosphorites 359, 365
  phosphorus 347, 351, 352
  recycling 94, 101, 348, 351, 357, 360, 368
  reversion to superoligotrophy 355
  salinity stratification and 355
  scavenging 348ff.
  secular increase in biomass and diversity and 362ff.
  submesotrophic levels 354, 355
  superoligotrophic conditions 348; *see also* energy and evolution
NWC site (Long Island Sound) 114, 115
Nyquist frequency; *see* artifacts, taphonomic

obrution 14, 54, 199, 204, 237, 238, 239, 251, 255, 258, 289; *see also* storms; tempestites; turbidites
Occam's Razor (principle of parsimony); *see* methodology in historical sciences
occurrences, lowest and highest of fossil datums; *see* biostratigraphy

oceanic anoxic events (OAEs) 253, 359; *see also* anoxia
oceanic circulation 310, 314, 316, 349, 357
  Antarctic bottom water (AABW) 42, 123, 320
  caballing 359
  chemocline 347
  estuarine 346
  halocline 255
  halothermal 346, 348
  hungry mode in Permian 357
  lagoonal 346
  North Atlantic deep water (NADW) 123, 320
  oligotaxic conditions 316
  overfed mode in Permian 357
  overturn rates, deep ocean 355
  oxycline 250
  oxygen minimum zone 216
  polytaxic conditions 316
  psychrosphere 312
  pycnocline 243
  salinity stratification and 355
  surface gyres 348, 349
  thermohaline 312, 348
  see also anoxia; glaciation; nutrients; oxygenation; productivity; upwelling
oligotrophic conditions; *see* nutrients
Olson E.C. 3
ooids 314, 317
oolitic ironstones 314, 317
oozes 271
  calcareous 301, 314, 315, 321, 362, 366, 367
  calcite compensation depth (CCD) and 354
  foraminiferal nanofossil 121ff.
  incipient calcareous 354
  red clay 121
  transfer of $CaCO_3$ to deep ocean 338
  siliceous 121, 143, 147, 362, 367
    expansion of grasslands and 325, 361

Paleozoic cherts 143
  phosphorites and 147
  see also biogeochemical cycles; calcite compensation depth (CCD); diagenesis; megabiases; plankton
opal 140, 142
  alteration sequence 143
  amorphous 142, 143
organic matter 40, 41, 47
  acid production 114
  bacterial oxidation of 114, 129, 144, 148
  decay of angiosperm versus gymnosperm litter 360
  decay of silica-rich grasses 325, 361
  degradation of 137, 343
  labile 114, 215
  localized organic decay 150
  matrix 47
  metabolizability 137, 339, 341
  oxic degradation 115, 208
  productivity and 217
  refractory 96, 112, 137, 243, 244, 347
  sediment accumulation and concentration 343
  see also anoxia; nutrients
orientation, shell
  concave-up 52, 238
  concave-down 52, 59
  convex-up 52
  cross-current 66, 76
  current direction indicators 53
  current, parallel 54
ostracodes 43–45, 252, 253, 259, 385
  ornamentation 44
  shell strength 44
  transport 44
otoliths 91
outcrop (sampling) area; see artifacts, taphonomic; megabiases
oxycline; see oceanic circulation
oxygen isotopes 183, 350
  bones and 385
  datums 301, 302
  sea-level change and 384

stacking of records 197
  see also artifacts, taphonomic; graphic correlation
oxygenation 24
  atmospheric 366
  banded iron formations (jaspilites) 338
  oxygen minimum zone 216
  progressive oxygenation of the oceans 341
  see also bioturbation; organic matter; dissolution and precipitation of $CaCO_3$

paleontological interest units (PIUs) 336, 337; see also artifacts, taphonomic; counting procedures
paleophysiology; see applications of taphonomy; artifacts, taphonomic
paleosols 221, 279, 282
  estimates of stratigraphic completeness in terrestrial sections, using 279ff.
  see also soils; stratigraphic completeness
pan-African orogeny 350
particle model of fossil populations 378–379; see also applications of taphonomy
peat 3, 93, 100, 157
  bogs 205, 264, 266, 267
  mires 232; see also coal
permafrost 157, 264; see also lagerstätten
permineralization 133, 139, 155, 160, 257, 258, 264, 386; see also petrifaction; replacement
petrifaction 133, 139, 141, 155, 157; see also permineralization replacement
petroleum source rocks 314, 360; see also anoxia; nutrients
petroleum wells 301, 302, 305
  caving 394
  composite ditch samples 108, 127, 394
  use in taphonomy 370, 394
pH 110, 138, 141, 144, 153, 155–157
pH-Eh conditions 135, 153, 177
phenols 143

phosphate
    calcification mechanism and 327, 350;
        see also nutrients; poisoning of
        $CaCO_3$, precipitation
phosphatization 145, 150, 240, 243, 246,
        252, 256
    apatite 240
    association with siliceous oozes 147
    francolite 240
    phosphorites, formation of 146, 147, 210
    vivianite 240, 261
phytoliths 156
plankton
    acritarchs 321, 349, 351, 354, 355, 360
    bacteria 351
    coccolithophorids 42, 125, 180, 253, 301,
        321, 332, 354, 360, 366, 367, 376
    conodonts 43, 44, 45, 146, 201, 237, 328,
        377
    diatoms 42, 143, 144, 147, 206, 159, 206,
        261, 263, 360, 367, 368, 382
    desmids (freshwater) 157
    diatom-like structures from late
        Proterozoic 368
    dinoflagellates 43, 44, 147, 157, 360
    dinosteroids 360
    diversification and evolution of 359, 368
    graptolites 237, 321, 349
    masked by terrigenous debris 143
    phytoplankton 349
    pteropods 123
    radiolaria 43, 44, 143, 147, 253, 321,
        348, 349, 367, 368
    red clay and 121
    response to diatoms 368
    silicoflagellates 367
    transport of 42
    zooplankton 349
    see also megabiases; oozes; nutrients;
        plankton
plants, land 93ff., 244
    abscission of leaves 94–96
    amber 90, 255, 264, 265, 266, 332
    Abies amabilis (conifer) 95

allochthonous leaf mats 230, 231
Alnus (alder) 96
Ambrosia (ragweed) 104
angiosperms 94, 95, 97, 98, 229, 230, 360
Annularia (sphenopsid) 257
Araucariaceae 265, 332
Avicennia (mangrove) 141
bark 229, 257
beech 96
Beta vulgaris 156
Betula (birch) 104
Calamites (sphenopsid) 102, 227
Charophytes 356
"climatic filters" and 94
coal balls 258
coal swamps 102, 256, 258, 314, 316,
    347, 353, 356
common oak 96, 97, 104
cones, fruits, and seeds 229, 231
conifers 94, 95, 98, 229, 253
conifer needles 97, 262
Copaifera 265
cuticle, destruction of 156
cycads 253
evergreens 95
Fagus sylvatica (beech) 96
ferns 230
flowers 227
fruits 231, 237, 257, 263
Ginkgo 253
gymnosperm forests 356, 360
hackberry 156, 159
horsetails 253
Hymenaea 265
leaves 94ff., 229, 230, 237, 257, 261, 262
Lepidodendron (lycopod) 102
lignin 96, 142, 229, 240, 347, 353, 365
log jams 230
lycopods 102, 230, 356
macrofossils 93ff., 244
macrofossils and pollen 227, 232
Magnolia 262
mangrove leaves 94
multiple elements of 93

Index

Neuropteris (seed fern) 257
Paleozoic floras 142
petrifaction in carbonates and sulfides 141
phytoliths 156
Pinus (pine) 102, 103, 104, 265
pollen 3, 227, 232
propagules 315
Pseudolarix 265
Quercus (oak) 96, 97, 103, 104
Rhizophora (mangrove) 94, 141
Rhododendron 94
root-shoot ratios 93
rootlets 85
Salicornia europaea 156
Sawdonia (lycopod) 102
seed accumulations 230, 231
settling behavior of plant parts 95
Sphagnum 267
Spenophyllum (sphenopsid) 257
sphenopsids 257
Stigmaria (lycopod) 102
Thalassia (turtle grass) 48, 214
Ulmus (elm) 104
vascular land plant debris 90, 137
vascular plants, rise of 347
wood 90, 133, 229, 230
see also Lagerstätten; lignin; paleosols; peat; pollen; soils
Plattenkalk 239, 253; see also Lagerstätten
Poincaré, Henri 387
poisoning of $CaCO_3$ precipitation
    by Mg 150
    by phosphate 327
pollen 3, 28, 102, 157, 221, 227, 232, 240, 263, 266, 373, 382
    Ambrosia (ragweed) 104
    anthropogenic impacts and 102
    experiments 103
    focusing of (sorting and redeposition) 103
    sporopollenin 240, see also plants; pollen
polychaetes 213, 257; see also annelids; worms and worm tubes
polysaccharides 143
Popper, Karl 8, 9

positive feedback 358
potash 334, 338
Potonié, H. 3
power laws; see fractals; methodology in historical sciences
prairie dog "towns"; see bioturbation
preadaptation 363
predation and scavenging 53, 57, 58, 64, 65, 66, 72, 87, 105, 241 322, 330, 333; see also megabiases
prediction of preservational quality 286, 289; see also Lagerstätten
predictive value, of taphonomic models 14, 24, 289, 300, 387, 388
pressure solution 20, 145, 188; see also dissolution and precipitation of $CaCO_3$; stylolites
principle of parsimony (Occam's Razor); see methodology in historical sciences
productivity, surface 125, 217, 255, 261, 340, 356
    alternative interpretations of marine 365ff.
    Ba, Cd, and Sr as indicators 128, 351, 356
    marine carbon-to-Phosphorus (MCP) burial ratios 351, 352, 354, 359, 360, 362, 364, 365
    see also energy and evolution; nutrients; oozes; plankton
progress; see teleology
provinciality, biogeographic 307, 337; see also biostratigraphy
pseudo-steady-state conditions 163; see also actualism; methodology in historical sciences; steady-state view of the earth
pseudomorphs 117, 241
    bone 385–386
    see also permineralization; petrifaction; replacement
pteropods; see plankton
pycnocline; see oceanic circulation
pyrite, and other sulfides 64, 114, 115, 134, 152, 153, 215, 252, 253, 257, 259
    amorphous FeS 135, 137
    ancient occurrences 139

pyrite, and other sulfides (*contd*)
  black coloration of sediment 137
  bladed 139
  chamber linings 139
  framboidal 135, 251
  galena 252
  greigite 134, 135, 137
  mackinawite 134, 135, 137
  marcasite 134, 139
  Mississippi Valley-type ores 139
  modern analogs and ancient settings 338
  overpyrite 139
  pyritization 24, 133, 215, 240, 250, 251, 253, 257, 338ff., 347
  pyrrhotite 135
  sequence of formation 135, 139, 144
  silicification and 144
  sphalerite 257
  stalactitic 139
  see also megabiases; permineralization; replacement

R-sediment model; *see* Kidwell models of assemblage formation
r-strategist (opportunistic) benthos 94, 250, 363
racemization, amino acid; *see* amino acids; *see also* radiocarbon dates
radiocarbon dates 5, 100, 132, 177, 186–188, 190–204, 206–219, 221, 223, 226, 227, 229, 231, 232, 376
  bulk samples 219, 220, 232
  shells 388, 389
  see also bivalves
radiolaria; *see* oozes; plankton
radiometric dates 203, 283, 377
radiotracers
  beryllium-7, 178
  carbon-14, 177
  cesium-137, 177
  decay 163, 172
  estimation of sedimentary parameters and 175ff.
  half-life commensurate with sedimentary parameters 175

lead-210, 175, 178, 179
mobility of 177
plutonium-239/240, 178
profiles 175
thorium-234, 178
random walk (Brownian motion)
  fossil population dynamics and 382, 383
  model 165ff.
  see also bioturbation; particle model
rare earth elements (REE) 199, 386
  rates of sedimentation and sediment accumulation 269
red clay 121; *see also* calcite compensation depth (CCD); oozes; plankton
Redfield ratios 113, 146; *see also* biogeochemical cycles
  energy and evolution; nutrients; productivity
redox chemistry; *see* pH-Eh conditions
redox surface conditions of the earth; *see* carbon and sulfur isotopes
reduction fronts 138
reductionism; *see* methodology in historical sciences
Refugia, oligotrophic; *see* energy and evolution
replacement 117, 133, 139, 243; *see also* permineralization; petrifaction
reptiles 72, 85
  Compsognathus 254
  crocodiles 160
  dinosaurs 85ff., 155, 385
  dinosaur skin 263
  eggshell 90
  hadrosaurs 90
  ichthyosaurs 253
  lacertilians 160
  marine 90
  nesting sites 90, 222
  plesiosaurs 253
  pterosaurs 252, 253, 254
  Saurischia 253
  skin 92
  therapsid 87

Index

turtles 160
residence time at sediment-water interface (SWI) 11, 15, 23, 32, 270, 371
return time 195, 270, 310, 317, 388
    erosion depth and 270; *see also* methodology of historical sciences; stratigraphic maturation
reworking, physical 13, 194, 195, 198, 206, 210, 213, 214, 217, 219, 221, 268, 269–271, 299, 303, 304, 307, 376, 377, 386
    advective mixing 179
    conodonts and 45
    cryoturbation 162
    dimensionless reworking rate (DWR) 269
    dinosaur teeth 374
    gambler's ruin model of 269
    graviturbation 162
    karst and 162
    pollen and 103
    random walk model of 270, 273
    recognition 198
    remanié 198, 203, 206
    subsurface flow in hot springs 162
    subsurface flow in tar pits 162
    uprooting of trees 162
    versus bioturbation 198
    see also bioturbation; biostratigraphy
Reynolds number (Re); *see* sediment transport
Richter, R. 7
Röt Event 359; *see also* sulfur isotopes
Rubey's settling curve; *see* sediment transport
rules of taphonomy 12ff., 389ff.

sampling 300
    aliasing 197
    analytical time-averaging 108, 190, 229
    choosing 233
    cyclicity induced by sampling regime 197
    Nyquist frequency 197
    pooling (resampling) 193, 198
    sampling programs and processing 190, 197, 233, 299, 393

Signor–Lipps effect 288, 296ff., 373, 377
    spacing 198
    stacking of oxygen isotope records 197
    see also artifacts, taphonomic
scale; *see* hierarchy theory; methodology in historical sciences
scavenging; *see* predation and scavenging; *see also* megabiases
Schäfer, Wilhelm 7, 55, 56, 57, 66, 92
sea-level, CO_2, lithology, and biotic response 310ff.
sea-level change 146, 185, 271, 380
    base level 22
    cratonic sequences 313
    holocene changes 383ff.
    oxygen isotope curve and 384
    second-order cycles 313
    sediment accumulation rates and 371, 373
    supercycles 310ff.
    third order cycles 313; *see also* sequence stratigraphy; unconformities
sea surface temperature (SST) 121, 123; *see also* artifacts, taphonomic; oceanic circulation
secular trends in preservation; *see* megabiases
Sedgwick, Adam 6
sediment transport 28ff.
    angularity 28
    Bernouilli effect 31
    bioclasts, nature of 28, 47, 65
    bottom traction 28
    buoyancy 33, 34, 51
    bypassing 22, 62
    capacity 28
    colloids 28
    competence 28, 51
    Corey shape factor (CSF) 43
    critical threshold (traction) velocity 30
    density 28, 30
    dimensionless parameter (h) 27
    dimensionless shear stress 32
    drag coefficient 34
    drag force 31
    dynamic viscosity 28, 30

sediment transport (*contd*)
  eddies 29, 30, 34
  eddy velocity 33
  eddy viscosity 30
  effective density 36, 43
  entrainment 30
  fluid velocity 30
  fluid viscosity 28, 30, 34
  fossil transport and 28, 76ff., 95
  friction velocity 32
  gravity 31, 33, 34, 35, 40, 80
  Hjulström diagram 31
  hydraulic behavior 36
  hydraulic equivalence 36, 77
  intercept sphericity 42
  kinematic viscosity 30, 32, 43
  laminar flow 29, 30, 34
  maximum projection sphericity (MPS) 36, 42
  mode of transportation 28
  nominal diameter of grains 36
  organic matter content 40
  Reynolds number, grain 30, 31, 43, 103, 104
  roughness, grain 40
  Rubey's settling curve 35, 39
  saltation 28
  settling velocities 33ff., 44–45, 119
  shape 28
  shear stress 28, 30
  shear velocity 28
  shell interference during transport 53
  Shields diagram 31, 33
  specific gravity 33, 40
  sphericity, operational 36
  Stokes' Law 34, 42, 43, 45, 103, 104
  supply 28
  suspended load 28, 34
  traction (threshold transport) velocities 36, 37, 42
  transportability 80
  turbulent flow 29, 30
  unusual equilibrium orientations 53
  viscosity 30, 34
  weight 36, 38, 43, 95
  see also orientation
sediment (soil)–air interface 75, 89
sediment–water interface (SWI) 28, 37, 41, 48, 49, 54, 57, 91, 110, 148, 163, 179, 181, 185, 238, 241, 250, 252, 254, 269, 372
sedimentary parameters, estimation of 175ff.; *see also* bioturbation
sedimentation (accumulation) rates 110, 163, 178, 194, 195, 196, 208, 215, 238, 269, 289, 333, 350, 371, 372, 390
  apparent versus true 177
  deep-sea 197, 304
  destruction of bones and 83
  focusing 234
  levee effects 177
  over short versus long intervals 275, 372
seed accumulations 230, 231; *see also* plants; pollen
seismic reflectors 127; *see also* taphonomic signals
septarian structures 147
sequence stratigraphy 284
  accommodation space 306
  bioherms, occurrence of 286, 314
  biostromes, occurrence of 286
  condensed sections 286, 289, 290, 305, 372
  coquinas, occurrence of 289
  cratonic sequences 313
  depositional sequence 284, 289
  distribution of datums in relation to systems tracts 287
  downlap and 292
  highstand systems tract (HST) 284, 286
  lowstand systems tract (LST) 284, 286, 287
  maximum flooding surfaces (MFS) 289
  occurrence of systems tracts 286
  onlap 293
  parasequences 276, 284, 286, 289, 291
  sequence boundaries 284, 289, 291, 303, 305
  taphonomic gradients and 288

Index

toplap 293
transgressive systems tract (TST) 284, 286, 287
see also graphic correlation; hiatuses; unconformities
shelf-slope experimental taphonomy initiative (SSETI) 393
Shields diagram; *see* sediment transport
siderite 135, 147ff., 239, 257, 258, 261, 263; *see also* calcite and related minerals; concretions
sieving; *see* artifacts, taphonomic; counting procedures; sampling
signal-to-noise ratio (S/N); *see* filters; *see also* bioturbation
Signor–Lipps effect 288, 296ff., 373, 377; *see also* artifacts, taphonomic; bioturbation; mass extinction; reworking; sampling
silica 157, 338, 348, 368
  beekite 145
  biogenic 143
  chalcedonite 139, 140
  chalcedony 139, 140, 141, 145
  cristobalite 143
  cryptocrystalline 145
  crystalline opal (opal-ct) 142, 143
  expansion of grasslands and effect on microfossils 325, 361
  lutecite 140
  megaquartz 140, 145
  microcrystalline 145
  microcrystalline quartz 142
  microquartz 140
  morphologies 144
  quartz 140, 152
  quartzine 140, 145
  sources of 140
  tridymite 143
  see also siliceous oozes; plankton
siliceous oozes; *see* oozes; plankton
silicification 139, 150, 240, 263, 264
  concurrent carbonate dissolution and 141, 144, 145

  of calcareous fossils 144
silicoflagellates; *see* oozes; plankton
simulations, computer 198; *see also* experiments, taphonomic; numerical modeling
size, effects on dissolution 118ff.
skeletons, origin of; *see* metazoa
Sloss, Lawrence 284; *see also* sequence stratigraphy
small-number systems; *see* methodology in historical sciences
soda (sodium carbonate-rich) lakes 141, 323
soda ocean 141
soils 76, 89, 141, 152ff., 221, 233, 266, 349, 353, 374
  acidic oxidized 159
  alkaline 74, 75, 153
  entisols 158, 159
  gley minerals 153, 159, 199, 207, 208
  inceptisols 159
  podsols 153
  rendsinas 156
  spodosols 159
  ultisols 159
  see also paleosols; weathering
sole marks 244, 247
Sorby principle 47, 118, 389
Souris event 351; *see also* sulfur isotopes
sponges 48, 141, 367, 381
  clionid 48
  spicules 143
sporopollenin; *see* pollen
stable isotopes 259, 275, 362, 376, 377, 384, 385, 386; *see also* biogeochemical cycles; carbon isotopes; oxygen isotopes; strontium isotopes
steady state view of the earth 5, 157, 163ff., 218, 219, 388
  geochemical models and 6, 337
  pseudo-steady-state conditions 163
  steady-state conditions 112, 157, 330, 335, 368
  see also methodology in historical sciences

steinkerns; *see* internal molds
Steno, Nicolaus 1
stenotopic taxa
   biotic turnover and 382; *see also* particle model of fossil populations
"stepping stones" for deep-sea faunal elements 91
Stokes Law; *see* sediment transport
storms 24, 54, 95, 133, 135, 182, 201, 202, 208, 213, 214, 229, 237, 238, 244, 247, 253–255, 259, 269, 271, 276, 319, 333, 347
   fairweather wave base 54
   hurricanes 178, 243, 381
   insurance companies and frequency 182
   intensity related to atmospheric CO_2, 319
   megamonsoons 316
   northeasters 178
   northers 2
   tempestites 18, 59, 249, 319, 347
   washovers 20
   see also obrution; tempestites; turbidites
stratigraphic
   acuity 269
   completeness 216, 269, 269, 271, 309, 310, 312–316, 318–323, 325, 326, 327–329, 334ff., 371, 373
   disorder 194, 198
   erosion depth and 270
   gaps, duration of 275
   maturation 269, 271, 282
   see also graphic correlation; sequence stratigraphy; temporal resolution
stratigraphy of shell concentrations 289ff.
stromatoporoids 145, 361; *see also* Cnidaria and associated biota; sponges
strontium isotopes 199, 301, 315, 354, 355, 356, 359, 360, 376, 377
   pan-African orogeny and 350
   runoff and 354, 359
stylolites 20, 188; *see also* pressure solution
submesotrophic conditions; *see* nutrients
substantive uniformitarianism; *see* methodology in historical sciences

sulfate reduction, bacterial; *see* alkalinity; anoxia
sulfides, oxidation to produce acids 114ff.
sulfur isotopes 314, 341, 343, 350, 354, 359
   Röt Event 359
   Souris event 351
   Yudomski event 327, 350
superoligotrophic conditions; *see* nutrients
superposition, principle of 11
swamps 75, 100, 102
   black(fresh) water swamps 93
   coal 102, 243, 314, 316, 347, 353
symbiosis 50, 119, 129, 328, 329, 348, 349, 354, 360

taphocoenosis 7; *see also* assemblages, fossil
taphofacies 24ff., 41, 62, 83, 89, 90, 257, 390–391; *see also* gradients
taphonomic
   bias 3, 72, 236
   clock 49, 208, 386, 389
   feedback 20, 22, 57, 64, 190, 200, 205, 206
   grades 10, 49, 132, 143, 157, 202, 206, 208, 211, 371, 372, 389
   hierarchy 306
   histories (pathways) 5, 10, 11, 49, 282, 387, 389
   mode 15
   prediction 286, 289, 237, 244, 387, 388
   signals 127
   see also artifacts, taphonomic; equifinality; experiments, taphonomic
taphonomic active zone (TAZ or surface mixed layer); *see* bioturbation
taphonomy
   as a predictive science 91, 289, 300
   differences in methodologies 392
   foundations of 1ff.
   positive contributions of 4, 188ff.
   rules of 12ff., 389ff.
   see also applications of taphonomy; information; time-averaging
teeth 1, 77, 78, 146, 152, 155, 160, 199, 201, 206

cave bear and oxygen isotope composition 386
dinosaur, reworking across K–T boundary 373
tooth (gnaw) marks 74, 83
tooth/vertebra ratio 78
see also bioerosion
teleology 5, 6; see also methodology in historical sciences
temperature 75, 92
    ambient water 66, 366
    see also gradients
tempestites 18, 59, 249
    as function of sea-level 319
    escape burrows 200, 204, 238, 249, 347
    graded beds 43
    see also obrution; storms; turbidites
temporal
    acuity 279, 391
    cause-and-effect and 268
    estimation using isotopes 269
    marine record and 275ff.
    resolution 268, 392
    scope 268, 279
    terrestrial record and 279ff.
    tradeoffs between acuity, sedimentation, and thickness 279, 292
    see also biostratigraphy; bioturbation; stratigraphic acuity
terpenes 265, 266
terrestrial floras, development during paleozoic 313, 349ff.
terrigenous regimes, versus carbonate 28ff.
thanatocoenosis; see assemblages, fossil
tidal pumping 150, 232
time-averaging 4, 5, 13, 22, 25, 66, 94, 186ff., 237, 269, 270, 279, 282, 313, 369, 375, 379, 380, 390, 392, 394
    abundance 191
    actualistic criteria 202–203
    advantages 188
    age, weighting 188
    allochthonous 192
    analytical 108, 190, 229

    autochthonous 192
    cohorts 188
    computer simulations of 226
    consequences of 191
    criteria for ancient settings 203ff.
    criteria for single beds (plants) 230
    durations 21, 188, 189, 199ff., 201, 206, 223, 269, 350, 361, 388, 394
    enhancement of ecological signals 369
    estimates of 202ff.
    experiments; see experiments, taphonomic
    exponential model of age distributions 188
    linear superposition, principle of 186, 207, 209, 390
    minimum durations 193
    paleobiological studies and 190
    positive aspects 4, 188ff.
    recognition 202
    skewed ages 187ff.
    stratigraphic disorder 194
    taxonomic composition and diversity 192
    terrestrial 223
    trophic and life habits 194
    types of 199ff.
    within-habitat 200, 201, 205
"tonguestones" ("glossopetrae") 1
trace element and stable isotope compositions 117; see also rare earth elements
trace metals 347
transport 27ff.
    Lagerstätten and 238, 247
    transport groups for coyote and sheep bones 76
    transported and reworked pollen 103
    see also sediment transport
transportation effects; see exposure effects
trilobites; see arthropods
trilobite-rich fauna 352, 361, 362
trophic and life habits 194
turbidites 45, 52, 54, 199, 237, 238, 244, 246, 247, 249, 251, 255
    degree of articulation and 54
    mudflows 30
    see also obrution; storms; tempestites

unconformities 138, 148, 276, 284, 287, 289, 300, 303, 306, 306, 373; *see also* diastems; graphic correlation; hiatuses; sea-level change; sequence stratigraphy; stratigraphic completeness
uniformitarianism 392
  geochemical 337
  relation to present 391
  taxonomic 55
  see also actualism; aktuo-paläontologie; methodology in historical sciences
unique objects; *see* methodology in historical sciences
upwelling 44, 50, 123, 143, 190, 322, 323, 351, 359; *see also* anoxia; glaciation; nutrients; oceanic circulation; productivity

varves 90, 231, 233, 243, 261, 262, 263
  Milankovich cycles and 262
vendobionta (vendozoa) 244ff.; *see also* Lagerstätten
vertebrates 72, 223, 255
  microfossil concentrations 223
volcanism, and preservation 93, 139, 140, 142, 161, 163, 168, 180, 183, 230, 237, 244, 302, 313

ash 302
tuffs 262
Walther, Johannes 1
Wasmund, E. 7
water table, perched (stagnant) 100
water/rock ratios 117
weathering
  acceleration by climate change 336, 367
  carbon dioxide drawdown by 326, 367
  clay minerals and 333, 353
  continental 310, 315, 349, 360, 361
  function of proportion of exposed lithologies 316
  function of rock hardness 316
  hydrothermal 315, 318, 323, 334, 338
  microbial 349
  terrestrial floras and 349, 354, 356
  volcanism and 140, 367
weathering stages of bone; *see* bone
Weigelt, Johannes 2, 11, 72, 87, 92
Whewell, William 6
wind 42; *see also* pollen
worms and worm tubes 48, 162, 179, 241, 257, 331; *see also* annelids; polychaetes

x-rays, core 175; *see also* bioturbation

zeolites 152, 153